AF905488

Applied Ecology

Jan Frouz • Jaroslava Frouzová

Applied Ecology

How agriculture, forestry and fisheries shape our planet

Springer

Jan Frouz
Charles University
Praha, Czech Republic

Biology Centre CAS
České Budějovice, Czech Republic

Jaroslava Frouzová
Charles University
Praha, Czech Republic

Biology Centre CAS
České Budějovice, Czech Republic

ISBN 978-3-030-83224-7 ISBN 978-3-030-83225-4 (eBook)
https://doi.org/10.1007/978-3-030-83225-4

© The Editor(s) (if applicable) and The Author(s), under exclusive license to Springer Nature Switzerland AG 2022
This work is subject to copyright. All rights are solely and exclusively licensed by the Publisher, whether the whole or part of the material is concerned, specifically the rights of reprinting, reuse of illustrations, recitation, broadcasting, reproduction on microfilms or in any other physical way, and transmission or information storage and retrieval, electronic adaptation, computer software, or by similar or dissimilar methodology now known or hereafter developed.
The use of general descriptive names, registered names, trademarks, service marks, etc. in this publication does not imply, even in the absence of a specific statement, that such names are exempt from the relevant protective laws and regulations and therefore free for general use.
The publisher, the authors and the editors are safe to assume that the advice and information in this book are believed to be true and accurate at the date of publication. Neither the publisher nor the authors or the editors give a warranty, expressed or implied, with respect to the material contained herein or for any errors or omissions that may have been made. The publisher remains neutral with regard to jurisdictional claims in published maps and institutional affiliations.

This Springer imprint is published by the registered company Springer Nature Switzerland AG
The registered company address is: Gewerbestrasse 11, 6330 Cham, Switzerland

Acknowledgment

We are very grateful for the valuable comments and advice we received from the reviewers of the Czech version of the book, namely Prof. B. Moldan, Ing. J. Urban, Assoc. Prof. J. Remes, and Assoc. Prof. J. Matěna, and for the help and advice we received from other colleagues. Among them, it is necessary to mention in the first place a number of employees of the Central Institute for Supervising and Testing in Agriculture of the Czech Republic, namely Ing. R. Šafaříková, Mgr. Š. Poláková, and Dr. S. Malý, who provided us with valuable comments, comments on the text, and made it possible for us to use the unique data of the institute. We also received valuable advice and information from employees of the Ministry of Agriculture and employees of the Czech Statistical Office. The recommendations of Dr. Oulehle and Prof. Hruška from the Czech Geological Survey, Prof. Petrusek from Faculty of Science of the Charles University are greatly appreciated. Among many colleagues who provide us valuable information, we would like to mention Prof J. Franklin from the University of Tennessee (USA), who provide invaluable input on the proportion of particular tree genera in US forests. We would also like to mention Prof. P. M. Curry (France), Dr. Kazumichi Fujii from the National Research and Development Agency Forest Research and Management (Japan), and Prof. J.M. van Mourik from the University of Amsterdam (Netherlands), and we are obliged to many other colleagues who provided us valuable information opinions and discussed various issues. The book has excellent illustrations by Mgr. L. Buchbauerová; we also thank many of our colleagues and students for technical assistance in editing and processing the text. In particular, Prof. J. Franklin, Dr. T. Flynn, and L. Cameron are thanked for language improvements in the English version of the book. Our work has been supported by a number of grant projects, of which at least the following projects supported by the Ministry of Education, Youth and Sports should be mentioned: LM2015075, EF16_013/0001782, and 8I20001; TL03000752 supported by TACR; and last but not least Strategy AV 21 Academy of Sciences of the Czech Republic.

Introduction

Mankind substantially affects ecosystems of our planet. The biomass of all humans exceeds 10 times the biomass of all wild animals (terrestrial vertebrates) and the biomass of domestic animals exceeds the biomass of wild animals more than 20 times (Smil, 2011). It is estimated that ecosystems on more than three-quarters (according to some estimates, even more) of non-glaciated land are to some extent affected by humans. In contrast, man-untouched ecosystems, or wilderness, not only cover just about a quarter of the terrestrial land but are largely situated in places where primary production is naturally low (tundra, deserts, high mountains, etc.) and thus produce only about 11% of the planet's total primary production (Ellis & Ramankutty, 2008). Conversely, human impacts in the seas and other waters of the world are less visible; however, the FAO reports that a third of fish stocks show signs of overfishing. And so we could keep on. Human impact on key functions of our planet, such as global climate change, the nitrogen cycle, and biodiversity, may potentially exceed safe ecosystem boundaries, which could lead to unpredictable system behavior that could result in a global catastrophe (Rockström et al., 2009).

Interestingly, most current ecology textbooks do not talk much about human impact on ecosystems, but rather explore how individual species, populations, or communities will interact and evolve on their own. This approach is certainly justifiable, because most ecological relationships evolved from an evolutionary point of view before the rise of human civilization, and so they somehow do not need man for their natural functioning. The problem is that this natural functioning is now limited to an ever smaller and smaller part of our planet's ecosystems, and not even random selection in ecosystems but mainly less productive ones. Therefore, if we want to understand the functioning of the current global ecosystem, we cannot ignore human influence.

At the same time, agriculture, forestry, and fisheries play a key role in this influence. Understanding the role of these activities in the functioning of our planet is not just an interesting academic exercise, the well-being and the very survival of humanity depends on it. That is why we organized the book from a more anthropocentric perspective. We explain that mankind needs some goods and services, which are either obtained by or affected by agriculture, forestry, and fisheries. Then we describe the main ecological and biological principles on which modern agriculture, forestry, and fishing are based, to clarify how these principles are used to increase the production of food and other raw materials (wood, biofuels, building fibers, and other materials). At the same time, we explain how the intensification of production of these goods changes the affected ecosystems. Their structure, energy, and material flows affect the functioning of these ecosystems and subsequent provision of non-productive ecosystem services (clean water supply, runoff regulation, climate regulation, etc.), on which the well-being of human society also depends. Last but not least, we describe the ways in which current science and society strive to increase the sustainability of agriculture, forestry, and fisheries. We do not try to replace ecology textbooks, but we try to remind readers the very basic relevant ecological rules and relationships in order to make the book understandable to a wider audience and to point out the broader context.

Although the book is not a textbook of the above disciplines (agriculture, forestry, and fishing), it must, to some extent, acquaint the reader with the principles of operation and development of their technologies because the impact on ecosystems largely depends on the technological procedures used. We believe that this approach from empirical knowledge to underlining principles will make the book readable for those without a background in theoretical ecology, and for those who have such a background, it will make the book an interesting supplement to their existing knowledge.

The book's ambition is to serve as an introduction to the problem of a wider educated and interested public, especially those who deal with the environment in some way. At the same time, we tried to write the book clearly that it could serve as an introduction to the problem for interested students, decision makers, and the general public. Therefore, in addition to a basic description of the main principles, we tried to demonstrate these phenomena on illustrative examples including specific trends of key variables. This factual data serves both for a better understanding of the described mechanisms, as well as for a basic orientation in individual problems. Therefore, in addition to the main text, the book is accompanied by a number of boxes that explain the principles of the methods used or develop the described mechanisms on more detailed illustrative examples. Boxes also explain some basic concepts from other disciplines needed to understand the text. This should allow the study of a text with minimal demands on the knowledge of related disciplines. We also tried to bring links to key literary sources, which should facilitate further more detailed studies of selected aspects. We have tried to keep the use of mathematical approaches to a minimum in order to make the book available to readers less skillful in the mathematical description of reality. This led to the omission of a more detailed description of some methods based on complex algorithms, where we usually limited ourselves to a brief mention and reference to the relevant literature.

Contents

Author Biographies

Jan Frouz studied agronomy at the Czech University of Life Sciences and biology at Charles University in Prague. He received his PhD in entomology from the Czech Academy of Sciences in 1995. He held several postdoc positions in the USA, Italy, and Germany. Since 1992, he has been working at Biology Centre CAS, where he is now director of soil and water research infrastructure. Since 2013, he has been working also at Charles University where he is at present Professor of Environmental Sciences in the Faculty of Science and director of the Environmental Centre. His scientific interests include soil science, soil biology, and ecosystem ecology. His research addresses the question of how interactions between plants, soil, and soil biota determine gradual succession changes of soils and consequently whole ecosystems, and how we can use this knowledge to restore ecosystems. He has published more than 300 scientific papers, with more than 200 articles in WoS-registered journals, and has authored or edited 3 books and 17 book chapters.

Jaroslava Frouzová studied agronomy at the Czech University of Life Sciences in Prague. After completing her MSc in 1989, she gained extensive life experience as a manager of agricultural companies working in both agronomy and animal husbandry; later, she worked for The State Land Office of the Czech Republic. After that practical experience she returned to academia and in 2005 received her PhD in ecology from the University of South Bohemia in České Budějovice. She worked at Fisheries Research Laboratory of the Fish and Wildlife Conservation Commission at Florida, USA. Currently, she is employed by the Biological Centre at the Academy of Sciences of the Czech Republic as a researcher, and by Charles University, Faculty of Science, and the Institute for Environmental Studies as assistant professor, where she teaches methods of abundance and biomass estimation, special methods in ichthyology, and applied ecology. Her research focuses on the development of hydroacoustical methodologies and acoustical data interpretation. She conducted hydroacoustic survey in freshwater ecosystem in many European countries and also in Philippines, Thailand, and Siri Lanka. She has published about 100 scientific contributions including more than 60 papers in WoS.

Chapter 1
General Background in Ecology and Environmental Sciences

1.1 Ecosystem Services and Human Impact on Them

In their classic work, Costanza et al. (1997) asked themselves how much ecosystems contribute to the human economy and human well-being. They estimated that ecosystem goods and services contribute to the human economy $32 trillion annually. Later, Constanza et al. (2014) corrected the number to $145 trillion and the figure is often criticized and considered undervalued. However, the number itself does not say much, and its significance only stands out when compared to the gross domestic product of all the countries of the world, which Costanza et al. (1997) state as $18 trillion. Thus, ecosystems provide key services that, in principle, enable human society to exist and normally operate. Ecosystem services can be divided into several categories, such as production, regulatory, cultural, and social. Mankind directly and indirectly uses these three types of services; however, it is necessary to mention that their existence depends on various natural processes, in an earlier concept these were called supporting services (Fig. 1.1).

Provisioning services include all goods and services produced by ecosystems that can be used for the material consumption of human society. This includes the production of plants providing food, fibers, fuels and other goods, the production of drinking water, etc. Regulatory services regulate key natural processes, such as the water cycle, the production of greenhouse gases and thus climate change, etc. Regulatory services therefore to a large extent create a framework of environmental conditions in which our society operates. Many people tend to have them for granted. We usually realize their importance only when there is some problem with their functioning and we start to suffer from that in a way of floods, droughts, dust storms or climate change, etc. These services are generally not subject to the market, although some market mechanisms such as carbon credits have emerged recently, but far more often, regulatory services that have been identified as key to human society are subject to regulations by various legal regulations. Cultural and

© The Author(s), under exclusive license to Springer Nature Switzerland AG 2022

J. Frouz, J. Frouzová, *Applied Ecology*, https://doi.org/10.1007/978-3-030-83225-4_1

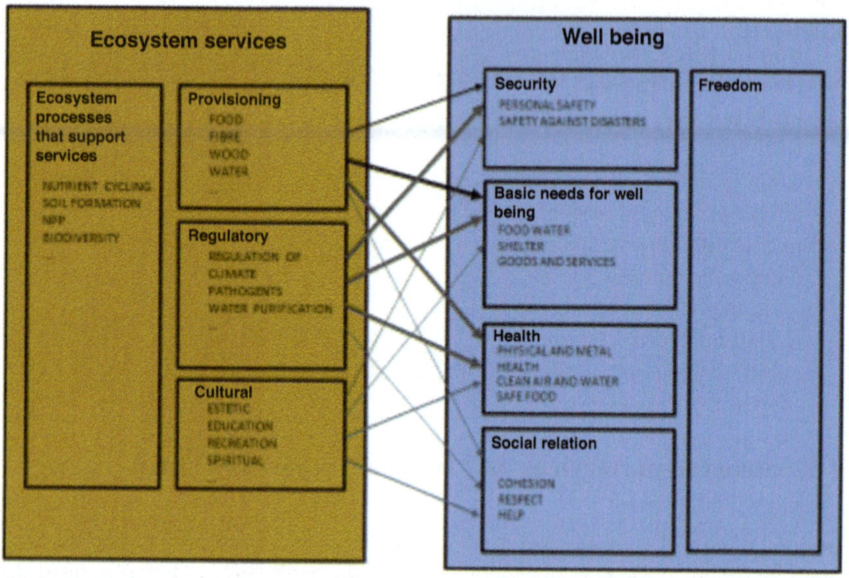

Fig. 1.1 Overview of the main types of ecosystem services and their relationship to the well-being of human society according to the Millennium Ecosystem Assessment 2005 (adapted)

social services represent all of this, what nature means to our psychical well-being, culture, artistic, spiritual inspiration, etc. For example, there are a number of works showing that vegetation improves people's mood, lowers blood pressure, and so on (Ulrich, 1986). Although these services are generally not marketable per se, they play a large role in the economy of many human activities, such as the tourism and entertainment industries, literature, and arts. For example, the natural beauty of some places may be subjected to (over)exploitation by tourism, which degrades their attractiveness and have an important consequence on the economy (Box 1.1).

Supporting ecosystem processes that provide services such as primary production, pollination, element cycle, etc., are usually not a direct subject of the market, which complicates their valuation. There are rare exceptions of some supporting processes such as pollination, where in some crops, farmers directly pay for pollination (Aizen & Harder, 2009; Burgett, 2010).

The utilization of nature by human often reduces the ability of ecosystems to provide ecosystem services. Constanza et al. (2014) estimated that global change, land-use change, and other influences have reduced the ability of the world's ecosystems to provide ecosystem services around the turn of the century by $4–20 trillion a year.

We used this approach here to illustrate value of nature for human society. However, this concept also suffers from criticism. One should keep in mind that the fact that we are able to value the economic benefit that nature brings to mankind does not mean that this is the whole value of nature. Nature has value in its own

Box 1.1 Life Cycle of Tourist Destination

One of the cultural services of ecosystem is the recreational potential of the landscape. We can express this service relatively well economically as the willingness of people to pay for having recreation in a given location, in a given landscape, etc. On the recreational potential of the landscape (or recreation potential of ecosystems), we can also clearly show how its use by human may affect this potential (Butler, 1980). The concept applies similarly here to natural and cultural values.

In the beginning, the destination is discovered, then visitors start to come to a place. These visitors are looking for the authenticity of the natural (and or cultural) values of the place. Their number is limited, which is due to the lack of tourist infrastructure. As a larger number of visitors are informed about the new destination, it becomes the destination for other tourists. Subsequently, as the destination's development phase takes place, the infrastructure is developing simultaneously. The original facilities serving the first visitors are modernized, providing more capacity, and a higher standard so that the destination can receive more visitors. At this stage, however, the first impacts on the environment are already becoming apparent and the number of visitors seeking authenticity, natural beauty, etc., starts to decrease; however, their number is balanced by an increase in "mainstream" visitors and the total number of visitors grows. Subsequently, in the consolidation phase, the increase in visitors stagnates, most visitors want to visit the place, because all the other people have already been there, but they are not really interested in the place as such.

On the one hand, the locals are highly dependent on income from tourism, and on the other hand, they are harassed by tourists and the existence of tourists displaces their everyday life and the natural beauty of the place is often deteriorated by infrastructure development. The phase of consolidation thus gradually passes into a phase of stagnation, when the negative impacts on nature and local communities intensify. As the number of visitors and the length of their stay is declining, the tourism industry is stagnating, there are no funds for the maintenance of tourist infrastructure and thus, the place becomes less attractive for clients. Further development in the post-stagnation stage may lead either to a complete decline or to some stabilization related to the discovery and appropriate marketing of new destination options.

right independent of human uses (Rea & Munns, 2017). Consequently, we will deal in the following text with much broader range of effects of agriculture, forestry, and fisheries, than just those related to goods and services provided to mankind.

Over time, the human population grows and the technical capabilities of mankind increases (Fig. 1.2), we produce more and travel faster. It can be seen that this increase was gradual and linear for most of the reported history then accelerates at the beginning of the industrial revolution around 1850. Another major turning point

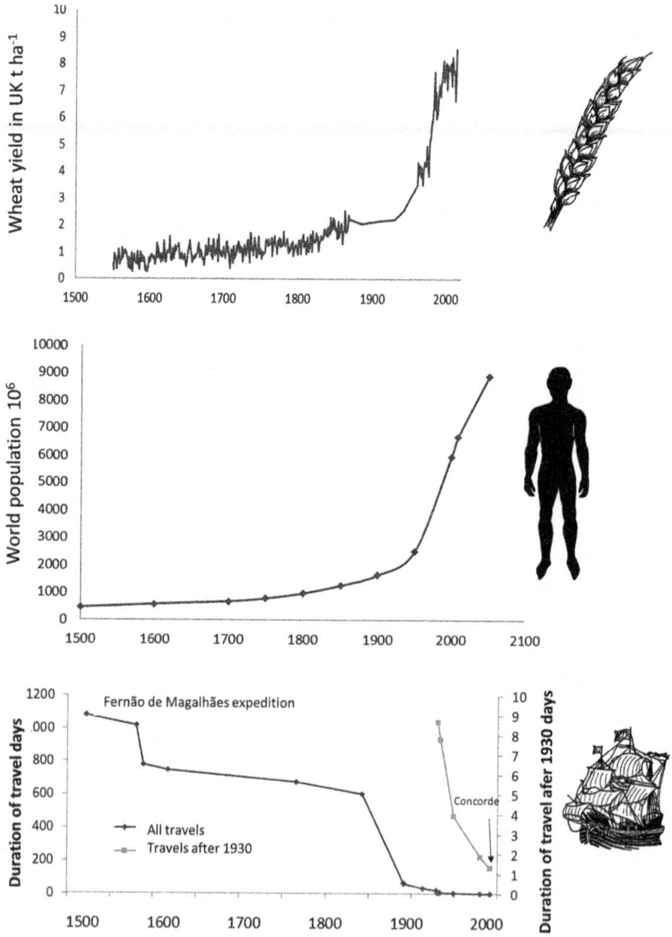

Fig. 1.2 Long-term changes in wheat yields in the United Kingdom, world population growth, and a minimum time required for a suborbital journey around the world (the gray line corresponds to the right axis)

came around 1950 when nitrogen began to be obtained from the atmosphere on an industrial scale by Haber–Bosch synthesis (Fig. 1.2). At the same time, human influence on the functioning of ecosystems and geochemical processes is growing. Based on geographical indicators such as population density, roads, agriculture use, etc., it is estimated that 83% of the global terrestrial biosphere is under the direct human influence (Sanderson et al., 2002). Certainly, not all biomes have been affected equally. It is certainly not surprising that Mediterranean and temperate biomes (i.e., areas where complex civilizations developed a long time ago) are most effected (Fig. 1.3). Given that ecosystem production services are most tangibly

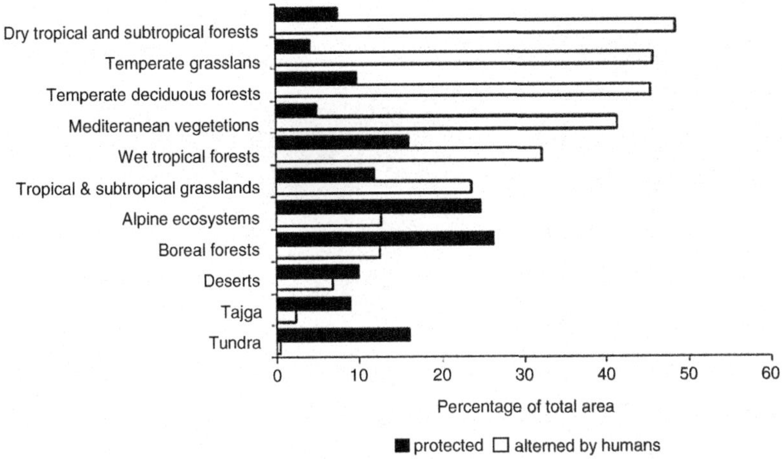

Fig. 1.3 The share of the original area of individual biomes that has been transformed by human (converted into fields, meadows, plantations, cultural forests, build-up, etc.) and the share of the biomes area that is protected in some way. (Adapted from Hoekstra et al., 2005)

linked to the well-being of society (they are the subject of a market), it is not surprising that ecosystems are often changed to increase their production capacity at the expense of other processes and services.

In many biomes, the area transformed by human is so large that only small fragments of the original ecosystems remain (Figs. 1.3 and 1.4). In these areas, the flow of energy, water, and matter is largely controlled by the behavior of human-made ecosystems. Ellis and Ramankutty (2008) estimated that more than 75% of the non-glaciated land is to some extent affected by humans. At the same time, human has influenced more biomes and ecosystems that have the ability to provide greater production. In contrast, untouched ecosystems are largely located in places where primary production is naturally low (tundra, deserts, high mountains, etc.) and thus, these relatively unaffected ecosystems produce only about 11% of the planet's total primary production (Ellis & Ramankutty, 2008). Some researchers even speak of a new geological epoch of the anthropocene (Box 1.2). Anthropogenic disturbance affects virtually all ecosystem functions, ecosystem structure, diversity, etc. (Fig. 1.5). These mostly negative effects persist even if the anthropogenic effect disappears and the ecosystem returns to its natural state, either through natural processes or assisted restoration. During this recovery period, the average annual deficit for species diversity was 27–33% and for the carbon cycle, 32–42% (Moreno-Mateos et al., 2017).

Looking at natural ecosystems, we can notice that in terms of matter and energy flows, they are very closed systems. Most of the minerals released by the decomposition of the dead bodies of plants and animals enter the soil and are used again by other

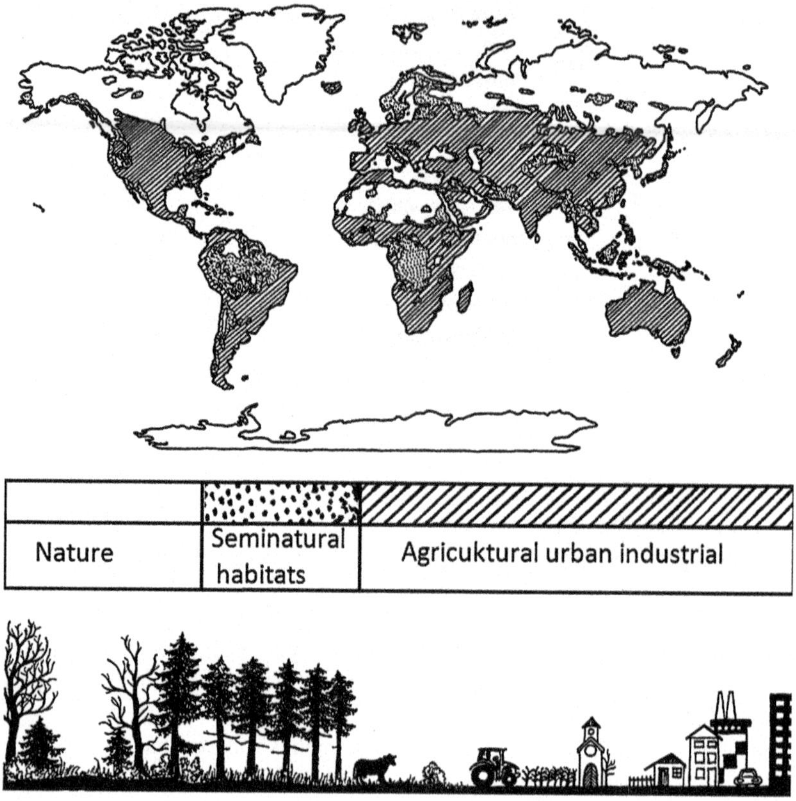

Fig. 1.4 The estimated human impact on global ecosystems according to Ellis and Ramankutty (2008)

organisms. The losses are relatively small and are supplemented by the supply of nutrients by weathering, by the supply of nutrients dissolved, suspended in water, or brought as an aerosol in an aeolian deposition.

In human-made ecosystems, humans often appropriate a significant part of primary production. This human appropriation of net primary production, abbreviated as HANPP, leads to a reduction in energy (food) available for other organisms. This reduction of input into natural food webs can have negative effects on biodiversity. Moreover, it also affects the flow of water, carbon, and other elements (Box 1.2). In many anthropogenic ecosystems, typically those on arable land, humans supply every year large amounts of nutrients to compensate for the biomass removed. These nutrients are then removed by harvest and a large part also leaves as a loss from leaching and other processes (e.g., nitrate by denitrification, ammonia volatilization, and so on). If we carry out farming on large areas, then these effects are magnified at the landscape level. In the case of nutrients, this leads to an increase in

Box 1.2 Anthropocene

The name Anthropocene comes from the Greek words Anthropos, which means human and cene comes from Greek kainos which means new or recent. These concepts appear at the beginning of the century. The term itself is attributed to biologist Eugene Stoermer and was subsequently developed by the atmospheric chemist Paul Crutzen (Crutzen & Stoermer, 2000).

Currently, a proposal has been prepared to the Stratigraphic Commission for the recognition of Anthropocene as a separate geological epoch at the Holocene level. Traces of nuclear explosions in the 1940s and other global tracers were proposed as the stratigraphic boundary of this era. However, the official recognition of the Anthropocene as a separate geological epoch has not yet taken place.

Human impacts on biodiversity are one of the frequently mentioned attributes of the Anthropocene. Human activity has greatly accelerated the extinction of species (Andermann et al., 2020). The rate of extinction is indeed close to the rate of extinction during the so-called large extinctions. However, it must be said that the time for which extinction takes place is geologically short and recent. There have been other periods in the past when extinction was at least as rapid as it is today. On the other hand, if the current rate of extinction continued for a significant period of time from a geological point of view, it could become the sixth largest extinction.

One of the geological symptoms of the Anthropocene is an increase in the content of carbon dioxide (CO_2) in the atmosphere. During the glacial-interglacial cycles, CO_2 fluctuated by approximately 100 ppm (from 180 to 280 ppm). At present, the CO_2 concentration has reached approximately 400 ppm, with an ever-increasing trend. Most of this increase is due to the burning of fossil fuels such as coal, oil, and gas, as well as changes in land use. Combustion of fossil fuels has also caused an increase in the concentration of black carbon (tiny carbon particles caused by the imperfect combustion of fossil fuels), inorganic ash, and spherical carbonaceous particles in recent sediments around the world. The concentration of these components increased significantly and almost simultaneously around the world in the period beginning around 1950. The Anthropocene is also associated with various changes in the concentration of trace elements. For example, between 1952 and 1980, tests of thermonuclear weapons left a clear global signal of an increase in ^{14}C, ^{239}Pu, and other artificial radionuclides.

Human activities such as deforestation, agriculture, and large transport infrastructure have made surfaces more susceptible to erosion and thus greater sediment removal. On the other hand, the construction of dams on many rivers around the world has caused the deposition of sediments, so it cannot be said that the supply of sediments to the seas increased during the Anthropocene. However, there has been a greater shift in some elements, such as phosphorus, whose entry into the seas has almost doubled.

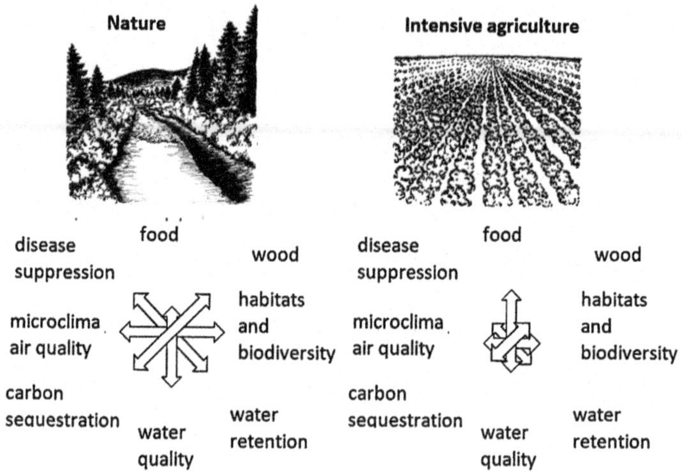

Fig. 1.5 Schematic expression of the ratio of production and nonproduction ecosystem services on the example of a natural ecosystem and intensively cultivated field

the outflow of nutrients into water courses and reservoirs. Here this excess of nutrients can cause eutrophication and impair the quality of the water usable for treatment in drinking water. Increasing in an ecosystem service (food production) can thus lead to a reduction in another ecosystem service (drinking water supply). More generally intensification of production often leads to a reduction in other ecosystem services (Fig. 1.5).

This has implications for global flows of matter and energy in nature. For example, the global application of mineral fertilizers has doubled the input of nitrogen from the atmosphere into ecosystems and the outflow of phosphorus from land to the sea. These changes then have a major impact on biodiversity and ecosystem processes.

Thus, humans often support the production of ecosystems at the expense of other services (provision of drinking water, regulation of water flow, climate regulation, etc.). Additionally, globalization has magnified this effect through the global displacement of land use (Weinzettel et al., 2013). As transport costs become cheaper, crops are exported over long distances. For example, the increase in soybean production in Brazil is mainly due to an increase in exports of this commodity, initially to the EU and later, mainly to China. Increasing soybean production in Brazil is one of the main causes of deforestation in the rainforest and, in particular, the decline of the shrubby forest-steppe ecosystems called Cercado, which is being converted to soybean fields. This has major consequences not only on the loss of biodiversity but also on the global carbon cycle. For example, soybean fields hold less carbon than the original forest and changes in land use will also affect the water cycle from erosion, etc. (Box 1.3).

1.2 Metabolism of Human Society

We usually consider processes in nature in physical units describing the amount of substances, energy flows, etc. In the previous section, we mentioned that the functioning of natural systems can also be expressed in money, or in the financial benefit for human society. On the contrary, when we talk about human society, we usually talk about the productivity of the economy, gross domestic product, well-being and similar indicators, the measure of which is mostly money in one form or another. However, human society also consumes real physical goods, raw materials, crops, water and conversely, deposit the products of its metabolism in the environment in the form of waste, contamination, etc. (Fig. 1.6). A discipline that seeks to view human society from the perspective of energy and material flows is called Metabolism of Human Society (Brunner & Rechberger, 2002; Hák et al., 2015). These energy and material flow through human society are important for natural processes because they represent a significant proportion of the total flows on the planet and also because human significantly changes the structure and function of ecosystems to best meet the requirements of social metabolism. These flows can be viewed in different ways, we can relate them to a specific area, or municipality. We can also deal with the ecological footprint of products, or the entire complex production process. Given that the exchange of goods is an important feature of the human economy, territorially conceived flows include both flows of substances and energy from local sources and flows from imports as well as those that are intended for exports. Similarly, if we look at the consumption of material, we can distinguish between direct material inputs. In other words, the consumption of material on site. Indirect material inputs are materials, which have been consumed in the production of the imported products. In addition, it should be kept in mind that besides the

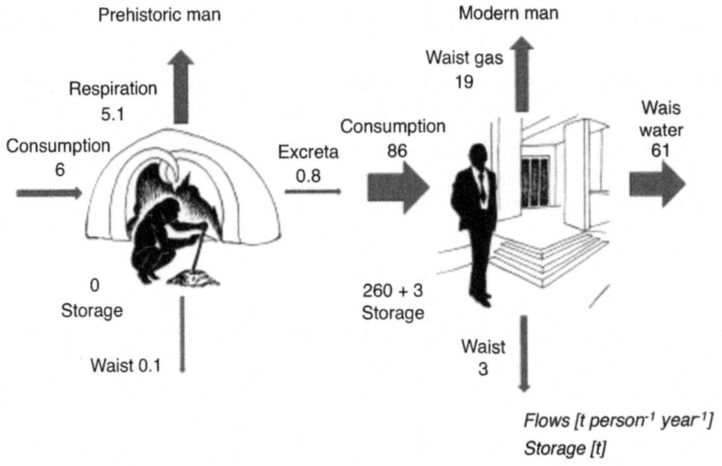

Fig. 1.6 Main material flows, consumption of goods, and production of waste in t person^{-1} year^{-1} according to Brunner and Rechberger (2002) edited

material (and energy) flows directly used by human society (used flows) such as coal for heating, there are also so-called unused material flows, such as material flows that are directly related to extraction and using the given product, but we will not use them directly, for example, heap of overburden generated by mining. Many of these indirect and unused materials flows originated in places that are spatially separated from the consumption of products. Thus, most of the environmental impacts also remained in these places, which is why we often do not even realize them. Therefore, a number of indicators have been created such as the so-called carbon footprint, water footprint, etc. These indicators describe how demanding of a certain type of product is for individual energy and substance flows.

The synthetic indicator of human consumption of natural goods is the so-called ecological footprint. This is defined as the total area of ecologically productive land and water area used exclusively to secure resources and assimilate waste produced by a given population (or individual), with a locally normal way of life using common technologies. However, this area may not only be the area of local ecosystems where the individual or population lives but also the use of ecosystems in other parts of the planet that are exploited as a result of imports of products, raw materials, energy or services used by that individual or population. This includes, for example, an area of rainforest in Southeast Asia that has been transformed into an oil palm plantation to meet our demand for palm oil. In connection with this, we express the ecological footprint as the consumption of so-called global ha (g ha). In 2016, the average ecological footprint of the average inhabitant of the planet was 2.7 global hectares (g ha). The differences between countries are considerable, highest ecological footprint being in 2016 was Quatar (13.5 g ha) and the lowest in Timor-Leste and Eritrea (0.5 g ha), while the USA had a footprint of 8.1 g ha and Europe 4.5 global g ha (https://data.footprintnetwork.org). These figures are effected not only by the amount of directly and indirectly consumed goods and services but also by how the material flows are modified. In nonindustrial societies, for example, most of human excreta or so-called night soil has been used as a fertilizer. The excreta are spread in large area and their concentration is in the range when natural processes can handle them and consequently, nutrient contained in them become part of natural nutrient cycles. For example, nitrogen mineralizes into ammonia and nitrate, which can be used by plants (Oinam et al., 2008). This locally increases nitrogen availability but globally causes only small alternation of N cycle. It largely mimics N flows in nature where most of the N in biota recycle between plants, other organisms, and soil.

In industrialized societies, where human feces becomes part of wastewater, the feces gets mixed with other waste and polluted by other contaminants which made their fertilizer use more complicated. In the end, we struggle to remove N in wastewater treatment plants, where instead of being used by plants, N, to a large extent, turns into N_2 through denitrification (Ni et al., 2017). Although denitrification allows us to get rid of excess N, it also produces NO_x, which are important greenhouse gases (Wunderlin et al., 2012). On the other end, we use Haber–Bosch synthesis to increase the amount of reactive N which enters into the human food chain. This example shows that we not only change the intensity of nutrient cycling but also change alternate major pathways of the cycle, which as will be shown later, that has many consequences on ecosystem functioning.

Human society obtains a variety of goods from nature (this includes the production services of nature described above) and at the same time takes care to eliminate the products of social metabolisms, which mostly falls under the regulatory ecosystem services. At the same time, human is changing ecosystems, as described above, to affect their ability to provide different types of services. At the same time, a number of social influences determine what products we obtain from nature and how we do it. For example, the increase in demand for palm oil (used as a biofuel and raw material, especially in economically developed countries in a temperate climate) has led to a significant increase in the size of oil palm plantations at the expense of native ecosystems (mostly in tropical countries). The emergence of cities during the development of human civilization enabled the concentration of people in one place and thus, specialization in different professions and division of labor. This has accelerated technological development and brought greater prosperity to people, especially the urban population. Over time, this has led to more and more people living in cities and relatively fewer people in the surrounding countryside. This is possible, among other things, because the development of technology allows fewer people to produce the same or more food. These technologies such as the use of machinery, fertilizers, pesticides, etc., then change the way we produce food and thus the impact we have on ecosystems.

Box 1.3 The Appropriation of Net Primary Production and Its Impact on Ecosystems and the Shift of the Environmental Burden by International Trade

Human appropriation of net primary production (HANPP) was introduced by Vitousek et al. (1986) as a comprehensive indicator that reflects both the amount of land used by humans and the intensity of land use. The appropriation of plant production by removing ecosystem production that serves as human food, as feed for livestock, or raw materials for further use is one of the oldest human activities. Subsequently, human began to change ecosystems to increase the part of production he wanted to appropriate, but this often reduced the total production. Human changes the amount of primary production in ecosystems that are available for ecosystem processes in two ways: (1) due to a change in the type of ecosystem, it changes its total primary production (e.g., if we dry a wetland and transfer it to a field, production will decrease); (2) by appropriating part of the primary production (e.g., harvesting, logging, etc.), it reduces the part of the total production available for ecosystem processes. HANPP, as has been said, is therefore equal to the primary production contained in the harvest (NPPh) and the change in production due to land-use change (ΔNPPLC):

$$HANPP = NPPh + PNPPLC$$

(continued)

Box 1.3 (continued)

In order to be able to estimate ΔNPPLC, we must therefore know or estimate the net primary production of potential natural vegetation NPPpot and further measure the current net primary production of NPPact.

$$PNPPLC = (NPPpot - NPPact)$$

Not surprisingly, the greatest rate of appropriation of primary production is in densely populated areas (Fig. 1.7). HANPP changes the input of energy into food webs in ecosystems. Based on the so-called energy hypothesis (Wright, 1983), it is expected that less energy flowing through the ecosystem means that fewer individuals will feed and food chains will be shorter (have less trophic levels), which in turn can reduce the number of species in these ecosystems. These effects are expected to be particularly significant in soil ecosystems, as in natural ecosystems most primary production flows through the detritus food web in soils and the supply of organic matter to soil also affects soil organic matter, which in addition to diversity largely effect ecosystem functions, such as water and nutrient retention in ecosystems (Miko & Storch, 2015). However, empirical data so far suggest that NPP and stock of biomass act is key in this regard rather than HANPP. Thus, it is not the amount of energy taken, but the amount of energy that remains in the system (Haberl et al., 2009). HANPP is directly or indirectly relevant in the context of global biogeochemical flows (Steffen et al., 2004; Gerten et al., 2005).

Moreover, products derived from the use of appropriated biomass do not have to be consumed locally but are often exported. Therefore, some countries or communities may have high HANPP values on their territory, but this may not reflect their HANPP consumption because much of the appropriated primary produce is exported. However, the resulting environmental impacts remain. An illustrative example of such a phenomenon is the loss of rainforests in connection with the transfer to crop plantations, especially soybean fields or oil palm plantations. These crops serve only partially to satisfy local consumption and are largely exported. Many of these crops are the reason for the eradication of natural habitats. A typical example is the expansion of oil palm plantations at the expense of rainforest. Palm oil is the most widespread vegetable oil in the world, in 2014/2015 (market year from October to September) its production was almost 62 million tons. A large number of products (with a slight exaggeration, half of the products in the supermarket) from food to cosmetics, candles, and biofuels contain palm oil or products derived from palm oil. In total, 27 million hectares (an area roughly the size of New Zealand) have been transferred to oil palm plantations (Fig. 1.8), leading to a significant loss of biodiversity (Fitzherbert et al., 2008).

(continued)

Box 1.3 (continued)

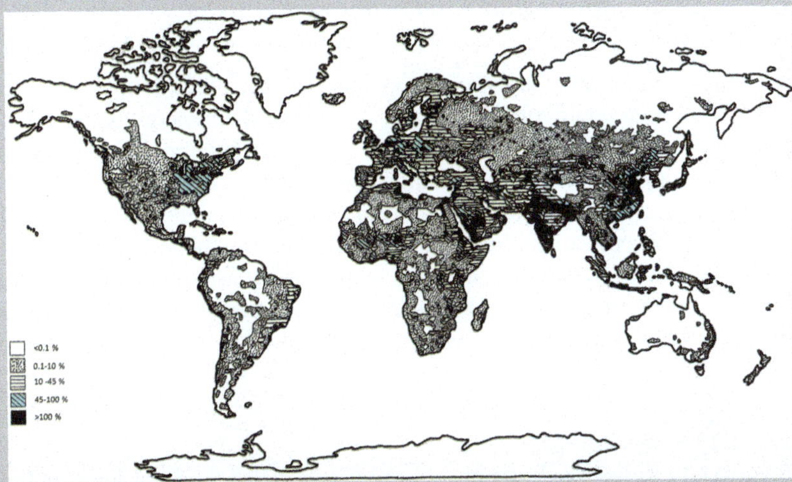

Fig. 1.7 Human appropriation of net primary production HANPP (Modified from Haberl et al., 2009)

Fig. 1.8 Oil palm plantations and related logging of the original forest in Borneo

1.3 Primary Production and Factors Influencing It

1.3.1 Photosynthesis and Energetics of Plants

Plants and some other photoautotrophic organisms convert solar energy into chemical energy through photosynthesis (Box 1.4). Part of their energy is immediately released by plant respiration and this energy is used for basal metabolism, renewal of membrane enzymes, etc. The remaining energy can be used for growth. In order for a plant to grow, the plant needs not only energy but also minerals and water. The energy needed to extract ions from the soil is a significant item of energy expenditure associated with growth, which is why we often set it aside separately. There are differences in the use of energy accumulated by photosynthesis between plants of different growth strategies.

Box 1.4 Photosynthesis

Photosynthesis has two basic phases, the first is called the light phase and is used to capture the energy of photons and convert it into chemical energy in the form of Nicotinamide adenine dinucleotide phosphate (NADPH) and Adenosine triphosphate (ATP). The second called the dark phase, uses NADPH and ATP in the set of chemical reactions, in which CO_2 is incorporated into organic molecules.

Light is captured by leaf pigments at two wavelengths about 400 nm (blue area) and about 700 nm (red area). Chlorophylls A and B are used in both areas and carotenoids are important in the blue area. The synthesis of one mole of glucose requires $6 \times 8 = 48$ moles of photons, which corresponds to an energy of 8440.6 $kJmol^{-1}$. However, only 2884.5 $kJmol^{-1}$ is released from glucose, so the theoretical efficiency of photosynthesis is 34.2%. However, there are other factors causing that only some photons reaches the reaction center and part of their energy is lost along the way, this and other constrains, causes the actual efficiency is much less (Table 1.1).

NADPH and ATP are further used in subsequent reactions, also called the dark phase of photosynthesis. The basic cycle is the so-called Calvin cycle that occurs in C3 plants (Fig. 1.9). The name C3 comes from the fact that the first stable compound formed by the incorporation of CO_2 has three carbons. It is formed by the binding of CO_2 to ribulose-1,5-bisphosphate (by the enzyme Rubisco-ribulose-1,5-bisphosphate carboxylase/oxygenase). However, Rubisco also has an oxidative activity that leads to the cleavage of photosynthetic products called photorespiration. The C3 cycle is mainly used by plants of temperate and cold regions, as the temperature in these areas is not high and thus photorespiration does not predominate over photosynthesis. Rubisco is also very discriminating against ^{13}C, so C3 plants are relatively depleted in ^{13}C.

(continued)

Box 1.4 (continued)

Table 1.1 Efficiency of photosynthesis

Out of sunlight, 47% is outside the applicable wavelength
Of the remaining 53%, (400–700 nm) 30% of the photons are lost due to imperfect absorption (they collide with something else than the chloroplast, turn to heat, or radiated back by fluorescence)
From the 37% (absorbed photons), 24% is lost due to the degradation of the energy of shortwave photons to the energy level of 700 nm photons
The remaining 28.2% (solar energy retained by chlorophyll) is converted to ATP-α-NADPH- and subsequently to glucose with 32% efficiency (this give 9%)
9% of energy is the accumulated in sugars, out of which 35–40% is consumed by respiration and photorespiration of the plant
Maximum theoretical effectiveness is thus about 6%
But the actual efficiency in field conditions is 0.1–2%

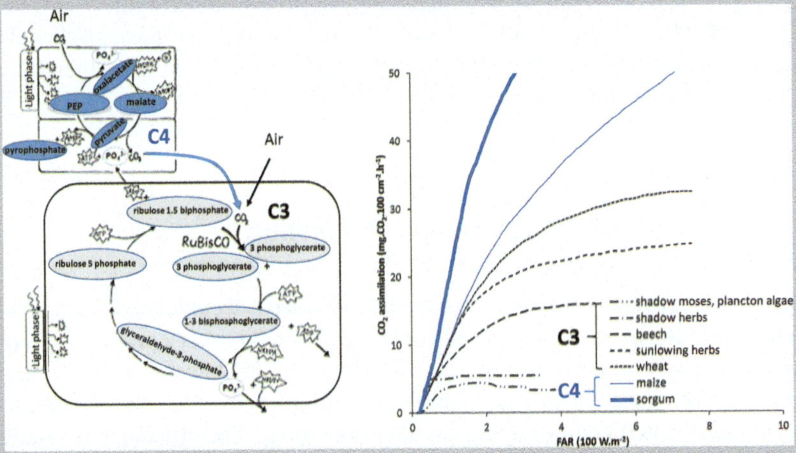

Fig. 1.9 Scheme of Calvin and Hatch–Slack cycle and dependence of rate of photosynthesis on the amount of light for C3 and C4 plants

Another widely distributed cycle is the Hatch–Slack cycle or C4 cycle (substances with four carbon atoms are formed by CO_2 incorporation). In this cycle, the CO_2-catalyzing enzyme is phosphoenolpyruvate carboxylase (PEP), which catalyzes the carboxylation of pyruvate to oxaloacetate. It further changes to malate and CO_2 is split from it, this CO_2 enters the Calvin cycle and pyruvate is formed and the whole cycle is repeated. The C4 cycle thus forms a kind of extension of the Calvin cycle. The C4 cycle is mainly used by thermophilic plants because at elevated temperature photorespiration is more pronounced, and thus the efficiency of C3 photosynthesis may decrease. C4 plants are well adapted to high light intensity and make better use of water.

(continued)

Box 1.4 (continued)

These are mainly subtropical grasses. Corn, sorghum, and sugar cane belong there. In addition to these two cycles, there is a crassulacean acid metabolism (CAM cycle) used by desert plants.

The rate of photosynthesis is affected by light, CO_2 concentrations, heat, and water. Water is important as a hydrogen donor distributes assimilates and introduces ions. In the absence of water, plants close their stomata, which limits the availability of CO_2. Photosynthesis increase with increasing temperature, reaching an optimum rate at temperatures between 15 °C and 25 °C. However, above 30 °C, there is a significant decrease in the rate. The reaction to temperature and water availability differs between C3 and C4 plants (Fig. 1.9).

C3 and C4 plants differ in their requirements for CO_2 and light. In general, increasing the concentration of CO_2 increases the rate of photosynthesis. In C4 plants, this increase stops around 500 ppm, in C3 plants around 750 ppm. In terms of light, C4 plants have a higher compensation point (i.e., the value of light when photosynthesis prevail over respiration), but they reach light saturation at a much higher light intensity than C3 plants. C4 plants, therefore, perform well a high degree of light intensity (Fig. 1.9).

1.3.2 Factors Affecting Primary Production

The plant immediately respires some of the assimilates accumulated during photosynthesis (Foley et al., 1996). What is left is called net primary production (NPP). The efficiency in which a plant produces NPP is measured as the energy contained in biomass (produced in a certain area over a period of time) divided by the energy of sunlight, (which comes on a given area over time). The efficiency is relatively small, usually tenths of a percent, at maximum of few percent. NPP is given in units of biomass per area and time. The amount of biomass produced can be given either as mass (i.e., g m^{-2} $year^{-1}$) or as energy (J m^{-2} $year^{-1}$). In order to convert mass into energy, we use the so-called combustion heat. This is the amount of heat released by the complete combustion of the material in an oxygen atmosphere and determines the total amount of energy contained in the material. In most plant tissues, the combustion heat value is around 17 kJg^{-1}.

In addition to NPP, we sometimes consider the so-called gross primary production (GPP), which we can calculate by the addition of respiration for a given period to the NPP (see Box 1.5). In the following text, we will deal with the factors influencing primary production and its role in ecosystems where primary production is the basis of food chains and the functioning of ecosystems. See basic ecological concepts, which are briefly explained in Box 1.6.

The main resources necessary for primary production are light, water, CO_2, and mineral nutrients (Chapin et al., 2011). In nature, light availability is often associated with higher temperature because areas with higher solar radiation are usually warmer. In addition, heat increases the rate of enzymatic reactions. The most important factors determining the global distribution of primary production in terrestrial ecosystems are solar radiation, water, and nutrient availability. Solar radiation, as mentioned, affects both the availability of energy for photosynthesis and temperature. Therefore, it is not surprising that primary production in terrestrial ecosystems shows the closest correlation with humidity and temperature. This is especially true with the annual actual evapotranspiration (AET), which combines both of the above parameters. AET is the amount of water that evaporates from ecosystem during the year that depends on both temperature and humidity. For example, in a system where there is enough water, but has a low temperature (such as tundra), less water evaporates because the lack of heat, as a considerable amount of thermal energy is needed to evaporate water. Conversely, in a system with a large amount of available thermal energy but with a lack of water (such as a desert), the AET will also be small because it will be limited by the availability of water. In order for the AET to be large, both the temperature and water available must be high.

In addition to radiation and water, nutrients are another important parameter determining terrestrial primary production. The most limiting nutrient in terrestrial ecosystems is usually nitrogen, relatively often phosphorus (especially in the tropics), and less often other nutrients. This is because relatively few organisms (only those that fix atmospheric N) are able to use large reserves of nitrogen that occur as a gas in the atmosphere. Most others rely on recycling nitrogen from dead plant and animal bodies. Once nitrogen from dead biomass is mineralized to ammonia or nitrates, it becomes relatively mobile in the soil and, if not taken up by plants or microorganisms, can be removed from the soil by leaching (Vitousek et al., 1991).

The availability of nitrogen and phosphorus varies with latitude (Huston & Wolverton, 2009) (Fig. 1.10). In the tropics, rapid weathering occurs due to high

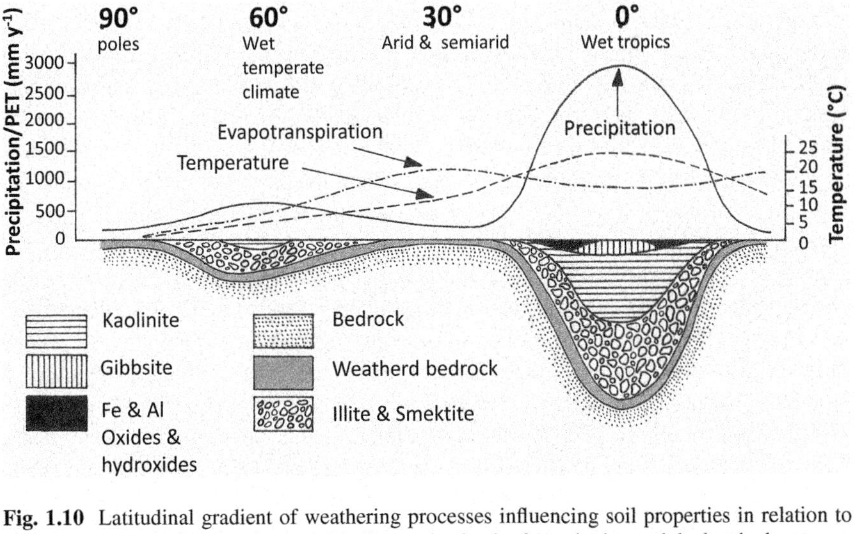

Fig. 1.10 Latitudinal gradient of weathering processes influencing soil properties in relation to climate. The depth of the underground indicates the depth of weathering and the hatched areas are the predominant products of weathering. (According to Strakhov, 1967, modified)

temperatures and precipitation, causing a significant loss of nutrients that plants obtain from rocks. In particular, phosphorus, basic cations (Ca, Mg, K), and many other microelements are lost in this way. This loss of mineral nutrients is accelerated by the fact that the soils of the tropics are often very old, i.e., weathering takes place here over a long period of time. In addition, clay minerals with lower sorption, such as kaolinite, form in tropical climates, which means that tropical soils are less able to retain nutrients by sorption. This increases the loss of basic cations and soil acidification (Huston & Wolverton, 2009), which in turn reduces the availability of phosphorus and some other mineral nutrients. On the other hand, temperature and precipitation are lower in the temperate zone, and thus there is slower weathering and nutrient loss. Moreover, a significant part of the temperate zone was directly affected by glaciation. The movement of the glaciers exposed the layers of unweathered rocks. In the areas where glacier retreated, there were large erosions. Eroded material transported by water and wind erosion formed massive layers of sediment (e.g., loess), which is often rich in mineral nutrients. As a result, temperate soils are richer in phosphorus and other minerals than tropical soils (Huston & Wolverton, 2009).

Nitrogen, on the other hand, is obtained primarily from the atmosphere; this source is equally available in the temperate zone and in the tropics. Biological nitrogen fixation is closely correlated with evapotranspiration and is greater in tropical than in temperate regions (Cleveland et al., 1999). Moreover, nitrogen is constantly lost from the ecosystem by leaching and is therefore limiting in many ecosystems (LeBauer & Treseder, 2008). Due to the fact that temperate soils are well supplied with phosphorus and nitrogen fixation is lower than in the tropics, nitrogen is the most limiting element in a temperate climate. Conversely, in the tropics, where the availability of P is severely limited, phosphorus may be more limiting than nitrogen in many places.

As we will show in the following chapters, mankind has now doubled the nitrogen input to ecosystems (Vitousek et al., 1997). This is done both by the application of nitrogen fertilizers obtained by Haber–Bosch synthesis and by the combustion of fossil fuels which release nitrogen oxides. The increase in nitrogen concentration will be reflected more in the productivity of temperate ecosystems, which are more limited by nitrogen than, e.g., in tropics limited often by phosphorus. In addition, most economically developed countries, which supply relatively more nitrogen to ecosystems, are in the temperate zone. As a result, the production of temperate ecosystems is growing. If we look at agroecosystems where the supply of nutrients through fertilization is most intense, then the production is significantly higher in temperate agroecosystems than in tropical ones (Green et al., 2005). In other words, the opposite direction than for natural ecosystems. This is due to a number of factors. As mentioned above, temperate soils are younger and therefore naturally better supplied with nutrients, including microelements. The most significant is the limitation by nitrogen and its delivery has a great influence on the yield. In the tropics, it is necessary to supply phosphorus and other nutrients, including microelements. Clay minerals of temperate soils (illite and smectite) have better sorption properties

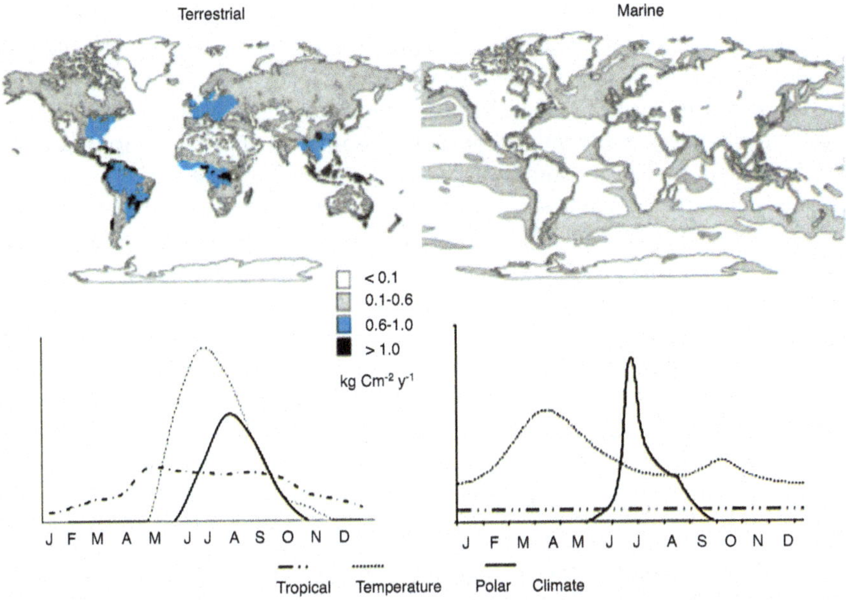

Fig. 1.11 Net primary production on land and in the oceans as well as seasonal changes in primary production in tropical, temperate, polar terrestrial ecosystems, and seas. The lower graphs show the relative distribution of primary production during the year. On the x-axis, there are the individual months of the year and on the y-axis the primary production in individual months. The scale is adjusted so that the maxima of land and sea are the same. (Adapted from Huston & Wolverton, 2009; Falge et al., 2002; Rifai et al., 2018)

and retain nutrients in soil better than kaolinite, which is an important clay mineral in the tropics (Strakhov, 1967) (Fig. 1.10).

Production, especially the ratio of production to existing biomass, is also influenced by the type of plants. It is also related to how nutrients are available and how the plants use them. Some plants have a large perennial biomass and the annual increase (production) is much smaller than this biomass supply. While other plants that emerge from seeds each year, produce relatively large biomass production during the year and overwinter again as seeds.

To summarize (Foley et al., 1996), most production in terrestrial ecosystems is provided by vascular plants. Mosses, lichens, and algae have only minor involvement especially in places where growth is low due to low temperatures or low humidity and vascular plants are restricted. Radiation (temperature), water, and nutrients are the main factors controlling primary production in terrestrial ecosystems. The largest values of primary production are in the equatorial regions. The production gradually decreases toward deserts with a lack of water and toward the poles with a decrease in radiation and temperature (Fig. 1.11). The supply of nutrients significantly limits primary production. Production in some fertilized temperate ecosystems may be higher than in warmer and wetter ecosystems with a lack of nutrients (Fig. 1.11). Of course, the decrease in radiation toward the poles should be perceived as an average value during the year. In temperate and polar climates, there is a large seasonal fluctuation of production during the year. In some part of the year production is close to zero but maximal values are higher than in tropics. In contrast, in the tropics, production is more balanced during the year, resulting in higher total production. However, the anthropogenic fertilization of temperate ecosystems with nitrogen may cause many of them to produce more than tropical ecosystems. Some authors even argue that a better nutrient supply in temperate terrestrial ecosystems means that current temperate biomass production may be comparable or larger than in the tropics (Huston & Wolverton, 2009). These overall comparisons are influenced by the fact that data on the production of whole ecosystems in the tropics are scarce.

Unlike land, most of the production in the oceans is provided by phytoplankton floating in the water column (Table 1.2). In the seas and oceans, the spatial distribution of production is also different. Places with intensive production are located at the estuaries where nutrients are brought from the land and where there is also a strong upwelling. On the high seas, the largest production is in subarctic and

Table 1.2 Comparison of selected characteristics of primary production in terrestrial, freshwater, and marine ecosystems[a]

Parameter	Terrestrial ecosystems	Freshwater ecosystems (lakes wetlands)	Marine ecosystems
Major producers	Vascular plants	Vascular plants and or phytoplankton	Phytoplankton
Light competition	The individual plants are usually rooted in the soil, grow from the same area, in competition for light; it is advantageous to grow higher, gain more light, and shade the others. To achieve greater height, supportive tissues with a high proportion of cellulose hemicelluloses and lignin are needed	Wetlands are more similar to terrestrial ecosystems, lakes to sea	Individual organisms float in the water column, their position depends on the mixing of water
Body composition and food web	Plant bodies are mostly less digestible to indigestible (a large proportion of cellulose, hemicellulose, and lignin), a large proportion of 90–50% production dies and the dead mass is decomposed by decomposers The biomass of consumers is small, on the contrary, the biomass of decomposers, including microorganisms, is large	Wetlands are more similar to terrestrial ecosystems, lakes to sea	The bodies of primary producers are easily digestible, a large part will be eaten by consumers, there will be less left for decomposers (approx. 25%) Consumer biomass forms a significant part of total biomass

(continued)

Table 1.2 (continued)

Parameter	Terrestrial ecosystems	Freshwater ecosystems (lakes wetlands)	Marine ecosystems
Approximate level of primary production t ha^{-1} year^{-1}	Tropical rain forest 20 Temperate forest 12 Tajga 8 Savanna 9 Steps and prairie 5 Semidesert 1	Plankton 2 Submerged plants Tropical 17 Temperate 6 Wetland vegetation Tropical 38 Temperate 75	Coral reef 20 Estuaries 18 Upwellings 2.5 High seas 0.5
Biomass (standing stock) t ha^{-1}	Tropical rain forest 440 Temperate forest 300 Tajga 200 Savanna 40 Steps a prairie 20 Semidesert 10	2–40	Coral reef 2.5 Upwellings 0.06 High seas 0.04
Biomass turnover time Years	Tropical rain forest 22 Temperate forest 25 Tajga 25 Savanna 4 Steps a prairie 4 Semidesert 14	0.02–2	Coral reef 0.125 Upwellings 0.024 High seas 0.080
Geographic distribution of production	Highest on the equator, descending to the poles and toward dry areas (deserts)	Higher in the tropics decreases toward the fields and with higher altitude depends on the input of nutrients from the basin	Highest in cold subarctic and subantarctic waters and estuaries
Limiting macronutrients	N ≥ P N in boreal and temperate climate P in tropics	P > N	N > P
Limiting micronutrients	–	–	Fe

[a] According to Chapin et al. (2011) and other authors, modified

sub-Antarctic waters (Fig. 1.10). This is because the spatial distribution of primary production in the oceans is determined by the availability of nutrients, especially nitrogen and phosphorus. Seawater must not only contain sufficient nutrients but must also have an adequate ratio of the main biogenic elements N and P. This optimal ratio is 16:1 and results from the ratio of the main macroelements C:N:P in algae 106:16:1, the so-called Redfield ratio (Redfield et al., 1963).

The content of macroelements and a suitable N:P ratio occur around continents and at the mouths of large rivers, in the open ocean, and in areas where deep sea currents emerge through the upwelling to the surface. These deep-sea currents are part of the so-called thermohaline cycle (Box 1.7) and enriching surface waters with nutrients originating from the decaying dead bodies of organisms falling to the ocean floor. These currents are cold but less salty than surface waters. Surface waters are saltier and warmer, and higher temperatures make them lighter. However, as water reaches cold areas closer to the poles, it cools, increases in density, and begins to sink to the bottom pushing nutrient-rich deep currents upward a process called upwelling. This is even more supported with sea currents flowing parallel to the coast. Thanks to Coriolis forces, they are diverted from the coast and this surface water is then replaced by deep water rich in nutrients.

Interestingly, on the map of global production, we can also notice thin bands of high production directly on the equator. These are caused by the action of eastern winds on the sea surface. Winds blowing from the east push the water west. In addition to the wind, the so-called Coriolis forces also act on the water masses. These are forces caused by the earth's rotation because gas and water tend to move clockwise in the northern hemisphere and counterclockwise in the southern hemisphere. As a result, in addition to the east–west direction, the water also acquires the direction going in both hemispheres toward the poles. As a result, the water on the surface of the ocean seems to spread out from the equator to the poles pushing on the lower layers around the equator. The water around the equator is thus, under greater hydrostatic pressure and at the equator under less hydrostatic pressure. This causes a rise of nutrient-rich deep water toward the surface. This is further enhanced by the wind pushing warm surface waters toward the western side of the Pacific Ocean. Due to the lower density of warmer water, the sea level is about 60 cm higher on the western side of the Pacific than on the eastern side when the wind blows at full strength. This tilts the thermocline which is deeper on the west side of the Pacific, but closer to sea level on the east side from where the wind shifted the surface layers of water to the west. This further facilitates the upward movement of cold water in the eastern equatorial Pacific on the principle described above. This will then increase the productivity of these waters (see also Sect. 1.7 and the boxes contained in it).

There are also areas of the oceans where there are enough macronutrients, and yet there is little production. It is believed that production here may be affected by a lack of microelements, especially iron, or possibly in combination with zooplankton predation pressure. Iron is crucial for a number of enzymatic reactions including nitrogen fixation. While on land, there is usually enough iron (Fe is the fourth most common element in the earth's crust) in the seas it can be deficient. The main sources of iron in the oceans themselves are ferrous manganite crusts and hydrothermal springs at the ocean floor. These are located mainly in the open ocean, often separated by a thick layer of water column from the euphotic layer on the surface of the ocean where there is enough light to support photosynthesis. There are exceptions, some studies show that in some parts of the ocean, especially at higher latitudes, these deep-sea sources can be a significant source of Fe (Tagliabue et al., 2017; Ardyna et al., 2019). However, the main source of iron in the seas and oceans, especially in lower latitudes, is iron coming from land. It comes in sediments suspended in rivers, but mainly as dust transmitted by wind deposition. This wind-borne dust comes mainly from parts of the continents with little vegetation cover, especially deserts. This may be one of the reasons why the production of the seas and oceans is higher in the northern hemisphere than in the southern one. In the northern hemisphere, there is a larger proportion of land and therefore there are more seas that are in contact with the land. More land also gives more desert area which is the main source of aerosols bringing Fe but also other nutrients (P) to the seas. This supply of iron by aerosols and erosion of land is crucial, especially in lower latitudes, where it significantly promotes bacterial nitrogen fixation (Tagliabue et al., 2017).

As we have seen, the productivity of the world's oceans depends to a large extent on the supply of nutrients. This is largely due to the mixing of the lower and upper layers of the ocean. This creates upwelling areas especially in the cold subpolar seas and around some coasts. Subtropical seas water is mixed only occasionally, usually as a result of some catastrophic events, such as hurricane passages (Babin et al., 2004). These nutrient-poor areas of the oceans are also called oceanic deserts. The size of these deserts can be increased by factors that limit the mixing of seawater between the surface and deep layers. One of the factors, that reduces the likelihood of mixing, is the increase in sea surface temperature. As the surface temperature of the seas increases, in connection with global climate change, the difference in densities between the sea surface and the deep sea increases which reduces the chances of colder deep nutrient-rich waters rising to the surface (see also Sect. 1.7 and boxes there). A NOAA study showed that between 1997 and 2008, as sea surface temperature increased, ocenic deserts increased by 15% amounting to 6.6 million km^2 (Polovina et al., 2008).

The mixing of the seas also indirectly affects the availability of light. Algae occurs in the mixed layer of the sea. Its depth ranges from tens to hundreds of meters and is generally greater in colder waters (i.e., it rises toward the poles and in the winter months). A larger mixed layer can also mean a larger algae biomass overall. However, algae can only photosynthesize if they have light available. We are talking about the so-called euphotic layer which is a layer where light reaches 1% of the value of light on the surface. The depth of the photic layer depends on the transparency of the water, but in the open ocean, it is about 200 m. When the depth of the mixed layer falls below the limit of the euphotic layer, the total production begins to decrease because part of the algae spends part of the time outside the photic layer. In the tropics, primary sea production is very stable during the year but due to a lack of nutrients, the values are much lower than in the temperate zone. Maximal values in the polar regions are even higher due to sufficient nutrients and a long day in the summer (Fig. 1.11).

Even more complex are the factors controlling production in freshwater. For many freshwaters especially running waters, not only the primary production inside aquatic ecosystems (autochthonous) but also the organic matter coming from the surrounding terrestrial ecosystems (allochthonous) may be important for secondary production. In general, with some simplification, standing water bodies have a larger primary production than streams or rivers under similar conditions. The production of comparable habitats is usually higher in the tropics and decreases toward the poles and to higher altitudes. The main producers can be very diverse, in streams and upper course of rivers, algae attached to stones (periphyton) will predominate.

In lower courses of rivers and some reservoirs submerged macrophytes or nonsubmerged macrophytes become important if the water layer is shallow enough. While in river reservoirs, phytoplankton predominates. Unlike terrestrial ecosystems, phosphorus is more often a limiting element in freshwater.

Box 1.5 Measurement of Production and Biomass

By primary production (or net primary production NPP), we mean the increase in biomass, the mass of plants (or more often the dry mass of plants), over time. In order to be able to measure production, we must therefore measure the biomass of producers at least two times (but preferably several times) or continuously measure a parameter related to production (e.g., oxygen production or CO_2 consumption). In exceptional cases, it is sufficient to measure biomass only once (e.g., harvesting annual plants) because we can make some reasonable assumptions about what the biomass was in the beginning. Some methods used to estimate production and biomass are summarized below.

The simplest method of measuring NPP is the harvest method. It can be advantageously used on annual plants. Its principle is that we harvest plant biomass in a given area at a time when the biomass has reached its maximum and we assume that the initial biomass was equal to the seed biomass which we either somehow estimate or neglect. For annual plants, we can thus estimate above-ground and belowground biomass. If we consider only above-ground production, we can also use this method for perennial plants that regrow every year (e.g., meadows) and where we can assume that above-ground winter biomass was zero and then the harvest at the time of maximum biomass is an estimate of annual production. For example, in meadows, this method is one of the most widely used methods for determining NPP. As we have already said, it usually does not include belowground biomass (root production), the production of root exudates, parts of the plant that died before harvest, or get decomposed or were eaten by herbivores. It is estimated that these uncounted items usually represent more than half of the NPP. We describe it in such detail here because we can demonstrate that the seemingly simple method has a number of pitfalls and a source of error. The same applies to other methods where we may not go in such detail. These errors may not be a big problem if we are, for example, comparing NPP in different places where we can assume that errors are similar everywhere. However, the situation is different when, for some reason, we need exact numbers.

The disadvantage of the harvesting method is that it is laborious and destructive. Therefore, a number of methods have been developed to monitor biomass nondestructively using some indirect indicators. Here belong group of methods that consist of repeated estimations of biomass by some indirect method. Methods based on allometric equations are often used. These equations are based on a number of empirical measurements that estimate the biomass of individuals on the basis of some easily measurable parameters. For

(continued)

Box 1.5 (continued)

Fig. 1.12 Examples of different methods of measuring primary production: eddy covariance (**a**), measuring CO_2 production by respiration, comparing CO_2 production by respiration and its consumption by photosynthesis (**b**), the dark and light bottles method (**d**), the harvesting method (**e**), measuring tree dimensions and its calculation on the basis of allometric equations (**f**), and their comparison with NDVI (**g**)

example, in forestry, methods for estimating a tree's biomass based on its trunk height, diameter, or a combination of these parameters are common. This will allow us to measure parts of the forest repeatedly over time and see how it grows. These methods can be extended to the determination of some parameters of the stand and estimation the biomass of the stand, based on correlation with these parameters, for example, from its height and cover. The advantage of such a procedure is that the height and foliage coverage can be measured by remote sensing. For example, from an airplane or a combination of lidar and multispectral scanners and thus, we can (repeatedly) estimate biomass over a larger area (Fig. 1.12).

The next group of methods we will mention here are methods based on the exchange of gases between the ecosystem and the environment. These methods are based on the fact that biomass production (i.e., photosynthesis) consumes CO_2 and produces O_2. Conversely, biomass consumption during respiration produces CO_2 and usually consumes O_2. By measuring gas exchange, we can quantify these processes. One of the old and very elegant methods of this type is the black and light bottle method used in hydrobiology. Using this method, one fills three bottles with water, water in one bottle is used to determine the

(continued)

Box 1.5 (continued)

actual oxygen concentration in the water, the other two bottles are closed and incubated in the water column (Fig. 1.12). However, one of these bottles will be transparent and the other is not. After some time, bottles are collected and the oxygen content in both bottles is determined. If we performed the experiment during the day, then photosynthesis took place in the transparent (light) bottle by the action of phytoplankton and the oxygen content increased. Simultaneously with photosynthesis, respiration takes place here, so the observed increase is the result of the balance between an increase in O_2 produced by photosynthesis and O_2 consumed by respiration, or in other words, is proportional to NPP. On the other hand, only respiration takes place in the dark bottle, so there is a loss of oxygen. If we add this respired oxygen to the NPP, we get the GPP.

In terrestrial ecosystems, we can measure the production and consumption of CO_2 using chambers of different sizes, which cover part of the soil and vegetation. Using transparent chambers, we can measure the total gas exchange between the ecosystem and the environment (balance between respiration and photosynthesis). If the chamber is dark, then we can measure the overall respiration of the system. However, chambers can cover only limited area, to cover whole tree or part of the forest would be difficult. For tree-based ecosystems, or over large areas so-called Eddy covariance can be used to monitor the gas exchange between the atmosphere and the ecosystem. They are based on monitoring the turbulent flow between the atmosphere and the surface. During the day, thanks to solar energy the surface is warmer. Vortices (eddies) of warmer air are then moving off the surface, they rise upward and are replaced by vortices of colder air from the atmosphere. Eddy covariance measurements are based on placing instruments measuring the speed of air movement in 3D, the concentration of CO_2 and water vapor in these eddies. Upward eddies will be depleted of CO_2 during the day compared to downward eddies. At night, it will be the other way around. If we measure often enough (ordinary Eddy towers measure nine times per second) with a little math (which goes beyond scope of this text), we can calculate how much CO_2 was absorbed or emitted in a given time period (day, month, year, or how long we measure) by the ecosystem. It should be emphasized that both chambers and towers measure the total exchange of CO_2 between the atmosphere and the ecosystem, which is given by the difference between NPP and heterotrophic respiration. Because most production in terrestrial ecosystems ending in soil, where it has been decomposed, we often neglect other components of heterotrophic respiration and estimate NPP as the total gas exchange after deducting soil heterotrophic respiration. Soil respiration can be measured using chambers similar to those mentioned above. The problem is that this soil respiration often consists of both heterotrophic respiration (decomposition) and root respiration. Root respiration can be very variable and represents 10–90% of total soil respiration. Separation of roots and heterotrophic respiration is possible

(continued)

Box 1.5 (continued)
by a number of methods such as measuring the respiration of soil without roots or using stable isotopes, each of which has its pitfalls.

Using the procedures described above, we can measure NPP in one or several places. But how to proceed if we want to estimate the NPP for a larger area whole country, or possibly for the whole world? The first attempts to somehow estimate the NPP were based on the use of climate data. Rosenzweig (1968) estimated NPPs using AET, which became the basis for a number of climate models estimating NPPs, who used precipitation and temperature data. Their disadvantage is that the dependence of NPP on climatic variables shows considerable variance as NPP values are affected by nutrients and other parameters. Models that try to predict NPP on the basis of radiation have a similar disadvantage as they work well in model systems not limited by water and nutrients, but the availability of these sources can be very heterogeneous in the landscape. Later, methods for estimating NPP based on plant biomass estimates made by remote sensing emerged. The most common tool for global biomass estimation is the normalized difference vegetation index (NDVI). NDVI is based on the fact that living vegetation absorbs well (and thus reflects little) in the red part of the visible spectrum (RED) and on the contrary, absorbs little and reflects more near infrared radiation (NIR). In contrast, bare surfaces and dead vegetation reflect relatively less NIR and much more RED. Thus, we calculate NDVI as the difference between the reflected NIR and RED divided by the sum of these two wavebands:

$$NDVI = (NIR - RED) / (NIR + RED).$$

NDVI correlates very well with the total area covered by vegetation. Therefore, NDVI corresponds well with vegetation biomass in stands with simple spatial architectures, ideally having only one vegetation layer. However, NDVI estimates plant biomass less reliably in stands with complex architecture. For example, it is difficult to capture differences between stands with similar cover but different trunk heights, in multiple layers of canopy, etc. The advantage of NDVI is that it can be measured using satellites for every place on earth every few weeks and allows us to calculate the real growth of aboveground biomass. However, this can be influenced by the fact that the dependence between the change in NDVI and the change in biomass is not linear. For higher NDVI values, a smaller change in NDVI means a relatively larger change in biomass for the reasons described above.

In order to address the abovementioned shortcomings of NDVI, production efficiency models (PEMs) have been developed that combine a leaf area estimate obtained by remote sensing of NDVI-like parameters with empirical estimates of maximum light use potential and possible limitations of other environmental factors, such as intensity of photosynthesis and respiration. Currently, models based on the use of data from MODIS sensors (MODIS—Moderate Resolution Spectrophotometer), so-called MODIS models, are used to estimate NPP using remote sensing. The most commonly used is MODIS17 (Zhao et al., 2005).

Box 1.6 A Brief Summary of the Main Ecological Concepts and Principles

The main ecological principles important for understanding the impact of agriculture, forestry, and fisheries are elaborated in more detail in individual chapters. However, they use terms common in ecology, which we will briefly repeat here.

All organisms need resources, energy, and building blocks from which to build their bodies. For plants, the sources of building materials are water, carbon dioxide, oxygen, mineral nutrients, and the source of energy is light. In addition to water and oxygen, food is important for animals, food serves as a source of both substances for building animal bodies and a source of energy. In addition to *resources*, each species requires a set of suitable *environmental conditions*. Examples of such environmental factors are temperature, water current, pH, salinity, nature of the bottom, etc. Some entities can be both a source and at the same time determine the environmental conditions. In general, it can be said that for each source or environmental factor, each species has a set optimal value that suits it best. If the amount of a given factor is lower or bigger, the performance of the organism decreases, and individual species are able to exist within a certain range of resources and environmental conditions. This set of factors that allow the existence of certain species is called a niche. A *niche* is defined on the one hand by resources and environmental factors (so-called fundamental niche) and on the other hand, by interactions (competition, predation, occurrence of diseases, etc.) with other organisms (so-called realized niche), which limit species distribution. If the conditions are favorable, the populations of individual species grow until they reach a value that is in balance with the conditions, this value is called the carrying capacity of the environment. Its size is determined by the fact that the amount of limiting resources covers the needs of only a certain number of individuals. Thus, individuals compete for resources. *Competition* takes place not only between individuals of the same species but also with other species that use similar resources; however, competition within species is more intense because individuals of the same species have very similar, almost identical requirements while different species differ more and so may use resources differently. For example, different plants can obtain water from different depths of soil allowing species to coexist. Coexistence of species is easier if the environment is heterogeneous and each species can use a microsite that suits it better. A set of species in a certain area is called a *community*. The community can be characterized by a number of parameters such as the number of species, species composition or various diversity indices, etc. In most natural communities, most individuals or majority biomass consists of a small number of species, while most species in the community reach much smaller numbers or biomass. The community, together with the abiotic environment in which this community lives, is called the ecosystem. There is an

(continued)

Box 1.6 (continued)
exchange of matter and energy, inside the ecosystem as well between ecosystem and its surroundings. Both biotic and abiotic components of the ecosystem participate in these processes. Biotic energy flow begins mainly with photosynthesis when the energy of solar radiation produces biomass, which is then used by herbivores that eat plants. Herbivores can be eaten by predators. Network of these interactions describing who eats who is called a food web. When energy passes through the food web, it is gradually lost. We distinguish two main branches of food webs: (1) forage one, which begins with the consumption of living plant biomass and (2) detritus one, based on the consumption of dead plant tissues. The detrital food web is usually more important in natural terrestrial ecosystems.

Abiotic matter and energy flows include the water cycle and other processes driven largely by solar radiation. They are usually quantitatively larger but the biotic component can significantly affect and regulate them.

Natural ecosystems are usually in balance with their environment. If an ecosystem is disturbed or if we abandon an ecosystem that we have transformed earlier for some purpose (e.g., we abandon an agricultural field), the ecosystem will gradually change. We call this process *succession*, it involves changes in plant and animal communities, which affect the abiotic environment especially the soil, and these changes in environmental conditions further change communities. The succession is gradually moving toward a stabilized community, which we call *climax*, which is in balance with the soil and climatic conditions of the place. These changes usually last for tens to hundreds of years and are similar for similar conditions so we can predict them to some extent. Climax communities are often prone to disturbances (fires, pest gradation, windstorms, etc.). This allows species specialized in the early successive stages to persist in the natural landscape for a long time. If the climax community persists in one place for a long time (thousands of years or more), it slowly deteriorates due to depletion of key nutrients especially phosphorus, a process we call *retrogression* (Wardle et al., 2008). Disturbance usually reduces the number of species (diversity) in the community; however, over time, the number of species grows and reaches its maximum in the early and middle stages of succession. With advancing time without disturbance, the number of species then decreases again and the climax or retrogression stages already host fewer species than the middle stages of succession (Wardle et al., 2008).

1.3.3 Plant Growth Strategies, Allocation to Production of Different Types of Plants, Changes in Production, and Biomass During Succession

In addition to external conditions such as temperature, water, nutrients, plant production is also affected by plant traits, which we collectively describe as growth strategies (Table 1.3). Clearly, growth strategies are not completely independent of external conditions. Plants with a fast-growth strategy tend to grow in conditions suitable for plant growth and conversely, in less suitable conditions, we usually find slow growing plants with conservative growth strategy. We have already mentioned in the previous chapter that plants with different growth strategies allocate assimilates to growth at different intensities. It turns out that many plant traits correlate with each other and are related to the tendency of plants to grow faster or slower. In general, plants with higher growth rates (larger NPPs) have thinner leaves and have more nitrogen and last shorter, while the opposite is true for plants with slow growth strategies.

Primary production, the amount of biomass, and the rate of biomass turnover (ratio of biomass to NPP) vary depending on the availability of nutrients and the prevailing growth strategy of plants, and the succession age of the ecosystem.

If we look at the development of biomass during succession, it is perhaps not surprising that the total biomass of vegetation increases with time since disturbance. The initial rate of this increase will depend on the availability of nutrients and water. In older stages which are close to climax, the rate of biomass growth will be slower over time than in the middle stages of succession until eventually, the biomass stabilizes at some, more or less, constant value when the climax is reached. In areas that have not been disturbed for a long time (thousands or rather tens of thousands of years), or only less drastic disturbances have been applied, there is a decrease in total biomass (retrogression) which is related to the loss and poorer availability of key nutrients, especially phosphorus.

While total biomass gradually increases during succession, the rate of biomass turnover decreases. In young ecosystems, annual plants predominate, and the turnover time is therefore very short. In contrast, the lifespan of trees in climax

Table 1.3 Plant properties associated with a conservative and fast-growing strategy

Fast-growing strategy	Conservative strategy
Thin leaves large specific leaf area	Thick leaves
Short living leaves fast turnover	Slow foliage turnover often evergreen
Higher N content in leaves	Lower N content in leaves
High rate of photosynthesis	Low rate of photosynthesis
Grow well in nutrient-rich soils, suffer when nutrients are limited	Grow even in low amount of nutrients
No mycorrhiza or arbuscular mycorrhiza	Ectomycorrhiza or ericoid mycorrhiza are more common

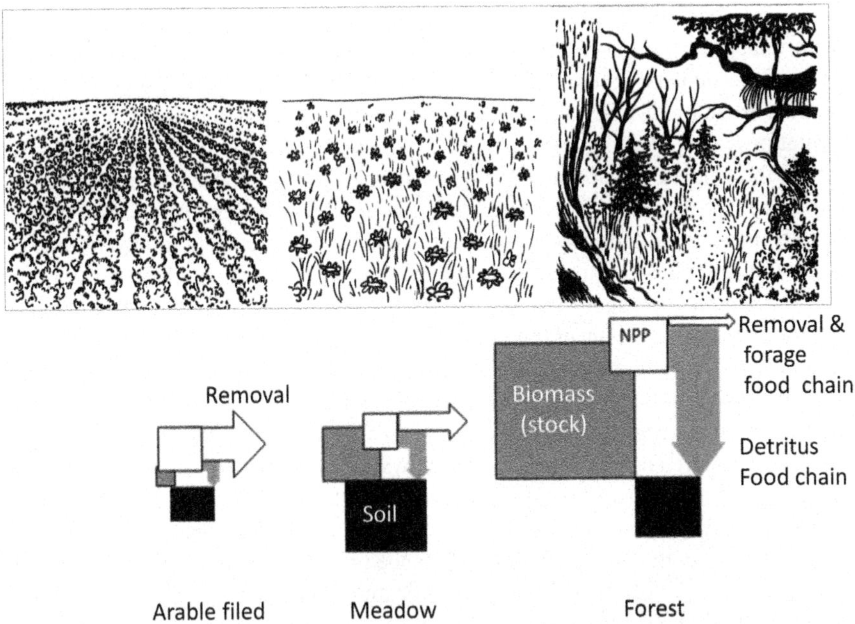

Fig. 1.13 Main reservoirs and flows of organic matter in the forest, meadow ecosystem, and in agroecosystem. Gray squares indicate annual average biomass, white annual biomass production, and black average soil supply of soil organic matter (data from Lal, 2004; organic matter is calculated as twice the soil carbon). Arrows indicate the approximate distribution of NPP between decomposition (gray) and grazing (white) food chains

ecosystems is much longer (decades or even centuries) and therefore the turnover time is longer.

The combination of these two trends shows that at the beginning of succession, the total production will be small because only some plants are able to colonize the newly released area. With the advancing succession, the production will increase and will peak in the middle phases of the succession, latter toward the climax it will decrease again. In the middle stages of succession, most of the production is easily available for herbivores. Therefore, many preagricultural societies burned the landscape to maintain the large area of habitats typical of the middle stages of succession (Native Americans and Australians still do so today). It is believed that in North America, the area of long-grass prairies has been greatly shifted east because of the burning of these ecosystems by the Native Americans.

In Fig. 1.13, the main flows and reservoirs of organic matter in ecosystems of various land uses are summarized. It is evident that in agroecosystems, especially on arable land, production is maximized, but the volume of biomass and the amount of organic matter in the soil is reduced. In other words, with increasing intensification more production is deviated to forage food chain (a food chain starting with the consumption of live plant biomass by consumers, including livestock and humans) and decrease the share of NPP going to the decomposer food chain (food chain starting with dead organic matter, more about food chains in Sect. 1.7).

As mentioned above, many ecosystem services depend directly or indirectly on the primary production available for natural processes. For other ecosystem services, in particular regulatory services such as climate control, protection against erosion, retention of water and nutrients in ecosystems, and support services such as habitat provision, diversity support, etc., standing biomass is essential. Finally, for others, the amount of carbon entering the decomposer food web in the soil is crucial. For example, it can be expected that the capture of rain energy and soil protection against erosion will depend more on the amount of vegetation present than on the rate of vegetation turnover. From this point of view, a low-yielding meadow can be a better protection against erosion than a neighboring field. In field biomass production is high but is concentrated in a short peak during vegetation, with several months to half a year completely without vegetation-or with only sparse vegetation. If we follow the example of erosion, then resistance to erosion can be supported by a larger supply of organic matter in the soil. Soil organic matter supports the formation of soil aggregates and thus improves water infiltration and mechanical resistance of soil to erosion. From this point of view, the amount of organic matter that enters the soil but also the disturbance of the soil, which affects the retention of organic matter in the soil affects the resistance of the soil to erosion.

In summary, with increasing intensification, the rate of biomass turnover increases but biomass standing stock as well as the amount of and organic matter retained by the ecosystem decreases. The ratio of NPP to biomass or the ratio of NPP to total organic matter retained by the ecosystem increases. Many ecosystem functions may depend on biomass (or total organic matter) in the ecosystem and this intensification typically reduces these functions.

1.3.4 Soils and Their Effect on Primary Production and Suitability for Agriculture and Forestry

The soil creates the basic conditions for plant growth, provides support for the roots, and retains water and nutrients that the plants can use (Box 1.7). The soil is formed under the influence of several soil-forming factors namely geological parent material, climate, terrain topography, and vegetation. All of these factors act gradually over time so, older soils differ from younger ones. During soil development, rocks are first weathered, followed by enrichment of the upper horizons of soils with organic matter produced by vegetation. At the same time, a number of transport and transformation processes take place in the soil profile. According to the nature of these soil-forming processes, we distinguish between different soil types, which are grouped into different soil classes. Weathering of rocks changes with the progressive development of soils. In bare rocks, weathering is relatively slow because little water is retained on the bare surface, which could contribute to chemical and physical weathering (using solutes and volume changes). Once a thin layer of soil develops, weathering accelerates because the soil retains water. The substances dissolved in it and the volume changes of water during freezing promote weathering. As soil

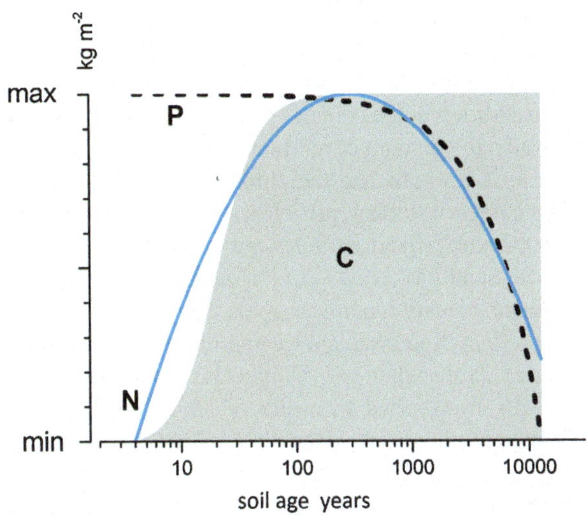

Fig. 1.14 Long-term development of reserves of total carbon (C), available nitrogen (N), and phosphorus (P) during soil development. (According to Vindušková et al., 2019, adapted)

thickness increases, the rate of weathering first increases, the amount of water retained increases, the influence of roots increases, etc. However, as soil thickness continues to increase, the original rock moves away from the soil surface, becomes isolated from temperature fluctuations, water and plant roots, and weathering slows down again. This affects the nutrient content of the soils.

With the exception of nitrogen, which enters soils as organic nitrogen due to the fixation of nitrogen from the atmosphere by free living or symbiotic microorganisms, other nutrients come from the bedrock. Thus, in young soils developing freshly on the exposed geological parent material, the N content in the soil will be small (Fig. 1.14). Conversely, the content of P and other mineral nutrients (e.g., K, Ca, Mg, and microelements) will be the highest possible, as determined by the quality of the geological parent material. The content of P and other mineral nutrients can affect the intensity of N fixation from the atmosphere, which is the main source of N. Over time, organic matter will accumulate in the soil and with it, N.

Phosphorus released from rocks will be recycled between dead biomass, soil, and plants. Its possible losses will be supplemented by weathering. In old soils, where the soils are very separated from the unweathered geological substrate, especially in the tropics, where weathering and leaching of P and other nutrients are supported by high temperatures and precipitation, there are large losses of P leading to its deficiency. In addition, most of the remaining P is bound in unavailable forms. The soils of the tropics, especially the geologically very old ones, are therefore often poor in mineral nutrients. In contrast, in the temperate zone, soils are richer in nutrients mainly due to the action of glaciers, which from time to time (meaning after some tens or hundreds of thousands of years or so) caused removal of existing soils, large erosion, and thus "reset" soil development and prepared new soil-forming substrates relatively rich in nutrients. This is also the reason why, especially in the tropics, the geologically younger soils around the volcanoes are so popular for agriculture. Most tropical soils are heavily weathered (Ultisols, Oxisols), with washed-out nutrients, while volcanoes bring in geologically younger material with more nutrients, thus helping to create more fertile soils (Andisols).

In addition to the content of elements in the soil-forming substrate and the age of the soils, the amount of nutrients in the soils is also affected by the grain size distribution of the soils and the climate. According to grain size, we distinguish three main categories of soil particles: sand (2–0.02 mm), silt (0.02–0.002 mm), and clay (<0.002 mm). Sand particles are formed by physical weathering and are mainly quartz. Sand has a low sorption capacity and thus relain less nutrients. Sand supports the formation of macropores. Sandy soils dry easily, they are well aerated. Silt is a product of mechanical weathering. It has the consistency of flour, smooth when dry and scrubs when wet. Unlike clay, which is sticks, silt retains the hand imprint, is easily transported by water or air as dust, and is placed in river depositions (Brahmaputra, ancient river states). Dust carried by the wind in the Ice Ages is the basis of loess. Clay is largely made up of clay minerals, which were formed by chemical weathering and have small particle size. These play a significant role in water retention, nutrient sorption, and soil adhesion and cohesion. Clay particles themselves, consist mainly of silicate layers. In some clays, water may penetrate between these layers. This causes significant volume changes during wetting and drying. Soils with a high content of clay are difficult to cultivate when wet, they are sealed and or muddy for a long time and they crack when drying. Due to the fact that they retain a lot of water, they slowly heat up and are cold. Due to their high sorption capacity, they often retain large amount of nutrients.

The grain size distribution is determined by properties of the geological substrate. Soils that have developed on quartz-rich parent rock, such as granite, sandstone, or loose sand, lead to the development of sandy soils. In contrast, quartz-poor rocks, such as shale or basalt, form soils with little sand. Soils are sometimes formed on unconsolidated deposits, such as sand or loess, where grains of a certain size have been sorted by transport processes.

In addition to the geological substrate, the amount of nutrients in the soil is fundamentally affected by the amount of precipitation. In general, in places with a high amount of precipitation, nutrients are leached from the upper layers of the soil. Clay minerals and organic substances are also washed out. This decreases the sorption capacity of these soils thus, they have a lower ability to retain nutrients (Box 1.7). Conversely, in soils where rainfall is very low and most of the infiltrated water evaporates, upward water flow predominates due to the combination of evaporation and capillarity. The water which infiltrates to soil dissolves the minerals and transports them to the soil surface, where they remain after the water has evaporated. In extreme cases, saline soils can be created that limit plant growth.

Soil development is strongly influenced by the type of vegetation. The effect of plants on the soil corresponds well with the plant growth strategies described above (Sect. 1.3.3). Based on that we distinguish three types of humus: Mor, Moder, and Mull (Ponge, 2003). Slow-growing conservative plants, such as conifers or heath (Gymnosperms and Ericaceae), tend to grow naturally on soils with low mineral nutrient content and also tend to retain nutrients in their tissues. Often, these species

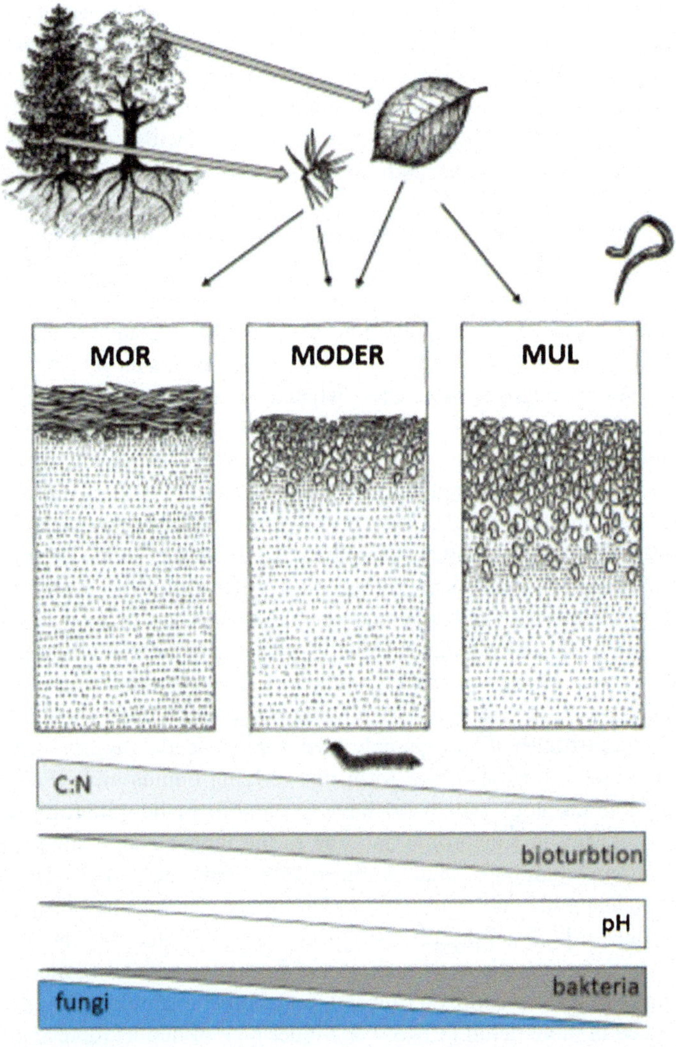

Fig. 1.15 Scheme of the main forms of humus and their relation to the main properties of the soil, according to Frouz (2018a)

are evergreen, with thick leaves or needles. The litterfall typically has a high C/N ratio, a high lignin content, and a high polyphenol, which significantly slows down their decomposition. As a result, a layer of partially decomposed litter with a low pH accumulates on the soil surface. Decomposition is provided by fungi. Communities of soil fauna are mainly formed by mesofauna (springtails, mites, etc.), there is a lack of larger soil fauna that would ensure significant bioturbation (mixing of the soil, e.g., by earthworms). The combination of this type of litter and the absence of bioturbators, leads to the formation of a humus-type mor (Ponge, 2003; Fig. 1.15). The quality of the litter contributes to the formation of fulvic acids, which act as acidifying and chelating agents, promote weathering and at the same time migrate down the soil profile, thus removing both organic matter and other nutrients from the topsoil layers. Fulvic acids precipitate together with other substances (sesquioxides, Al, Fe) and are immobilized in the lower part of horizon B (Lundström et al., 2000). The presence of plant species with a conservative growth strategy, often growing on poor sandy soils, thus leads to the formation of Spodosols (podzols).

In communities of fast-growing plants on mineralogically richer substrates, N is also enriched in natural conditions thanks to the fixation of atmospheric N_2, mainly by symbiotic fixation bacteria on the roots of Fabaceae and other plants. Nitrogen fixation may be limited by soil conditions, availability of P, and other nutrients. N_2 fixation is energy expensive for plants, so N_2-fixating plants most often dominate in the early or middle stages of succession, when the N supply is low and light is abundant. Once enough nitrogen has accumulated in the soil, especially in the soil organic matter, N_2-fixing plants are often replaced by other plant species whose ecophysiology focuses on rapid growth and better competition for light. Fast-growing plants have thinner leaves that are short and have a higher rate of photosynthesis and corresponding to higher N content in the leaves. They also invest less in producing secondary metabolites and produce litter with a lower C/N ratio. As a result, a large amount of easily degradable litter enters the soil. Compared to slow-growing plants with a conservative growth strategy, they produce litter that decomposes quickly and releases nutrients into the soil. The chemical composition of this litter and the properties of the soils provide a suitable habitat for the soil fauna involved in bioturbation of soil (namely earthworms), leading to the formation of a massive A horizon, typical of the form of overlying humus mull (Ponge, 2003). Formation of humic acids, which are much less mobile in the soil profile than fulvic acids and thus stay in soil profile and due to the higher clay content, soils have higher sorption, which helps maintain most of the nutrients in the soil (Fig. 1.15). An easily weatherable substrate with a pH close to neutral, the presence of fast-growing plants, a high supply of easily decomposable litter, and the bioturbation that mix the litter into the soil leads to the formation of soils called Mollisols. Chernozems belong to the most fertile soil. Some other examples of main soil types are introduced in more detail in Box 1.8. Moder type of humus represents intermediate stage between mull and mor and is typical for broadleaf forest.

From the point of view of agricultural use, both the nutrient content and the physical properties of the soils, which determine their response to cultivation are important. If we start cultivating the soil, the cultivated layer of soil will mix. In soils formed under temperate grasslands or deciduous forests, where there is a deep A horizon (e.g., Mollisols), cultivation is concentrated in this horizon and changes in soil properties are not so extensive. Conversely, in forest soils, where most of the organic matter is in the overlying horizon (e.g., Spodosols), these horizons are mixed with deeper soil layers (with the B horizon, often depleted), which causes dilution of soil organic matter in the depleted B horizon. Cultivation will increase the susceptibility of the soil to erosion and accelerate the mineralization of organic matter, which leads to its loss. These are again somewhat smaller when the organic

Table 1.4 Distribution of global soils in terms of their suitability for agriculture

Suitability for agriculture	% of all world soils	Proportion of world population that live on them %
Suitable soil without limitation	12	25
Soil useable for long-term cultivation with some limitations	34	51
Soils not suitable for cultivation but useable as pastures	26	13
Soils not useable for agriculture	28	11

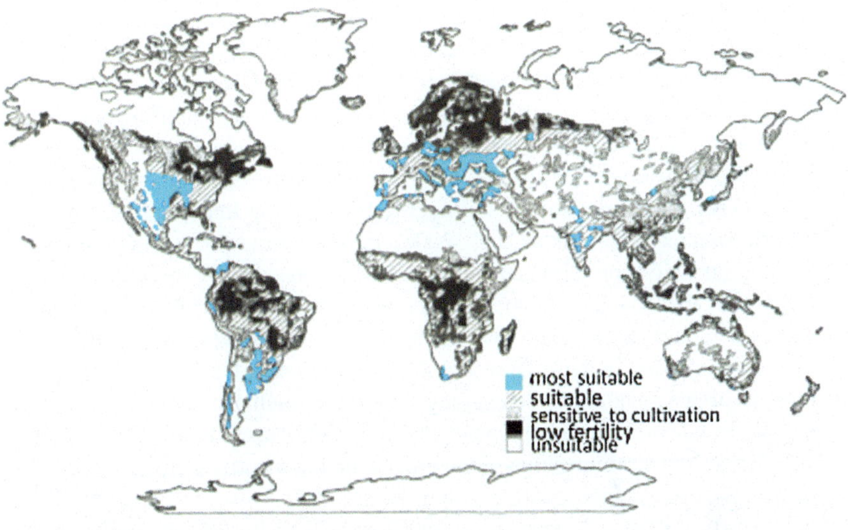

Fig. 1.16 Suitability of world soils for agriculture according to Blum (2013), modified

matter was bound to the mineral matrix in the humus A horizon (e.g., Mollisols) than when it was on the soil surface in the overlying horizons (e.g., Spodosols). Sensitivity to the negative consequences of cultivation described above is also increased in shallow, stony soils, and sandy soils. In addition, shallow and stony soils are poorly cultivated because the stones damage the tools used.

To summarize, in order for a soil to be suitable for arable land, it must not only have favorable chemical and physical properties but must also have the ability to maintain these properties during cultivation. There are a few soils (soil types) that meet all these conditions and are directly usable as arable land without much difficulty. They represent 12% of all land in the world. Most of these soils occur in the temperate zone and only about a third in the tropics. This 12% of soils also support almost a quarter of the world's population (Table 1.4, Fig. 1.16). Land that can be used as arable land, with some limitations, accounts for another 34% of the world's land. Another roughly 25% of the world's soils are land unusable for tillage-based

Box 1.7 How Soil Retains Organic Matter Water and Nutrients?

The organic matter entering the soil in the form of litter gradually decomposes and becomes partly mineralized to CO_2, water, and nutrients. At the same time, the remaining organic matter is stabilized. Historically, humification has been thought to be the main process responsible for this stabilization. Formation of humus by a series of polycondensation reactions, in which simpler substances, which are decomposition products such as amino acid sugars and phenolic substances, condense together to form more complex macromolecular humic substances.

Today, it turns out that the situation is a bit more complicated and humic substances make up only a small part of the stabilized organic matter. Much of this organic matter in the soil is formed by the dead bodies of microorganisms, especially extracellular cell walls (Lützow et al., 2006). Due to the large content of various aliphatic substances, they are hydrophobic (water-repellent) and are, therefore, difficult to decompose. However, what largely stabilizes soil organic matter are the various interactions with the mineral component of the soil, especially clays. The formation of soil aggregates plays an important role in the stabilization process (Lützow et al., 2006). In their formation, soil microorganisms produce glycoproteins cementing soil aggregates, fungi connecting aggregates with their hyphae, or the production of specific substances cementing aggregates, such as glomalin. Last but not the least, earthworms and other soil fauna are involved in the stabilization of organic matter. Earthworms incorporate this organic matter into soil aggregates, where small pieces of organic matter (particulate organic carbon) are coated with a thin layer of clay. This coating slows down the decomposition of the organic matter incorporated for a number of reasons. For example, coating by a water-saturated layer of clay causes oxygen to diffuse inward through the water saturate layer. Its diffusion is slow and oxygen can be consumed by microbes in the surface layer of aggregates, and thus anaerobic conditions often occur inside, which in turn, slows down organic matter decomposition. In addition, soil bacteria grow on the clay minerals around this tiny piece of coated organic matter, as they die, most of their cell walls and other necromass remain attached to the clays. This dead microbial biomass eventually forms most of the organic matter that accumulates in the soil. The activity of earthworms is essential here, if we only mechanically mix the sediment into the soil, we do not achieve the same degree of carbon accumulation in the soil as with the help of earthworms (Frouz et al., 2014).

In addition to light, which is a source of energy, plants need water and nutrients to live. If we omit hydroponic cultivation systems, where plants obtain water and nutrients from an aqueous solution, plants obtain water and nutrients from the soil. Water in the soil is affected by a number of forces. Probably the simplest is the idea of gravity's effect, where water flows down as if it were flowing through a water pipe. In addition to gravity, a number of other forces affect water in the soil, probably the most important of which are capillary forces. In small pores (due to capillarity), water tends to rise through

(continued)

Box 1.7 (continued)

these pores, even against gravity. This can easily be demonstrated on a sugar cube, which when immersed in a teaspoon of coffee, one can see the dark liquid rising up the sugar cube.

Capillary forces are responsible for most of the water retention in the soil. The amount of water that a soil can hold for the long term (mainly due to capillary forces) is called water-field capacity. The amount and size of capillary pores are determined by the grain composition of the soil. In general, there are smaller voids (pores) between the smaller particles, and so soils with a higher clay content have a greater ability to retain water. However, the predominance of clay particles and the absence of large pores lead to poorer infiltration of water into the soil. In contrast, in sandy soils with a predominance of macropores, there is good infiltration, but the water retention capacity is much lower. In addition, part of the water may be so tightly bound in the soil that it cannot be used by plants. The moment when only water which is not accessible for plants remains in the soil is called permanent wilting point (Fig. 1.17).

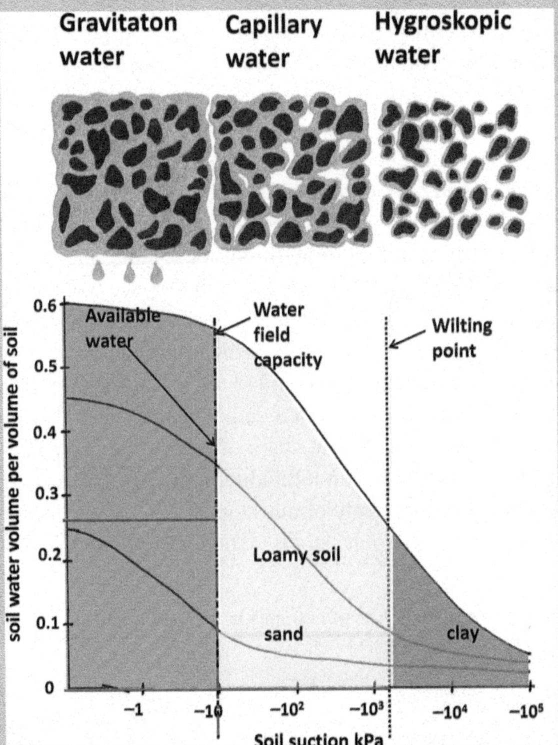

Fig. 1.17 Indicative values of the relationship between the volume of water in the soil and the suction pressure required to obtain this water for different soil types and the distribution of water in different soil microenvironments

(continued)

Box 1.7 (continued)

In addition to the proportion of soil particles of various sizes (soil texture), the behavior of water in the soil is also due to the arrangement of these particles (soil structure). If the soil particles are cemented into small lumps or soil aggregates, then capillary pores predominate inside the aggregates, which can retain water, while noncapillary pores predominate between the aggregates, which allow water to soak in.

In addition to the clay content and the formation of soil aggregates, the retention of water in the soil also affects the content of organic matter. Rough estimates suggest that soil organic matter can bind up to ten times as much water as its own weight. An increase in soil carbon content by 1% at a depth of 0–30 cm would therefore mean an increase in water retention in the soil by 39 L m^{-2}. Organic matter contributes to water retention mainly indirectly by influencing the formation of soil aggregates and increasing the number of capillary pores.

In addition to water, nutrient retention in the soil is important. The largest supply of nutrients is in minerals and organic matter. Organic matter is especially important for nitrogen. Minerals play an important role in most other nutrients. Minerals (primary or secondary), gradually weather and release nutrients into the soil solution. Soil organic matter is another important pool of nutrients, its mineralization can release nutrients into the soil solution. Finally, some of the nutrients in the soil solution can be bound by sorption to soil organic matter and clay minerals (sorption complex). Due to the fact that both humus and clay minerals carry mainly a negative charge, the sorption bond mainly concerns cations. In addition, these ions bound to the sorption complex can be exchanged for another ion, so we sometimes speak of exchange sorption. For example, when CO_2 is released during the respiration of the roots, root surroundings acidify which leads to dissociation of water and the released H^+ ions, which can bind to the sorption complex and displace other cations such as K^+. The released cations enter the soil solution and the plant roots can uptake the ion. The share of individual reservoirs in the total supply is highly variable between individual nutrients (Table 1.5). In general, in natural ecosystems, the supply of nutrients in the soil solution forms only a very small part of the total stock. The second smallest reservoir is

Table 1.5 Content of selected types of nutrients in individual reservoirs on the example of US forest soils (Dahlgren & Singer, 1994)

Pool/nutrient	Ca	Mg	K	N	P
	Kg ha^{-1}	Kg ha^{-1}	Kg ha^{-1}	Kg ha^{-1}	Kg ha^{-1}
Soil solution	16	4.3	8	0.2	0.018
Exchangeable sorption	31,900	8000	1105	10	135
Organic matter	2337	1440	1168	10,700	4400
Minerals	789,700	488,100	46,700	0	10,500

(continued)

Box 1.7 (continued)
exchangeable sorbed nutrients. As already mentioned, the role of this reservoir is greater for cations than for anions.

Plants can directly use nutrients in soil solution and exchangeable (bound by exchangeable sorption), or nutrients in some easily degradable, usually secondary minerals. Wild plants in nature use a number of methods to get to less degradable forms of nutrients. They can stimulate the decomposition of minerals by means of weak acids and root exudates, they can also support symbiotic microorganisms, which then provide nutrients in various ways, mostly in exchange for energy-rich organic substances provided by the plant. However, most cultivated plants rely mainly on available forms of nutrients. Therefore, from the point of view of plant nutrition, it is not only the total content of nutrients that is important, but especially the content of available forms.

The content of available forms depends not only on the total content of the nutrient in the soil but also on other factors. The amount of nutrients in the soil solution is limited. In addition, nutrients in the soil solution that are not taken up by plants, microorganisms, or bound to the sorption complex can be easily leached out of the soil. Thus, the content of available nutrients is largely determined by the amount of the exchangeable nutrients and hence, the capacity of the sorption complex. It depends, as we have already said, on the content of clay and soil organic matter.

Another important factor that affects the amount of available nutrients is the pH of the soil as we can show an example with P. Phosphorus is readily available at neutral or weakly acidic pH, ideally in the range of 6.5–7. At a more acidic pH, P binds to Al and Fe to form insoluble complexes. In contrast, at basic pH, it binds to Ca to form insoluble apatite (Fig. 1.18). However, it is

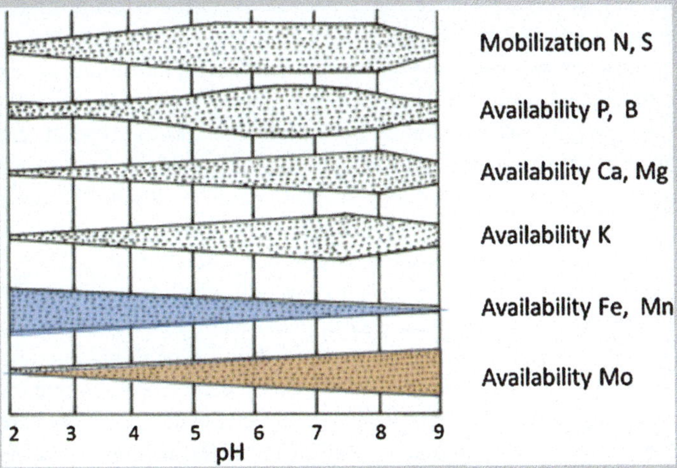

Fig. 1.18 Influence of pH on the availability of selected nutrients for plants

(continued)

Box 1.7 (continued)
not just P. Most nutrients are best available at a slightly acidic pH. An exception is most microelements, whose availability increases with decreasing pH. Among microelements, on the other hand, Mo is an exception, the availability of which increases with increasing pH. Thus, in addition to fertilization, pH adjustment is an important factor that can improve the availability of nutrients and thus, their uptake by plants. In addition to affecting the availability of nutrients, pH also affects the activity of the soil microflora and a number of other soil processes that directly or indirectly affect plant growth.

Box 1.8 Soil Types and Their Main Representatives
Soil classification is largely based on how soils form (IUSS Working Group WRB, 2015). We deduce how the soils were formed from the characteristics of the soil horizons. We distinguish several main soil horizons. As we have already seen in the text on forms of humus, a layer of more or less decomposed litter may lie on the soil surface. This layer, we call O and like other horizons, can have a number of sublayers and different types. Below it is a layer in which soil organic matter accumulates. This horizon is usually significantly darker than the rest of the soil profile and is called A. Below it lies horizon B, which is formed by weathered and variously transformed mineral rock. In the B horizon, a number of transports, redox, and other processes occur, which are important for determining the type of soil. Below it is a weathered soil-forming substrate, which we call C.

As we have already mentioned, the development of soil is determined by the soil-forming substrate, the climate, the topography of the terrain, the vegetation, and all this works over time. As a result of the effect of time, young soils look different from old soils and we refer to them as a different soil type. Over time, one soil types may pass into another. As for the influence of vegetation, climate, and substrate, these are often related, because the natural occurrence of vegetation is determined by the climate and the type of substrate. Throughout the history of soil classification, a number of schools have emerged that classify soils in different ways. Internationally, the World Reference Base for Soil Resources is used, which distinguishes 32 reference groups of soils; however, many countries and organizations use different systems.

In the following overview (Fig. 1.19), we will mention only some types of soils (more, e.g., Němeček et al., 1990). In the section on forms of humus, we mentioned the type of mull humus. Soils with this type of humus formed on loess under steppe vegetation are called chernozem. They are characterized by a very deep A horizon, which is rich in organic matter and carbonates, and the C horizon (loess), which is one of the most fertile soils. Luvisols (also forest soils) are common under deciduous forests with a form of moder-type humus. These soils usually formed on quaternary sediments, loess clays, and

(continued)

Box 1.8 (continued)

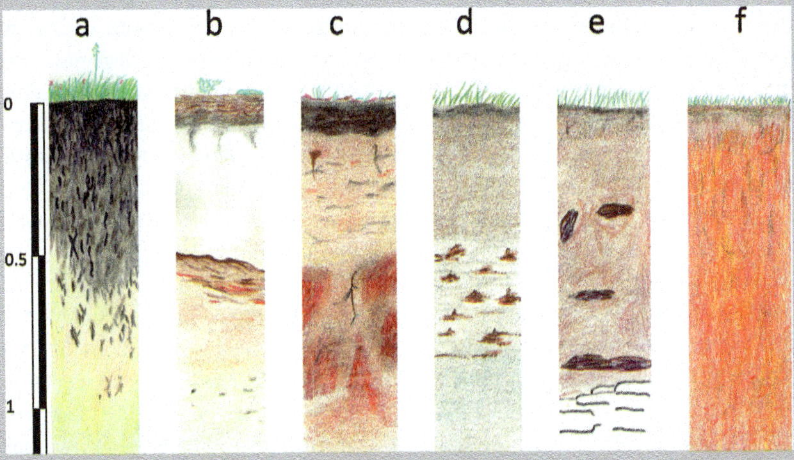

Fig. 1.19 Examples of soil types. (**a**) chernozem, (**b**) podsol, (**c**) luvisol, (**d**) gley soil, (**e**) cambisol, (**f**) laterite (drawing J. Frouz)

other unconsolidated sediments. In areas with higher precipitation, this led to the leaching of carbonates and, after acidification, the subsequent leaching of clays. Due to this, the upper part of B is depleted (eluviated) by clays and the lower one, on the contrary, is enriched by clay (illuviated).

Podsols are formed on even poorer substrates, sands, or weathered sandstones, and with even higher precipitation. Typically, they occur under coniferous forests, with the mor-type humus. These are characterized by an acidic subsoil of more acidified litter layer. The absence of bioturbation leads to the formation of a shallow A horizon. Not only carbonates and clay are washed away but also other cations, especially iron, manganese, and others. As a result, the upper part of the B horizon is bleached, below which precipitated metals accumulate into a colored horizon, which can sometimes take the form of a hard, low-permeability layer. These soils are characterized by intense weathering, which means that there are few weatherable minerals left. They are dominated by kaolin, quartz, and iron oxides, which give them a red or yellow color, and soils poor in basic cations and mineral nutrients.

The movement of water in the soil profile plays an important role in soil formation. In soils where the A horizon is flooded by water, peat soils (Histosols) are formed. Flooding of the B horizon starts the reduction processes, which gives the reduced parts of the B horizon a green to blue-gray color and leads to the formation of gley soils. When B horizons are affected by a periodic water table, reduction and oxidation processes alternate over time giving part of the B a mosaic rusty (places where oxidation prevail) and gray–blue (reduction prevail) colors (Fig. 1.19).

(continued)

Box 1.8 (continued)

Many soils have very little development of the soil horizons. This includes, for example, alluvial soils formed on the alluvium of large rivers, which are constantly affected by sedimentation or the erosion of eroded material. Cambisols (brown soils) are the common soils in many parts of Europe, they are characterized by the presence of stones and internal weathering, during which, nutrients and iron are released, which gives the soils a characteristic brown color. These soils are very diverse with many subtypes.

It is noteworthy that mankind is also an important factor in soil formation. There is a whole group of antroposoils, soils that are largely determined by human activity. Those include not only various soils developing on backfill and dumped substrates but also soil altered by incorporation of some additives. Examples of the latter are so-called terra preta, dark fertile soils, formed in ancient civilization probably by incorporation of large amount of charcoal in soil (Fig. 1.20).

Fig. 1.20 Terra preta as an example of ancient anthropogenic soil, note anthropogenically altered dark layer with numerous ceramics fragments, near Manaos, Brazil

agriculture, but some of them can serve as pastures. The remaining 28% of soils are deserts and other barren places (Blum, 2013).

1.4 Secondary Production

1.4.1 Energy Balance of Heterotrophic Organisms

Heterotrophic organisms obtain energy as well as chemical compounds, and they need to build their bodies from food. If we look at what happens to food (Fig. 1.21), we see that part of the consumed food or consumption (C), leaves the body as feces or defecation (D), and the rest was assimilated (A). From this assimilated part, the part that leaves the body by excretion (E), in the case of mammals, is urine. However, in many cases (e.g., birds or insects), it is difficult to distinguish between excretion and defecation. Therefore, we may for simplification, assume it is as a loss from assimilation. Similarly, we reduce assimilation for the energy that gets lost with gas in some animals. For example, the loss of ammonia in terrestrial isopods or the loss of methane in ruminants. Assimilation (after deduction of excretion and energy of gases, if any) is further divided into the part used to build the body called production (P) and the part that the animal respires to cover its energy called respiration (R).

$$C = A + D\left(+E + G\right)$$

$$A = R + P$$

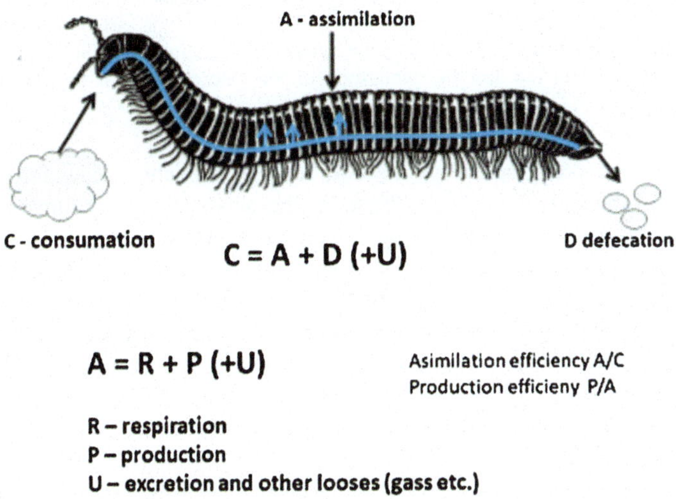

Fig. 1.21 Scheme of energy balance of animals

Table 1.6 Individual components of the energy balance in different groups of animals in % of the total energy contained in the food consumed (energy consumption = 100%)[a]

Species of group	Assimilation	Defecation+ Excretion	Respiration	Production	P/A
Domestic rumninants[1]	55	45	35	20	36
Elephant (*Loxodonta africana*)[1]	33	67	32	1	3
Kob antelope (*Kobus kob*)[1]	85	16	83	1	1
Rodent (*Microtus pensilvanicus*)[6]	85	15	68	17	20
Yellowhammer (*Emberiza citrinella*)[1]	90	10	89	1	1
Chicken (*Gallus gallus*) broiler[8]	74	26	52	22	30
Snake (*Heterodon* sp.)[2]	84	16	16	68	81
Corn snake (*Elaphe guttata*)[2]	89	11	9	77	86
Frog (*Bufo terrestris*)[2]	74	26	38	36	49
Common bream (*Abramis brama*)[3]	76	24	22	54	71
Prey fish[4]	80	20	60	20	25
Daphnia (*Daphnia pulex*)[5]	31	69	14	17	56
Herbivorous invertebrates[7]	40	60	24	16	40
Predaceous invertebrates[7]	79	21	56	24	30
Invertebrate decomposers[7]	20	50	13	9	43

[a]Assimilation is the sum of respiration and production, therefore the sum of the line does not give 100%. In total, 100% gives either the sum of assimilation and defecation + excretion, or the sum of defecation + excretion plus respiration plus production. The last column shows production efficiency as a percentage of production from assimilation. Values are rounded to whole percentages if the original range of the studies indicated a mean value. According to various authors: [1]Klimo et al. (1994), [2]Smith (1976), [3] Sorokin and Panov (1966), [4] Welch (1968), [5] Richman (1958), [6] Golley et al. (1965), [7] Heal and MacLean (1975), [8]Calculated on the assumption that the gross energy of the feed mixture is 18,918 kJ kg^{-1}, the metabolizable energy is 14,000 kJ kg^{-1}, the conversion factor is 1.7 and the heat of combustion of the chicken is taken from Wolynetz and Sibbald (1990)

In order to express the energy flow, we must have energy values for individual members of the equations above. These can be obtained from dry matter that has been consumed, defecated, assimilated, etc., and then multiply them by the appropriate energy value (dry matter of consumption by energy value of food, dry matter of excrements or excreta energy, content of excrements and excreta, etc.). The energy content can be determined, for example, by burning a sample of the material in an oxygen atmosphere and determining the amount of energy released in a combustion calorimeter.

Equations are also used to describe two types of efficiencies. The first is assimilation efficiency which is basically assimilation in percentages or in other words, the percentage of assimilated food from the food consumed. The second is production efficiency which indicates the share of production from assimilation. Table 1.6

gives examples of the individual components of the energy balance for different species or groups of animals. Differences in assimilation efficiency are large. Assimilation is high in predators, smaller in herbivores, and smallest in decomposers that feed on dead biomass. There are also large differences in production efficiency, which will be explained in the next chapter.

1.4.2 Factors Influencing Secondary Production

From Table 1.7, it can be seen that the production efficiency is generally greater in invertebrates and ectothermic vertebrates (those who do not maintain constant body temperature) than in endothermic vertebrates (those that maintain constant body temperature similarly to humans). This is due to the fact that maintaining a constant body temperature requires a substantial amount of energy. Maintaining a constant temperature allows endotherms to occupy niches inaccessible to ectotherms, to be active even in adverse climates, to provide long-term energy-intensive performance, but at the cost of even higher energy expenditure. A more detailed comparison of the advantages and disadvantages of ecto and endotherms is in Table 1.7.

From the point of view of production efficiency, ectotherms or endotherms are key factors. While for wild endotherms, the production efficiency is in single percentages, while for ectotherms it is tens of percent and often more than 50%. The main reason for this difference is the high loss to thermoregulation of endotherms and also to cover their overall greater metabolic activity. The more active metabolism of endotherms allows for greater energy expenditure in foraging, reproduction, and other activities. Reducing energy expenditure on movement and thermoregulation (restricting movement while providing suitable microclimatic conditions) are important factors that increase productivity in domestic animals (compare yellowhammer and chicken in Table 1.7).

Larger endotherms tend to have greater production efficiency than smaller ones on average (Tables 1.7 and 1.8), although there is considerable variability. This is due to the fact that the surface volume ratio decreases with body size. Thus, larger organisms have a relatively larger body volume, a larger volume of muscle, other organs that produce heat energy, and a relatively smaller body surface area through which heat loss occurs. As a result, larger endotherms, relative to their weight, lose relatively less energy on thermoregulation and may have greater production efficiency. The influence of body size on metabolic activity is described by the so-called Kleiber's law, named after Max Kleiber, who derived it in the 1930s. According to this relationship, the metabolic activity of the animals corresponds to their mass to the ¾ power (Fig. 1.22). The metabolism of an animal that is 100× larger than another similar animal will, therefore, be not 100× larger but only 31.6× larger. The cause of this phenomenon lies, as already mentioned, in the allometric change in the proportions of the body. As the size of the body increases, its volume changes by the cube of the length (l^3), while the area changes by the square of the length (l^2). At the same time, some processes in the metabolism, such as heat

Table 1.7 Comparison of ecological and ecophysiological consequences of ectotherms and endotherms

Parameter	Ectotherms	Endotherms
Thermoregulation	Their body temperature and metabolic rate vary depending on the ambient temperature	Keep constant body temperature in large range of ambient temperatures
Metabolisms	Overall metabolism is lower	Overall metabolism is higher
Ecophysiology	The optimal temperature has comparable speed with endotherms. It slows down in suboptimal temperatures and they may have difficulties to conduct energetically costly behavior for long term (e.g., run a long distance)	They are able to conduct energetically demanding behavior for long term in large range of ambient temperatures
Energetic loses	Low energy loses	Energy cost and water demands are up to 17× higher in comparison with ectotherms of the same size
Relation to body size	As the surface volume ratio increases with decreasing size, smaller animals, at the same temperature, have greater heat loss and therefore it is more difficult for them to maintain a constant temperature. Therefore, ectotherms are more advantageous for smaller forms	More suitable for larger organisms because heat retention is easier in larger forms, which have a larger body volume and thus a greater metabolism compared to a smaller body surface area and thus less heat loss. It is stated that due to their size, dinosaurs were probably mesothermic (their temperature was at least partially independent of the ambient temperature)
Ecological consequences	They have advantageous in the tropics and subtropics but in the colder regions of the world their activity is limited by lower temperatures	They can be active at night and in bad weather or even in winter, they can occupy ecological niches inaccessible to ectotherms (e.g., polar wastelands)
Energetical cost of reproduction	Production form large part of whole metabolism and consequently, the amount of energy invested in production of next generation, the production of eggs or offspring is relatively large	Production form relatively small part of whole metabolism and consequently, the amount of energy invested in production of next generation, the production of eggs or offspring is relatively small
Lifespan	Live longer than endotherms of the same size	Live shorter than ectotherms of the same size

Table 1.8 Average values of production efficiency P–A for selected groups of wild organisms

Group	P/A %
Insectivores	0.9
Other small mammals	1.5
Other mammals	3.1
Birds	1.3
Fish	9.7
Insects—Herbivores	38.8
Insects—Predators	55.6
Insects—Durophages	47.0
Social insects	9.7
Invertebrates—Herbivores	20.6
Invertebrates—Predators	36.2
Invertebrates—Durophages	27.6

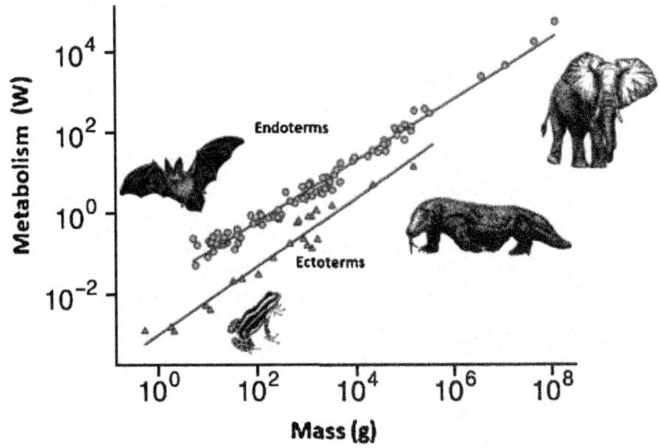

Fig. 1.22 Relationship of energy needed for the metabolism and body size

production, are more related to volume, while others such as heat loss are more related to body surface area.

The mechanisms that codetermine the phenomenon, however, are more complex. For example, younger and smaller individuals of the same species may have a greater metabolism than older (larger) individuals due to the higher energy costs of growth. Adults of smaller species will have a faster metabolism than adults of larger species, due to the fact that smaller individuals will have a larger proportion of active tissues and their maintenance requires energy. Conversely, larger individuals will have relatively more reserves, which are not so energetically costly to maintain. Another interesting consequence of this is that there is a linear positive relationship between the logarithm of the animal's weight and its lifespan. In other words, larger animals live longer. However, it is suspected that this is related to the rate of metabolism, especially the stress on the circulatory system.

In addition to the thermoregulation strategy, the metabolism is also directly affected by temperature. Lower temperatures reduce the production efficiency of endotherms because it leads to more thermoregulation expenditure. Ectotherms have a higher growth rate and efficiency at higher temperatures. However, ectotherms growing at lower temperatures grow longer and mature later than the same species at higher temperatures and thus, individuals growing at lower temperatures tend to be larger (Angilletta et al., 2004).

As already mentioned, in addition to thermoregulation, a significant part of the energy, animals spend, goes at the expense for movement related to food retrieval, defence against predators, and other similar activities. Different types of movements are differently demanding on energy. In general, digging in solid material (soil or sediment) is the most difficult, followed by walking, then flying. Swimming is the least energy-intensive. The movement of a large animal costs quite more energy than the movement of a smaller animal. However, if we calculate the energy costs of moving the unit weight of the body, then these costs are lower for larger animals than for smaller animals. In other words, the movement of larger animals is relatively cheaper from an energy point of view. In addition to the type of movement, different animals also have different strategies for doing this or that, resulting in different energy costs of the outcome. In particular, in the case of obtaining food, we can express the energy efficiency of a given behavior by comparing the energy obtained in the food and the energy expended in searching for it.

Food is another key factor in determining production efficiency. The energy content is important as well as the usability (digestibility) of individual food components. In addition, the content of key nutrients and essential substances is crucial (Table 1.9). As we have already mentioned, the energy content of food can be measured with a burning calorimeter. However, the heat of combustion of the main components of plant biomass with the exception of fats (Table 1.10) is similar for most of the plants around 17 KJ g^{-1}. Equally or perhaps more important than the energy content of food is its digestibility, which can be expressed as assimilation

Table 1.9 Combustion of heat of various natural substances and materials in dry matter material

Substance	kJ g^{-1}
Cellulose	17.5
Sugars	17.2
Proteins	17.2
Fat	38.7
Wheat grain	17.5
Straw	16.1
Potatoes	16.7
Clover	15.4
Rape seed	25.1
Maize for silage	16.1
Wood	17.1
Whole chicken	24.8

Table 1.10 Distribution of biomass at harvest time among the main parts of the plant in different crops and the share of yield (in bold) in total biomass in %[a]

	Letuce[1]	Sugar beat[2]	Wheat[3]	Kernza[4]	Cesnats[5]
	Lactuca sativa	*Beta vulgaris*	*Triticum aestivum*	*Thinopyrum intermedium*	*Castanea sativa*
Seed	–	–	22	2	2
Other aboveground parts	79	33	42	48	78
Roots	21	67	36	50	20
Share of yield in whole biomass	**79**	**67**	**22**	**2**	**2**

[a]The part used for human consumption is marked as yield, therefore other parts that can be used as forage (crumbs or straw for beets or wheat) are not included. Rounded average values by various authors, [1]Kuzyakov et al. (2002), [2]Špaldoň (1982), [3]Siddique et al. (1990), [4]Culman et al, unpublished data., [5]Battisti et al. (2013), Fonseca et al. (2004), Patrício et al. (2005)

efficiency. In general, animal food that serves as food for predators and parasites has a high assimilation efficiency (60–90%). Of the various types of animal food, insects have the lowest assimilation efficiency due to the large proportion of indigestible exoskeletons. Plant foods such as seeds, nuts, and other fruits have a higher assimilation efficiency (80%). Among other parts of the plants, young tissues with a low proportion of cellulose, hemicelluloses, and lignin have a higher assimilation efficiency (60%). Older leaves and twigs have worse assimilation efficiency (30–40%). Wood and dead parts of plants (litter) have the lowest assimilation efficiency (5–20%). In general, the digestibility of plant tissues decreases with increasing content of cellulose, hemicelluloses, and lignin.

In addition to the amount of assimilated energy, the content of key macro- and micro-elements is important. It might seem to be ideal if the ratio of the individual elements in the food is the same as in the body of the animal that feeds on the food. In many cases this is true. The most significant exception is the ratio of carbon to other elements, especially nitrogen. Carbon or carbon-rich substances (hydrocarbons), in addition to being used for building new animal tissues (i.e., to produce new biomass), also serve as a source of energy. Thus, if the C:N ratio of both the food and the consumer is the same, for example, 10:1, and if the consumer used half of the energy received in the food for maintaining their metabolism, then of the 20 moles of carbon molecules received, the consumer respires ten moles and uses another ten moles to build its body. It receives two moles of nitrogen molecules per 20 moles of carbon. To build its body, it uses only ten moles of carbon. Of these ten moles of carbon needed to produce the body, it consumes only one mole of nitrogen (out of two taken). The remaining one mole of nitrogen molecules are in addition to the loss of nitrogen, also costs it some extra energy. This can be a problem especially if one tries to accumulate some energy reserves, for example, for the winter. Therefore, bears if they have enough salmon, eat only the fat-rich brain and eggs and leave the protein-rich muscle intact. Conversely, if an element is deficient in food, the animal may receive more food to make up for the element's deficiency or try to obtain it in some other way. Typical examples are sodium and potassium. Potassium is the main electrolyte of plant cells, while sodium is the main electrolyte of animal cells. Thus, in general, animals that eat only or mainly plant foods often lack sodium. Therefore, salt is added to livestock feed. Salt was relatively rare even in the African savannas, where our ancestors evolved, so we look for it and we like salty chips, salted peanuts, and a lot of other salty treats.

1.5 Diversity

1.5.1 Factors Affecting Diversity

The next chapter will deal with the relationship between productivity and diversity at the local scale. The aim of this chapter is to mention major patterns of diversity at global and regional scales and elucidate its relationship with local-scale diversity. Probably the most striking pattern of global biodiversity, for many very diverse groups of organisms (although not all of them), is that the highest diversity can be found in tropical areas and declines toward the poles (Gaston, 2000). There are a large number of partly nonexclusive hypothesis, which explain this pattern. Tropical areas represent the largest area on earth as the circumference of the equator is larger than any other parallel. Conditions in the tropics are very stable as temperature is very similar between the 20 north and south parallels. Moreover, conditions around the equator have been more stable during geological history compared to more northern and southern latitudes, where ecosystems were almost completely erased in large areas (e.g., during ice ages). This spatial and temporal stability may allow long-term speciation and eventually finer divisions of available energy between more species. At the same time, productivity of natural terrestrial ecosystems in tropics is larger than in natural ecosystems in more southern and northern latitudes. This larger net primary production means a larger amount of energy is available for all other levels of the food web and more energy can support more individuals. More individuals can mean that more species can coexist (Gaston, 2000). This more individuals hypothesis is also supported by the general pattern that species numbers often tend to decrease with increasing elevation and decreasing evapotranspiration. Consequently, diversity is lower in higher altitudes or in very cold or very dry areas. Very cold or dry climates as well as very high elevation reduces primary productivity in terrestrial ecosystems and then potential the number of individuals in these ecosystems.

At more regional scales, one of the most prominent patterns of diversity is the species–area relationship, in other words, recognition that a larger area can harbor more species. More specifically, it is often mentioned that when we plot the log of species number against the log of area, we get a straight line with a slope of about 2. The practical implication is that if an area of a given habitat is reduced ten times, then the number of species decreases to half. Recently, we know that the pattern is more complex and these rules do not apply in very small or very large areas (Allen

& White, 2003). However, this simplistic rule has often been applied to estimate the potential effect of habitat loss on species diversity. The mechanisms responsible for species–area relationships may vary across spatiotemporal scales. At large spatio-temporal scales, processes like speciation and species extinction may be important, while at smaller scales, habitat heterogeneity and species migration may be important. The theory of island biogeography explores this smaller to medium scales, based on recognition that smaller and more isolated islands harbor less species than larger ones closer to the mainland. Proximity of islands to the mainland affect its accessibility by migration from the mainland and thus, potential sources of new species, while Island size indirectly effect species persistence on the island (probability of extinction). Species are more likely to go extinct in smaller islands where populations will have less individuals, than on large islands with larger populations, which would be more resistant to random fluctuations of population dynamics or environmental disturbances. Moreover, a larger island is likely to be more diverse, containing larger numbers of various habitats, which can support more species (MacArthur & Wilson, 1967). Island biogeography principles are often applied in nature conservation, where an isolated patches of natural ecosystems can be seen as "islands" in the "sea of" cultural landscape. Biodiversity at regional scale strongly affect diversity at local scale as species occurring in given habitat are subset of species pool in the surrounding landscape.

There is increasing concern that human activity may substantially contribute to the ongoing biodiversity loss. Many of these processes are directly or indirectly related to agriculture, forestry, or fisheries and will be discussed in detail in the next chapters. Some species, namely, many megafauna' species can be hunted to extinction. However, indirect effects are generally assumed more important. Among them, habitat loss is the most commonly mentioned. Natural habitats get lost in terms of area and replaced by urban and agricultural areas, at the same time habitat can lose its ability to support species by changes in habitat structure (e.g., conversion of natural forest to planted monocultures), increasing human appropriation of net primary production (which decrease the amount of available energy for rest of the food web and thus number of individuals) and contamination. This can decrease habitat area and affect species by species–area relationships or decrease the suitability of habitats for some species (e.g., removal of certain tree species from large areas will affect many insect herbivores specialized on that tree). Remaining habitats may be fragmented in small islands, which again may increase risk of local extinction as explained earlier. Species invasion is often mentioned as another reason for the reduction species numbers. Introduced species can cause the extinction of native species by predation, competition, or habitat alternation. They may also alter natural succession and cause that disturbed habitats may not follow their natural succession trajectories.

Finally, human affects biogeochemical cycles including cycles of carbon and major nutrients. Nutrient availability effects on diversity will be explored, in detail, in the next chapter. Effects of the carbon cycle is associated with global climate changes. Global change affects species in many ways, it shifts climate, which may cause changes in distribution of habits. The effects on species performance can be, however, more complex, for example, climatic changes can alter phenology, which may cause mismatch between seasonal parent resources and species life cycles. Moreover, effects of these factors on biodiversity are not isolated, but largely magnified by their joint influence. For example, species can response to latitudinal climatic shifts can be followed by a shift in species range, this can be, however, hampered by loss or degradation of potentially suitable habitats in the "new" species range by human activity. Loss of biodiversity can in turn, feedback in changes of ecosystem functioning, human health, and wellbeing (World Resources Institute, 2005; Hooper et al., 2012).

1.5.2 *The Relationship Between Production and Diversity*

Although there are other components of diversity, in the following text, we will deal with species diversity given the number of species and their representation in the community. If we examine the relationship between production and diversity, we can proceed from two different ends. We can change production by changing the productivity of the environment and observing how diversity will react, or we can manipulate the number of species under the same environmental conditions and observe what happens to production. On a global scale, the primary production of terrestrial ecosystems is highest in the tropics, where there is also the greatest diversity. Thus, on this scale, productivity appears to be positively correlated with diversity (Lepš, 2005). Factors affecting diversity are complex (Sect. 1.5); however, if we look at productivity at the landscape level, it is most affected by the availability of nutrients. If we look at the diversity of plants in areas with different nutrient supplies (Fig. 1.23), which can potentially achieve different production, we see that in areas with very little nutrients, the diversity is low. With an increasing supply of

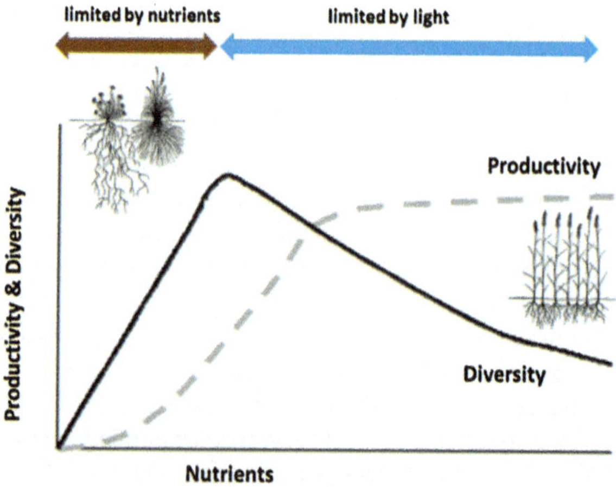

Fig. 1.23 The relationship between nutrient availability, primary production, and species diversity, schematized by Lepš (2005)

nutrients, the number of species increases, but only to the extent that the amount of nutrients is limiting (Lepš, 2005).

At some point, the number of species will begin to decline again as the productivity of the environment increases. This is due to the fact that, when nutrients are limiting, underground competition plays a major role. Underground competition can be very variable, as plants can obtain nutrients from different soil layers and can also receive nutrients in different forms. In such condition where the growth of plants is limited, it is also likely that facilitation in terms of cooperating between various plant species will be more important than competition. This is conceptualized in the stress gradient hypothesis, which depends not only on nutrients but also climatic and other conditions. The hypothesis predicts that in environmental conditions that limit plant growth, facilitation becomes relatively more important and competition weakens (Maestre et al., 2009).

As the amount of nutrients increases, nutrients are not limiting anymore and competition for light becomes the most important. Unlike variable underground competition, which takes place in 3D space, competition for light takes place in 2D, and the ways in which successful competitions apply are similar (greater height and larger leaf area). Unlike nutrients in the soil, which plants could easily share between themselves, light is thus, easy to monopolize. Plants that are larger and taller with a larger leaf cover get most of the light, overshadow the other plants below them, and there is not much light left for them. Due to the fact that only a few species can monopolize light as a limiting source, only a smaller number of species will occur in nutrient-rich conditions, and the diversity of the community is declining (Lepš, 2005). Here, we would like to point to the fact that as explained earlier, most of the natural ecosystems that are left today are less productive (Ellis & Ramankutty, 2008). Consequently, we know much less about the functioning of highly productive natural ecosystems (because those were already more severely affected by humans in past) than less productive ones. Moreover, many of the studies of plant response to nutrient availability were done on sites where nutrient availability was affected by anthropogenic influence.

On the other hand, as we have said, there is a possibility that we will examine the relationship between diversity and production by random selection of some species from the species pool found in a given landscape and growing them either alone or in mixtures with a different number of species (Fig. 1.24a, b). Such an experiment was tested in two classical experiments based on varying the number of species of grass communities in the USA and Germany (Cedar Creek and Jena experiment). The results show that single-species stands have on average, less biomass than multispecies stands, and the biomass increases with increasing number of species (Tilman et al., 2001).

How to explain such a trend? With a little simplification, we can say that this trend has two main explanations. The first explanation is that natural communities are usually dominated by only a few species, which produce most of the biomass or cover (Fig. 1.24c). On the contrary, most species which are represented in natural communities occur with only low cover and biomass (Lepš, 2005). These species, which do not form dominants in nature, are usually unable to occupy space and

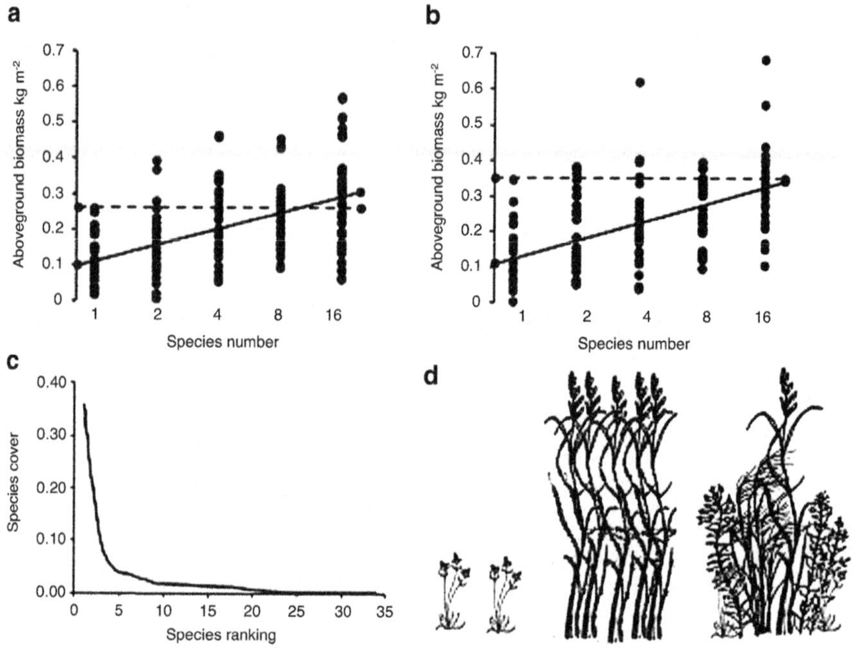

Fig. 1.24 Relationship between the number of species in the community and aboveground biomass in the first year (**a**) and subsequent years (**b**), an example of the share of species cover in the community (**c**) shows the distribution of species cover in the community shoving that only a few species dominate the community while other species are underrepresented, and the demonstration of the occupancy of the space by one small species, one species with high biomass production and a mixtures of species (**d**) according to Tilman et al. (2001) and Lepš (2005) modified, more detailed explanation in the text

create a large production of biomass, even if we sow them separately without the competition of other species. Therefore, if we sow one randomly selected species, there will be a small probability that it will be one of the few species that naturally forms the dominants. On the contrary, there is a high probability that we will select some species that grows naturally in the shade of others, produces little cover and biomass, and even if it is grown alone, it will not produce much biomass (Fig. 1.24d). The more species we select from the list of local species, the more likely it is that just by chance, we will fall on a naturally dominant species which, produce a lot of biomass and thus, the biomass of the whole community will increase significantly (Fig. 1.24d). The first reason why stands with more species achieve more biomass is the fact that there is simply a better chance that they will contain species that have a large biomass production. There are few species that achieve high cover in the local species pool (Fig. 1.24c), and therefore, the chances of randomly selecting them if we select only one species from the list of species are small. The more species we add to our experimental community, the greater the chance that we will chose a growing species. If we selectively choose the most productive species, their monocultures will achieve similar or larger biomass than mixtures of species. This strategy is used in agriculture, where monocultures of high-producing species are often grown (e.g., as forage crops).

However, this cannot explain all the increase in biomass and we will see later in the section about monocultures and polycultures that mixture of species have in average higher biomass than the individual component of the mixture. Another reason why mixtures of species can achieve greater biomass is the fact that species not only compete but also cooperate, so mixtures of more species can make better use of locally available resources. This benefit of species interaction increases over time, as has been nicely demonstrated in an experiment at Cedar Creek (Tilman et al., 2001). Polycultures, as we have already said, produce more biomass than monocultures at the beginning of the experiment (although this is due to the fact that there is a statistically greater chance of containing growing species), but after 16 years of experiments, this difference is much greater than at the beginning. This is due to a number of factors, including the fact that polycultures use the soil more evenly than monocultures, for example, each species obtains nutrients a little differently (uses different forms of the same nutrient, in different parts of the soil profile, etc.). In addition, this diversity means that some species may use resources unavailable to other species (some species may mobilize nutrients from great depths, others may fix nitrogen, etc.). This not only reduces competition but also makes these resources mobilized by one species available to others. For example, N fixed from the atmosphere or nutrients brought up from deeper layers may become available for other plants. When the litter of these plants comes into the soil, it contains some of the nutrients that the plant has accumulated, and these are released to other plants by the decomposition of the litter. Through a number of mechanisms, similar to those described in the previous sentences, polycultures change the soil so that it is more beneficial to the community over time, while the opposite is true of monocultures. This is because monocultures use the land unilaterally, they have no one to provide the nutrients they get in excess, and no one to get excess nutrients from which they obtain less efficiently. In monocultures, diseases and pests also accumulate over time, much easier than in polycultures. Rather than the higher productivity of polycultures, which is due to the fact that one of the more randomly selected species will be one with a large biomass production, the stability of this product over time is important due to the species cooperating and using resources differently.

To summarize, monocultures of randomly selected species have less biomass than polycultures. But monocultures of species that usually dominate can achieve the same or often higher biomass production compared to polycultures, especially if these species are specially bred for this purpose and nutrients are not limiting (e.g., due to intensive fertilization). However, if we do not supplement nutrients, at low nutrient availability species cooperation (higher diversity) often increase productivity. In such conditions, the advantages of polycultures and the disadvantages of monocultures increase over time. By constantly adjusting the conditions and supplementing nutrients, we can, however, maintain the high production of monocultures of selected species. At the same time, increasing production by adding nutrients often leads to a decrease in diversity.

1.6 Competition Between Organisms and Its Relationship with Production and Yield

If organisms will not be limited by competition, their number would increase by some constant factor (growth rate) in each generation. This growth rate is given by natality and mortality. In such a case, the bigger population would grow more, in absolute terms, as it has more individuals that can reproduce, such growth is called exponential (Fig. 1.16a). However, populations cannot grow indefinitely because resources are limited, and organisms compete for resources. If the same amount of resources (e.g., the same area) is shared by more individuals, there will be greater mortality, natality of surviving individuals will be lower, and also their development may also take longer. All of this (higher mortality, lover natality, and longer developmental time) leads to the fact that if there are more individuals per unit area (higher density of individuals), the growth rate of the population will be slower than at a lower density.

As the population grows, population growth rate will decline and eventually population growth stops completely. This means that the next generation will have the same number of individuals as the previous one. The value of the population density where this cessation of population growth occurs is called the carrying capacity or K. When the slowdown of growth rate is linearly proportional to population density we get the so-called logistic growth (Fig. 1.25b). We get such a smooth slowdown to the value of K only if the reduction of population growth rate due to competition is in direct proportion to the population density. This is the case, for example, with species that have territories, the growth rate decreases already in low densities with any increase of population. However, for many species, an increase in competition may not be observable at all in low densities, as the amount of resources per individual will still exceed their needs. If we have two or maybe four Daphnia in the pond, the increase in numbers will not be reflected in any way because there will be enough food for everyone. Sometimes, it can even be easier to obtain food with increasing density. So, population growth rate may even increase with increasing density. For example, if one bark beetle attacks a tree, the tree can easily kill it by producing a drop of resin. If many beetles attack the tree, the tree will be weakened, it will not be able to defend itself and it will become a better source of food for bark beetle.

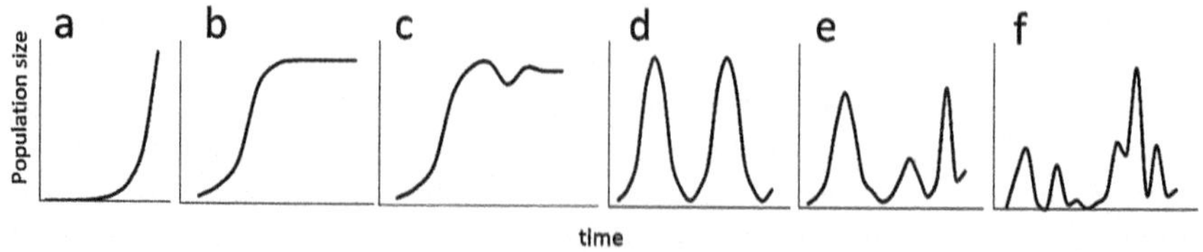

Fig. 1.25 Different types of population development over time, (**a**) exponential growth, (**b**) sigmoidal logistic growth, (**c**) damped oscillations, (**d**) regular cycles, (**e**) irregular cycles, (**f**) chaotic changes

In such cases, when the slowdown in the population growth rate of with increasing density is not linear, the population does not smoothly stops at its carrying capacity, but it can overshoot it and reach densities higher than K. This overshooting will be greater, the faster is the population growth rate, and the looser is the dependence between the population density and the competition. Larger overshoots may mean that the population size does not stabilize at K but will be subject to regular or irregular cycles or even chaotic oscillations over time (Fig. 1.25d–f).

According to the nature of population dynamics, we distinguish between the so-called r and K strategists. K strategists live long, reproduce slowly, mostly repeatedly, have a smaller number of offspring, which they often care for, their population grows slowly, often have territories or other interactions that share resources. Their population dynamics is similar to that in Fig. 1.18b, c. In contrast, r strategists are usually smaller species that reproduce rapidly, have many offspring, their population dynamics is similar to that in Fig. 1.25d–f.

If we consider a monoculture of one species that we will grow in an area, the total biomass in a certain area will be the sum of the biomasses of individuals living in that area. If we have two carrots in the flowerbed, their total biomass will be larger than if there is only one carrot. In other words, total biomass can be expected to grow with the growing number of individuals. This increase may be almost linear at first, but then it begins to slow down, as with an increasing number of individuals they begin to compete for resources (light, water, nutrients), and the size of individuals decreases. This decrease in the biomass of individuals will lead to the fact that at a certain density of individuals, the increase in total biomass stops, and with a further increase in the number of individuals, total biomass no longer increases or even decreases (Fig. 1.26).

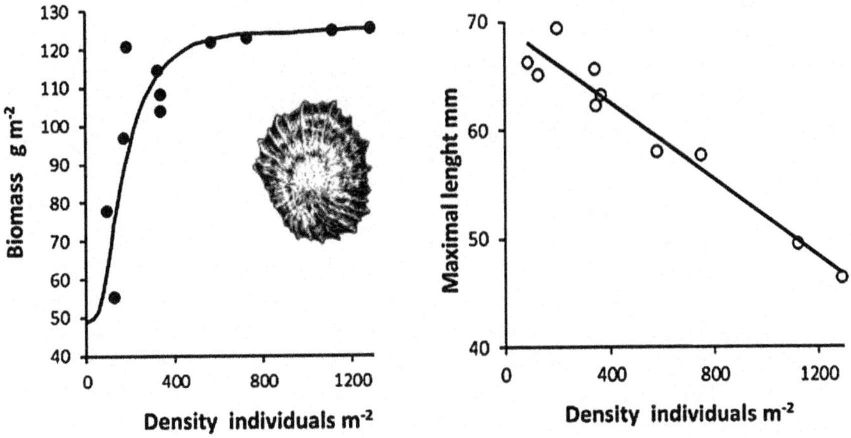

Fig. 1.26 Dependence between individual density individual size and total biomass, as reported by Begon et al. (1986)

So far, we have talked about biomass production and compared it under different conditions. But for farmers or foresters, the production of total biomass is not as important as its usable part (i.e., yield). For example, the total biomass of cereals includes the roots, leaves, stalks, seeds, etc., while the yield is made up only of seeds. The share of yield in the total biomass is very variable, it is higher for crops where we harvest and use the entire aboveground part (e.g., lettuce or beets) and smaller for crops where we use only seeds or fruits. In plants where we harvest seeds, there is a larger share of yield in total biomass in annual crops (wheat) than in perennial crops (e.g., perennial cereals—kernza or chestnuts tree) (Table 1.6).

The fact that annual plants allocate a larger proportion of biomass to seeds than perennial plants is the reason why many crops are annuals. Thus, the increase in yield is related not only to the increase in total biomass production. An increase in yield can also be achieved by redistributing biomass between the parts important for yield and other parts of the plant. For example, in the case of cereals, varieties with a shorter stalk that have a greater allocation of assimilates to grain may have a higher yield than plants with a lower allocation of assimilates to grain with the same total biomass. Sometimes even an increase in total biomass can lead to a decrease in the proportion of the part that generates the yield. Focusing on dwarf (now common) wheat varieties was one of the significant contributions to increasing wheat yield made by Norman Borlaug (Nobel Prize winner 1970).

The economic value of a yield often depends not only on its total quantity but also on some of its specific qualities. At the same time, quantity and quality can often go against each other, and producers often optimize them in order to achieve maximum profit. For example, in the case of grape yield, not only the total quantity but also the sugar content and other factors are important. It is easier to produce quality (and thus more expensive) wine from grapes with higher sugar content than from grapes with low sugar content. Therefore, winemakers often reduce the total number of grapes per plant, thereby reducing the overall yield. The assimilates are then concentrated in the remaining grapes, thus achieving their greater sugar content. Similarly, in forest production, it may be more valuable to produce a slightly smaller volume of larger diameter logs than greater small diameter wood biomass, which will not be suitable for many technological processing options.

So far, we have talked about the density of individuals grow in one even-aged generation (one cohort). In the end, we harvest all individuals and the population is not able to recover itself, we must ensure this by sowing or planting it again. In this way, we manage a number of crops, from wheat fields to clear cut forests. We can imagine another situation. Imagine that we have a self-sustained population of individuals in a certain area (e.g., trees in the rainforest, fish in the ocean, or wildlife in landscape), which is large enough to maintain themselves at the carrying capacity without our intervention. The aim is to remove individuals in a way that allows the original population to maintain itself. How is it possible? If we remove individuals from the population, we will reduce the number of individuals that can reproduce. However, at the same time we will reduce competition and thus increase the chances of other individuals for survival and successful reproduction. This is, after all, the principle of the influence of predation, the predators (with the exception of humans, as will be shown below) usually removes mainly those individuals who would otherwise perish as a result of competition and so their effects on the population may not be very pronounced.

As described above, the growth of the population limited by competition can be described by the logistic curve. In the beginning, the population grows slowly because it has few individuals who are able to reproduce and thus contribute to population growth. Later, as the number of individuals increases, the number of reproducing individuals increases, and population growth increases and enters the exponential phase. However, as the number of individuals increases, so does competition. At the low population densities, absolute decrease of growth rate due to competition is small. However, later one at higher population densities effect of competition becomes more pronounced and this reduces individuals' chances for reproduction as the birth rate decreases and mortality increases.

When the birth rate of the population equals its mortality, the overall growth of the population is zero and the population stabilizes at carrying capacity. Let us look once again at the population growth curve as a function of population density. The whole curve is given, as we explained, by the interaction of the number of individuals in the population and the competition. The curve has an ascending and descending part. In the descending phase, the removal of individuals will increase the growth of the population due to the decrease of competition. In the ascending phase, the removal of individuals will reduce the resulting population growth due to the decrease in the number of individuals. If we want to maintain the same value of the population, we must operate in descending part of the curve where the removal of individuals increases the population growth. However, when one removes only a small number of individuals, yield will be small.

The maximum yield, which leads to a stable population, is achieved if we take individuals in a way that the resulting density gives the maximum population increase. This value is called the maximum sustainable yield (MSY) (Fig. 1.27). If competition were linearly dependent on the number of individuals, the MSY would be equal to half of K. However, in many cases, it is not so easy to determine because of many reasons. For example, competition often depends on density in a not linear way, K, and the rate of population growth change over time depending on the

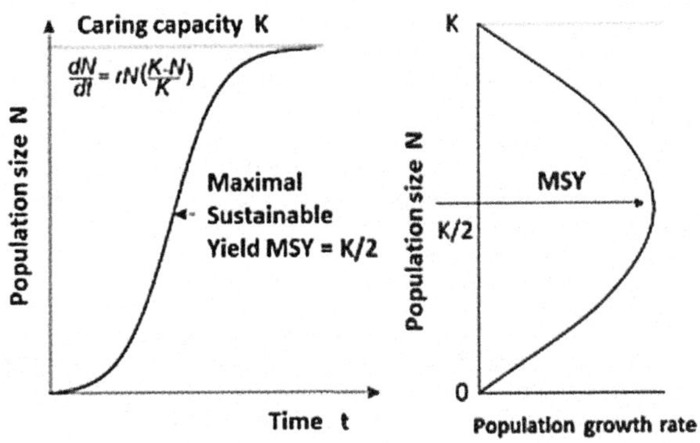

Fig. 1.27 Maximum sustainable yield. Logistic population growth, and the relationship between growth rate and population size

availability of food, long-term climate changes, and many other factors. The practical determination of MSY and its use for the exploitation of populations is, therefore, more complicated.

Another problem is that if we take a number of individuals other than MSY, we can get two balanced states of the population. One of them is at population densities higher than would correspond to the state of the population at MSY and this state is stable. If there is a sudden decline in the population, for example, a natural disaster, the population will return to its original level. The second steady state is below the MSY, which is unstable, and the population may collapse in a sudden disturbance.

1.7 Flow of Energy and Matter in the Ecosystem

1.7.1 Energy Flow

When we talk about the flow of energy through the ecosystem, we can mean very different energy flows. For the purposes of this text, we divide them into two categories: (1) energy flows that have been bound to the bodies of living organisms by photosynthesis and possibly in other chemoautotrophic processes and are subject to energy flows in food chains and (2) all other (abiotic) energy flows (Boxes 1.9 and 1.10). The latter one is quantitatively more important as we can clearly demonstrate this in the diagram of the energy balance of solar radiation in Fig. 1.28. The flux of

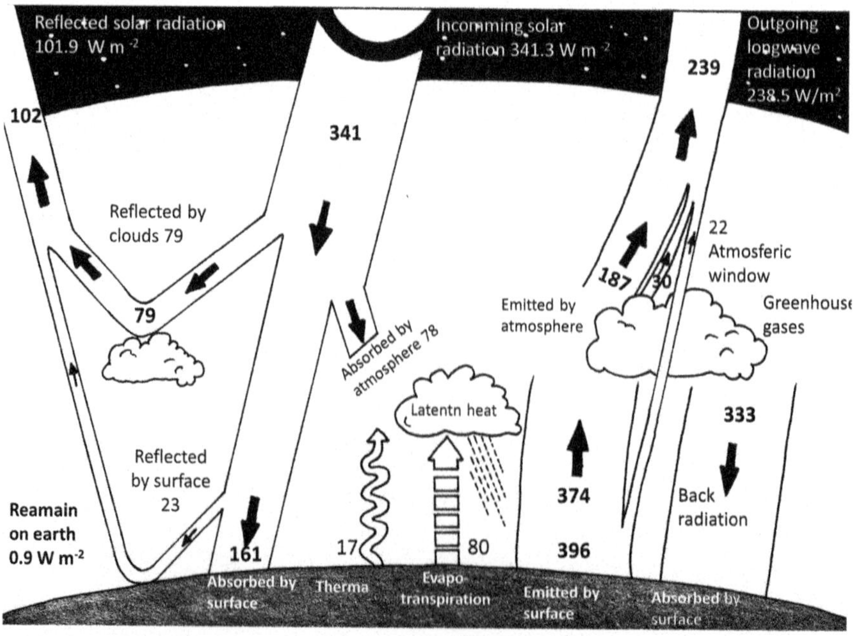

Fig. 1.28 Radiation balance of the earth's surface based on Trenberth et al. (2009) and Blankenship et al. (2011)

Box 1.9 Thermohaline Cycle

The thermohaline cycle, also called the Global Ocean Conveyor or Great Ocean Conveyor Belt, causes the circulation of deep-sea waters. The wind drives the waves only to a depth of about 100 m, but the thermohaline cycle sets the water in motion at much greater depths. It is controlled by differences in seawater density given by temperature and salinity (Fig. 1.29). Water from the warmer parts of the ocean passes into the polar regions. This surface warm water is already relatively salty due to the fact that part of the water has evaporated in the warm seas. It cools in the polar regions, which increases its density even further. In the polar regions of the Earth, the ocean waters are very cold and form sea ice. In the formation of sea ice, most salts are concentrated in seawater under ice, although some salts may be partially trapped in small pockets between ice crystals. The water under the ice is cold and has a higher salt concentration and it becomes denser than the surrounding ocean water. This dense water begins to sink to the bottom of the ocean and pushes the deepwater upward. Water dropping from the surface downward brings oxygen to the depths of the ocean, while water rising from the bottom of the ocean brings nutrients especially N and P. These nutrients got into the deep layers of the ocean due to the constant rain of debris and dead bodies of organisms gradually falling to the bottom, where they slowly decompose. Due to the fact that there is darkness, there is no photosynthesis, and nutrients are not built into the bodies of algae, they are carried away by currents and rise to the surface in subpolar regions.

The bodies of water that have sunk in the North Atlantic are slowly moving south. This flow is driven by cooling water at higher latitudes and heating closer to the equator. Deep-sea currents will reach south and become part of the current flowing around Antarctica. Beyond South Africa, this current branch into a branch extending into the Pacific and a branch into the Indian

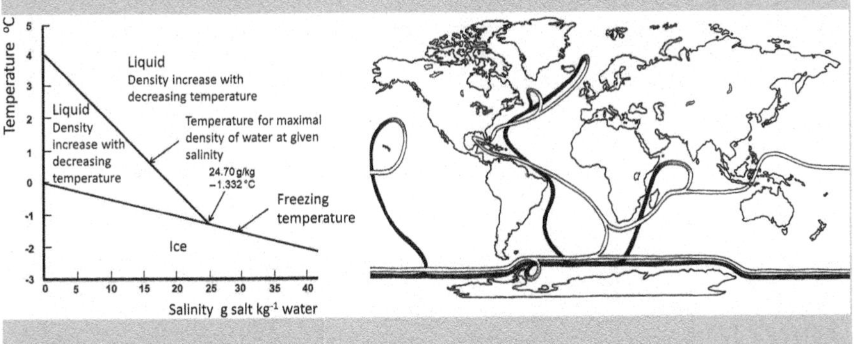

Fig. 1.29 Diagram illustrating factors influencing seawater density (left) and thermohaline cycle diagram (right), surface currents white, deep black, based on NOOA, modified

(continued)

Box 1.9 (continued)

Ocean. In the Indian Ocean, due to the fact that the Indian Ocean is closed from the north and the cold and saltwater of the Atlantic meet the opposing flow of warmer and less saltwater from the tropical Pacific, a number of complex phenomena occur. This results in the equatorial circulation of sea currents within the Indian Ocean waters in which deep-sea currents reach the surface. In the Pacific Ocean, the rest of the cold and saltwater from the Atlantic undergoes an upward gradient on the salinity and becomes warmer and faster. In the North Pacific, these deep-sea currents are rise off the west coast of North America. The surface currents then move to the southwest. The sea level of the Atlantic is slightly lower than the Pacific and the salinity of the water in the Atlantic is higher than in the Pacific. It generates a large but slow flow of warmer and less salty ocean waters from the tropical Pacific to the Indian Ocean through the Indonesian archipelago, replacing the cold and saltwater of Antarctica. This warmer, less salty water from the Pacific flows through the South Atlantic to Greenland, where it cools and sinks to the bottom, as described above, thus closing the thermohaline cycle (Fig. 1.30).

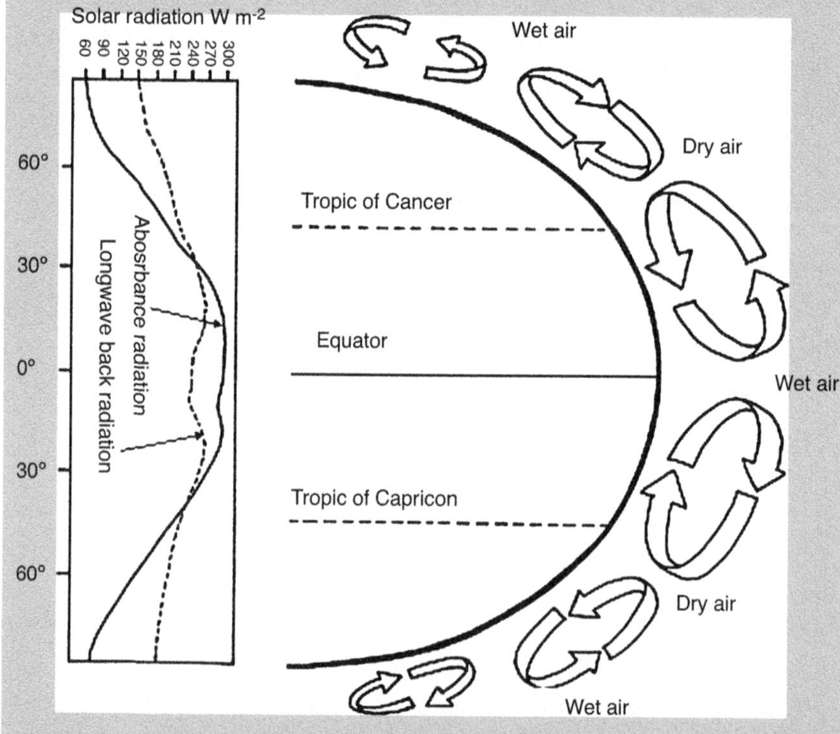

Fig. 1.30 Distribution of solar radiation on Earth and its impact on the global atmospheric circulation

(continued)

Box 1.9 (continued)

As already described, thermohaline cycle is powered by temperature differences between polar areas and the equator, wind movement, and earth's rotation. With global warming, the polar region warms faster than the equator and the temperature difference between them decreases (Baling, 2005). Some scientists believe that climate change could disrupt thermohaline circulation (Schiermeier, 2007) by warming waters in the polar region and create an influx of freshwater from melting glaciers into subpolar waters. Since freshwater is less dense than saltwater, significant freshwater infiltration would reduce the density of surface waters, moreover warmer polar region makes the water warmer, which will reduce its density even more. Therefore, reduced water density in the polar region will prevent water from sinking, which drives thermohaline circulation. There are also opinions that a similar disturbance of thermocline cycle has already occurred in the geological past of the earth and was related to some of the mass extinctions (Zeebe, 2012).

Box 1.10 Radiation, Greenhouse Effect, and Global Circulation

The main source of energy on earth is solar radiation. If we were located on a perpendicular line between the sun and the earth, in the distance of the earth from the sun, but outside the influence of the atmosphere, the solar radiation would be about 1360 W m^{-2} (this value is called the solar constant, although it also oscillates slightly and changes over geological time). The earth is round, so the actual value of the energy falling on the upper limit of the atmosphere is equal to a quarter of the solar constant (342 W). This is due to the fact that only half of the planet is always illuminated, the other half is in the dark. The length of daylight on the equator is the same for about 12 hours each day while in higher latitudes the length of daylight is more variable.

In the polar regions, there is a period of continuous daylight and continuous darkness lasting several weeks or months (polar day and polar night). Despite this variability, the total annual time of daylight in all places on earth is very similar and is slightly greater than 50% of the total time (this time is even a few percent greater in the polar regions than at the equator). The roundness of the earth causes only a small portion of the rays at the equator to fall perpendicular to the earth. At higher latitudes, the rays fall obliquely, so the radiation spreads over a larger area and its power input per unit area is smaller (Fig. 1.30). As a result, the energy input of sunlight at the equator is much higher than at higher latitudes. As mentioned in Sect. 1.7.1, solar radiation heats the earth and it then emits radiation back into space. However, while the sun's radiation is shortwave (corresponding to the higher temperature of the sun), the earth's radiation is longwave (corresponding to the lower temperature of the earth).

(continued)

Box 1.10 (continued)

The ratio of incoming shortwave radiation to emitted longwave radiation is not the same everywhere on earth. In the vicinity of the equator, the radiation balance of the earth (the ratio of the incident shortwave radiation from the sun and the longwave radiation emitted by the earth) is positive, and at the poles it is negative. This uneven distribution of radiation sets in motion air and sea currents that distribute heat from the equator to higher latitudes. The movement of sea currents given by this temperature gradient from the equator to the poles allows the existence of a thermohaline exchanger (Box 1.9). The global air circulation conditioned by this gradient, is then responsible, among other things, for the formation of the main climate zones on Earth (Fig. 1.30). Their more detailed description can be found in Goosse et al. (2010).

To illustrate, we describe the behavior of air masses around the equator. In equatorial areas where there is a high input of solar radiation, there is also a large amount of evaporation, the humid air rises and after it reaches the level of condensation, it water condensate and rain out. The air deprived of water vapor cools in the upper atmosphere, at the same time moving away from the equator and returning to the ground in the region of Tropic of Cancer (in the northern hemisphere or Tropic of Capricorn in the southern hemisphere) and starts returning toward the equator. The fact that the incoming air is free of water vapor, which condensate and rain out around equator, is the cause of the dry climate in the desert zone around the Tropics of Cancer or Capricorn, and the humid climate around the equator.

The input of solar radiation is affected not only by changes in solar activity but also by small deviations in the earth's rotation. The Earth's rotation is more circular in some periods, sometimes more elliptical, the inclination of the Earth's axis (23.5°) fluctuates slightly by 1° on both sides and there are some other small fluctuations. These oscillations of the Earth's rotation described by the so-called Milakovich cycles cause small changes in the input of solar radiation. However, in the consequences they can, for example, cause oscillations between glacial and interglacial periods.

In contrary to shortwave solar radiation, the earth's long-wave radiation that returns energy to space is partially absorbed by greenhouse gases. The most important greenhouse gases include water vapor, carbon dioxide, and methane. The greenhouse gases subsequently radiate the captured energy in all directions and also partially back to the Earth's surface, which causes the earth cool more slowly (Fig. 1.28). We call this process the greenhouse effect (National Research Council, 2005). The greenhouse effect raises the temperature of the earth and is one of the basic conditions of life on earth. It is estimated that without the greenhouse effect, the average earth temperature would be −18 °C.

(continued)

Box 1.10 (continued)

We have already mentioned that the changes in the input of solar radiation given by the Milankovich cycles trigger the alternation of glacial and interglacial periods. However, it would be wrong to assume that the change in energy input itself was the only driver of this change. Rather, it can be imagined that a small change in radiation triggered other interrelated processes. A small increase in temperature will increase respiration and thus, the concentration of CO_2, which will further increase the temperature. This process will increase the evaporation and again increase the greenhouse effect and consequently the temperature. This can lead to melting glaciers, however, exposed soil absorbs more energy than snow (albedo decreases), this increases the amount of solar energy absorbed by the Earth and this again increases the temperature, etc. Therefore, changes in solar radiation interact with changes in the greenhouse effect and other factors.

The amount of solar energy and the intensity of the greenhouse effect has changed in the geological history of the Earth (National Research Council, 2005). The luminosity of the Sun increases significantly over geological time (more energy comes to Earth now than in the past). These changes are negligible during human history. However, during geological development, these are much larger changes than those caused by Milakovich's cycles. How is it possible then that the temperature of the earth was, for example, higher in the Proterozoic than today, when less solar energy was coming? This is due to the different intensity of the greenhouse effect (Fig. 1.31). In the Proterozoic, the CO_2 content in the atmosphere was about ten times higher than today, and this

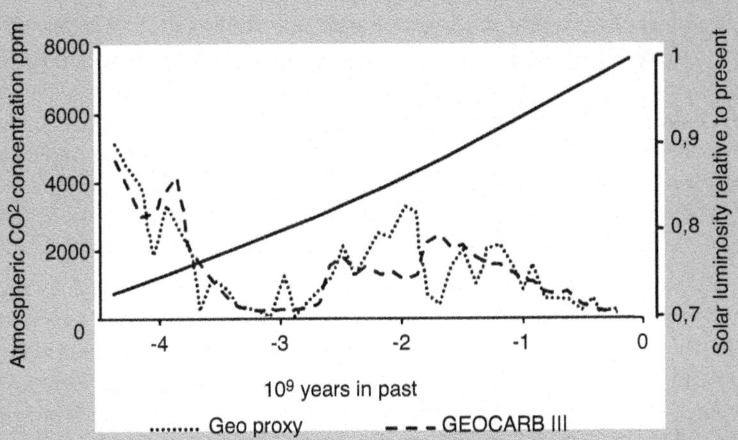

Fig. 1.31 Evolution of CO_2 concentration and solar luminosity (proportional to solar energy input) during the geological evolution of the Earth from the National Research Council (2005), adjusted

(continued)

Box 1.10 (continued)
corresponded to a greater intensity of the greenhouse effect, which then ensured a similar temperature as today, or even slightly higher even with lower solar radiation (National Research Council, 2005).

How is it possible that the strength of the greenhouse effect has changed so much? There are a number of mechanisms that affect the concentration of CO_2 in the atmosphere. One of them is plate tectonics, as described in the chapter on the cycle of elements. Plate tectonics is involved in the carbon cycle. The CO_2 produced by volcanoes enters the atmosphere here, among other things helps weather rock to form carbonates, which accumulate as sediments on the seabed and with the displacement of tectonic plates over millions of years, can reach subduction zones where sediments become part of magma, which closes the cycle. Weathering, like any chemical process, is temperature dependent, more weathering occurs at higher temperatures and drains CO_2 from the atmosphere. Conversely, when it is cold, weathering is slower and volcanoes replenish CO_2 over time. This process is very slow (CO_2 fluxes mediated in this way are negligible when compared, for example, to the exchange of CO_2 between the atmosphere and oceans, or the atmosphere and soil) but could help balance the planet's temperature already in the abiotic part of the Earth's development.

In addition to plate tectonics, there are other large carbon reservoirs today that can affect the concentration of CO_2 in the atmosphere, including water in the ocean, vegetation, and soil. Soils can support or slow down the greenhouse effect e.g., at the end of the ice ages, permafrost melted, and released a considerable amount of methane, which promoted faster warming. At an optimal temperature, vegetation fixes a considerable amount of CO_2 and this is stored in soils and sediments (peat) and thus, draws CO_2 from the atmosphere. Thus, the ecosystem of the planet has a number of feedback mechanisms that affect the greenhouse effect. However, mankind can also influence these mechanisms by effecting the global cycle of elements, especially the carbon cycle, including various ways of land use by agriculture and forestry.

solar radiation coming to earth from space is 1360 W m^{-2} (so-called solar constant); however, because the earth is curved not every meter of this energy falls vertically and thus the energy captured is reduced (the largest is of this reduction is in the polar regions). In addition, light does not fall on any part of the planet for some time, so the average amount of energy falling on 1 m^2 at the upper limit of the atmosphere is less than about 342 W (for more details, see Box 1.10).

Part of this energy is absorbed and dispersed in the atmosphere, some are reflected back by clouds, about half of it hits the planet's surface and some of this energy is reflected from the earth's surface. The ratio of incident and reflected energy is called albedo and depends on the properties of the surface. Snow reflects most of the light and has high albedo darker surfaces effect less light have lover

Table 1.11 Global size of selected energy flows on earth (Makarieva et al., 2008)

Category	Energy flow TW (10^{12} W)
Solar radiation	8×10^4
Sensible heat	4×10^4
Latent heat	2×10^4
Thermohaline conveyer	1000
Atmospheric circulation	1000
Photosynthesis total on earth	100
Mankind energy consumption	15

albedo. For example, pine trees remain darker in Alaska even in the winter, which absorbs more heat than the surrounding snow and therefore, are warmer. On the contrary, deforestation and its replacement by dark soil will reduce albedo and contribute to a higher temperature of areas without vegetation. Energy that is not reflected is absorbed by the surfaces. Only a small part of this energy is used in photosynthesis and it is the subject of energy transfers in food chains (Fig. 1.28) (Table 1.11). Nevertheless plants substantially energy budget by effect on many other energy flows, e.g., by changes in albedo and transpiration.

Most of the solar energy that has reached the surface is converted into either so-called sensible heat or so-called latent heat. Sensible heat is responsible for the temperature of the surfaces and it also causes the air above the surface to heat up, which then rises. This heating is different in different places, which also causes horizontal displacements of the air masses, which affect the weather and climate. In addition, air movements are responsible for wind erosion, aerosol and nutrient transfer, and other important processes. Heating surfaces causes volume changes and promotes weathering. Water heating then affects the bulk density of water, which contributes to the stratification of water bodies in reservoirs and oceans, and the heterogeneity of this heating leads to heterogeneity of densities that support the movement of sea currents. Latent heat is the heat needed for the transformations of water from liquid to gas (2257 kJ kg^{-1} is needed to evaporate the water; this value may vary slightly depending on the temperature). This heat is consumed when water evaporates (or melts the ice), is released again when the water condenses or the ice solidifies, and this leads to buffering of temperature fluctuations.

The presence of vegetation can affect the ratio between sensible and latent heat. If we use the example of a clear-cut forest, which we convert into a field, then when there is no vegetation in the field, its temperature will be higher not only because it will have a smaller albedo but also because the forest will evaporate the soil. In the case of vegetation, a large amount of water transpires and thus the air above the forest will cool.

Water evaporation has a number of other consequences in essence it allows for the water cycle in nature, the transfer of water from the oceans over land, and related processes such as water erosion, chemical weathering of rocks, etc. In addition to the energy coming from the sun there are other forces related to the rotation of the earth, the gravitational action of the earth and the moon, and the movement of tectonic plates, which then determine the position of the continents, affect the movements of water and air masses. A description of all these massive and crucial energy flows for our planet would be beyond the scope of this book. We will limit ourselves to this brief introduction and in the individual chapters, we will deal with processes that either affect some human activities or may be affected by human activities. Activities such as changes in vegetation composition, which affects albedo or the amount of evaporated water and thus the proportion of sensible to latent heat, etc.

Let us go back for a moment to the flow of energy that has been bound to the bodies of living organisms in photosynthesis and possibly other chemoautotrophic processes and is the subject of the flow of energy in food chains. The efficiency of photosynthesis after deducting plant respiration is about 0.2–2%. Part of this energy can be consumed by consumers and transferred to the biomass of their bodies. As a rough estimate, it is often assumed that biomass at a higher trophic level (i.e., in organisms that have consumed organisms on lower trophic level) will represent about 10% of biomass at a lower trophic level (i.e., those organisms that have been consumed). This is given by the combination of assimilation and production efficiencies we talked about earlier, including the fact that not all biomass from a lower trophic level can be eaten, etc.

It follows that the higher is an organism in the food chain, the more space is needed for its livelihood. For example, the average annual territory of roe deer (*Capreolus capreolus*) which are herbivores, is about 0.5 km^2 (Kjellander et al., 2004), while the territory of the wolf (*Canis lupus*) which are predators that feed on roe deer among other is on average, about 210 km^2 (Jedrzejewski et al., 2007).

This certainly applies on humans as well. Let us imagine a hypothetical example of two different human communities to illustrate this. The first, which would feed on rice and legumes (like many people in many Asian countries), will consume about 100 kg of rice and 20 kg of legumes per person a year and would need an area of 200 m^2 to make a living for one person, 150 m^2 for growing rice, and 50 m^2 for legumes based on the estimated yield of 7.5 tons of ha^{-1} rice and 4.5 tons of legumes ha^{-1}. There will still remain some rice and legumes left to sow the field the next year. Our second hypothetical community would feed only on pike, which one would hunt in small lakes and ditches in a grassland type of landscape. Pike would feed on frogs and small fish, and they would feed on herbivorous insects (e.g., grasshoppers) that would feed on grass (admittedly this is quite simplistic and not entirely realistic, but let us allow this possibility for a while). Pike meat contains more water, but let us say one fisherman will consume 150 kg of pike a year, so he needs a pike population at least twice the size to feed (see maximum sustainable yield, in fact a little more, because the catchable population is just a few years old) or 300 kg of pike. They would then need to feed 3 tons of frogs, small fish, and in turn, 30 tons of insects, which would need 300 tons of grass. Assuming a grass yield of 5 tons ha^{-1}, which would represent 60 ha or 3000 times more than in the first case. In other words, an area that will feed one pike hunter would feed 3000 vegetarian farmers. This is a very hypothetical but hopefully illustrative example.

Table 1.12 Primary production, ecological efficiency, and number of trophic levels in selected biomes or ecosystems

Ecosystem	Primary production t ha^{-1} $year^{-1}$	Ecological efficiency %	Number of trophic levels
Open Ocean	0.5	25	7
Estuaris	18	20	5
Temperate grassland	5	10	4–5
Tropical rainforest	20	5	3

Surely you have noticed that in the hypothetical fisherman case, we used someone who feeds on predatory fish as an example of a human at a very high trophic level. This allowed us to push more trophic levels in front of the consumer in this artificial example (at least in a somewhat realistic way) than if he ate sheep, for example. In such a (more realistic) case, a shepherd would need only about half a hectare or about "only" 30 times more than a vegetarian farmer at the same yield of grass (but a six-member shepherd family would still need the same land as a 180-member vegetarian village, again rough illustrative estimate).

This leads us to the interesting observation that the length of food chains varies between different types of ecosystems. Grazing food chains in terrestrial ecosystems tend to have fewer trophic levels than those in aquatic ecosystems. This is related to the fact that ecological efficiencies differ between individual types of ecosystems (Table 1.12). This is mainly because the speed of biomass turnover and digestibility varies between ecosystems. The share consumed by consumers is higher in aquatic ecosystems and smaller in terrestrial ecosystems. This is due to the fact that in aquatic ecosystems, producers (algae and cyanobacteria) have a rapid turnover of biomass and little or no supporting tissue. In contrast, terrestrial plants build a lot of supportive tissues that are difficult to digest for consumers and have a slower biomass turnover rate. With higher turnover speeds and greater digestibility, more food is available per unit area in aquatic ecosystems. This can be limiting because, as we have shown above, as the trophic level increases, the area that the organism needs for its life increases. In addition, in order for the organism to survive, it needs to maintain a reasonably large population, which increases the area. In addition, organisms must be able to meet in this area with others in order to reproduce and at the same time find places with enough food where they can raise young, etc.

1.7.2 Water Cycle

As we have already mentioned in the chapter on energy flows, substantial part of the energy that falls on the Earth's surface is used for changes of state of water, most often for water evaporation. Evaporated water enters the atmosphere, where it can condense or even freeze, creating precipitation that fall on the earth's surface. In this

Table 1.13 Main reservoirs and flows of water on the Earth

Pool	Approximate volume km^3	Mean retention time
Oceans and seas	1348 mL	3200 years
Glacier	27 mL	20–100 years (20,000 years)
Groundwater	8 mL	100–200 years (10,000 years)
Lakes	190,000	50–100 years
Soil moisture	140,000	1–2 months
Rivers	10,000	2–6 months
Atmosphere	13,000	9 days
Annual evaporation from seas	430,000	
Annual evaporation from land	70,000	
Annual precipitation over seas	390,000	
Annual precipitation over land	110,000	
Annual runoff from land	40,000	

way, we could briefly describe the water cycle. Most of the water on earth is stored in the oceans (Table 1.13) and most of the water evaporates over the oceans. Some of it also returns to the oceans in the form of precipitation, but some is transferred by air currents over the mainland, where it precipitates. This cycle of water, including evaporation of water from the oceans, moving water vapour over land, their precipitation over land, and water runoff back into the ocean in rivers is called the large water cycle. On the contrary, the cycle of water evaporating from the land and returning to the land in the form of precipitation is called a small water cycle. A large water cycle is typical, for example, of deserts and semideserts, while a small water cycle dominates over rainforests and is important in well-vegetated areas.

Let's Let us take a closer look at what happens to precipitation. If the surface is covered with vegetation, part of the precipitation (about 30%) is captured by vegetation, so-called interception. Part of the water captured by interception then evaporates, part flows, falls on the ground, and part is taken up by plants. The water that reaches the surface is partly infiltrates the soil, partly evaporates, and partly becomes surface runoff. The water that has infiltrated becomes part of the soil water, which is retained mainly by capillary forces or becomes part of the groundwater. Soil water can either evaporate or be taken up by plants, become part of an underground runoff, or emerge from soil bit downhill and become part of surface runoff. Interestingly, most of the water that flows out of the catchment is the so-called old water. In other words, water that was already part of the soil or groundwater and the newly infiltrated rainwater "pushes it out" and takes its place. Besides, runoff water is lost from the soil by evaporation and be taken up by the plants. Plants also evaporate most of the water taken from the soil. This plant-mediated water evaporation is called transpiration. We often talk about evapotranspiration, which is the evaporation of water from soil and plants.

1.7.3 Cycles of Major Biogenic Elements

In this chapter, we describe the global cycles of the main biogenic elements and note how they have changed under the influence of mankind in comparison with the preindustrial period (i.e., the state before 1850). We will pay attention to the main features of individual cycles, which may be affected in a locally significant way by the activity of organisms. The effects of plants were also discussed in the chapter dealing with plant growth strategies and the effect of animals in Box 1.11.

If we look at the largest carbon reservoirs (Fig. 1.32), we see that the largest amount of carbon is stored in rocks, especially in carbonates, and a small amount in organic sediments. However, rock reservoirs naturally have only a very slow turnover (Houghton 2007). Thanks to plate tectonics, sedimentary rocks containing carbonates are shifted until they reach subduction zones, where they sink under other tectonic plates. This leads to high pressures and temperatures, and some of the submerged rocks are converted to magma, which is released by volcanic activity. The carbon originally contained in the sediments is then released into the atmosphere as CO_2. CO_2 then dissolves in water, forms weak carbonic acid, which participates in the weathering of rocks producing carbonates. Carbonates become part of

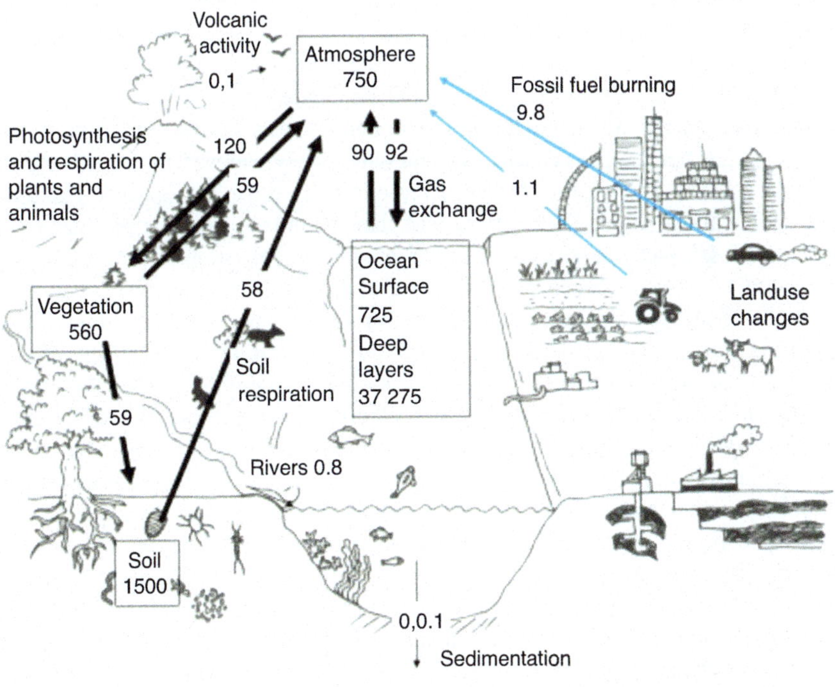

Fig. 1.32 C cycle according to Houghton (2007) and Global Carbon Project (2008), modified. The contents of the reservoirs are in Pg, the flows are given in Pg year^{-1}. Values are approximate mean values. The main natural flows are in black, the main anthropogenic influences in blue

sedimentary carbonate rocks formed at the bottom of the sea, which, thanks to plate tectonics, move to subduction sites, and the whole cycle is repeated. However, this cycle is really a cycle only on geological timescales. If we consider the values of annual flows, the flows associated with the processes described above are almost negligible (Fig. 1.32).

Other large carbon sinks are carbon in the form of CO_2 dissolved in water, especially in ocean water, carbon in the atmosphere, in soils, and biota. Water absorbs a significant amount of CO_2 because dissolved CO_2 reacts with water to form carbonate and bicarbonate ions. This exchange, despite its overall large volume, is almost balanced. The atmosphere and the world's waters exchange large amounts of CO_2, but the total amount going in both directions is almost the same. At present, due to the increasing concentration of CO_2 in the air, the flow from the atmosphere to the oceans is greater than back, which, among other things, leads to the acidification of the oceans.

As we have already said, terrestrial biota and soils represent another important carbon sink. Biomass contains about the same amount of carbon as the atmosphere and the world's soils 2× to 3× more. CO_2 is taken up by plants from the atmosphere by photosynthesis, a significant part is released back by plant respiration, the rest remains as plant biomass. Plant biomass can be partly used by herbivores and becomes the basis of forage food chains, but most of the carbon comes in the form of dead biomass to the soil, where organic matter decomposes and releases CO_2. The amount of CO_2 released by soils and biota is somewhat less than the amount of CO_2 entering the soil. In other words, soils sequester some carbon and reduce the current increase in CO_2 concentration in the atmosphere.

We have already mentioned the anthropogenic increase in CO_2 in the atmosphere several times, but the flows described so far only slowly remove CO_2 from the atmosphere. So, what is the cause of the increase in CO_2 in the atmosphere? There are two main causes: burning fossil fuels and changes in land use. The mechanism of action of fossil fuel combustion is easy to imagine. Combustion of coal or gasoline releases CO_2, which escapes into the air. So, what is there imagine under changes in land use? These are a number of different influences that increase the flow of CO_2 from the soil to the atmosphere compared to the state in undisturbed ecosystems. Here we list just a few of them. Probably the most straightforward is the idea of felling and burning rainforests and other ecosystems, the burning of biomass releases CO_2 into the atmosphere. After the transition of the original ecosystems to the arable field, organic matter is often lost from the soil and CO_2 goes back into the atmosphere, which is caused by accelerated microbial respiration. Drainage of wetlands and peatlands has a similar effect. Due to anaerobic conditions, peatland bound a large amount of organic matter, when the soils are oxidized due to drainage, the organic matter begins to decompose rapidly. Although CO_2 fluxes caused by fossil fuel combustion and land-use change do not seem large at first glance, they are a constant addition into an otherwise well-balanced system and are the main causes of the observed increase in atmospheric CO_2 concentration (Houghton 2007).

Nitrogen also has large reservoirs in rocks; however, rocks contribute little to the overall cycle (Lavalle et al., 2005). The largest reservoir of N is in the atmosphere

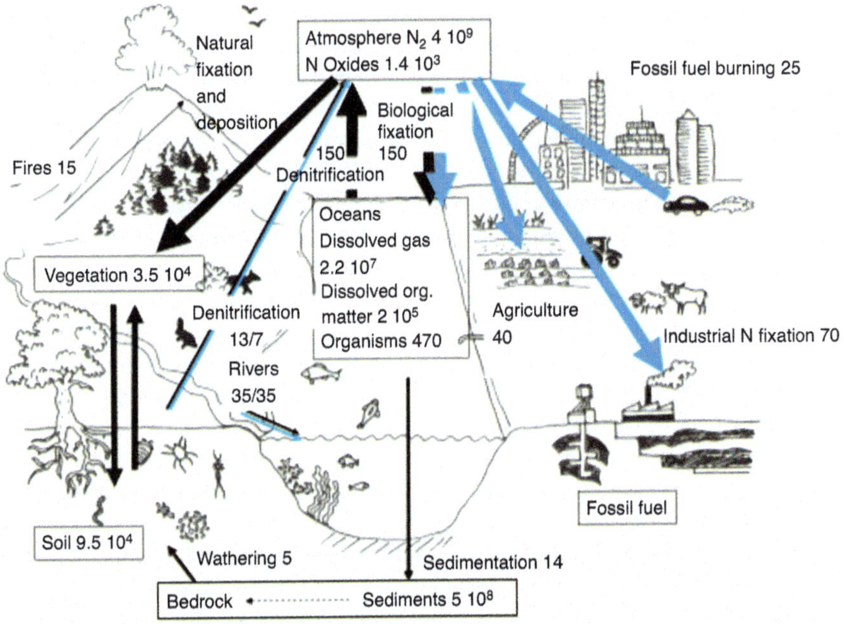

Fig. 1.33 N cycle according to Lavalle et al. (2005), modified. Contents of containers are in Tg, flows are given in Tg year^{-1}, and values are approximate mean values. The main natural flows are in black, the main anthropogenic influences are in blue, if the anthropogenic influences have led to an increase in natural flows, this increase is given as preindustrial values/increase, so the current values are the sum of both numbers

from where N can enter the soil by means of electric discharges (lightning) or into organisms by means of nitrogen fixation (Fig. 1.33). Nitrogen bound in the body of plants is released by their decomposition as ammonia. Ammonia can be taken up by plants or can bind to the soil sorption complex. Ammonia is usually oxidized in two stages to nitrite and then to nitrate. In this step, bacteria that perform this activity gain energy by oxidizing ammonia. Nitrates can be absorbed by plants and are very mobile in the soil, so it is easily leached out of the soil. Some bacteria can, in the absence of air, use nitrates in redox reactions as an electron donor and thus reduce them back, by denitrification, producing N_2, which escapes into the atmosphere (Lavalle et al., 2005). In this process, nitrogen oxides NO_x are formed, such as N_2O, which is an important greenhouse gas.

The fixation of N from air is an energy-demanding process, and the amount of N fixed from atmosphere is thus limited. Since N only physically dissolves in water but does not react with it, the volume of dissolved N is relatively smaller than that of C. For these reasons, N in the atmosphere remains predominantly in the atmosphere, and N bound to soil and organisms is recycled between soil and organisms. There are transitions between these two parts of the N cycle, but quantitatively these flows are small relative to the volume of the reservoirs. The persistence of the N-reservoir in the atmosphere can also be demonstrated by the fact that while you read this chapter, you inhaled, with a probability bordering on certainty, at least one of the nitrogen molecules that were in Julius Caesar's last breath, so by other words, most of the N molecules in air are the same as in Roman times. This would not be the case for carbon, as carbon molecules are exchanged more quickly with other reservoirs, and carbon molecules in the atmosphere have changed considerably since Caesar's time.

Human has significantly influenced the global N cycle through a number of activities. From the beginning, human accelerated the turnover of N in ecosystems, for example, by plowing. Plowing accelerates the decomposition of organic matter and thus the mineralization of nitrogen. Similarly, the supply of animal excrement or manure has increased the proportion of mineral forms of N compared to the situation where dead plant biomass enters the soil. This resulted in faster mineralization of N and part of the N made available in this way could be used by plants. However, the consequence of such a greater access to N was its loss from ecosystems. In natural ecosystems, most of the nitrogen is bound in the dead organic matter and the available N very low, so available N is immediately utilized by plants and microorganisms.

The acceleration of N mineralization and a higher availability of mobile forms of N have caused not all N being used and part of N has been removed from the ecosystem by leaching (NO_3), volatilization (NH_4), or denitrification (N_2). This could lead to some parts of the landscape being gradually depleted of N. This changed with the industrial fixation of N (more also in Chap. 2). This activity globally doubled the N input from the atmosphere to ecosystems. At the landscape level, the supply of N by fertilization and atmospheric deposition causes a substantial increase in the availability of N. In addition, as the intensification progresses, the nitrogen cycle becomes less and less closed. In natural ecosystems, almost all N is cycled in one place. Conversely, intensification of land use leads to N losses, not only in the ways described above but nitrogen is also removed with biomass removed from the area (such as crops, logging, etc.). Nitrogen in food becomes part of wastewater and is removed in wastewater treatment plants, mostly by denitrification. At the same time, nitrogen uptake by plants and entry into ecosystems (e.g., by atmospheric decomposition, fertilization, etc.) do not necessarily coincide (Galloway et al., 2008).

Compared to the C and N cycles, the P cycle is a slower process. The main source of P is rocks (Lavalle et al., 2005). Phosphorus occurs in many rocks, the most abundant is in phosphorites containing 5–55% apatite (calcium phosphate), which is mined and used as a P fertilizer. Most other rocks contain a small amount of P, in the order of hundreds to thousands of ppm. Among the volcanic rocks, there are lower contents of P in acidic silica granites (hundreds of ppm) and higher contents in alkaline basalts (thousands of ppm). In sedimentary rocks, the P content decreases with increasing grain size, the largest is in claystones (>1000 ppm) and the smallest is in sandstones (500 ppm) (Porder and Ramachandran, 2013). Phosphorus is released from rocks by weathering. It is taken up by microorganisms and plants and then animals. Mycorrhizal fungi play a major role in the uptake of P by plants. In organisms, P is part of a number of key molecules such as phosphoglycerol, which is part of membranes, as well as forms part of DNA, ATP, and ADP.

When organisms die, P is mineralized, returned to the soil, and can be reused by plants. In this process, part of the P from the ecosystem is lost mainly by erosion. Therefore, in very old and very weathered soils, phosphorus is depleted and often, the remaining phosphorus is bound in hardly available forms. Phosphorus, which leaves ecosystems by leaching and erosion, enters rivers and with rivers enter into the sea, P can also enter the seas by wind erosion and dust carried by the wind, especially from deserts and semideserts (Lavalle et al., 2005). As a result, ecosystems on younger soils (e.g., in the temperate zone where soil development has been reset by glaciers) are better supplied with P than ecosystems on old weathered soils (e.g., many tropical ecosystems), where P is often limiting. In the sea, P then becomes part of marine sediments over geological time. These can then be lifted during geological processes (over millions of years), become part of the rocks, and the process repeats itself. However, while the first part of the process is relatively fast, the return of P in the form of rocks that could weather is very slow.

Mankind has substantially accelerated the P cycle – firstly, by removing permanent vegetation (e.g., by tillage) and thus supporting erosion and movement of P to rivers and into the sea, and secondly, through the mining of phosphorus-rich rocks and using them as fertilizer and for other industrial applications. This again increases the entry of P into rivers and subsequently into the sea (Lavalle et al., 2005), which has almost doubled compared to preindustrial times (Fig. 1.34).

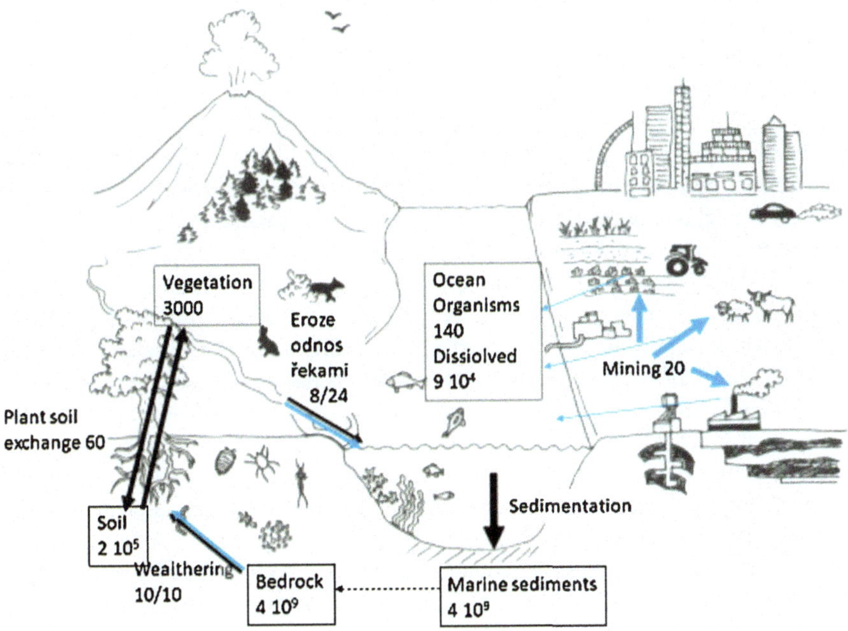

Fig. 1.34 P cycle according to Lavalle et al. (2005), modified. Contents of containers are in Tg, flows are given in Tg year^{-1} and values are approximate mean values. The main natural flows are in black, the main anthropogenic influences are in blue, if the anthropogenic influences have led to an increase in natural flows, this increase is given as preindustrial values/increase, so the current values are the sum of both numbers

If we compare the main biogenic cycles, we see that the C cycle is closely linked to the flow of energy in ecosystems. From the atmosphere where it was in the form of CO_2, C is bound to organic compounds by photosynthesis, which converts the energy of radiation into the energy of chemical bonds. All living organisms then gradually break down these chemical bonds and oxidize organic matter to CO_2 and water (in anaerobic environments also to other end products), thus releasing the energy they use to cover their energy needs. Similarly, in the N cycle, energy is needed to fix N from the atmosphere, or in other words, to convert it from N_2 to HN_4^+ (16 ATP molecules per N_2 molecule). This energy needed during biological fixation is at the expense of the chemical bonds formed in photosynthesis, as described above. Energy can also be supplied in other ways, either industrially in the form of high temperature and pressure in the presence of catalysts (Haber–Bosch process) or by some natural processes, especially lightning. Organic N from the bodies of dead animals is converted into ammonia. Oxidation of ammonia to nitrates then leads to the release of energy stored in chemical bonds. Denitrification also serves to obtain energy, but in a different way. Here, under anaerobic conditions, microorganisms use nitrate as an electron acceptor instead of oxygen. In contrast, P, which is part of the ATP–ADP energy transfer system in cells, does not oxidize or reduce in any way during its cycle. Both the C and N cycles have an active gas reservoir, while the P cycle has no gas reservoir. The C and N cycles also have large reservoirs in the rocks. However, they are very slow in terms of exchange with other reservoirs and so they are used in the cycle in geological rather than ecological time. Similarly, the P cycle is a cycle in geological time, while in ecological time it represents a rather one-way flow of P from weathering rocks to the sea, which is slowed down by the recycling of P between organisms and dead organic matter.

The interactions of individual biogeochemical cycles with each other or their interactions with water flow and other key ecosystem processes are less explored. At the same time, it is clear that there must be close ties. As mentioned above, C in soils and plant biomass is a significant global carbon sink. Here, C is bound in living or dead biomass, which contains other biogenic elements in addition to C. As shown in the chapter on plant growth strategies, the rate of biomass turnover is related to the rate of decomposition and thus affects the availability of other elements, especially N and P. In contrast, the plant community is influenced, among other things, by the availability of nutrients. In natural ecosystems, the main source of N is the fixation of N from the atmosphere. It is an energy-intensive process that correlates well with the degree of evapotranspiration (Cleveland et al., 1999), which reflects the availability of water and solar energy. In addition, N fixation is dependent on the availability of P and other microelements (Huston & Wolverton, 2009). In contrast, the availability of N and other mineral nutrients affects primary production, as described above. These are just some examples of the interactions of the main biogenic elements, but they show the complex interconnectedness with biogeochemical cycles.

Box 1.11 Animals and the Nutrient Cycle

Animals can significantly affect the flow and distribution of nutrients in ecosystems and their spatial distribution. An important way of moving P and N from the sea to freshwater and land is the transfer of these nutrients by fish-eating birds and fish that live in the sea but spawn in freshwater. Doughty et al. (2016) estimate that fish migrating to freshwater have historically been able to bring from the seas to land 1.4 10^8 kg P $year^{-1}$ and seabirds 6.3 10^6 kg P $year^{-1}$. However, with the decrease of seabirds and migratory fish, the total amount of P brought onto land by both groups is 5.6 10^6 kg P $year^{-1}$, which means a decrease of more than 96%.

Animals could also be involved in the horizontal and vertical distribution of nutrients in the oceans. In the ocean, nutrients are constantly falling out of the photic layer as the excrement of crustaceans and fish and their dead bodies gradually sink to the bottom. Also, the migration of crustaceans into deeper layers can speed up this process. Whales and other large marine mammals often hunt under the photic layer and defecate when breathing near the surface, which can bring N, P, and other nutrients into the photic layer (Roman & McCarthy, 2010). Roman and McCarthy (2010) estimated that whales and other large marine mammals bring into the euphotic layer of Maine Bay 2.3 10^4 t N $year^{-1}$, which is more than all rivers that enter into the bay. Doughty et al. (2016) estimated that intensive whaling reduced this vertical distribution of nutrients to 23% of the original value (approximately for P from 34 10^7 kg P $year^{-1}$ before the intensive hunting period to today's 7.5 10^7 kg P $year^{-1}$).

Similar to large marine mammals, terrestrial mammals may be involved in the horizontal transfer of nutrients, however, to a lesser extent. Doughty et al. (2016) estimated that extensive hunting of large mammals and whales by humans has reduced this transport capacity in the seas and terrestrial ecosystems to 5–8% of the original values, respectively. The movement of large mammals affects not only the movement of nutrients on a macroscale but also its distribution on a microscale. In Africa, it has been observed that gazelles graze often on areas of grass that are better supplied with nutrients. They also deposit their excrement near these places. This creates places that will be better supplied with nutrients in the future. A better supply of nutrients favors grasses in competition with tree seedlings and helps keep these areas free of trees. This again increases the attractiveness of these sites for gazelles, as a clear open space reduces the risk of predation (van der Waal et al., 2003). At the same time, the deposition of game or even cattle excrement can leave its mark in the form of an increased amount of nutrients even after 40 years (Augustine, 2003).

Small-scale distribution of nutrients can also be affected by smaller species of organisms. For example, forest ants, thanks to the accumulation of food and the accumulation of needles in nests, which also decompose rapidly here, support both the accumulation of nutrients in anthills and their transition to accessible forms. Frouz et al. (1997) estimated that about one-third of the total supply of available P in a spruce forest is stored just below the anthills of forest ants, however, this occupied less than 1% of the area.

Chapter 2
Agriculture

2.1 Definition, Development, and Types of Agriculture

Agriculture is one of the human activities with large effect on our planet. It is an activity aimed at obtaining food, animal feed, and other goods through the cultivation of plants and animal husbandry, usually domesticated and bred for this purpose. Through various agrotechnical measures, farmers affect various environmental factors especially the soil, in order to improve the conditions for the growth of cultivated plant species and thus increase their yield. The land used for agriculture is called agricultural land, it includes arable land, permanent crops (orchards, vineyards, and hop gardens), meadows, and pastures. Worldwide, it forms slightly less than a third (Table 2.1) of the total land, but its proportion varies substantially among various countries. Given that agricultural land and arable land in particular, is the main tool for the livelihood of the human population, it is interesting to know how much land we have per person. The world average is 0.38 ha per person. Among the countries with a larger amount of arable land per person are, for example, the USA having 0.46 ha (Table 2.1), while the EU average is 0.21 ha. On the contrary, in Papua New Guinea, where the original forms of shifting agriculture are still practiced, there is only 0.04 ha per person, but here, the land is used only for a short time (1–2 years) and then abandoned to restore its fertility.

2.1.1 *Development of Agriculture*

Agriculture probably developed independently in several places around 11–8000 years ago. One of the centers of agricultural development was the so-called Fertile Crescent in the Middle East. Recent studies show that even in this region, agriculture has probably developed in two isolated cultures, one living in the area of today's

© The Author(s), under exclusive license to Springer Nature Switzerland AG 2022

J. Frouz, J. Frouzová, *Applied Ecology*, https://doi.org/10.1007/978-3-030-83225-4_2

Table 2.1 Area of agricultural and forest land, as a proportion of total area, and amount of arable land per person in selected countries and territories according to FAO stat and World Bank data as of 2015

Country or territory	Arable land and permanent cultures %	Meadow and pastures %	Agricultural land %	Forest %	Arable land per person ha
World	12.3	25.2	37.4	30.7	0.38
USA	16.9	27.4	44.4	33.9	0.46
EU	28.1	15.4	43.5	38.0	0.21
Germany	34.6	13.4	48.0	32.7	0.14
Czechia	41.6	13.0	54.6	34.5	0.29
Poland	36.8	10.1	46.9	30.8	0.28
Papua New Guinea	2.2	0.4	2.6	74.1	0.04

Jordan and Israel, while the other occurred on the other side of the Zagros Mountains in western Iran. These two cultures then began to spread to Europe and Asia. In addition, agriculture is considered to have developed independently in South America, Papua New Guinea, and China (Box 2.1).

Box 2.1 Neolithic Revolution

The domestication of cultivated crops and animals was a great step in the formation of human society. As a hunter and gatherer, humans created permanent settlements in the places of game migration and made plans for hunting prey. However, the Neolithic revolution meant a fundamental shift from the nomadic to the sedentary way of life and the building of complex and often fortified settlements. Compared with hunters, the farmer's life was based on a much longer-term and more complex planning, which led to the emergence of complex structures, physically, culturally and socially. This led to a comprehensive change in human society.

This entails a number of other phenomena not directly related to agriculture leading to the development of a complex hierarchical society.

One of the important centers of agriculture was the already-mentioned area of the so-called Fertile Crescent around the mountain ridges of Zagros and Taurus in the present-day Iraq, Iran, Syria, Turkey, and Palestine, where agriculture began to develop probably in the ninth and tenth centuries BC. During this time, there has been a gradual domestication of animals (sheep, goats, and cattle) and plants (single-grain wheat, two-grain wheat, figs, and barley). The great expansion of grain cultivation and the expansion of related agriculture took place around 5000 BC. During this time, the rise in sea levels slowed significantly, and a dry and warm climate could have helped the grass to spread. The existence of stable river floodplains and riparian plains, possibly with floods, also created suitable conditions for agriculture. There was also significant progress in tool manufacturing technologies. Humans learned to

(continued)

Box 2.1 (continued)
drill hole in a stone to put on a shaft. Also, sickles with many thin blades embedded in a joint in wood or bone as well as tools for grinding grain were created. People learned to process bronze.

An interesting question is whether the original hunters and gatherers first became shepherds and only then settled, or whether pastoralism developed at the same time as cultivating the land and specialized pastoralists arose only as a result of more intensive and specialized agriculture. The second theory assumes that the domestication of animals occurred simultaneously with traditional agriculture. Only with the emergence of irrigation systems, the development of more intensive irrigation agriculture occurred on better irrigated land and specialized herders, and nomads in peripheral areas.

Due to the independent emergence of agriculture in different areas, it is possible that the natural conditions and cultures of the original farmers differed considerably. The barley fields of the first Neolithic farmers may have resembled the stands of wild barley that we can still come across in Central Asia and the Middle East. The forest gardens of the people of Papua New Guinea may have looked similar to the gardens of some forest tribes to this day.

2.1.2 Types of Agriculture

Agriculture can be divided according to a number of aspects. According to the degree of specialization in mixed agriculture, where farming on one farm includes crop production (often a combination of several different crops) and animal production. According to intensity, we can classify agriculture into subsistence agriculture (often mixed) and intensive or commercial agriculture, the aim of which is to produce crops for sale (often specialized). A special category is the so-called organic agriculture and other approaches specifically focusing to achieve higher sustainability.

Most agricultural activities are fixed to a permanent location in the landscape, but there are also other types of agricultural activities that periodically change their location that includes in particular, the so-called shifting agriculture, or slash and burn agriculture. Slash-and-burn agriculture is followed when part of the forest is clear cut, burned, and agriculture is operated on the place for one or more seasons, the area is abandoned and then allowed to regrow to forest again. This type of farming still occurs in some tribes in the Amazon or Papua New Guinea. Similarly, nomadic pastoralists are constantly shifting, often over large areas of land.

Agriculture is closely linked to the possibilities of the environment, especially the climate and soils in the area, however, it significantly affects the characteristics of the environment. These relationships will be described in the following chapters, but in addition to them, agriculture is significantly influenced by a number of social, economic, cultural, and religious influences. Probably, the most striking of the religious influences is the exclusion of the meat from certain species of animals from the diet of various religions. Socioeconomic factors that influence farmers' behavior and its environmental impacts are discussed in Box 2.2.

For the purposes of this book, we have divided agricultural activities into several groups according to cultivated plants or farmed animals, which show a certain similarity in the technologies used and have a similar impact on the environment.

Crop production includes various ways of growing crops, both growing different types of usually annual crops in the fields and growing permanent crops, grasslands (meadows and pastures), orchards, viticulture, and hop growing.

Horticulture mainly combines the cultivation of vegetables, medicinal and aromatic plants, fruits, and other crops on cultivated land. What sets it apart from other types of agriculture is the smaller spatial scale, which allows for much more intensive tillage. We also traditionally include mushroom growing and ornamental gardening.

Animal production includes animal husbandry either in the form of nomadic and non-nomadic pastoralism or the breeding of housed animals.

Box 2.2 Socioeconomic Factors Determining the Impact of Agriculture on the Environment

Agriculture is a human activity used to earn a living or to make a profit. As such, it is greatly influenced by a variety of social, economic, political, and cultural influences. As a matter of fact, similar factors affect forestry and fisheries, where we mention them more briefly. Ancient agriculture may have had little effect on the environment. The study by Kukla et al. (2019) showed that the disturbance of the lowland rainforest in Papua New Guinea by traditional slash and burn agriculture was low in intensity due to the small area of agriculture. Moreover, the gardeners were often unable to completely remove large trees and besides planting which happen locally, did not disturb soil by cultivation. This low intensity was combined with low population density, due to which it took at least several dozen years for the natives to return to the same spot. Therefore, the overall impact of traditional agriculture on ecosystems was small and sustainable in the long run. However, as population density increased in mountain valleys due to the absence of malaria, agriculture removing the original continuous forest using stone tools far before the arrival of the Europeans. The original forest was then replaced by a mosaic of secondary forest, fields, grassland habitats, and fragments of the original forest. Even so, the newly created ecosystem was sustainable in the long run.

The original farmers produced crops mainly for their own use leading to a relatively large diversity of crops grown in one place. As agricultural production increased, a surplus appeared that could be exchanged allowing people to be more specialized in certain activities. In addition to food surplus, specialization was also supported by the concentration of people in permanent

(continued)

Box 2.2 (continued)

settlements. The concentration of people in settlements enabled the accumulation of surplus, and at the same time, the concentration of people to whom could be offered products or services. As some people specialized in certain activities, they gained experience and accumulated technological improvements. This increased the efficiency of their work and they became specialized craftsmen. Over time, it was easier for craftsmen to obtain other goods in exchange, while it was more advantageous for others to buy their services or goods. Technological innovations, on the other hand, made it possible to increase the efficiency of agricultural production, while the development of trade brought new crops and also made it possible to specialize in certain commodities in suitable areas.

All this brought not only the growth of agricultural production but also another concentration of people in cities, where huge wealth accumulated. This led to the further development of crafts, trade, a hierarchical society, bureaucracy, and the army. As a result, the relationship between rural food production and urban consumption became increasingly tense. At the same time, the services provided to peripheral rural areas deteriorated, central governments tried to compensate for the fall in budgets with higher taxes. During this constellation, the civilizations became repeatedly sensitive to external influences and failed wars or climate fluctuations led to the collapse of ancient states (Tainter, 1988).

Despite the decline of some civilizations, successful civilizations always led to the emergence and expansion of cities. Thus, even after the fall of the Roman Empire, there was a slow and gradual development of medieval cities. It was initially small, because medieval farmers were able to support only a small number of other, nonagricultural populations, but the importance of cities gradually grew. The growing class of urban merchants and craftsmen became important drivers of progress, especially at the beginning of the modern age, as they made full use of the potential of the newly discovered colonies.

The Industrial Revolution brought significant changes to agricultural rural life. During the Industrial Revolution, more sophisticated crop rotation systems started to be applied as well as new agricultural machines such as a seed drill appeared. The production of these machines or their components was concentrated in cities. A major breakthrough in this direction occurred with the discovery of the Haber-Bosch process, which enabled the industrial fixation of nitrogen (see Sect. 2.2.2.3). On the one hand, all these inventions increased the productivity of agriculture (increasing the surplus), thus enabling an even larger part of the population to engage in something other than agriculture. However, the importance of external inputs for agricultural production (tools, machines, and fertilizers) increased.

Securing these inputs directly required a larger part of the population than before to be engaged in activities other than agriculture. It required skilled and

(continued)

Box 2.2 (continued)

specialized workers to produce fertilizers and machinery. To produce both, mineral resources were needed, which had to be extracted, transported, and processed, and all of which required specialized workers. The farmer began to buy raw materials and sell his own products to a greater extent than before. This led to the specialization in certain crops for which farmers had the best natural conditions and at the same time, the best market. This effort maximized production, but farmers ceased to be self-sufficient, as they became producers of a limited number of commodities and bought the others. Since natural conditions are often similar for geographically close farms, entire areas were homogenized and focused on one or a few crops. This allowed farmers to acquire specialized machines and increase the efficiency of their production. The world would be divided into food producers living in rural areas and consumers living in cities. While this would create the conditions for population growth, the population of rural areas was gradually declining, as more money could be made in cities for a similar amount of work.

This outflow of jobs created even more pressure to replace human labor by agricultural machinery, but it also changed the landscape, so that the machines were used as efficiently as possible (Fig. 2.1). While this increased the food production and the general population, it would not stop the outflow of labor from the rural areas, quite the contrary. In 1900, there were 6.7 villagers living in agricultural areas per one city dweller. Conversely, it is expected that in 2025, there will be 1.5 city dwellers per one villager (Satterthwaite et al., 2010). This decline in the rural population (food producers) and the increase in the urban population (food consumers) will lead to an increasing need to

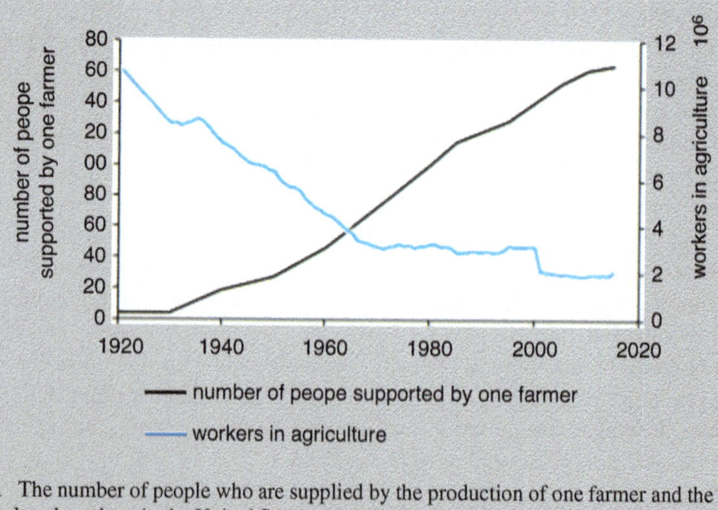

Fig. 2.1 The number of people who are supplied by the production of one farmer and the number of agricultural workers in the United States, according to Roser (2013) and the USDA

(continued)

Box 2.2 (continued)

increase the intensity of agricultural production, the replacement of human labor with mechanization and the use of chemical plant protection products. The intensification of agricultural production (the ability of one person to produce more food) and at the same time, the process of urbanization (and thus the departure of people to cities and the loss of people in rural areas) will lead to an increase in the agricultural holdings (Winsberg, 1980). This can be clearly seen in the increase in farm size in the United States in the 1950s and 1970s, which was accompanied and driven by an increase in the agricultural production while a decline in the agricultural workforce (Fig. 2.1).

This period was named the "green revolution" and is characterized mainly by the increased use of industrial fertilizers, pesticides, and new mechanization. It was a response to food shortages during the war with significant state support (introduction of subsidies). Subsequently (gradually, especially since the 1980s), these subsidies and the intensification of agriculture led to overproduction in industrialized countries, which had and still results in many problems.

Using the example of the concentration of agricultural production in former communist countries, we will show that political decisions can also have a significant influence. In the 1950s and 1960s, a process of collectivization started in communist lead Czechoslovakia, which led to a significant increase in agricultural enterprises. This process was not started by a sharp increase in the productivity of agricultural production (especially at the beginning of collectivization), but by political pressure. However, following some consolidation, it must be said that socialist agriculture achieved production growth. However, despite some economic benefits of concentrating agricultural production, the enormous rate of increase in the size of agricultural enterprises in Czechoslovakia was driven more by political decision than by economic necessity.

The average size of agricultural cooperatives and state farms significantly exceeded the average size of farms not only in Western Europe but also in the USA. In fact, the large associated companies were divided into individual centers, which operated more or less, independently. It can also be seen that the growth of agricultural production continued after the political changes at the end of the last century and a significant reduction in the size of agricultural holdings. Despite this reduction, the Czech Republic has by far the largest average farm size in the EU (Fig. 2.2). This enables a large size of individual fields and many other consequences.

For the sake of completeness, it should be added that similar trends are characterized by the rapid development of agriculture and concentration after World War II in Western Europe and the United States, as described above. This involved an increase in land units and a reduction in the number of farms and people working in primary agricultural production, while the degree of concentration did not reach such dimensions.

(continued)

Box 2.2 (continued)

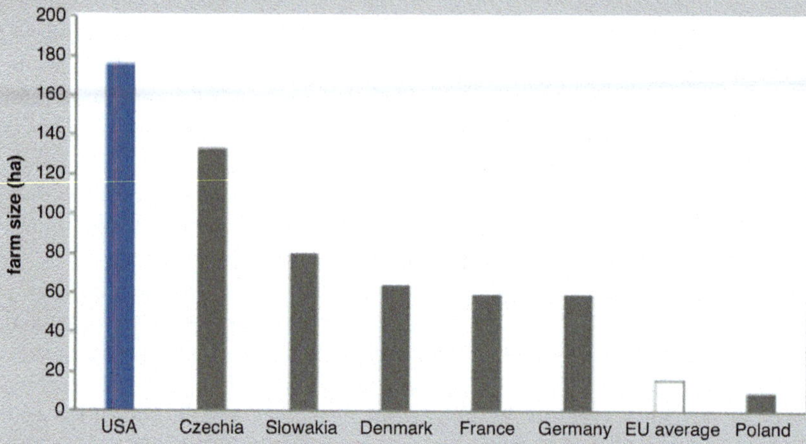

Fig. 2.2 Farm size (ha) in the USA and in selected EU countries. According to Eurostat and FAO

As we have said, urbanization has divided the world into rural food producers and their urban consumers. Let us take a closer look at the relationship between these two categories. While the self-sufficient producer has regulated his consumption by the possibilities of his production, the consumer perceives production only indirectly through the price of commodities, the production of which will be more demanding, will be rarer and will cost him more money. Urban dwellers, as can be well observed especially in developing countries, are shifting their dietary demands, consuming more meat, dairy products, oils and other "luxury" goods (Satterthwaite et al., 2010) than rural dwellers. With the growth of the economy and the growing purchasing power of the population, this difference is decreasing, rural areas cath up the cities. For example, in China, meat consumption in urban areas increased by 113% between 1981 and 2001, while in rural areas by 278% (Steinfeld et al., 2006). In developed countries, rural people behave the same way as consumers in urban areas.

The consumer influences the producer's decision on which commodities to produce, toward those on which the producer will make the greatest profit. However, this relationship between the producer and the consumer is becoming even more complicated. Initially, the farmer could certainly invest his time and bring the goods to market and sell them. Yet, urban dwellers are becoming increasingly interested in preprocessed products (Satterthwaite et al., 2010) and so instead of selling their products to a consumer, the producer sells it, increasingly, to the processor (millers who then continued to sell flour, pastries through bakers, a butcher who sells meat, etc.), who then sells the finished product to the consumer. Gradually, specialized traders began to emerge, who bought products from farmers or processors and then retail it. In many

(continued)

Box 2.2 (continued)

cases, other or even the same entities, could start offering farmers inputs, fertilizers, feed, etc. This has two consequences, leading to even greater separation of the consumer from the producer, even though the consumer still influences the producer's decision.

Another consequence of the arrival of traders is the internationalization of trade, which meets the efforts of consumers to buy exotic goods (i.e., those that cannot be produced locally). In developing countries, the tendency of consumers is again evident, especially among urban dwellers (Satterthwaite et al., 2010). While in contemporary developed countries, the purchase of exotic goods has become a common standard throughout the society, and this was not always the case. In the paintings of the Dutch Golden age in the seventeenth century, one can often observe lemons or other exotic fruit as a reflection of fascination with the possibility of buying something exotic. As we will show below, from the import of something exotic, it is a step toward the import of something that is not so exotic (such as apples), but which can produce elsewhere cheaper, out of season, in a different quality, etc.

If we look at the total added value that arises from the primary agricultural production to the product that the consumer buys, how this added value is divided between the producer, the processor and the trader, then the share of added value per producer ranges from 4% for bread to 40% for milk. It is important that most of the profit from a product that is bought in a supermarket does not go to the producer (i.e., the farmer), but from the processor and mainly retailer (Peltoniemi & Niemi, 2016). If we previously said that the consumer influences the behavior of the producer, then now both the processor and the retailer enter this chain with their interests, and together they influence the behavior of the producer (farmer). Peltoniemi and Niemi (2016) also showed that with the increasing concentration of the food trade in a smaller number of large (often nationwide or multinational) companies owning a chain of stores, the share of these large retailers' profits increase. Large retail chains thus become important players determining the character of the entire supply chain and can to a large extent, determine the behavior of producers.

Gereffi (1994) distinguishes between two types of supply chains. The first is managed by the manufacturers who have their own technologically advanced solutions and basically who "control" the distribution network. This includes, for example, the car market or some brand of electronics. In contrast, the food market (with the exception of truly luxury items) is rather driven by sellers who make decisions and sales strategy, price structure, etc. Therefore, the behavior of large trading companies is largely determined by the behavior of customers.

Despite the fact that customers choose according to a number of criteria, price is one of the key factors determining their decisions. The richer is the country, the smaller the share of their income citizens spend on food. Of

(continued)

Box 2.2 (continued)

course, this only applies relatively in terms of income, in absolute scale, citizens of richer countries spend more on food than those in poorer countries (Tilman et al., 2011). This is, among other things, due to the higher proportion of meat in food purchased. Also, people in richer countries purchase standard commodities that would be considered luxury in poorer countries. The customer is therefore guided mainly by the price and the price–performance ratio, and less so by assessing how environmentally friendly the production of the product was. However, in many cases, the customer may not pay attention to this or does not even have enough information to judge it.

Customers can certainly buy, for example, BIO products, for which it is very likely that their production was more environmentally friendly. However, the possibility of knowing whether the chicken they buy was fed locally grown feed, or soybeans grown on fields replacing cerrado and causing environmental damage on the other side of the world or is at least very limited. Sellers are aware of the key importance of price and price–quality ratio in order to offer the most advantageous products. Due to the limited possibility for the customer to assess the environmental impacts, a price reduction may be associated with less environmentally friendly products. At the same time, these measures certainly do not have to be taken by trading companies themselves, but they can force producers and processors to do so by putting pressure on the price and other attributes of the product.

One of the possible ways to reduce the price is to buy food or even feed used for food production in countries where natural conditions are suitable for the production of such commodities or production is just cheaper.

Colonial era was an important step in this globalization of food market. Agriculture transformation in colonies allow to increase the production of goods that were exported from colonies to Europe and also increase the self-subsistence of colonial economy needed for mining and other branches that support Europe by raw materials and other goods coming from colonies (Hippert, 2018). On the way, domesticated animals and plants commonly grown in Europe were introduced massively to the colonies, together with some technologies of European agriculture.

Recently, in addition to the above-mentioned imports of soybeans, is the import of vegetables, fruits, and herbs from sub-Saharan Africa, as described by Dolan and Humphrey (2000), can be an example of such a practice. The study points out that imports of these commodities are not typical trade between countries; instead, transnational chains creating their own partially or fully internalized structures that decide on production, quality control, and prices of delivered products. It is therefore a supply chain fully managed by the seller, who determines who enters it and who does not. Among other things, the chain prefers large suppliers of individual commodities, mainly due to the difficulties associated with working with small local suppliers.

(continued)

Box 2.2 (continued)

For small suppliers, it is much more difficult to ensure sufficient supplies and consistent quality throughout the year. Quality control is much more difficult because checking dozens of small orders is more time-consuming and expensive than checking one large order. In addition, large chains seek basic preprocessing of products, washing, peeling, etc., therefore, it is much easier to ensure the hygiene standards of the destination country and control them if it is concentrated. Furthermore, large producers are more economically stable, give a greater guarantee on possible investments, while small producers may have problems transporting their product to their destination due to poor transport situations. Although these observations are derived from specific commodities in Africa, some of the observations are certainly applicable even on the behavior of large retail chains elsewhere. In addition to what was described above, international food trade is also affected by various legislation and market barriers which modify the market in many ways (Friel et al., 2020).

Farmers' behaviors are also determined by the structure of ownership. Farmers can farm either on the land they own or on land they rent for farming purposes. The reasons that shaped the structure of ownership are various. For example, in many post-communist countries, political changes in recent history have been a major intervention in property relations. As shown earlier in the example in the Czech Republic, with the rise of socialism, agricultural cooperatives were formed that brought together landowners. At that time, the laws suppressed the rights of owners and transferred most of the rights to use the land to the cooperatives. However, in many cases, the property rights of the owners did not expire. After the political changes associated with the fall of socialism, it was possible to restore most of the broken property relations in the so-called restitutions. This created a large group of owners who lived in the cities and were not interested in doing business in agriculture, and these owners usually leased their land to farmers.

At present, Czech farmers farm about 25% of the land they own and 75% of the land they lease. This has implications for farming, especially in relation to land. Soil is a farmer's basic production tool and different uses of land wear the soil in different ways. Walmsley and Sklenicka (2017) showed that the biochemical properties of soil are better for organic farmers than for conventional farmers and among organic farmers, it does not differ between those who own the land and those who lease it. In contrast, in conventional agriculture, biochemical parameters (activity of soil enzymes) are statistically significantly better for those farmers who are also landowners compared with the tenants. These results show, among other things, that farmers are aware of the discrepancy between land-friendly and short-term profit-maximizing farming and if they own the land and have the prospect that they or their children will continue to farm it in the future, they treat the soil more gently.

(continued)

Box 2.2 (continued)

Finally, farmers' behavior can be affected by various economic incentives, the level of subsidy taxation and so on. An example of such subsidy titles is the Common Agricultural Policy (CAP), which was launched by the EU in 1962 (https://ec.europa.eu/info/food-farming-fisheries/key-policies/common-agricultural-policy/cap-glance). The CAP seeks to compensate for some of the natural problems in agriculture, such as a high dependence on natural conditions, especially weather and climate. The objective existence of a large time lag between consumer demand and farmers' production possibilities (growing more wheat or producing more milk takes time). These restrictions mean that farmers' incomes are significantly lower than for nonagricultural professions. This could lead to a loss of interest in agriculture, which would jeopardize food production and lead to the abandonment and gradual neglect of the agricultural landscape.

The aim of today's CAP is therefore to (1) support farmers and improve agricultural productivity so that consumers have a stable supply of affordable food, (2) ensure a fair standard of living for farmers residing and working in the European Union (EU), (3) help tackle climate change and sustainable management of natural resources, (4) preserve the character of the rural landscape throughout the EU, and (5) support the rural economy and thus jobs in agriculture, the agri-food industry, and related sectors.

The CAP is financed at the European level from the EU budget. It has three basic components. The first of these is direct payments to farmers, the aim of which is to ensure income stability and reward farmers for environmentally friendly farming and for the provision of public goods, which under normal conditions, are valued on the market such as rural care. Direct payments are mainly linked to the area of agricultural land farmed by the farmer (per hectare). Another component is market measures. The EU can take action to address difficult market situations, such as a sudden drop in demand due to health threats or a drop in prices due to a temporary oversupply in the market. Next are rural development measures, where national and regional programs address the specific needs and challenges facing rural areas in each country.

CAP payments accounted for around 39% of the total EU budget in 2013 and in the past, this percentage was even higher. In total, in 2018, the EU supported farmers through the CAP in the amount of €58.82 billion. Of this, direct payments amounted to €41.74 billion, rural development measures to €14.37 billion and market measures to €2.7 billion (EU website). The largest share of CAP payments are therefore direct payments to farmers. The farmer receives a subsidy for managing and fulfilling in particular, the criteria of environmental and food safety (Cross Compliance). It must be said that since

(continued)

Box 2.2 (continued)

its inception in 1962, the CAP has moved from an instrument aimed almost exclusively at securing and intensifying agricultural production to a more balanced instrument supporting, more than in the past, the provision of ecosystem services, biodiversity, and other nonmarket goods.

Despite the fact that the current CAP aims to support the so-called multifunctional agriculture with an important emphasis on nature protection, the impact of the CAP on biodiversity is perceived by many authors rather negatively. Donald et al. (2002) points out that in European countries that are part of the EU and thus in the CAP, there has been a faster decline in wild birds than in countries that were not members of the EU at the time. In addition, Reif and Vermousek (2018) point to a sharp decline in wild bird populations in Eastern European countries after joining the EU and applying the CAP. At the same time, the intensity of agricultural production increased during the same period. This only confirms the previous opinion of Donald et al. (2002) that despite all the environmental elements included, the CAP is still primarily a tool for increasing the intensity of agricultural production rather than promoting environmentally friendly farming.

There are several reasons for this. First, there is a lack of truly empirically validated information on what works at the landscape level to support biodiversity and ecosystem services. The level of individual measures is often designed to be more sustainable for farmers while maintaining the current intensity of production, rather than to actually have a measurably positive effect. Another big problem is trying to make the proposed measures easy to control, even at the cost of losing their effect. An example could be a subsidy for mowing meadows, when a deadline has been set by which the meadows are to be mowed so that the implementation of this intervention can be checked. As a result, the vast majority of farmers have mowed the meadows just before this deadline in order to check that they have actually been mown at that date. As we will show below, it is the mosaic mowing of meadows that is key for meadows to be able to fulfill their role as habitats for fauna and flora. Thus, the measures that were supposed to support habitats became an intervention damaging the habitats of many species. In this regard, the pioneering efforts of the British Society of Amateur Ornithologists RSPB, which in 1999 bought the Hope farm. On this 200-hectare farm, the society has more than double the population of wild birds in 10 years and improved other environmental indicators while maintaining the profitability and achieving the same or even higher yields. This example proves that there is a way to combine profitable agriculture with biodiversity of agriculture land.

2.2 Crop Production

2.2.1 *Crops, Origin, and Development*

There are about 250,000–300,000 plant species on earth, of which, about 10,000–50,000 species are edible and it is estimated that about 100–200 plant species are domesticated (FAO, 1998a, b). Some estimates speak of up to 500 partially domesticated species. Even so, it is about 1% of edible species. This is all the more interesting because many species are commonly collected in the wild, yet they have not been domesticated. For example, many European forest berries or herbs used as spices such as wild garlic (ramsons). It is possible that a small degree of domestication is associated with some natural barriers in individual plant species. On the other hand, it is also possible that greater efforts have been made to domesticate plants that can significantly meet human energy and protein needs, while in plants used as fruit, vegetables or spices, this pressure was less.

This reduction in the diversity of domesticated plants from potentially edible plants is even more pronounced when we realize how unevenly domesticated species are represented in food production. Only nine crops (wheat, rice, maize, barley, sorghum, potatoes, sweet potatoes, sugar cane, and soybeans) cover more than 75% of the human consumption. Of course, we do not only grow crops as a source of food, many crops are used as livestock feed, production of biofuels, fibers, and other uses. In addition to the fact that humans use only a small part of the diversity of edible plants for their subsistence, there has recently been a great reduction in cultural plant diversity even at the subspecies level. In areas where the original cultural varieties have been domesticated, there is still great variability of cultivars. In Peru, for example, there are now around 4000 potato cultivars. As a result of the mass spread of a smaller number of new varieties, historical local varieties are being abandoned and gradually dying out. The FAO estimates that about 75% of the variability of cultivated plant varieties has been lost over the last century. There are a number of efforts to stop this decline (Box 2.3).

Box 2.3 How Can We Support the Genetic Variability of Cultivated Crops?

The diversity of cultivated varieties is declining, leading to a loss of genetic diversity, which may be important as a reservoir of genes for further breeding. As mentioned, the FAO estimates that 75% of varieties and local cultivars of plants have been lost during the last century. Many national and international institutions are therefore trying to create genetic seed banks of various species and varieties. The Svalbard Global Seed Vault, built by the Norwegian government in Svalbard is far beyond the Arctic Circle in a place where the temperature is permanently −3 °C, is in a tectonically stable sandstone massif and at an altitude that should protect it against possible sea level rise, is very popular in this respect to protect against possible sea level rise.

(continued)

Box 2.3 (continued)

There are also several gene banks of plants for individual countries. However, you can also contribute to the support of genetic variability of cultivated crops in a relatively pleasant way by becoming a gourmet. Be aware that the taste, smell, color, and texture of the food you eat depend on the properties of the ingredients, and ultimately on the characteristics of the plants used to make it. Books have been written about different cultivars of wine, coffee or tea, and many people have found pleasure in enjoying the taste and aroma of drinks prepared from different cultivars and especially in finding opportunities for which this or that cultivar is suitable here or there. You can easily extend this art to things that look much more mundane but can affect your culinary experience.

Virtually every commodity we eat, from rice to tomatoes to lettuce, hides an inexhaustible variability of different forms suitable for different uses. When combined with the other two great fashion trends of today's culinary world, local and seasonal products, you can make a good and quite pleasant contribution to promoting the biodiversity of crops. Similarly, as described above, for individual forms of crops, we can also contribute to the return of minority crops (buckwheat, spelled, single-grain, two-grain, and millet, etc.) to our field and to our diet. If you have your own garden, or at least a box behind the window, you can over time, grow some less common forms yourself.

As mentioned, grains predominate significantly among the dominant crop species, as wheat, rice, and maize together represent more than 60% of human consumption (FAO, 1998a, b). This popularity of grains could have a number of reasons. The harvested part are grains that contain little water compared with the root crops, vegetables, or fruits, and are therefore easy to store and transport. The dominant grains are mostly annual species, which made it possible to use suitable areas created, for example, by flooding in the valleys of large rivers. This feature has also encouraged the colonization of new areas by grain crops, as the time required to obtain a crop to colonize a new area is relatively short. Moreover, the share of yield in harvested biomass is relatively large, especially compared with fruit-producing trees and shrubs.

Vavilov (1926) came up with the idea that although many crops are now cosmopolitan in distribution in almost all areas where climatic conditions are suitable for them, its wild ancestors usually grew in only one or a few places on earth. In these so-called genetic centers, the crop or group of varieties was domesticated and then spread around the world. Vavilov (1926) defined eight main genetic centers of origin of cultivated plant species (Fig. 2.3): East Asian center of origin of barley, rowan and seedless, sorghum, buckwheat, soybean, radish, hemp, pear, apple, apricot, and bean; Indian and Indo-Malaysian center of origin of rice, sugar cane, beans, cucumber, citrus, coconut palm, cruciferous plants, cotton, and banana tree; Central Asian

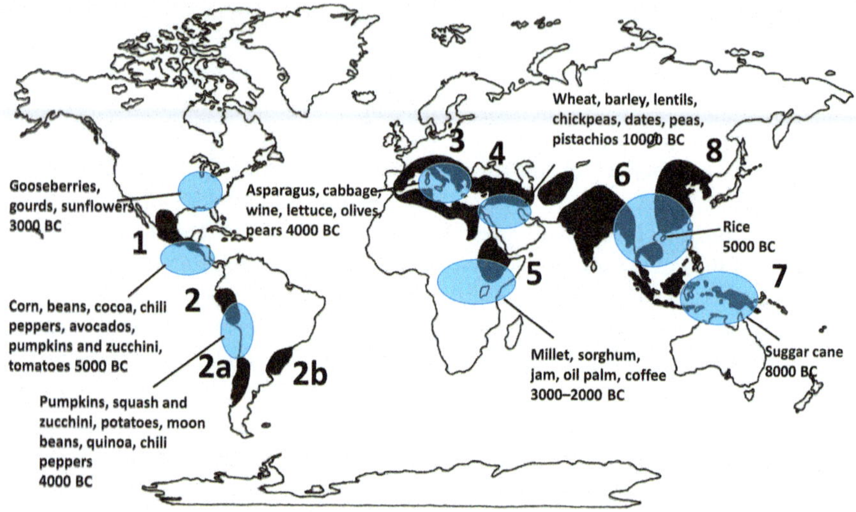

Fig. 2.3 Geographical origin of selected crops and main genetic centers of origin of crops in (1) Central American, (2) South American, (3) Mediterranean, (4) Middle Eastern, (5) Ethiopian (Abyssinian), (6) Central Asian, (7) Indian and Indo-Malay, and (8) East Asian. According to Vavilov (1926), modified

center from where wheat, peas, lentils, beans, small beans, flax, melon, radish, onion, spinach, apricot, almond, pear, apple, olive, vine, and walnut originate; Middle Eastern center which was home for wheat, rye, barley, onion, lettuce, beetroot, cabbage carrot, flax melon, cucumber, gourd, alfalfa, Persian clover, safflower, vetch, lupine, vine, pear, cherry, almond, sycamore, and walnut; Mediterranean center which was the origin of wheat, barley, peas, Alexandrian clover, clover incarnate, vetch, flax, mustard, cruciferous crops, beets, and olives; Ethiopian (Abyssinian) center from where wheat, barley, flax, peas, lentils, and beans originate; Central American center that brings beans, pumpkins, peppers, cacao, cotton, and tobacco; and finally South American center of origin of potatoes, corn, beans, tomatoes, pumpkins, cotton, peanuts, pineapple, and cocoa.

Cultivated plants often originated from multiple populations or different plant species from multiple centers. Many centers are located in mountains or foothill areas of the subtropics or tropics or on islands, where there were a number of separate geographical populations of the same species. This allowed the selection of suitable forms for domestication. Subsequent migration of domesticated forms with humans allowed them to hybridize with forms from different centers as well as the hybridization with both wild species and cultural forms. This led to the emergence of crops that are often very dissimilar to their wild ancestors. Examples of such crops that are very different from their wild ancestors are maize (*Zea mays* L.) and cabbage (*Brasica oleracea* L.). Particularly, cabbage has led to a number of very diverse vegetables, such as kale, brussels sprouts, curls, broccoli, cauliflower, kohlrabi, and Chinese broccoli (gailan).

The domestication of plants has improved the possibilities of their use by humans and facilitated their cultivation. Selections and gradual breeding increased the size of specific parts of plants that determine the yield (larger grains, larger fruits, etc.) and further improving the useful properties of these parts (increasing the sugar content in sugar beet tubers). At the same time, there is a reduction in the content of toxic and un-wanted substances (for example, in almonds), or the loss of thorns and other defense mechanisms. In a number of domesticated forms, the seeds are reduced, for example, in the case of melons and bananas; even completely seedless forms can be formed, which can multiply only vegetatively.

Another significant pressure on domesticated plants, especially those propagated by seeds, is the selection of lines and populations of plants which easily end dormancy and which emerge evenly, allowing the establishment of an even aged stand that also ripens and are ready for harvest at the same time. So, if we sow cultivated plants, all the seeds will germinate more or less at the same time. In contrast, the termination of dormancy and emergence within wild plant populations is highly variable as to prevent growing from unexpected fluctuations in conditions. When conditions suitable for germination and emergence are suddenly interrupted by some climatic fluctuations (frost, drought, floods, etc.), emerging plants can be severely damaged or completely eradicated. On the other hand, in the case of wild plants, only a part of the population germinates and emerges, the others wait as a seed bank and arise on another suitable occasion. Among most important crops (Table 2.2) is large proportion of annuals. There are also perennial plants that are grown as annuals, such as cotton (*Gossypium* sp.), or potato (*Solanum tuberosum*) whose wild forms are perennials and domesticated forms are annuals.

There is also a loss of natural mechanisms that ensured the spread of seeds. For example, in the wild ancestors of peas, the pods opened after ripening and placed the seeds into the immediate vicinity. At the same time, however, this mechanism made seed collection extremely difficult. Therefore, in domesticated forms of peas, the pod does not crack, and the seeds remain inside. In nature, such seeds would rot and the plants that did not reproduce would disappear. On the contrary, man preferentially collects such seeds and can subsequently ensure their spread. Similarly, the fruits of wild cotton's ancestors disintegrate, but they remain together in the domesticated forms.

With the help of humans, domesticated plants spread far better than their original wild ancestors did, both in terms of the number of individuals in a given place and the expansion of the area. However, domesticated plants often lose many of the properties needed for the survival in wild. They are less resistant to diseases, pests, weather extremes, often use less symbiotic interactions with other organisms, as well as have a limited ability to reproduce and spread. An extreme example of limited reproduction is vanilla (*Vanilla planifolia*), which is widespread in almost all tropical parts of the world. However, outside the original area (southeastern Mexico), it does not have suitable pollinators and the production of fruits (vanilla pods) in all these introduced areas, depends on artificial human pollination.

2.2.1.1 Main Types of Agricultural Crops

By crops, we mean cultivated plants that directly or indirectly provide a human benefit, such as food, feed, a source of fiber or other materials. It is interesting to note that what we often perceive as different crops, may be one botanical species (*Brassica oleracea* L., includes kohlrabi cabbage, cauliflower, etc.). On the other hand, if we buy a mixed salad in a supermarket, different salads are made up of a number of species, even belonging to different families. This has a number of implications for their cultivation, for example, closely related crop species often have similar diseases and pests. This means that if we grow closely related species, at least from the point of view of plant protection, we are still growing the same species, thus promoting diseases and pests to spread.

If we ask about the most important crops, we can choose different parameters to compare them, the total area on which the crop is grown, the total harvest or the total amount of the crop used for a particular purpose such as human nutrition. In terms of sown area, the most important crops are wheat, maize, rice, and soybeans (Table 2.2). However, in terms of yield, the most important crop is sugar cane, which is used partly to produce sugar, but especially to produce alcohol, which is used for both human consumption and as an industrial source of energy. The main crops that predominantly cover human consumption are maize, wheat, rice, potatoes, and soybeans. In the following overview, we will concisely introduce the main crops grown worldwide.

Cereals These are mostly annual grasses, grown mainly for grain. Cereals are among the most important crops used by humans. Cereal grain contains little water, so it is well stored and relatively easy to transport compared with other crops. Table 2.2 indicates the area, total harvest and average yield of individual crops. However, it should be noted that these values are highly variable between cultivation sites, as can be seen in the example of cereals (Fig. 2.4). The cultivated cultivars of maize belong to the species *Zea mays* and is a sturdy, 1–3 m tall true grass, grown for grain, which is found in characteristic sticks or is harvested as green forage or for silage. Most cultivated cultivars are F1 or F2 generation hybrids between maternal lineages. Cultivated rice belongs almost exclusively to the species *Oryza sativa* (*O. glaberrima* is grown in the Western Africa), and is usually grown as an annual in flooded paddy fields. It can also be grown in upland fields, but irrigation and weed control are usually more difficult here. It should be noted that the so-called wild Native American rice is not really a rice, but it is a grain of aquatic grasses, most often of the species *Zizania palustris* and *Zizania aquatica*. Wheat is an annual grass grown in winter or spring. Wheat contains various species of the genus *Triticum*, of which, at least six species are cultivated. The most common of them is *T. aestivum* (used, for example, to make pastries); *Triticum turgidum* subsp. *durum* (so-called semolina often used in pasta production) is also important. The old type of wheat is spelled *Triticum aestivum,* subspecies *Spelta*. There is also a cultivated hybrid of wheat with rye, which is called *Triticale*. Barley (*Hordeum vulgare)* is one of the earliest domesticated cereals. A wild ancestor of the same species still grows in the Fertile Crescent and elsewhere. It is grown as a spring, for example, for malt

Table 2.2 The overview of the world's major crops including the first 20 crops with the largest sown area supplemented by additional crops to include the first 20 crops that contribute the most to world production. Note that the order of most important crops differs substantially if sown area or world production is used. Data from 2016 according to FAOstat

Crop	Harvested area worldwide 1000 km^2	Average yield t ha^{-1}	Production total 10^6 t
Wheat	2201.1	3.4	749.5
Maize	1879.6	5.6	1060.1
Rice	1598.1	4.6	741.0
Soya	1215.3	2.8	334.9
Barely	469.2	3.0	141.3
Sorghum	447.7	1.4	63.9
Canoa (oil seed rape)	337.1	2.0	68.9
Millet	317.1	0.9	28.4
Cotton	302.1	2.2	65.4
Bean	293.9	0.9	26.8
Peanuts	276.6	1.6	44.0
Sugar cane	267.7	70.6	1890.7
Sunflower	262.1	1.8	47.3
Cassava	234.8	11.8	277.1
Oil palm	210.9	14.2	300.3
Vegetables	205.7	14.1	290.1
Potatoes	192.5	19.6	376.8
Chickpeas	126.5	1.0	12.1
Vigna	123.2	0.6	7.0
Coconut	121.7	4.8	59.0
Sugar beet	45.6	60.7	277.2
Tomatoes	47.8	37.0	177.0
Bananas and plantains	54.9	20.6	113.3
Sweet potatoes	86.2	12.2	105.2
Onion	49.6	18.8	93.2
Apple	52.9	16.9	89.3
Vine	71.0	10.9	77.4
Yam	74.5	8.8	65.9

or as winter, most often for feed purposes. Rye (*Secale cereale*) is a winter cereal pollinated by wind. It is used for the production of bread, for feed purposes and the production of alcohol (gin). It can be grown in colder climates than wheat. The most commonly grown oat species is *Avena sativa*. One of the youngest domesticated plants that came to Europe as a weed with wheat and barley, which is grown in the spring. Its grain is an excellent feed for livestock, an important food (used, for example, to produce oatmeal) and the whole plant can be used for green feed. Other, important groups of cereals is sometimes collectively called millet. They include

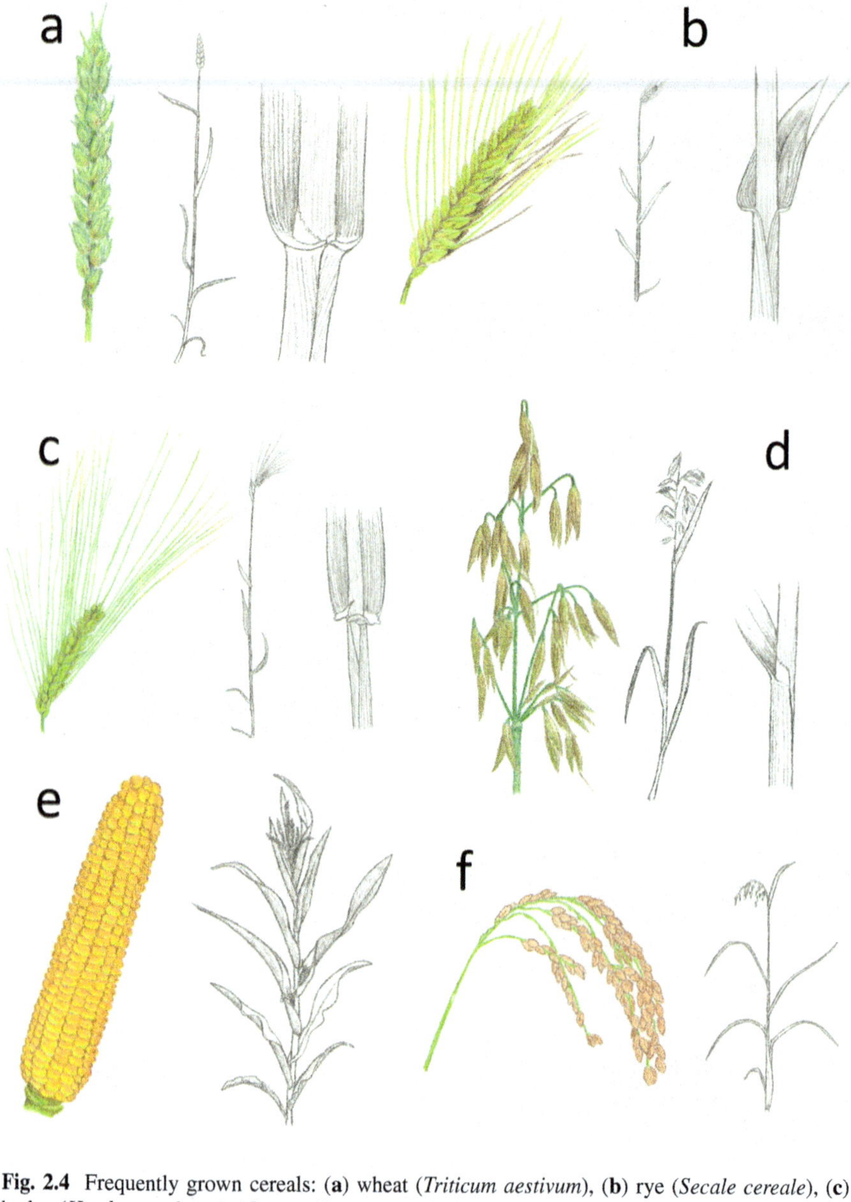

Fig. 2.4 Frequently grown cereals: (**a**) wheat (*Triticum aestivum*), (**b**) rye (*Secale cereale*), (**c**) barley (*Hordeum vulgare*), (**d**) oats (*Avena sativa*), (**e**) maize (*Zea mays*), (**f**) rice (*Oryza sativa*)

Fig. 2.5 Frequently grown millets, (**a**) proso millet (Panicum miliaceum), (**b**) sorghum (*Sorghum bicolor*)

sorghum, economically important are especially varieties, such as *Sorghum bicolor* and proso millet (*Panicum miliaceum*), and a number of other similar cereals (Fig. 2.5), of which, we name at least foxtail millet (*Setaria italica*), pearl millet (*Pennisetum glaucum*), and finger millet (*Eleusine coracana*).

Legumes They are a group of crops from the family Fabaceae and their fruit is a pod (Fig. 2.6). Like other vetches, they fix atmospheric nitrogen and have a positive effect on the soil. Soybean (*Glycine max*) is one of the most important legumes, it contains a high proportion of protein (up to 35%). It is a source of human food as well as feed for domestic animals. It comes from China and is one of the oldest cultivated plants. The genus *Phaseolus* is a bean with many species that mostly originated from the Americas, of which, at least four species have been domesticated. Today, the most commonly grown beans are the *P. vulgaris*, the others are the lima bean (*P. lunatus*) and the scarlet runner bean (*P. coccineus*). The genus *Vigna*, has a number of species similar to beans, but differ in their pollen structure and some biochemical features. *Vigna* is a large genus with more than 90 species, the most commonly grown include *V. angularis, V. mungo*, and *V. radiata*. The most commonly grown pea is *Pisum sativum*. It is a climbing herb grown both for human consumption and as animal feed. Lentil (*Lens culinaris*) is an annual vetch-like plant with small lentil-shaped fruit. Peanut or groundnut (*Arachis hypogaea*) grown for its seeds is also a legume, originally from South America. After flowering, the flowers form long stems and the pods are drilled into the ground, where they ripen. Faba (fava) bean or broad beans (*Vicia faba*) are grown mostly as a fodder crop, but some varieties are also grown for consumption.

Root Crops or Row Crops Unlike the previous groups, which usually consist of botanically related plants, root crops (similarly to other subsequent groups of crops) are a very diverse group from the point of view of phylogeny and are rather associated with the fact that the harvested part is usually below ground although it is not necessarily always the root. The term row crops refer to the fact that they are grown in wide rows that allow them to be cultivated by agricultural machinery. For root

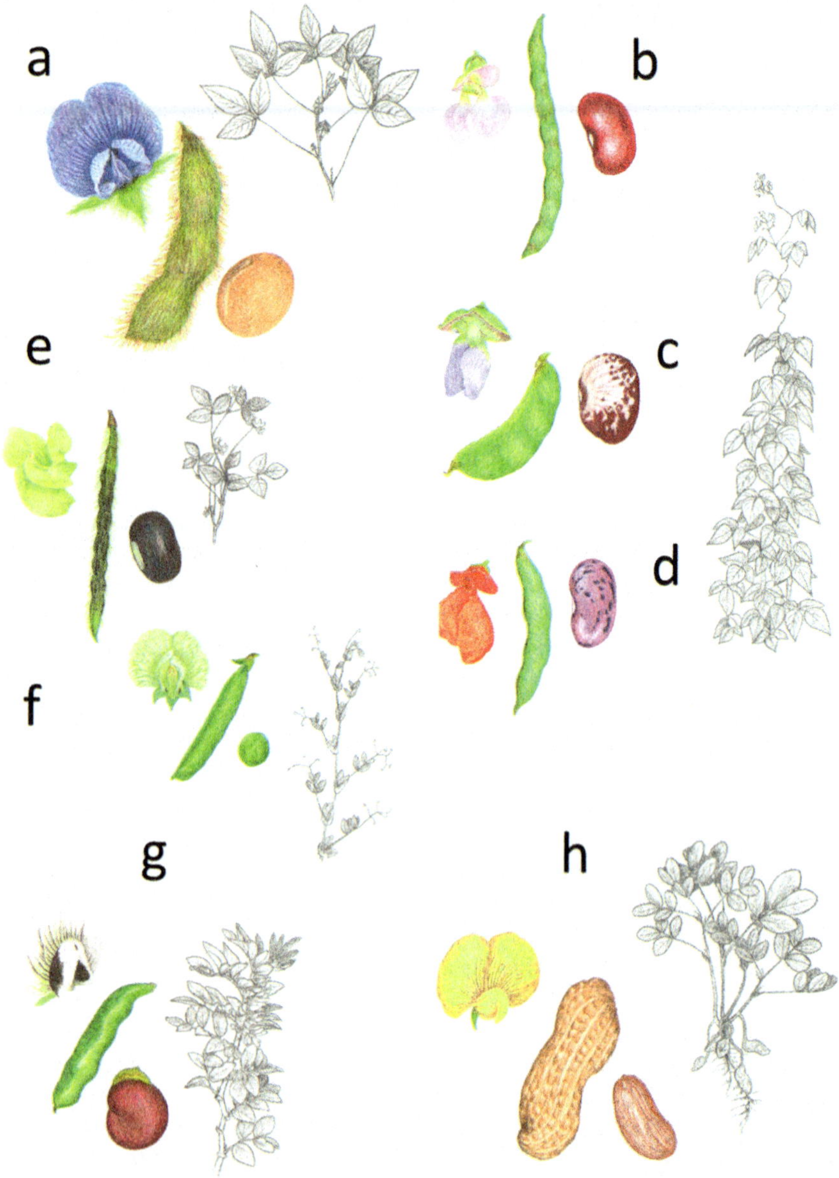

Fig. 2.6 Frequently grown legumes: (**a**) soybeans (*Glycine max*), (**b**, **c**, **d**) beans (*Phaseolus,* (**b**) *P. vulgaris;* (c) *P lunatus;* (d) *P. coccineus*), (**e**) vigna (*Vigna mungo*), (**f**) peas (*Pisum sativum*), (**g**) beans (*Vicia faba*), and (**h**) peanuts (*Arachis hypogaea*)

crops, tubers are usually harvested, which may contain part of the hypocotyl and usually a part of the stem or root, the tubers are located completely underground (Figs. 2.7 and 2.8). Some tubers are formed due to the thickening of the stems and contain leaves, at least in the form of buds (stitches). The others are formed by the thickening of the roots. The most famous root crops are potatoes *Solanum tuberosum*, they belong to the same genus *Solanum* (family Solanaceae) as eggplant and tomato. It was originally a perennial herb, but it is grown as an annual for stem tubers and comes from the Andes. Another root crop grown is the beet *Beta vulgaris* belonging to the Goosefoot family (Chenopodiaceae). It is grown for the bulb, which can be of different colors and can be used for various purposes. Sugar beet *B. vulgaris* var. *altissima* (white bulb) is an industrial crop cultivated commercially for the production of sugar, *B. vulgaris* var. *rapacea* (red or yellow bulb) is cultivated as feed; and *B. vulgaris* var. *conditiva* (red bulb) is cultivated as vegetable. Chard (*B. vulgaris* var. *cicla*) which is grown as leafy vegetables also belongs to the same species. At first glance, beets are similar to other root crops, which we also grow for pulpy tubers used as feed or food like Swedish turnip (*Brassica napus* var. *napobrassica*) and turnip (*Brasica rapa* var. *rapa*). But both belong to the cruciferous family (Brassicaceae). From a number of different types of root crops grown in the tropics and subtropics, we mention sweet potatoes *Ipomoea batatas*, a plant belongs to the family Convolvulaceae. Another important tropical root crop is the manioc (*Manihot esculenta*), also known as cassava, yuca, or tapioca (in Asia), is a tropical plant of the Euphorbiaceae family grown for its root tubers. Cassava contains glucosides, linamarin, and lotaustralin, which are broken down by the enzyme linamarase to form cyanide. Cassava varieties are divided into sweet (i.e., non-bitter which you can usually buy in supermarkets) and bitter. Bitter varieties, especially dry-grown, can cause poisoning if not properly treated. These must be cut or grated into small pieces before treatment, soaked thoroughly in water and boiled thoroughly. Jam or Yam is a group of many species of root crops of the genus *Dioscorea* (family Dioscoreaceae), are perennial herbaceous vines, and many species are grown for starchy tubers. The most important cultivated species include *D. rotundata* and *D. cayenensis* (yellow yam), *D. alata* (white yam), and *D. polystachya* (Chinese yam). We will end this brief overview of root crops with taro, an old cultivated crop from the Araceae family, it is grown for starchy tubers (*Colocasia esculenta*).

Oilseeds and Industrial Crops Oilseeds, as the name suggests, are grown in order to obtain oil (Figs. 2.8 and 2.9). Probably, the most important plant of this type in the world is the oil palm *Elaeis guineensis*. Its cultivation, as we have already mentioned, is given way by tropical rainforests in many parts of the world. There would be a long list of exotic representatives, but we focus on domestic crops of this category. An important oilseed crop is Canola or oilseed rape *Brassica napus*, a winter and spring crop from the Brassicaceae family grown for small black seeds. Canola is used for oil production, both for human consumption and as a raw material for the production of biofuels. Rapeseed is similar to several other plants of the same family. These are mustard (*Sinapis alba*) and black mustard (*Brassica nigra*), whose oilseeds are used among others for the production of mustard. Other oilseed crops are sunflowers and poppy. The sunflower is a large annual herb belonging to

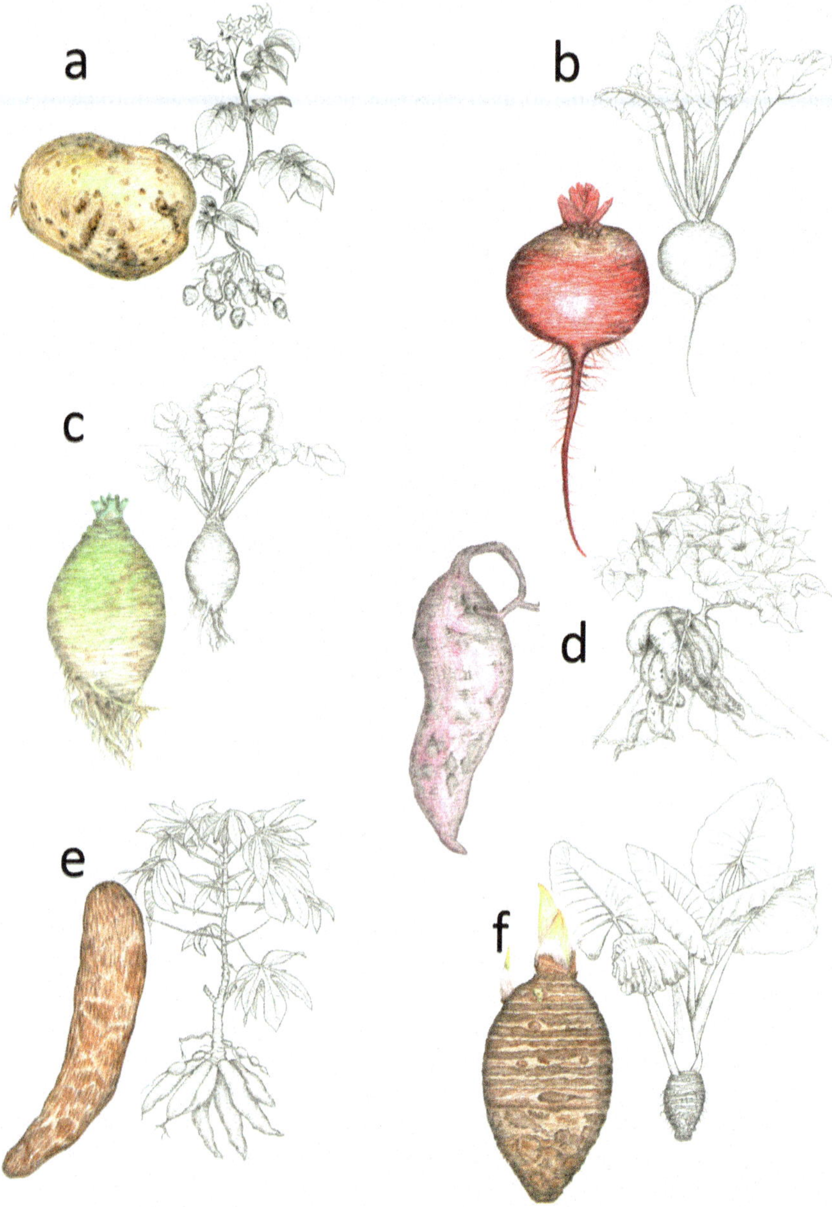

Fig. 2.7 Often cultivated root crops: (**a**) potatoes (*Solanum tuberosum*) (**b**) beets, red foma, there are also other forms with other colors (*Beta vulgaris*), (**c**) turnip (*Brassica napus* var. *napobrassica*), (**d**) sweet potatoes (*Ipomoea batatas*), (**e**) Cassava (*Manihot esculenta*), (**f**) taro (*Colocasia esculenta*)

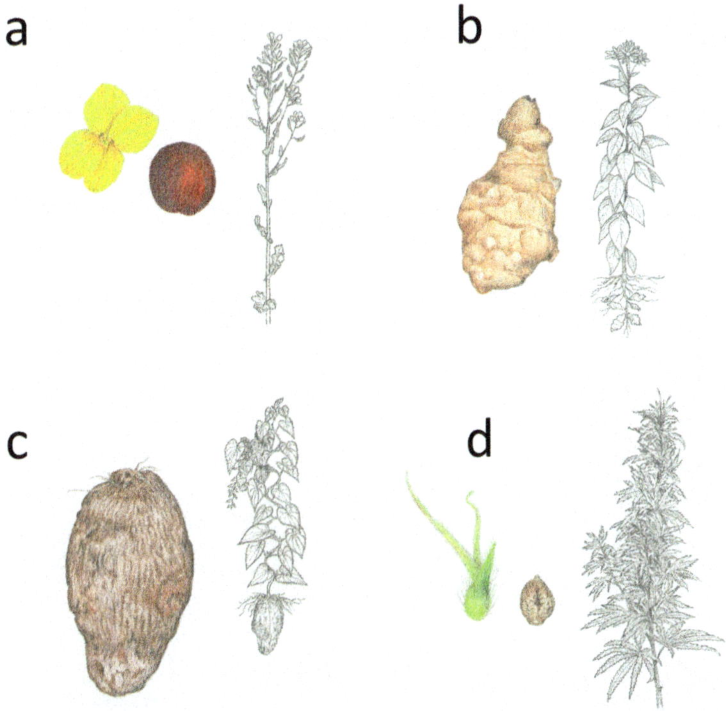

Fig. 2.8 Root crops, oilseeds, and industrial crops: (**a**) rapeseed (*Brassica napus* subsp. *napus*), (**b**) Jerusalem artichoke (*Helianthus tuberosus*), (**c**) yam (*Dioscorea alata*), (**d**) cannabis (*Cannabis sativa*)

the family Asteraceae, which is an annual *Helianthus annuus*. Another cultivated species of the same genus is the Jerusalem artichoke (*Helianthus tuberosus*), but it is grown for underground tubers. *Papaver somniferum* (Papaveraceae) is known as the poppy seed. Poppy seeds are popular for kitchen use in various cakes and pastries in many Slavic countries in Central and Eastern Europe, but its cultivation is regulated in many countries, because in addition to oilseeds, it can also be a source of opium. Some crops are grown for oilseeds and also as a source of fiber. Important representative of this group is cotton, which has many species of the genus *Gossypium* (Malvaceae), the most important of which is *G. hirsutum*. Fiber in cotton comes from the seeds. In other plants, namely, flax (*Linum usitatissimum*, family Linaceae) and hemp (*Cannabis sativa*, Cannabaceae) fibers are obtained from stems. Another group of industrial crops is used to produce sugar. The world's most important crop is sugar cane, which has several species of large tropical grasses of the genus *Saccharum* grown for sweet stems used for sugar production. Probably, the most important species of sugar cane is *S. officinalis* (and its hybrids) originating probably from New Guinea. We have already discussed about sugar beet in the section on root crops.

Fig. 2.9 Frequently grown oilseeds and industrial crops: (**a**) oil palm (*Elaeis guineensis*), (**b**) sunflower (*Helianthus annuus*), (**c**) poppy (*Papaver somniferum*), (**d**) flax (*Linum usitatissimum*), (**e**) cotton (*Gossypium hirsutum*), (**f**) sugar cane (*Saccharum officinarum*)

Fodder or Forage Crops These are grown for forage, that is, bulk animal feed (Fig. 2.10). As mentioned above, cereals harvested green can also perform this function. There is a wide range of fodder, some of the most important representatives of clover and grasses are listed. We mention alfalfa (*Medicago sativa*), red clover (*Trifolium pratense*), zigzag clover (*T. medium*), and white clover (*T. repens*) and from grasses, we mention ryegrasses, for example, *Lolium perene*, false oat-grass (*Arrhenatherum elatius*), cock's-foot (*Dactylis glomerata*), meadow foxtail (*Alopecurus pratensis*), and timothy grass (*Phleum pratense*).

Fruits and Vegetables The division of fruits and vegetables (Figs. 2.11 and 2.12) is to some extent, arbitrary. A good example of this arbitrariness is the tomato, which many of us would consider a vegetable, but according to the European Parliament's decision in 2011, it is a fruit. At the same time, there are several previous lawsuits, for example in the USA, which have alternately included tomatoes as vegetables or fruits. It should be noted that the very motivation for these lawsuits was not so much an attempt to assess the issue of principles, but rather different customs and other tariffs, which differed between fruit and vegetables. There is an inexhaustible amount of vegetables and fruits, so of the tropical species, we will only mention bananas. Many species of bananas of the genus *Musa* are grown, the most widespread and economically important are the varieties created by crossing the species *M. acuminata* and *M. balbisiana*. Plantains, also called vegetable bananas (*Musa* × *paradisiaca*) are grown for their use mainly in warm cuisine. A commonly grown **fruit** is the apple. The fruits of the apple tree (*Malus domestica*, family Rosaceae) were created by crossing several species of wild apple trees. The most important of which was *Malus pumilla*, which is also important as a rootstock in currently grown varieties. Its domestication took place in Central Asia. Another important fruit tree is the pear (several species are grown, especially *Pyrus communis*) belonging to the same family as the apple and its domestication also took place in Central Asia. Still another important group of fruits are stone fruits, especially the plum (*Prunus domestica*). This species has a number of subspecies, known as other plum-like fruits, such as pins, ryngle, half-plums, etc. These are hybrids of blackthorn (*P. spinosa*) and myrobalan plum (*P. cerasifera*), the last of which is still a cultivated fruit and is a significant rootstock of other stone fruits. Domestication probably occurred in the Caucasus and the Caspian Sea. Other important species of the genus *Prunus* are cherries (*P. cerasus*), sweet cherries (*P. avium*), apricots (*P. armeniaca*), and peaches (*P. persica*). An important small fruit is the currant fruit (*Ribes* sp., family Grossulariaceae). Two species of red (*R. rubra*) and black (*R. nigra*) currants are predominantly grown. As for the **vegetables**, onions (*Allium cepa*) are grown in large area in a number of color and taste variations. Onions are second in terms of total production after tomatoes. Other onion-like vegetables are garlic (*Allium sativum*), chives (*Allium schoenoprasum*), and a closely related leek (*Allium porrum*), which together form bulbs. The second most important vegetable in terms of area and the most important in terms of total production is the tomato (*Solanum lycopersicum*). Tomatoes are related to eggplant (*Solanum melongena*), peppers (*Capsicum annum*), cucumbers (*Cucumis sativus*), and pumpkins (*Cucurbita pepo*). These vegetables form a group of fruit vegetables. Another important group of vegetables are various cabbages, such as cabbage, curly kale, cauliflower, kohlrabi, broccoli, all of which belong to the species *Brassica oleracea*. Yet, another

Fig. 2.10 Frequently grown fodder: (**a**) alfalfa (*Medicago sativa*), (**b**) red clover (*Trifolium pratense*), (**c**) zigzag clover (*Trifolium medium*), (**d**) shite clover (*Trifolium repens*), (**e**) false oat-grass (*Arrhenatherum elatius*), (**f**) cock's-foot grass (orcharr or cat grass) (*Dactylis glomerata*)

Fig. 2.11 Frequently grown fruits, vine and hop: (**a**) apple (*Malus domestica*), (**b**) pear (*Pyrus communis*), (**c**) plum (*Prunus domestica*), (**d**) red currant (*Ribes rubrum*), (**e**) vine (*Vitis vinifera*), (**f**) hops (*Humulus lupulus*)

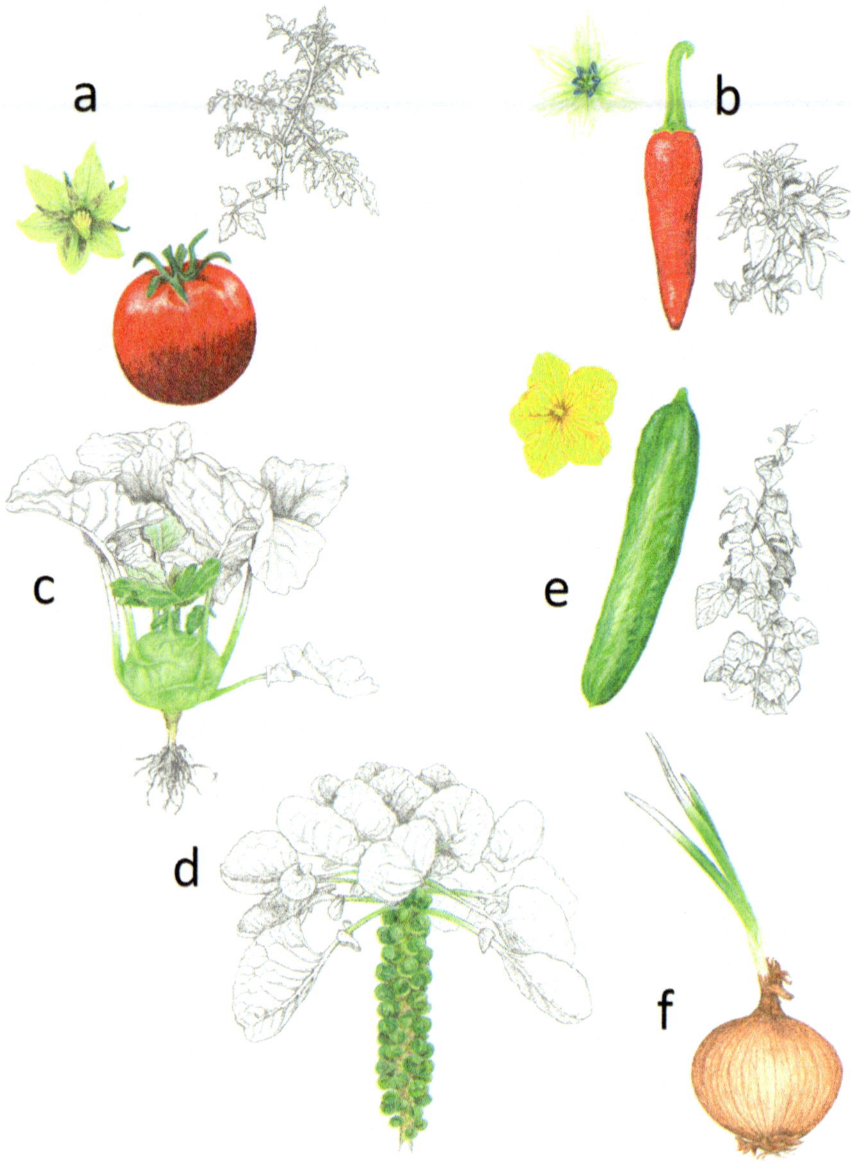

Fig. 2.12 Commonly grown vegetables: (**a**) tomatoes (*Solanum lycopersicum*), (**b**) pepper (*Capsicum* sp.) (**c**) cucumbers (*Cucumis* sp.), (**d**) Brussels sprouts (*Brassica oleracea* var. *gemmifera*), (**e**) kohlrabi (*Brassica oleracea* var. *gongylodes*), (**f**) onions (*Allium cepa*)

important group of vegetables are root vegetables, many of which belong to the Apiaceae family, such as carrots (*Daucus carota*), celery (*Apium graveolens*), parsley (*Petroselinum crispum*), and parsnips (*Pastinaca sativa*). Many root vegetables belong to the family Brassicaceae, the most important of which is radish (*Raphanus sativus*) grown in a number of forms known as radish, summer or winter radish, and daikon. Leafy vegetables grown for their leaves represent a taxonomically heterogeneous group of species belonging to a number of families. In addition to probably the most famous lettuce or locica (*Lactuca sativa*, Asteraceae) includes spinach (*Spinacia oleracea*, Amaranthaceae), valerian (*Valerianella locusta*, Valerianaceae), rocket (*Eruca sativa*, Brassicaceae), and many others. Finally, we mention a few vegetables that are grown as perennial, such as asparagus (*Asparagus officinalis*) or artichoke (*Cynara cardunculus*).

Wine and Hops European vines belonging to the species *Vitis vinifera,* have been domesticated in southern Europe and the Middle East. There are many other species of vines in Asia and America, such as *V. riparia, V. lambrusca, V. aestivalis* or *V. rotundifolia*. Many American vines are grown for wine production, for example, *V. rotundifolia* (muscatel), but they are also important as rootstocks for *V. vinifera* varieties. Since the invasion of the rhubarb aphid at the end of the nineteenth century, virtually all European grapevine varieties have been grafted onto American rootstocks that are resistant to this pest. Hops (*Humulus lupulus*) is a dioecious plant whose male plants produce cones that provide the bitterness used in beer production. It was domesticated from a wild plant of the same species.

For reasons of space, we have omitted spice and medicinal herbs in this overview. Firstly, because there are a large number of them, and secondly because they are grown on a small area.

2.2.1.2 Varieties and Breeding

Plant grown in culture are grown as a specific cultivar or variety. There are many definitions of variety, but they agree that a variety is a group of plants belonging to the lowest degree of botanical classification below species or subspecies. A particular variety is distinguishable from other varieties by certain traits resulting from an expression of a particular genotype or combination of genotypes, which are reproducible without change. A variety must be distinct from others, the distinctive characters are uniform and stable from generation to generation.

In orders for a variety to be marketed, it must be registered. The condition for registration is, among other things, compliance with the requirements for distinctness, uniformity, and stability. For important agricultural species, the utility value of the variety is also assessed (yield, resistance to biotic and abiotic stresses, technological parameters, etc.). The diversity, uniformity, stability, and utility value of the variety are verified in the relevant official authorities following specific regulation. In the EU, variety registration is managed by the Community Plant Variety Office (which delegate most of the administration in relevant bodies in individual member states). In the USA, the USDA is an authority that performs a similar task. These authorities may enforce other bodies to do actual variety registration and related administration.

Variety breeding is a long-term and financially demanding process. The author (Breeder) of the variety has the opportunity to protect the know-how invested in the variety by obtaining protection rights to the variety. Protection of a plant variety is a form of intellectual property protection, which provides the breeder with the exclusive right to use his protected variety. Another person may use the protected variety only with the consent of the rights holder, on the basis of a licensed agreement. License fees are an important financial source of reinvestment in breeding. Another condition for registration and protection of rights is the designation of the variety by name. In order for a name to be unmistakable, it must meet a number of criteria, be proposed by the breeder and be assessed by the relevant testing authority.

From the point of view of population biology, the variety can be formed by very diverse categories, it can be a vegetatively propagated clone, a line created by selection of self-pollinating plants, a population of heterozygotes in non-pollinated plants or a hybrid created by crossing maternal lines. Varieties such as vegetatively propagated clones occur, for example, in potatoes. In such a variety, all individuals of the variety are genetically identical. Because clones are the progeny of a single plant, often created by crossing, they are often heterozygous. The lineage is a generative progeny of one plant (single-line) or more sister plants of the same origin, selected from an almost homozygous line (multiline), resulting from repeated self-pollination. By crossing the lineages, a hybrid variety can be created in the first daughter generation, consisting of a population of heterozygotes that are very homogeneous. Thanks to the heterosis effect, these heterozygotes are usually more vital and more resistant. In addition, the production of varieties such as F1 hybrids are very popular with seed producers because the production of a given variety requires the re-crossing of maternal lines. If we use the seeds of F1 plants, only some of them will have the characteristics of the mother plant. The use of these hybrids is very common in maize, but it also occurs in other crops, such as cucumbers, tomatoes, and peppers. Finally, the variety may be formed by a population of cross-pollinating plants, such as rye, apple, pear, etc. Population varieties are more heterogeneous and predominantly heterozygous.

The activity leading to the creation of new varieties or to the improvement and maintenance of existing ones is called breeding. Humans have been choosing suitable plants probably since the beginning of their cultivation, and the selection of plants with suitable characteristics was certainly the oldest breeding technique used. In the early days of agriculture, plant selection was combined with random mutations and probably random crossing. Later, breeding developed as a targeted activity that combines different methods. In the first step, there is some targeted increase in the variability of plant material, and then there is its selection and testing.

The oldest classical method of increasing variability is crossing. People cross two or more varieties together to achieve some new combination of traits or improve existing traits. From then, with the required properties, plants are selected and then further tested. Imagine the whole breeding process based on crossing on the example of potato breeding. We will cross two or more clones for potato breeding. In order to cross plants, plants must bloom first. This may be a problem with potatoes, but breeders have various methods of forcing them to do so. For example, they grow plants so that they have access to their underground organs, and by constantly breaking off the tubers, they support the efforts of the potatoes to bloom. By pollinating each other, fruits (similar to small green tomatoes) and seeds are produced. One can get thousands of seeds from a few plants. We grow seedlings from the seeds, one to several thousand, from which we then select several hundred and plant them in the field. From here we will continue to propagate the potatoes as a vegetative clone. So, the good news is that our new variety may already be in our field, just select it (it may be even more complex when we try to produce a cultivar of plants, which are grown as heterozygotic population). However, even selection of the one, is not a simple task. Thus, breeders select several hundred to thousands of promising plants and further propagate and test them for taste, shape of tubers, yield, resistance to diseases, pests, etc., until they receive one or a few clones that have the new desired properties. Of course, not all tests can be performed in the first year, for some, one tuber is enough, for others, more is needed and it is possible to perform tests in the following years, when the material of the clone multiplies a bit more. Because clones can undergo random mutations during cultivation, can be infected with viruses, etc., we maintain selected clones in tissue culture, from which they can be replicated repeatedly.

Hybridization is not always that simple. We can cross several parents in different ways and we can cross different species with each other. An example of a crop crossed through interspecific crossing is triticale, a hybrid of wheat and rye. Interspecific crossing is often combined with modification (multiplication) of the number of chromosomes, the so-called polyploidy. Polyploidy is sometimes used on its own because plants with twice (or otherwise multiplied) the number of chromosomes have a larger nucleus and thus, larger cells and are larger overall. This is valuable, for example, in forage crops. Polyploidy also occurs in nature; in the laboratory, it can be caused by the action of some substances such as colchicine.

To clarify the principle of its action, we first look at the formation of germ cells (eggs or pollen) of plants. The formation of these buds occurs in a special way of division (meiosis) so that the resulting germ cells have half (haploid) the number of chromosomes. When germ cells fuse, a germ cell is formed which like all cells in the body, has all chromosomes twice. When colchicine is applied, germ cells having 2 N chromosomes are formed, and by fusing them, a cell with twice the number of chromosome is produced (4 N). Crossbreeding also uses artificial induction of mutations, chemicals, or radiation.

In addition to these classical methods, it is currently used to breed the production of genetically modified organisms (GMOs). (see Box 2.4).

For registered varieties, it is necessary to carry out the so-called maintenance breeding, which ensures the continued existence of the variety without change. In maintenance breeding, the variety is grown under strict supervision and plants that do not correspond to the description of the variety are removed. Material used for the further production of seeds or seedlings is obtained from maintenance breeding.

Box 2.4 Genetically Modified (GMO) Crops and Crops Produced by Modern Mutagenesis Methods

In principle, we can distinguish two types of genetically modified organisms (GMOs), which are used in agricultural practice. The first of these are organisms to which genome genes from another organism has been added. A typical representative of such crops is Bt maize, a maize in which genome, genes for the production of bacterial toxin from *Bacillus thuringiensis* have been inserted. This toxin then makes corn resistant to a variety of herbivorous insect pests. A natural analogue of such a situation is the so-called horizontal gene transfer (i.e., gene transfer between unrelated organisms). Such gene transfer is very common in bacteria and less common in multicellular eukaryotes. For example, some termites produce their own cellulose. This termite cellulose is believed to be derived from the cellulose of their symbiotic protozoa.

The second group consists of the so-called modern methods of mutagenesis or gene editing (genetic knockout) such as CRISPR–CAS. In this approach, nothing is added to the genome of the target species, but on the contrary, some genes are "turned off." One could call it a controlled mutation. This is a relatively new technique and attitudes toward it vary from country to country. Some regulators such as in the USA treat such crops as conventional crops, and bringing them to market is not subject to any special registration or additional testing beyond what applies to conventional crops on the grounds that nothing is added to the crop genome. In contrast, in the EU (by the ruling of the European Court of Justice of 25 July 2018), these techniques are considered to be GMO products with all the resulting restrictions (Callaway, 2018).

Methods based on genome modification are becoming more common and diverse. GMO production has its analogies in natural processes, the intensity with which we produce GMOs is incommensurable with these processes and leads to combinations of properties that would not arise in nature, or rather the probability of their occurrence would be very small. At the same time, there is a small but nonzero chance of spreading these genes to weeds and other wild plants. The scientific community sees this as one of the possible long-term dangers in the use of GMOs. On the other hand, it must be said that the plant varieties permitted so far have undergone very rigorous testing comparable to medical drug testing, but as critics point out, all these tests cover a relatively short period of use. A frequently cited benefit of using GMOs is the reduction in pesticide application. This is certainly true for some crops such as Bt maize. On the other hand, GMO varieties resistant to the Roundup herbicide are in many crops, which often leads to overuse of pesticides. In addition, seeds of GM varieties are mostly traded as part of a comprehensive supply together with pesticides and are relatively expensive, without the possibility of growing their own (farm) seeds.

Worldwide, 191.7 million ha of GMO crops are grown, the most abundant of which are soybean GMO varieties, which represent half of all the sown

(continued)

Box 2.4 (continued)
areas with GMO crops. The rate of GMO utilization in different countries varies, over 90% of the total area of GMOs is grown in five countries (USA, Brazil, Argentina, Canada, and India). In some of these countries, such as India, GMOs account for the vast majority of crops in a number of crops (ISAAA, 2018). Consumers of the EU countries and organic farmers have a negative opinion on the use of GMOs in agriculture due to precaution, which makes it difficult to grow GM crops in the EU, however, many commodities derived from GMO crops are imported into the EU.

2.2.2 *Agrotechniques: Adjustment of Conditions for Plants in Agriculture*

2.2.2.1 Additional Energy

Humans significantly change the conditions for the growth of cultivated plants at the expense of additional energy (Woods et al., 2010). The term additional energy refers to all the energy supplied by humans for the cultivation of plants. It includes, in particular, the energy of fuels needed to carry out agrotechnical interventions (plowing, sowing, etc.), energy needed for the production and application of fertilizers and pesticides, wear and tear of machines, energy for harvesting, etc. As an example, we can mention the energy balance in corn production in the USA (Fig. 2.13). During the growing season, about 29 TJ of solar energy comes per hectare field. Maize grown in this field can produce 7 tons of grain and another 7 tons of other biomass, which represents 289 GJ of energy. The efficiency of capturing solar energy during the season is therefore, about 1% (about 0.5% if we take into account the whole year). To achieve this production, man supplies about 47 GJ of energy, mainly in the form of fossil fuels. Both fossil fuels used directly within the field to powering machinery, etc. and also the energy used to produce fertilizers, pesticides, and other goods consumed to grow plants on the land. An important problem is energy for irrigation (which would be eliminated if we grew corn in a more favorable climate that does not need irrigation). Another important problem is the wear and tear of machines, or the energy needed to produce spare parts (e.g., spare plow blades) or new machines, which will replace those already worn. This is followed by fuel energy and energy contained in pesticides. It is worth noting that the energy contained in pesticides is very high, even though the weight of used pesticides is relatively low. This is due to the large amount of energy needed to produce them

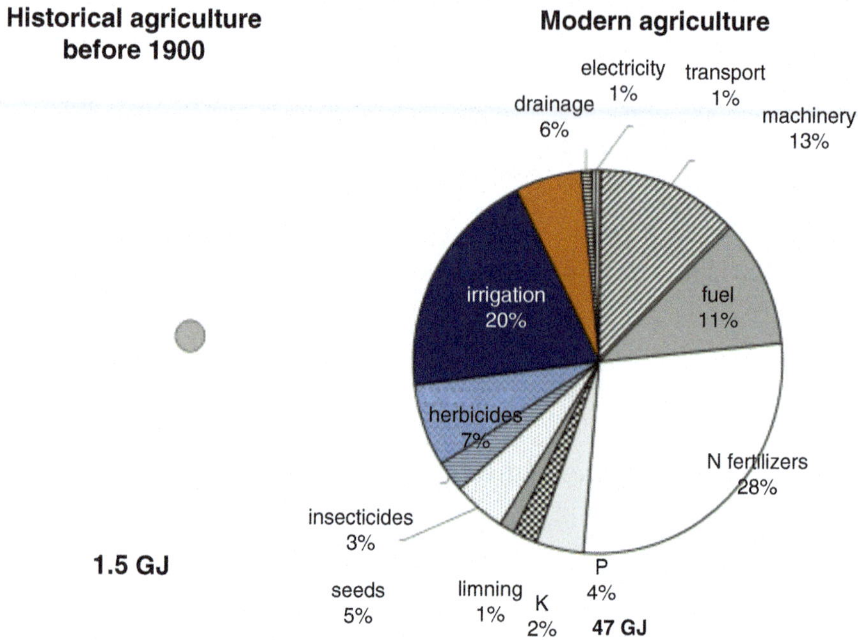

Fig. 2.13 Energy supplied by farmer to grow 7 tons of maize in the USA in historical and modern agriculture based on Pimentel et al. (1980, 1990)

(419 MJ kg^{-1}). Another highly energy-concentrated item is the replacement of machinery (109 MJ kg^{-1}) and is followed by the production of N fertilizers (88 MJ kg^{-1}).

In modern mechanized agriculture, the consumption of human labor is only about 25 MJ ha^{-1} $year^{-1}$. This is relatively small compared to labor consumption in conventional agriculture based on human and animal work (1530 MJ). However, the application of mechanization requires an additional energy in the form of machine wear and fuel consumption. Thus, with a similar yield, the efficiency of using additional energy in the original historical preindustrial management is much greater (Pimentel et al., 1980, 1990). The ratio of additional energy to the total energy accumulated in plant yield is the least favorable (highest) for annual plants harvested for grain (rapeseed, cereals). On the contrary, it is most favorable (lowest) for perennial crops, in which the whole aboveground biomass is harvested, such as meadow hay or switchgrass (*Panicum virgatum*), grown as energy crops (Fig. 2.14).

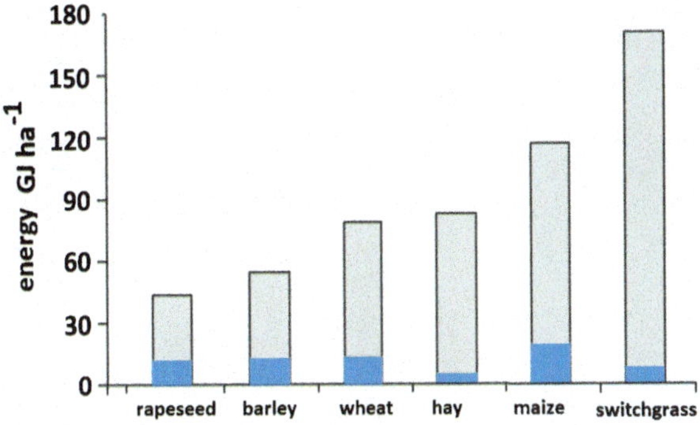

Fig. 2.14 Share of additional energy (blue) in total energy accumulated in yield. According to Pimentel et al. (1980, 1990)

2.2.2.2 Fertilization and Plant Nutrition

Plant tissue consists of more than 90% oxygen, nitrogen, and hydrogen. Only about 5% of the dry matter of organic matter is other nutrients. During the combustion of plant biomass, C, O, and H are released into the air in the form of water vapor and CO_2 while other nutrients generally remain in the ashes, which is why we sometimes call them ashes. The largest part of ashes consists of N and K, followed by other elements, Ca, Mg, S, P, and Cl (Fig. 2.15), and even in smaller concentrations of Fe, Mn, Cu, Mo, Zn, B, and Ni. The latter group consists of the so-called microelements (the division into micro and macroelements is somewhat arbitrary, Cl is often included among the microelements). These elements play a number of important functions in plant tissue. In the following overview, we will look at the main macroelements (Fig. 2.15):

Nitrogen (N) can be taken up by plants as NO_3^-, NH_4^+, or from the air by means of N_2 symbionts by symbiotic fixation. It forms a part of proteins, and thus is present in all enzymes. In addition, it is an important component of chlorophyll (more than 70% of leaf nitrogen) supporting the rapid growth of plants. In natural ecosystems, it is mostly bound in organic matter in the soil. The available forms are mobile in the soil and are easily lost by leaching.

Phosphorus (P) is taken up in the form of orthophosphates or hydrogen phosphates PO_4^{3-} $H_2PO_4^-$ HPO_4^{2-}. It forms a part of ATP and ADP, is essential for photosynthesis, energy transitions, supports flowering and growth. The availability of P strongly depends on the pH.

Potassium (K) is accepted as the K^+ cation. It supports the production of proteins, is important in osmoregulation. It is the major osmoregulatory cation of plant tissues in contrast to animal tissues, where sodium plays this role. The available K forms can be bound to a sorption complex.

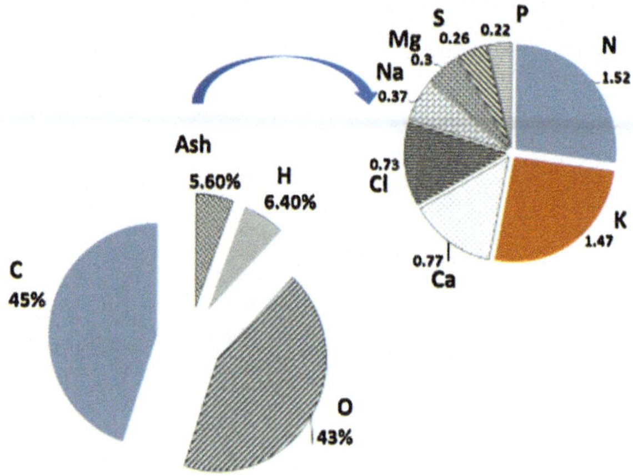

Fig. 2.15 Elemental composition of plant tissue

Calcium (Ca) is taken up as the cation Ca^{2+}. It is essential for cell wall formation and integrity; the available forms can be bound to a sorption complex. Magnesium (Mg) is accepted as the Mg^{2+} cation. It forms a part of chlorophyll and other enzymes. As with other cations, the available forms can be bound to a sorption complex.

Sulfur (S) is accepted as the SO_4^{2-} anion. It plays a role in the formation of proteins, the activity of certain enzymes and vitamins. In natural ecosystems, it is mostly bound in organic matter in the soil. The available forms are mobile in the soil and are easily lost by leaching.

The content of some elements in plants is much higher than their availability in the soil, especially if we consider the content of available forms of these elements. This difference is greatest for nitrogen. Which is also a reason why N is often a limiting element in most ecosystems. This means that the addition of N will increase the plant production, while the addition of other elements does not have such an effect. However, with N we often have to add other elements, especially P and K. This is because these elements may not be limiting at the beginning, but if we fertilize N, their lack can limit its use (Fig. 2.16). This condition is described by the law of the minimum, which states that a factor that is in the minimum (most deficient) limits the growth of plants. In other words, if there is one factor that is most in short supply, improvements in other factors have little or no effect. Only if we increase the availability of the limiting factor will production increase.

In practice, it is possible to assess the limitation of individual elements by fertilizing all elements in several doses that we suspect could be limiting and could measure the response in plant production. However, this procedure is relatively laborious and time-consuming. Therefore, the ratio of the elements in plant tissue is often used to estimate which element could be limiting. As previously described, the most common limiting elements in terrestrial ecosystems are N and P. In many practical applications, we use the ratio of these elements to roughly estimate the limitation. A N/P ratio <10 generally indicates a N limitation, while a N/P ratio >20 indicates a P limitation (Güsewell, 2004).

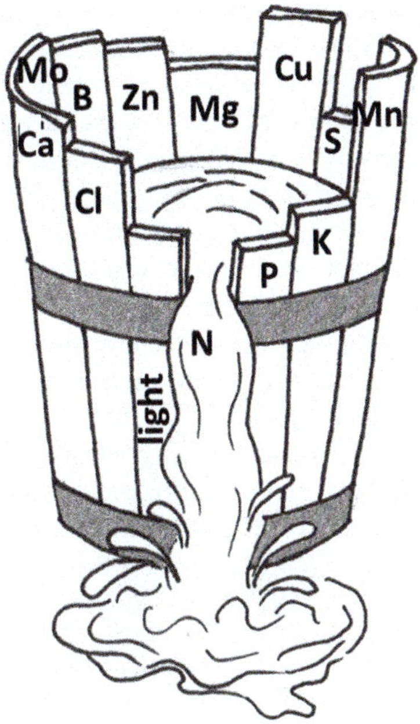

Fig. 2.16 Schematic representation of the concept of a limiting element. The resulting value of the process can be imagined as the height of the level in the barrel, which consists of wooden segments (pulp) of different lengths. If an element is limiting (its flesh is the shortest), then fertilizing with other nutrients (increasing other flesh) will not help. Only by increasing the limiting element (by extending the shortest pulp) will the yield (level) be raised. The limiting element does not have to be only factor but also water or light

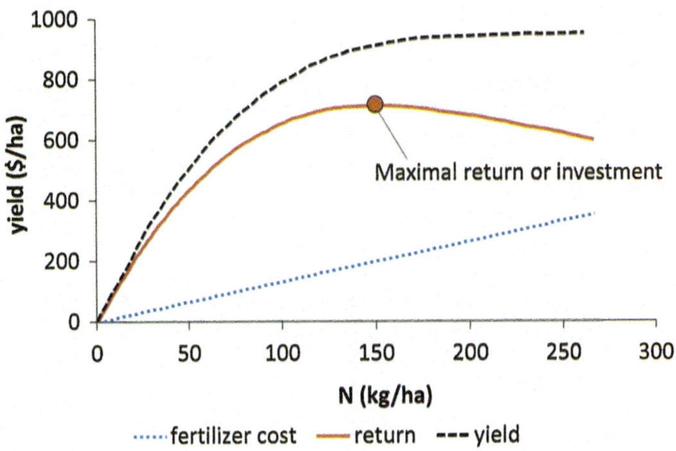

Fig. 2.17 Yield function, dependence of yield growth on the amount of nitrogen added

The increase in yield with the addition of a limiting factor is described by the so-called yield function that is not linear. From the beginning, the yield increases with the addition of the limiting nutrients (e.g., nitrogen) greatly, then with more addition, the yield increase is smaller until it finally stops (there is enough nitrogen, it is no longer limiting). In the example in Fig. 2.17, we see that if we add 20 kg of nitrogen, the yield increases by approx. 1.25 t. Another 20 kg (i.e., a total of 40) will bring an increase of only half a ton. If we fertilize less, return on investment is greater but we have lower yield, however, if we have enough area, we can get the same production with less investment. In our example, if we have 1.4 ha available, we can fertilize 20 kg N per ha (28 kg per whole area) and get a total yield increase of 1.75 t. To get the same yield from 1 ha, we have to fertilize 40 kg of N. From an economic point of view, we can also express the yield in monetary units and in the same way the price spent on the acquisition and application of fertilizers. We see that in the lower part of the curve, we get more return on the investment. This return gradually declines as we put more fertilizer. We can achieve the maximum profit at the point where the investment is equal to return, after that point, further increase of the fertilizer dose brings less profit then was the investment and it is no longer profitable.

The course of the yield function is different among various areas with different climate, agronomy practices and also among various soils (Fig. 2.18). Soils with higher organic matter content generally provide better yields (Fig. 2.19), with the same level of N fertilization. Higher proportion of N coming from natural N_2 fixation and less N coming from mineral fertilizer supports the higher and efficiency of N use (Quiroga et al., 2006).

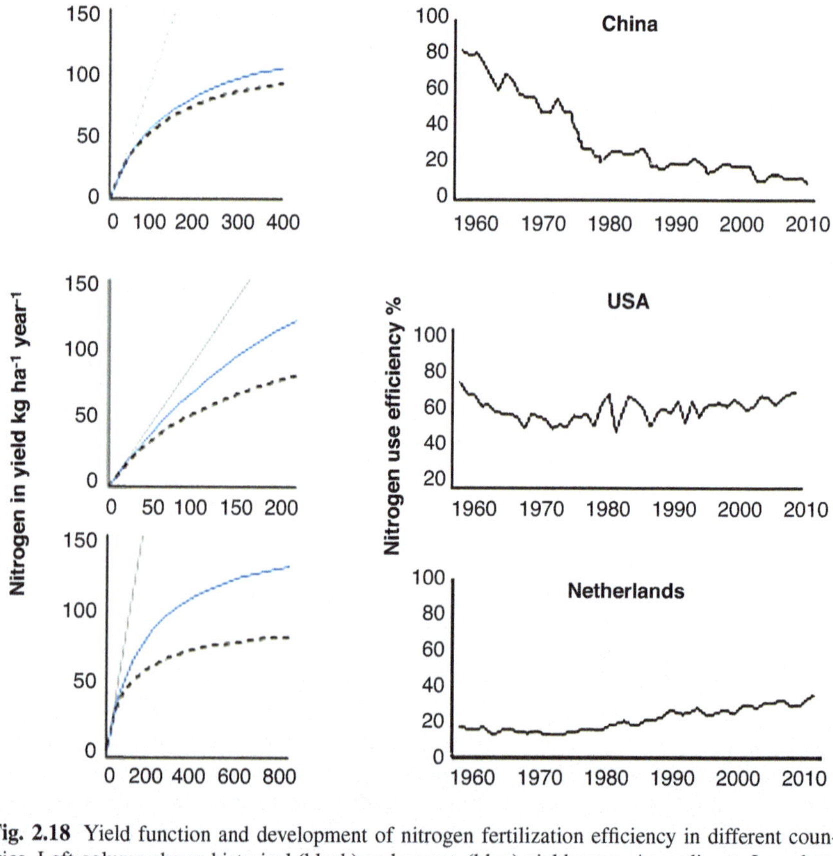

Fig. 2.18 Yield function and development of nitrogen fertilization efficiency in different countries. Left column shows historical (black) and recent, (blue) yield curve. According to Lassaletta et al. (2014)

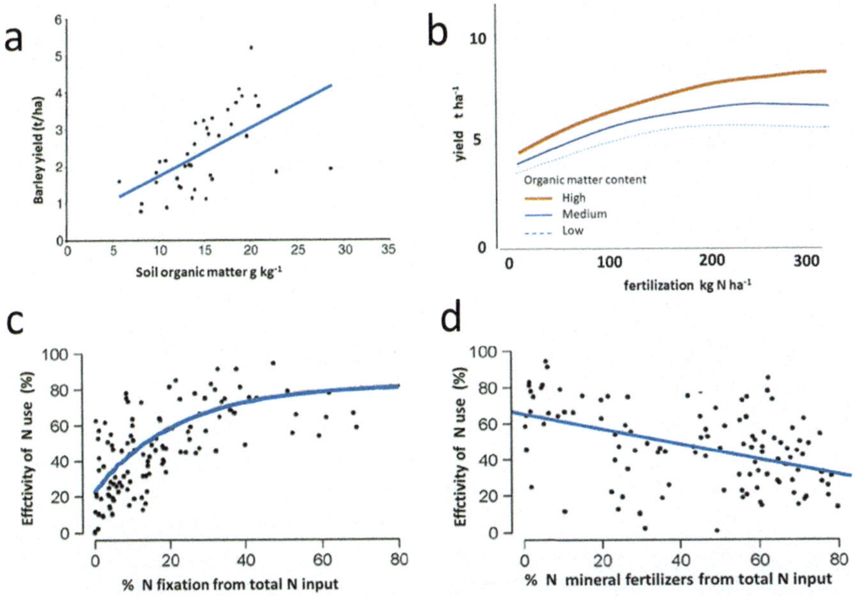

Fig. 2.19 Dependence of the yield on the content of organic matter in the soil (**a**, **b**) and the dependence of the efficiency of the use of nitrogen fertilizers on the share of nitrogen originating from its biological fixation (**c**) and from industrial fertilizers (**d**). According to Oldfield et al. (2019) and Quiroga et al. (2006)

Main Types of Fertilizers

Fertilizers can be divided into two main groups: organic fertilizers and mineral fertilizers. In this overview, we start with mineral fertilizers, which currently supply most of the nutrients to agricultural soils.

Nitrogen Fertilizers

Nitrogen fertilizers supply nitrogen in a number of chemical forms. Firstly, in the form of anion NO_3^- (there are various nitrates, we call them saltpeter), then in the form of ammonium cation NH_4^+ (ammonium saltpeter contains both these forms), then in organic form as NH_2 (urea) or as amide and cyano N (nitrogenous lime). Currently, the source of most industrially produced nitrogen fertilizers is Haber-Bosch synthesis. It is a process of obtaining ammonia from atmospheric nitrogen, which was developed by Fric Haber (for which he received the Nobel Prize in 1918). Later, Karl Bosch adapted this process so that it could be used on a truly industrial scale (Nobel Prize 1931). The principle of the process is the reaction of nitrogen and hydrogen at high pressure and temperature in the presence of magnetite, which serves as a catalyst. Subsequently, ammonia can be oxidized by the Ostwald reaction (Wilhelm Ostwald, Nobel Prize 1909) and is used to produce nitric acid (Fig. 2.20).

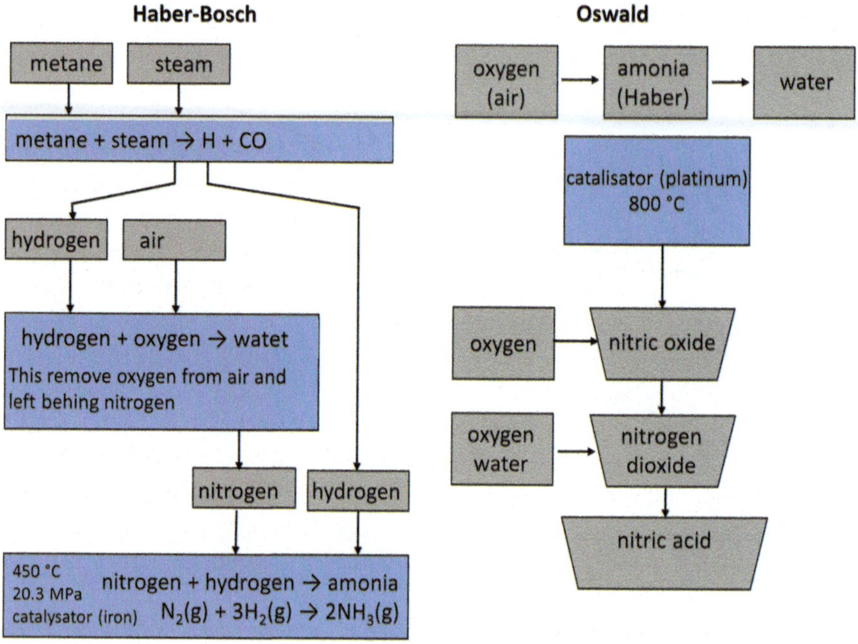

Fig. 2.20 Schematic of the Haber-Bosch and Oswald process, key chemical reactions used to produce nitrogen fertilizers

The main raw material for the production of nitrogen fertilizers is atmospheric nitrogen. Thus, it might seem that their extraction is not limited in any way, but it should be borne in mind that the recovery of ammonia in particular, is a very energy-intensive process. The theoretical energy intensity of ammonia production is 19.7 GJ t N^{-1}, however, the efficiencies of real chemical plants are usually in the range of 35–40 GJ t N^{-1}. Energy intensity depends very much on the type of fuel used to reach the required temperature. The efficiencies mentioned above relates to the use of natural gas. In third world countries, other fuels, such as diesel, fuel oil, and coal are used, which leads to the whole process taking place with 2–3 times lower efficiency (when using coal, it is up to 165 GJ t N^{-1}, so eight times less efficient than the theoretical efficiency).

Phosphorus Fertilizers

The main source of P fertilizers is the extraction of apatite minerals. Apatite is a calcium phosphate $Ca_3(PO_4)_2$ with a number of impurities. It can occur under a wide variety of conditions, from magmatic activity to sedimentation from accumulated organic residues. Therefore, it is a very abundant accessory mineral. Currently, the estimated recoverable reserves exceed 300×10^{12} t, (US geological Survey). Minable deposits were formed either during igneous segregation (5%) or by sedimentation of organic marine sediments (95% of all reserves). Mining in 2017 reached 263×10^6 t. The main producers of apatite are: China (53%), USA (11%), Morocco and other North and West African countries (10%) and Russia (5%). For geopolitical reasons, Morocco and other African countries are an important source of P for EU countries. This raises concerns about the future availability of apatite due to depletion of stocks and possible geopolitical fluctuations (Box 2.5). At the same

Box 2.5 Phosphorus as a Potentially Limiting Element of Intensive Agriculture

The behavior of P in the soils of natural ecosystems and in agroecosystems is very complex and the availability of P to plants is affected by a number of chemical, physical, and biological processes. As explained above, phosphorus is initially sufficient during soil development, as young soils are well supplied with unventilated geological subsoil. In old soils, P is gradually lost by erosion and leaching (Fig. 1.14). In addition, in these soils, most of the P is bound to hard-to-reach forms. This lack of P in tropical soils means that even P supplied by fertilizers can be easily immobilized (Oberson et al., 2006), which is limiting for agriculture in the tropics.

Even in young soils on which modern Western agriculture is practiced, P can be a limiting or co-limiting element, which can begin to be limiting especially if N is supplied in the form of fertilizers. P supplied in the form of fertilizers is primarily obtained by the extraction of apatite. This leads to a debate about whether the world's apatite reserves are sufficient to meet the ever-growing needs of agriculture and society as a whole. Although there is currently a sharp increase in the estimated world reserves of apatite, there is doubt if all declared reserves are actually recoverable (Edixhoven et al., 2013). In addition, it can be expected that the quality of the raw material may deteriorate and the content of undesirable impurities may increase. In addition, a large part of the reserves is in geopolitically problematic parts of the world.

In addition, the over-supply of P into ecosystems contributes to the eutrophication and degradation of natural ecosystems. Therefore, it is necessary to focus on greater recycling of P (e.g., from sewage sludge) and better use of P in agriculture. Current fertilization results from the principle of sufficient available P, which leads to the accumulation of P in soils and undesirable degradation and eutrophication of the surrounding ecosystems. This could be partially or completely avoided by increasing the efficiency of P use in agro-ecosystems. Through breeding and appropriate selection of cultivars, it would be possible to maintain sufficient yields even at reduced concentrations of P (Lammerts van Bueren et al., 2011; Schneider et al., 2019). In addition, a better supply of P can be ensured by more efficient recycling of P in agroecosystems. This can be enhanced by increasing soil organic matter content, the gradual release of P by mineralization of organic forms of P, optimizing symbiosis with arbuscular mycorrhizal fungi (AMF), and supporting other biota activities (e.g., earthworms) (Oberson et al., 2006; Schneider et al., 2019).

Table 2.3 Cadmium content in Apatites of various origins mg kg^{-1} (according to Man, 1994)

Origin	Cd content
Russia (Kola)	1
Florida, USA	6–15
Morocco	8–75
Israel	15–30
Tunisia	25–35
North Carolina, USA	36
Togo	42–80
Senegal	66–90
Idaho, USA	40–340

time, apatites from different areas have very different amounts of environmentally undesirable impurities, such as cadmium (Table 2.3). Although the content of undesirable impurities can be reduced by subsequent treatment, long-term use of apatite can affect soil contamination by these contaminants (Man, 1994).

Apatite after being ground can be directly used as fertilizer, but it is only very slightly soluble. The solubility of apatite is increased by the action of acids to form superphosphates, which is more soluble in water. Single superphosphate is produced by the action of sulfuric acid and double superphosphate by the action of phosphoric acid.

In addition to apatite, P occur as an impurity in iron ore, from which they are removed in some steelmaking processes in slag. It is the basis of the so-called Thomas slag, also usable as a fertilizer, and in addition to P, an important source of many microelements. Another marginally used P fertilizer is bone meal.

Potassium Fertilizers

Potassium fertilizers are obtained by mining and processing potassium-containing minerals. World reserves are estimated at 18×10^{12} t K_2O. In 2017, the production of potassium fertilizers was 56×10^6 t K_2O (FAO). The main producers are Canada (16%), Russia (12%), Belarus (8%), Germany (4%), China (3%) and Israel (3%). The most important K fertilizer is potassium chloride, which is mined from underground deposits, then purified to remove clay and sodium chloride. Another frequently used K fertilizer is potassium sulfate K_2SO_4, especially in crops which are less tolerant to additional chlorine from KCl. Potassium sulfate can be extracted from the mineral langbeinite or can be synthesized from potassium chloride by the action of sulfuric acid at high temperature.

Calcium and Mg fertilizers are important both to replenish the supply of the elements in question and to apply them in order to increase the pH of the soil. This is necessary because there are a number of processes that lower the pH, such as application of other fertilizers, leaching of cations, and atmospheric deposition. Moreover, keeping the right pH is important for availability of other nutrients.

Table 2.4 Average content of main nutrients in main types of organic fertilization in %

Parameters	Beef manure	Horse manure	Poultry manure	Beef slurry	Pig slurry	Urine fertilizer
Dry matter	24.0	25	31	7.5	7.2	1.4
Organic matter	17.0	20	25	5.5	6.0	1.0
N	0.48	0.65	2.82	0.40	0.60	0.23
P	0.11	0.13	1.25	0.10	0.13	0.01
K	0.51	0.52	1.23	0.40	0.27	0.33
Ca	0.37	0.21	0.62	0.10	0.20	>0.01
Mg	0.08	0.11	0.07	0.04	0.05	0.01

Other elements, such as Cl and S are rarely supplied by specialized fertilizers. This is because Cl and S are a common component of N, P, and P fertilizers, which usually covers more than the plant's requirements for these elements. In addition, there are a number of anthropogenic sources of these elements, such as road salting, or in the past, significant atmospheric S deposition which was associated with the combustion of fossil fuels, especially coal. At present, however, there has been a significant decrease in S deposition, and thus, under certain conditions, there may be a lack of S in the soil for some sulfur-intensive crops, such as rape. Therefore, it is recommended to use K and N fertilizers based on S, or autumn fertilization with gypsum. Furthermore, S fertilization lowers the pH, which is usually not desirable. Exceptionally, in some crops, such as raspberries and blackberries, S is applied for this purpose. Microelements are often supplemented with complex NPK-type fertilizers but there are also specialized fertilizers with a high content of microelements. One of the oldest of them is the above-mentioned Thomas slag.

Organic fertilizers serve both to supplement nutrients and, above all, to supplement the content of organic matter in the soil. Organic matter contributes to plant nutrition both directly by mineralization of the nutrients contained in it (Table 2.4), which becomes available to plants, and indirectly by adjusting soil conditions. This indirect effect, consisting in increasing the capacity of the sorption complex, supporting the soil structure, influencing the water regime, etc., is usually more significant than directly released nutrients. The most important organic fertilizers include manure, slurry, compost, digestates, and green healing. Manure is a mixture of solid and liquid feces together with litter bedding that has undergone a maturing process. It is usually applied in autumn, the usual dose is 40 tons of manure per ha, every four year (10 t ha^{-1} $year^{-1}$) and after spreading, manure is usually plowed. Slurry is produced through housing animals without bedding. It consists of a mixture of solid and liquid feces of farm animals. Urea is then fermented and diluted with the urine of livestock. These semi-liquid or liquid fertilizers are applied by spraying to the surface of the land or incorporated into the soil. Application of these fertilizers to the frozen surface brings the risk of their washing into watercourses.

With regards to the use of animal feces, it should be recalled that in the past and in some crops, the use of human feces as a fertilizer has been common to this day. Human feces represent a potentially interesting source of nutrients and organic matter, suitable as a fertilizer (Moya et al., 2019). The problem of their use is the danger of transmission of gastrointestinal diseases and parasites. However, this danger can be countered by appropriate sanitation, heat treatment, composting or vermicomposting (Moya et al., 2019). The use of human feces as fertilizer would both solve the problem associated with their disposal and would significantly contribute to greater closedness of biogeochemical cycles. The problem of their use lies both in social perception and in the real health risks described above, requiring a treatment and testing protocol that is lacking in many countries. In addition, in most cultures which use human feces as fertilizer (night soil) they are collected separately, so they will not become part of sewage and are not mixed with water and other waste which makes their use much easier.

In modern societies, feces are often collected in the form of wastewater, where the feces are mixed with other wastes and other wastewater, which complicates their use because the excrement is diluted and mixed with a number of other undesirable impurities. Wastewater is then treated in wastewater treatment plants, some nutrients get lost such as part of nitrogen, which is lost by denitrification. It is possible to use the resulting sewage sludge for fertilization, it still contains some original nutrients, but also contains heavy metals, microplastics, pharmaceuticals, and other pollutants that complicate the use of sewage sludge as fertilizer. Moreover, in addition to sanitation, the sludge must be stabilized, drained, and concentrated to improve their physical properties.

Compost is a partially decomposed organic matter created by mixing different organic matter so as to achieve an optimal C/N ratio of the mixture of about 15–30 and a moisture content of 50–55%. Composting takes place in well-aerated composters or in piles and has several stages. At the beginning, there is a rapid development of microflora, and its activity produces heat, the compost becomes heated in the thermophilic phase, and then the temperature of the compost can reach 50–70 °C. Due to the high temparature reached in thermophilic phase, the compost is sanitized. After thermophilic phase the compost cools. The microflora enter the mesophilic phase and then the maturation phase commences (de Bertoldi et al., 1983) for compost to mature. Composting is also possible with the help of earthworms where there is not as much intense heat, but the passage of matter through the gut of the earthworms leads to a reduction of pathogenic microorganisms (Edwards & Subler, 2011). It is also possible to compost biodegradable waste from households, urban greenery, etc. Compost is a valuable source of organic matter.

Green manure is a crop that we grow to incorporate biomass into the soil. Green manure replenishes organic matter in particular, because with the exception of growing vetch plants, which can promote symbiotic nitrogen fixation, all other nutrients that the plants return to the soil come from the land on which they grew. The most common crops used for green manure are white mustard, winter rape, bunch, peas, and spring wormwood. Mixtures are also popular, such as rape, mustard, and oatmeal.

Use of N-fixing plants as a source of nitrogen. Prior to the commercial boom of industrial nitrogen fertilizers, vetch crops grown either for human consumption or as feed were the main source of N, compensating for N losses (caused by harvesting, leaching, etc.). The amount of N that remains available for other crops is given on average, 113 kg ha^{-1} for monocultures of vetch crops, such as lupine, beans, soybeans, and chickpeas. However, the variability is considerable from several tens of kg to more than 200 kg (Evans et al., 2001; Carlsson & Huss-Danell, 2003). These authors also state that the alternation of vetch crops and wheat can completely cover the required dose of N without the use of industrial fertilizers. In clover grass stands, the amount of N that remains in the soil is estimated at 3–103 kg ha^{-1} with a median of about 50 kg (Ledgard & Steele, 1992). The use of symbiotic N fixation may be less energy-intensive than industrial fixation, and some studies show that it leads to lower NO_3^- leaching (Crews & Peoples, 2004).

Fertilizer Application

The amount of nutrients applied depends on a number of parameters, the type of crop grown, its technological purpose, the yield we want to achieve, the characteristics of the soil on which we grow, and its supply of nutrients (Box 2.6). We have already shown in the yield curve, the increase in yield depending on the added nutrient is different on different soils. In addition, the efficiency of the added nutrient (yield at the input cost) decreases with increasing fertilization rate. Thus, the farmers will optimize their fertilization for maximum profit rather than maximum yield. The application of fertilizers depends on their mobility in the soil. P and K are less mobile so, we can fertilize P and K by applying the entire dose in the autumn and plow the fertilizer into the soil. This is because the risk of leaching of these nutrients is relatively small due to poorer solubility and binding to the sorption complex. In contrast, for N, NO_3 is very mobile in the soil, so we usually apply only a small part before sowing and the rest is applied during vegetation (Box 2.6). Illustrative examples of the annual dose of essential nutrients for selected crops are given in Table 2.5.

Fertilizer application also affects the soil pH. The general rule of physical chemistry applies here as the salt of a strong acid and a weak base lowers the pH, and conversely, the salt of a weak acid and a strong base increases it. However, when applying fertilizers, we must also consider their removal by the plant. For example, KCl is a salt of a strong acid and a strong base, however, K will be taken up by plants more than Cl, and therefore more chlorine anions will remain in the soil solution and the resulting effect will therefore be a decrease in pH. The application of mineral fertilizers often has an acidifying effect, so the soils are limed with limestone or dolomite, or other calcium fertilizers that supply Ca or Mg to the soil. In particular, however, they adjust the soil acidity to slightly acidic or neutral values, which are favorable for plant growth and the intake of most nutrients.

Table 2.5 Examples of recommended doses of fertilization with main nutrients in kg ha^{-1} $year^{-1}$ for selected crops in central Europe, assuming that we want to achieve a high level of yield and the supply of soil with mineral nutrients is good (according to UKZUZ, 2020)

	N	P	K
Wheat	130	26	66
Winter barley	100	26	66
Spring barely for malt	60	26	66
Maize for silage	180	44	100
Canola	160	26	100
Beet	160	44	141
Potatoes	135	26	100
Clover	0	26	83
Alfalfa	0	26	83

Box. 2.6 Historical Development of Fertilizer Use and Some Broader Consequences

Figure 2.21 shows the historical development in the use of fertilizers. We can observe that the sharp increase in fertilizer consumption begins at the start of the last century. Its development was closely linked to the establishment of the Haber-Bosch process. Previously, there were industrial guano-based fertilizers imported from South America, but it did not become so important globally. We can observe that the use of K and P fertilizers essentially copies the use of N fertilizers.

Some researchers believe (Smil, 2004; Erisman et al., 2008) that the development of the Haber-Bosch process and the associated production of N fertilizers have significantly contributed to the population explosion of mankind over the past century (Fig. 2.22). These authors believe that the growth of the human population in this period was at least half due to the availability of N fertilizers and the associated increase in food production (Fig. 2.22). In addition, the Haber-Bosch process allowed for a significant increase in yields with almost the same amount of work, so that it was possible for part of the population not to be directly involved in agriculture. In addition, we have already mentioned that the Haber-Bosch process is very energy-intensive. This energy can only be obtained by extracting fossil fuels. Both the chemical synthesis of N and the extraction of fossil fuels require extensive technological support for other industrial productions and adequate human resources. It can thus be said that the Haber-Bosch process supported not only the growth of the world's population, but also the transfer of people from agriculture to other professions and migration from the countryside to the cities. However, the gradual process of urban population growth began much earlier and was certainly not caused by a single impulse. The development of agricultural mechanization and other intensification effects were certainly also significant stimuli in this direction.

(continued)

Box 2.6 (continued)

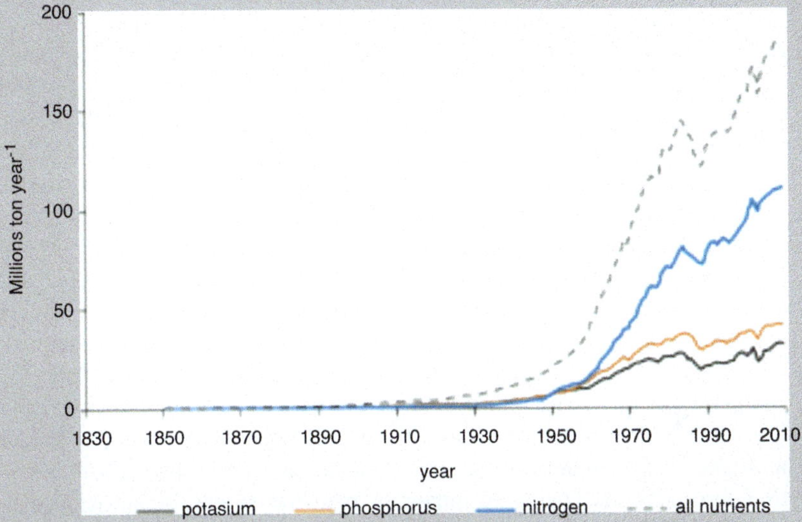

Fig. 2.21 Global production of industrial fertilizers. Data after 1961 FAO stat, data before 1961 Steffen et al. (2004)

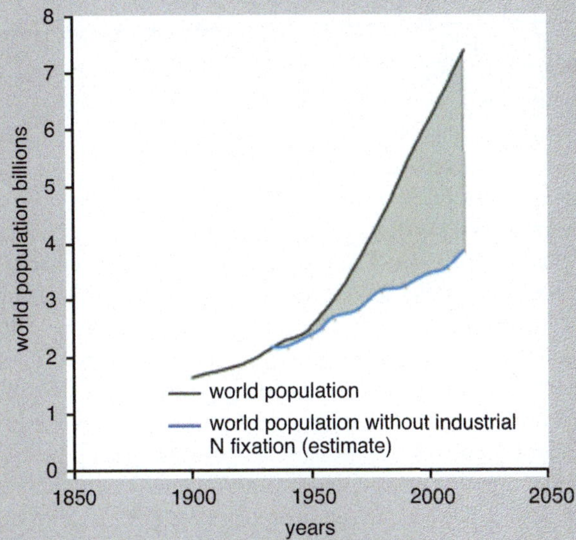

Fig. 2.22 The increase in the world's human population and its portion, which is hypothetical due to the increased availability of reactive N provided by the Haber-Bosch synthesis (gray area), according to Erisman et al. (2008), edited

(continued)

Box 2.6 (continued)

As already shown, many aspects of agriculture technologies are driven not only by economy and technological development, but also by policy. This is evident from the historical development of fertilizer use in the Czech Republic (Fig. 2.23). The broader picture is very similar to the global situation. Use of fertilizers increased during the last century and peaked at the end of the 1980s, when the consumption of pure nutrients contained in both organic and industrial fertilizers reached 197 kg ha^{-1} (N 100, P 30.5, and K 66.5 kg ha^{-1}). Following the political changes resulting from falling of communism in late 1980s, many agricultural cooperatives faced severe economic difficulties, and this was reflected in a reduction in the consumption of fertilizers. After this depression in 1990s, new economical subject formed and gradually caught up, which was followed by a gradual increase in mineral fertilizers application, namely, N ones. An example of the Czech Republic also illustrates the changes in the proportion of organic and mineral fertilizer in nutrient supply over time. Organic fertilizers have historically been an important source of nutrients. It was not until around 1965 that the supply of nutrients with industrial fertilizers became more important than the supply of nutrients with organic fertilizers. After political changes in the 1990s, the supply of nutrients with organic fertilizers decreased due to the decline in cattle breeding, and still remains at lower values (Fig. 2.23). This together with rise of mineral fertilizers have caused organic fertilizer now to play a minor role in nutrient supply.

The Czech example can be used also to monitor the effect of long-term application of fertilizers on the soil. Figure 2.24 shows the state of soil pH and the consumption of Ca fertilizers on the unique long-term data of UKZUZ. We can observe that since the second half of the 1970s, the share of strongly and extremely acidic soils has been increasing. This was matched by a large increase in the consumption of Ca fertilizers, which peaked in the late 1980s. It is difficult to explain unambiguously the causes of the increase in soil acidity, but certainly high doses of mineral fertilizers, many of which can acidify the soil, may have contributed to the acid deposition associated with the combustion of coal with a higher S content. In the 1990s, in connection with the political changes and economic transformation, the use of Ca fertilizers declined. However, despite the sharp decline in the consumption of Ca fertilizers in the 1990s, there was no significant acidification of soils, mainly because the consumption of mineral fertilizers decreased significantly (Fig. 2.23). Moreover, desulfurization of power plants took place during the 1990s, thus significantly reducing the acid deposition, which is one of the other causes of deteriorating soil acidity. In the years 1990–92, the share of acidic and extremely acidic soils was the smallest (Fig. 2.24). Subsequently, since 1992, there has been a slight increase in acidity, which may be related to a gradual increase in the doses of mineral fertilization and a lower rate of liming.

Box 2.6 (continued)

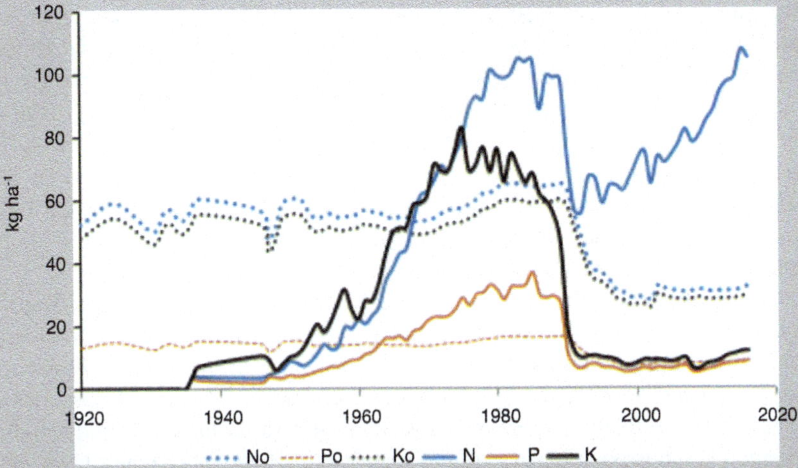

Fig. 2.23 Development of fertilizer use in the Czech Republic. N, P, K denote the supply of these nutrients in mineral fertilizers. No, Po, and Ko denote the supply of N, P, and K in organic fertilizers. Data about mineral fertilizers from Salusová (2018), the amount of nutrients in organic fertilizers estimated on the basis of the intensity of livestock breeding and the calculated manure production and nutrient content according to methods described in Decree No. 377/2013 Coll

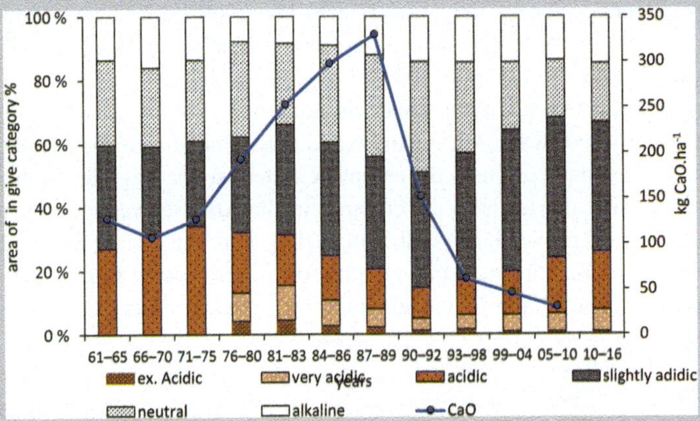

Fig. 2.24 Development of pH of calcareous soils in the Czech Republic, source Czech Institute for supervising and testing in agriculture

2.2.2.3 Irrigation and Drainage

In addition to nutrients and light, water is a basic condition for plant growth. Not surprisingly, plants cannot grow without water. On the other hand, excess water limits the availability of oxygen in the soil, impairs root respiration, and although many specialized plants can cope with this situation, it is undesirable for growing many plants. Therefore, farmers often try to regulate the amount of water in the soil. The FAO states that 20% of arable land was irrigated in 2012, however, 40% of agricultural production was produced on these irrigated areas. It should be noted that in many irrigated fields, irrigation provides water replenishment during the period of its deficit and the amount of water supplied by irrigation to field crops represents only a small part of their total needs.

Already, ancient civilizations addressed water shortages by irrigation. There are a number of irrigation systems. Probably, the most ancient method is flood irrigation, which basically mimics the flooding of soil. The irrigated land is flooded, the water is allowed to soak, and then the water is drained or the excess water dries out. Some crops, especially rice, are grown on flooded land. A similar technique to flood irrigation is sub irrigation by raising of water table. There are channels between the individual parts of the field, which are filled with water and raise the water table between channels it is a very simple way to irrigate. It requires a small amount of energy but is inefficient. In this, only about 60% of the supplied water is used by plants (Brouwer & Prins, 1989). Like flooding, it can only be applied on flat land. Another possibility of irrigation, probably the most common in field conditions, is sprinkler irrigation. It is more effective (75%) than the previously mentioned types of irrigation and can also be applied on sloping land. However, it requires a much larger amount of energy to supply the necessary pressure to spray the water. Finally, the most effective type of irrigation is drip irrigation (90%), where water flows directly to the individual irrigated plants. Drip irrigation is widely used in greenhouses or in the cultivation of permanent crops (Fig. 2.25).

Irrigation brings several potential environmental problems, which are associated with the collection of water for irrigation and may then be missing for other natural processes. An example of this is the extensive water use for irrigation in the Aral Sea basin, which basically leads to the collapse of the Aral Sea ecosystems as well as ecosystem in the Aral Sea basin (Micklin, 2014). This can be largely eliminated by using recycled water. Unfortunately, the use of treated wastewater is also not without problems. After all, the issues related to contamination have been resolved, the

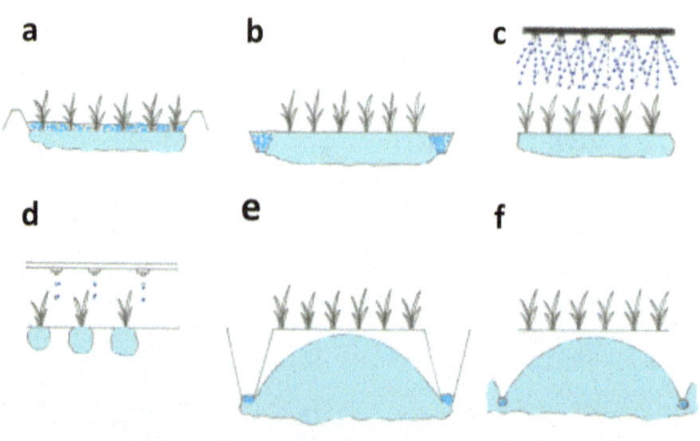

Fig. 2.25 Different types of irrigation and drainage, (**a**) irrigation by flooding, (**b**) sub irrigation by remodeling of water table, (**c**) irrigation by spraying, (**d**) drip irrigation, (**e**) drainage by channels, and (**f**) drainage by pipe drainage. The blue area is saturated with water

biggest problem at present, appears to be the hydrophobicity of the soil. Another significant risk associated with irrigation is soil salinization . Irrigation is usually applied in places where there is shortage of water, which in other words, means that there is usually a higher potential evaporation than precipitation. As a result, the flow of water toward the surface prevails in the soil. This water carries with it dissolved salts and these subsequently accumulate on the soil surface. Salinization promotes the use of salted water. Also, replacing the original deep-rooting arid vegetation with shallowly rooting crops promotes salinization.

In addition to water shortages, excess water is undesirable for many plants. High groundwater levels can be addressed through drainage. The simplest way to drain is ditches, which are deeper than the groundwater level and are sloping to drain water from the land. Instead of ditches, pipes or similar structures buried below the soil surface can be used. Drainage decreases the groundwater level and thus, the water content in the soil. Water has high heat tolerant capacity and water evaporation needs a lot of heat, soil with lower water content warms up quicker, namely, in the spring. This can result in an extension of the growing season.

Soil change promotes the mineralization of organic matter in it. This is especially important in the drainage of peat, where lack of oxygen is one of the main factors inhibiting the decomposition of organic matter. Along with the decomposition of organic matter, nutrients are released as a consequence of drainage. However, if the organic matter content decreases significantly, both the nutrient content and the water-holding capacity are reduced. Drainage has historically been one of the most common ways of increasing agricultural land in many countries at the expense of peatland and other wetlands. This has resulted in a significant decline in biodiversity, but also in the ability of soils to retain water and nutrients (Frouz et al. 2010a, b).

2.2.2.4 Plant Protection

A part of crop production is consumed or otherwise degraded due to the activity of various organisms, in the agricultural terminology of diseases, pests and weeds. Pests are herbivores, especially insects, nematodes, mites, molluscs, and vertebrates. Weeds are other competing plants that compete for light, water, and nutrients, but can also have other negative effects. For example, they make it difficult to harvest or technologically use the crop. Finally, pathogens are mainly fungi, viruses, and also bacteria. Even in nature, plants are exposed to the activities of various pathogenic organisms, herbivores, and competition from other plants. However, the goal of plants in nature is to reproduce enough to survive into the next generation. Due to the large overproduction of seeds or other reproductive organs, some loses are taken into account and it may not harm the population. In contrast, in agroecosystems, the loss of harvestable parts is a loss of the farmer's yield, and therefore profit. In addition, for many reasons, crops in agroecosystems are more sensitive to diseases and pests. An example of such an effect is the fact that most of the crops we grow are derived from r-strategists, annual plants of the early successive stages. Perennial plants that grow in stabilized ecosystems, we call them climax ecosystems, such as oak, invest large amounts of assimilates in the production of defensive substances, such as tannins or other polyphenols. Due to high amount of these substances, it is almost impossible for insects to decompose them enzymatically. However, during evolution, insects developed a number of strategies to feed on these plants, such as feeding on plants in the season when it has few defense substances. These strategies, combined with massive chemical plant protection, lead to a balance between plant and herbivore. It will allow the herbivores to consume a little bit of plant biomass, but not so much as to endanger the plants. Moreover, late succession plants are usually large, so they can sustain some biomass loss. Annual plants of the early succession stages, which specialize in growing on disturbed soil, benefit from the relative availability of nutrients and little competition. The strategy of these plants is therefore, to quickly colonize new disturbed areas, to grow quickly, and to produce large numbers of small seeds to colonize other suitable areas. However, these favorable conditions do not last forever, and so in nature these early successor species are soon displaced by other species, which are colonizing new areas more slowly, but are better competitors. Because these plants invest as much as possible in order to grow quickly, they invest little or nothing in the production of defenses. Thus, if they produce any defenses, there are only a small amount of them, although they can sometimes be very effective poisons, such as poppy papaverines, tobacco nicotine, or cruciferous glucosinolates. These poisons protect them well from nonspecialized herbivores. However, because there are small amount poisons, it is enough for specialized insects to develop an effective enzyme for its detoxification during evolution and can feast on the plants. Plants of early succession stages are protected from these herbivores by the fact that suitable habitats are randomly distributed in the landscape and last for short time. As a consequence these annual plant occur each year in different place in landscape. Consequently it takes an insect a while to find young plants. In the meantime, they grow a little and just produce enough seeds before they are completely eaten. Under natural conditions, this game of hiding between plant and a specialized herbivore can work quite well. But once we start growing such plant as a crops on thousands of hectares, the chances that pests from the last year's field will find this year's field is very high.

The potential yield can therefore be reduced by the activities of other organisms from which we try to protect our plants from (Heitefuss, 1989). Integrated plant protection approach is used for this. This includes (1) setting conditions that support plant defense and natural enemies to reduce pests (microclimate, nutrition, vegetation architecture, and timing), (2) the use of biological control agents, and (3) the appropriate use of pesticides. There are also systems that can only handle the first two points.

So how do you set conditions that support plant defense? Simply put, they will be the opposite of those that promote the occurrence of a harmful agent. For example, by the cultivation of sensitive cultivars, growing the same crop in the same place, especially with there is a history of previous disease, late sowing and N fertilization, high plant density and moisture support the occurrence of a number of fungal diseases. On the other hand, the cultivation of resistant cultivars, proper crop rotation, adequate nutrition, in particular, adequate fertilization with N, rapid crop start, adequate plant density and support for the natural occurrence of predators and parasites of pests can reduce the occurrence of harmful factors. Sometimes small adjustment of agriculture practice can be crucial. For example Olsen et al. (2012) show that random distribution of spring wheat substantially increase its ability to compete with weed in comparison with wheat seed in rows.

The use of natural enemies is also the basis of biological control. This method of plant protection has achieved significant success in horticulture, especially in greenhouses and also in permanent crops, orchards, etc. In this method, for some crops, there are well-developed comprehensive methodologies that, with the help of a number of types of natural enemies, protect against the main species of pests. However, the use of biological control in field crops is less often successful. In addition to methods based on the use of natural enemies, we include among the methods of biological control, a wide range of methods based on the use (and usually confusion) of some behavior of the target species (Heitefuss, 1989). The description of these methods would exceed the possibilities of this text, so we will mention only the principle of two of them. The first is the method of confusing the males. It is based on the use of the female sex pheromones. These pheromones are artificial analogues of pheromones produced by females to attract males to mate. They are often used to signal the occurrence of certain insect species. If we put too much pheromone into circulation, the pheromone trace will adhere to almost everything, on leaves, on twigs, etc. and for a male butterfly, almost everything smells like a female, and the male is then unable to find a suitable female to mate with. The second method we will mention is male sterilization. The principle is simple again. The method is particularly suitable for species that mate only once. We obtain a large number of males of the target species, sterilize them, for example, with the help of radiation, and release them into nature. Males dilute the natural population of males that mate with females. But because they are sterile, females that have mated with them will have no offspring. Although the principles of both methods look simple, their application in practice is no longer so simple, and again it is applied rather in agroecosystems, such as orchards and other more stable ecosystems.

In agricultural practice, a frequent way to control pests is the application of chemicals to control them (pesticides). Pesticide application is costly and brings negative effects on the environment. To decide whether we need to supply them or not, the concept of the so-called economic threshold applies as the pest population level which is likely to cause damage is equal in value to the cost of pesticide application. Figure 2.26 schematically shows the change in yield in relation to the degree of pest infestation. It is interesting to note that with a very small attack, the yield may not drop immediately and there may even be situations where the yield rises slightly. How is this possible? The infected plant tends to compensate for the damage by mobilizing reserves and promoting growth. If the infestation is small but still sufficient to trigger these compensatory mechanisms, the plant may overcompensate, and the growth of the infested parts may even be somewhat greater than without infestation. However, this does not always happen and only with a small level of attack. As the level of attack increases, the yield decreases and losses increase. When the losses exceed the costs of pesticides plus its application, the economic threshold is reached.

Of course, we want to know that we have to do something a little ahead of time, so there are a number of sophisticated methods to monitor the occurrence of the pests and predict the behavior of its population in the future. The threshold is often not produced empirically as shown in Fig. 2.26, but based on experience. Also there

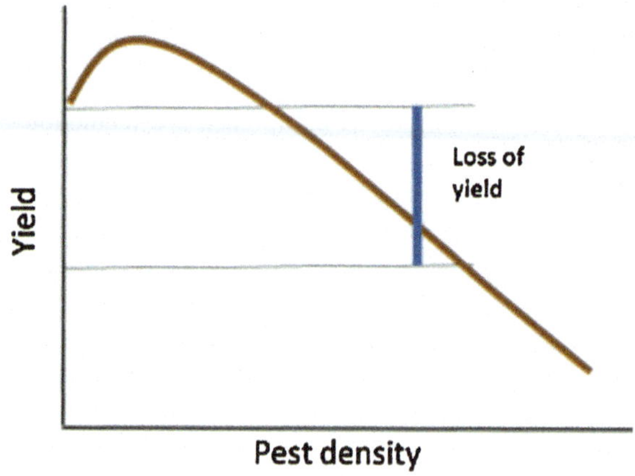

Fig. 2.26 Graphical representation of the harmfulness threshold principle as the magnitude of the yield loss that is greater than the cost of the intervention (e.g., spraying)

are other thresholds which follow different parameters such as quality, but the overall logic are similar. These thresholds are typically not set specifically to reduce negative environmental impact of pesticide. They justify economic sensibility of pesticide use but may also reduce chances for development of resistance against pesticide in pest population. However, any kind of regulation can indirectly help a bit also in the reduction of environmental impacts.

A number of plant protection products (pesticides) are available that work against a variety of pests. For example, fungicides act against fungal diseases and zoocides act against various groups of animals. More often, designations are derived from the specific group of organisms on which they act . For example, insecticides act against insects, rodenticides against rodents, acaricides against mites, nematicides against nematodes, and molluscicides against molluscs, etc. Finally, herbicides act against plants and are used against weeds. We also include desiccants, which speed up the ripening and drying of the crop and can simplify its harvest.

We apply pesticides in various ways, such as a powder, spray, some are used for dressing or coating seeds, etc. Some pesticides come into contact with the target organism on the surface of the plant, others, the so-called systemic herbicides, penetrate the plant. They are spread by conductive tissues and act on organisms that penetrate into the plant or, for example, suck on it. The actual application form of pesticides consists of the active substance, a filler that allows adequate dosing and application (in sprays it is often water) and substances that ensure good contact with the target organism (abrasives or detergents). Detergents reduce surface tension and allow good contact of water drops with the surface of the body, such as insects. The surface of the insect's body is covered with a thin waxy layer, while a drop of water would not stick to it at all, would run down or would have very little contact. Abrasives disrupt the wax layer and, on the one hand, allow the active substance to penetrate. Also, breaking the waxy layer can in itself cause the insects to die by drying out. There are even insecticides that kill insects only because of their abrasive effect.

Only the active ingredient kills or otherwise damages the target organism. The most frequently investigated property of pesticides is toxicity. We are interested in

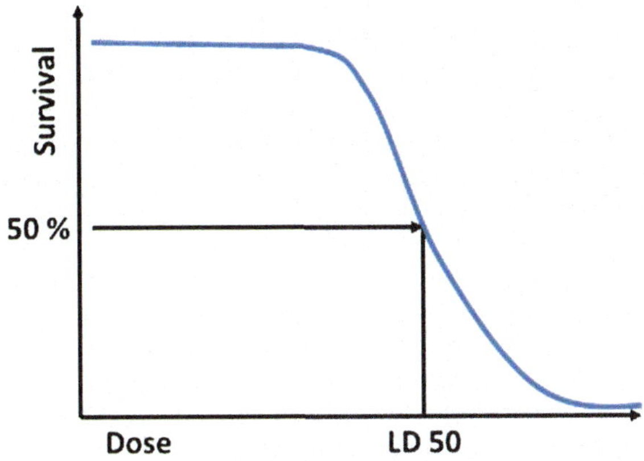

Fig. 2.27 Principle of determining the LD_{50} as the concentration of a preparation or application dose at which 50% of the affected individuals die

both toxicity to the target organism and toxicity to nontarget organisms. Toxicity to a certain organism is most often reported as the so-called LD_{50}. The LD_{50} is the dose (e.g., in mg kg^{-1}) at which half of the subjects die (Fig. 2.27) within 24 hours of being exposed to the substance. However, sometimes for practical reasons, it is not possible to determine the LD_{50} (it is difficult to introduce enough substance into an animal to kill it). The NOEC is used here, or in other words, the maximum dose at which no effect on the organism was observed. Although these toxicity tests are very sophisticated, their extrapolation to natural conditions is complicated. In nature, an active substance can bind to a variety of other substances and this can reduce its availability, it can be diluted or accumulated in the food chain, tests are performed on a limited number of species and each species reacts differently, a substance with small acute effects may have significant chronic effects, etc.

The use of pesticides makes it possible to reduce the occurrence of adverse factors relatively quickly. However, it has a number of environmental risks. A textbook example is the use of DDT (dichlorodiphenyltrichloroethane). Although the substance itself was synthesized as early as the end of the nineteenth century, its insecticidal effects were discovered in 1939 by Paul Hermann Müller (who received the 1948 Nobel Prize). DDT seemed to be an almost ideal insecticide as it was very effective against insects and almost nontoxic to warm-blooded vertebrates. However, the problem with DDT is that it decomposes very slowly and binds to fat tissue. This allowed its bioaccumulation (accumulation in food chains) and the top predators received much higher doses with food than organisms at lower level in the food chain. This was most pronounced in birds in which DDT interfered with the calcification of eggshells. Rachel Carson drew attention to this fact with her book Silent Spring (Carson, 1962). Subsequently, the use of DDT was gradually reduced and eventually banned.

The authorization of plant protection products is currently accompanied by very careful testing. However, even these do not have to detect and anticipate all possible environmental complications that the use of a given substance may bring. This is mainly because the applied substance can interact in a very complex way with the environment where it is applied. Most of the organisms used in testing are species that behave well in the laboratory but may not represent species abundant in nature due to their ecophysiology. Finally, it is very difficult in the laboratory to model complex long-term interactions between several different species and environments and the effect of pesticides on them.

New and new pest and disease control strategies are constantly being renewed, for many of them, we have only preliminary idea of how they can affect the environment. One example of recent developments is the use of the so-called RNA vaccination (Niehl et al., 2018).

2.2.3 *Cultivation of Field Crops*

The basic measure to modify conditions for plant growth on arable land is soil cultivation. As already mentioned majority, of most commonly grown crops are grown as annuals, this increae demand on cultivation. For annuals, soil has been commonly prepared by tillage and following cultivation. In fact growing annuals is a reason for widespread occurrence of bare soil in the landscape. Usually, immediately after the harvest of the previous crop the stubble is often incorporated in the soil by shallow tillage. Its goal is to break up the few top cm of soil and thus, break the soil capillarity and hence, interrupt the rise of soil water toward the surface and reduces its loss by evaporation. At the same time, it creates suitable conditions for weed germination. The emerging small weed plants can then be destroyed by subsequent medium or deep plowing in autumn. Plowing also brings deeper layers of the soil profile to the surface that helps to mineralize organic matter. This, on the one hand, increases the availability of nutrients, and on the other hand, it leads to a decrease in the content of soil organic matter. Probably, the best example of this decrease is the decrease in the content of organic matter in the soils of long-grass prairies after the start of tillage in this pristine ecosystem (Fig. 2.28).

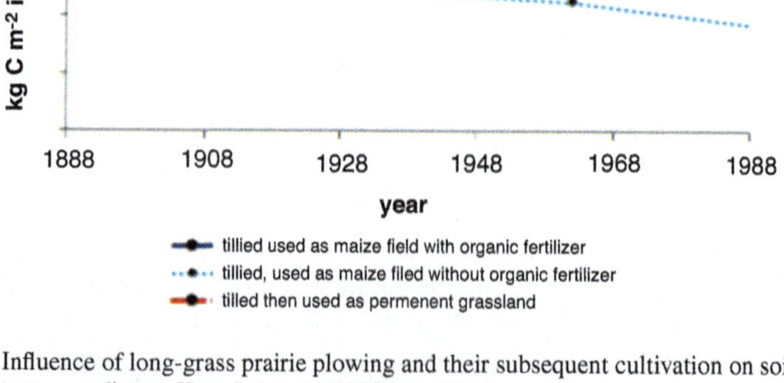

Fig. 2.28 Influence of long-grass prairie plowing and their subsequent cultivation on soil organic matter content according to Huggins et al. (1998), modified

After plowing, which takes place in the autumn, furrows remain in the field. These disintegrate for the most part during the winter, and in the spring, the soil is prepared before sowing. It can consist of several consecutive operations such as shearing, loosening, dragging, rolling, and sometimes these activities can be combined. These include the task of leveling the soil surface, loosening the top layer of soil, and compacting or leaving the compacted bottom layer of the soil. The goal is to create a good bed for sowing. The seed should lie on a low layer of soil, where the soil capillaries bring soil moisture to the seed. A loose layer of soil should lie above the seed, which makes it easier for the germinating plant to penetrate the soil and, at the same time, prevents water loss by breaking the capillary action (Fig. 2.29).

For some crops, especially the so-called root crops, we cultivate field not only before planting but also during the growing season. The aim is to remove weeds and, in some cases, especially potatoes, to add soil to plants. The addition supports the formation of stolons and tubers.

In arable fields, we usually grow annual crops, with a few exceptions, consequently sowing or planting is an important part of agricultural technology. We sow seeds in autumn (winter wheat, oilseed rape, rye, and winter barley) or in spring (spring barley, oats, spring wheat, corn, and beets). As already mentioned, the synchronization of emergence in a given population, is one of the distinctive features of domesticated plants. However, to further increase seed homogeneity, the seed can be sized. Alternatively, it can be coated, which gives the seeds a uniform shape and size and allows them to be sown with precision seed drills. The chances of seeds in

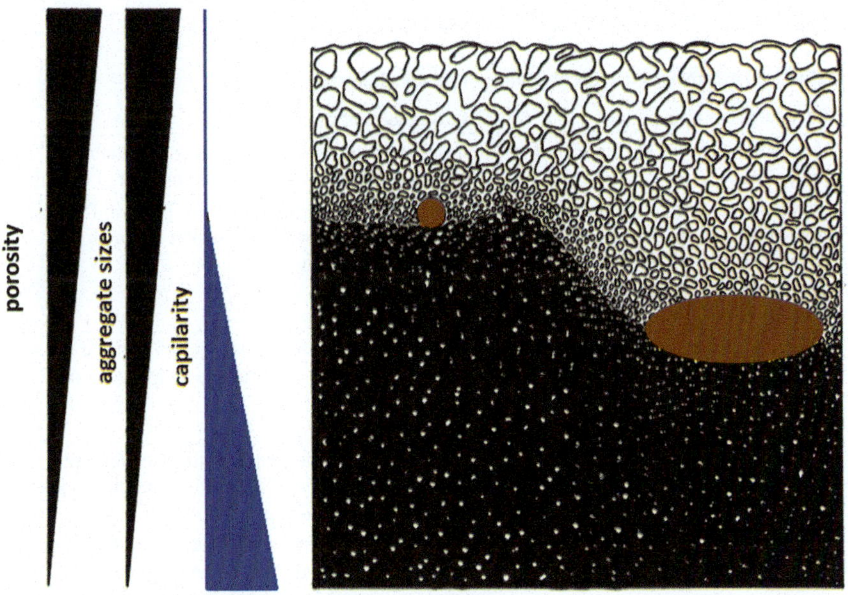

Fig. 2.29 Diagram of the optimal placement of the seed in the soil so that there is a non-capillary fluffy layer above it and a capillary water supply layer below it. A change in other soil parameters and the fact that the seed bed may be deeper may be indicated for larger seeds

successful emergence are increased by picking the seeds and or providing them with fertilizers or various inoculants.

Some plants can be sown under other plants, which is a common method of sowing clover, alfalfa or clover grass mixtures. For these plants, which are usually sown for forage production, their development is slow in the first year and therefore, we usually sow them in the form of the so-called undersowing. This means that we sow them together with some cover crop (for clover it can be oats or spring barley). After the cover crop is harvested, the already rooted undersown clover gets the opportunity to grow subsequently, it is grown on forage on the same plot for another 1 or 2 years. This means the field is not ploughed for 1 or 2 years.

On arable land, monocultures are usually grown, which are also grown as one cohort (a set of plants of the same age). There are exceptions such as the cultivation of clover grass mixture in the undersowing, as described above. However, although many attempts have been made in this direction, the simultaneous cultivation of mixtures of species at the same time in the same field is rather an exception in modern intensive agriculture (see Box 2.7), and is mainly used for fodder crops. Examples of this are mixtures of fodder legumes, such as fodder peas (peas) and cereals (e.g., oats), which are harvested before they are fully ripe and used for direct animal consumption or conserved (e.g., as silage).

Polycultures are used more in long-term perennial cultures (see Boxes 2.8 and 2.9 for more examples of comparison of annual and perennials). The reason for this may be the complexity of the current harvest of several species that would be harvested for seed, and perhaps the fact that the benefits of species-rich polycultures (Table 2.6) will be felt more in older stands (see Sect. 1.5.2 and Box 2.9).

Table 2.6 Advantages and disadvantage of monocultures

Advantages of monocultures	Disadvantages of monocultures
Monocultures avoid interspecific competition	Plants use resources such as nutrient and water in synchronized way which increased in peak of growth
Harvesting is easier, plant ripen in more synchronized way, it is easier to separate final product from other residues. This also made postprocessing easier	The use of soil may lead to negative plant soil feedback when the same crop will grow poorer when grown repeatedly on the same field
It is easier to optimize condition for plants and synchronize them with plant phenology using fertilization and other measures	Maintenance of monocultures require intensive modification of plant environment and weed control
Plan protection focus on one species which limits the number of potential pest species	Monocultures are more sensitive to pests, require intensive plant protection which cost energy and may affect nontarget species
	Biological diversity on level of individual fields or whole landscape is reduced this is even enhanced by fact that the most widespread crops belong to few families only, with grass Poaceae being the most common (Brassicaceae are another family in which many crops belong)

After sowing or planting, the vegetation is treated. This consists mainly of nitrogen fertilization and protection against diseases and pests, as described in detail for wheat in Box 2.8. However, the exact procedure varies from crop to crop. This is followed by harvesting, the process of which is again largely species-specific. To summarize the above, soil cultivation is one of the basic elements of agricultural technology on arable land. Cultivation has a number of meanings for adjusting plant's living conditions: it helps to remove weeds, helps manage water, loosens the soil and supports its aeration. Better aeration and decomposition of soil aggregates lead to the mineralization of organic matter and release of nutrients. On the other hand, cultivation also has its negatives: it leads to the formation of areas that are free of vegetation for a long time, which promotes erosion and changes the energy balance (e.g., reduces albedo, which leads to overheating of the soil), leads to soil disturbance and loss of organic matter, which further increases the susceptibility of the soil to erosion. Reducing the organic matter content also leads to other negative consequences, such as less water and nutrient retention in the soil. We try to mitigate these negative effects of cultivation in several ways. The first and most common of these, is the correct rotation of the crops. Another option is the application of no-till cultivation technologies. Finally, efforts can be made to breed new cereal crops that could be grown as perennials.

Crop rotation is a very old technique in which crops are rotated in different years to minimize the impact of unilateral land use and reduce the likelihood of diseases and pests. Ancient farmers used the so-called shifting agriculture, which is still practiced by some tribes in the rainforests of the Amazon and Papua New Guinea. In this method of cultivation, the farmers grub up part of the forest, burn most of the existing vegetation, which brings a considerable amount of nutrients and then establish a garden in which crops are planted (bananas, sweet potatoes, taro, etc.). After one or a few seasons, when the field loses fertility, it is abandoned and covered by forest again. In some areas where the forest has been completely removed, the abandoned fields are overgrown with grass only. This yield in alternating between fallow and field. This is a method of farming that is close to the three-field rotation that was common in Europe in the Middle Ages. It was practiced by dividing the land into three parts. One was sown in spring, the other in winter, and the other remained fallow. The next year, the fallow field was replaced by winter crops, winter crop by spring crops and the area overgrown by spring crops were left as fallow. The so-called Norfolk crop rotation system was developed from this three-field system. It consists of rotation of four crops: clover, winter cereal, root crops, and spring cereal which at the same time served as cover crop for clover seeded together with spring crop. The next year, there is clover again which closes the loop (Fig. 2.30). Norfolk crop rotation became the basis for modern crop rotation schemes. Norfolk crop rotation was later expanded by adding more crops, which made rotation longer. The general principle is that crops of different characteristics should alternate: shallow with deep-rooted, narrow-leaved with broad-leaved, less and more moisture-intensive, enriching the soil with N and N-intensive crops, with different demands on individual nutrients, with a different spectrum of pests and diseases that are well tolerated by pre-crops, etc. The practical context must also be considered, such as time needed to prepare land for the next crop

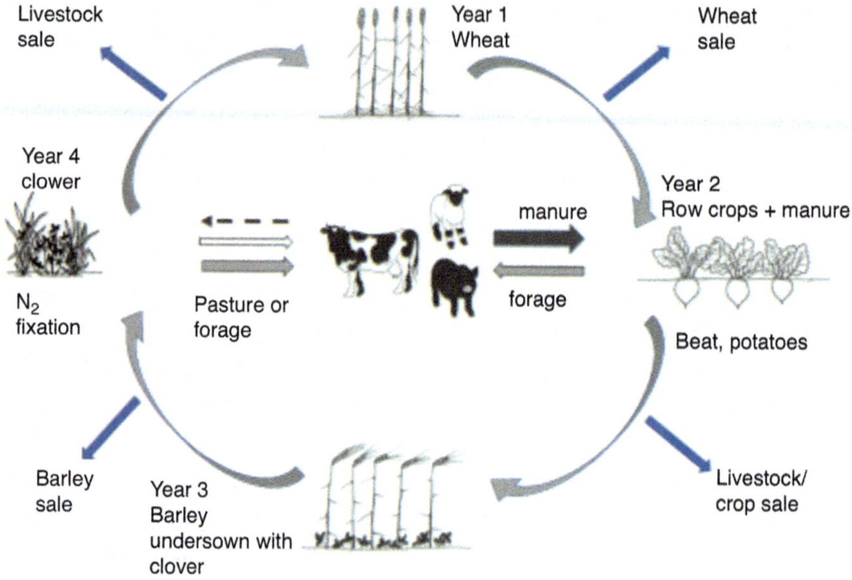

Fig. 2.30 Scheme of the Norfolk crop rotation system

after the pre-crop has been harvested. Therefore, crops with longer and shorter growing seasons and crops with different demands on soil preparation before sowing and planting alternate. An economic aspect must also be taken into account. Cultivated plants should bring a reasonable profit and the least possible business risk. Finally, crop selection needs to be adapted to local soil and climatic conditions. Crop rotation can help to combat soil degradation, to protect soil, one may use more often plants that improve soil condition, such as alfalfa, clover, or grass clover mixture. However, there may not always be economically sensible use for such plants. Another possibility is to include some intercropping options such as undersowing the main crop, helping to protect soil and using a crop in between major crops to improve soil quality, the so-called green manure. They provide organic matter and also cover soil in the time where soil will normally be without vegetation.

Another approach is to minimize tillage or the use of no-till technologies. No-till technologies abandon soil cultivation. Sowing is done by special sowing machines which can sow seeds into uncultivated soil. The reduction of water losses after harvest, which is achieved by the stubble plowing in classical cultivation, is ensured by a layer of postharvest residues left on soil surface in the case of no-till technologies. The control of weeds provided by cultivation is replaced by the application of herbicides between harvest and the emergence of another crop. This leads to an increase in the application of herbicides, especially glucosinolates. Due to the elimination of plowing, which is one of the most energy-intensive operation, no-till technologies lead to a significant energy savings.

Less disturbance of the soil surface and non-disturbance of soil aggregates reduces the possibilities of erosion and leads to greater retention of C and water in the soil. Higher soil moisture can result in lower soil temperatures, especially in

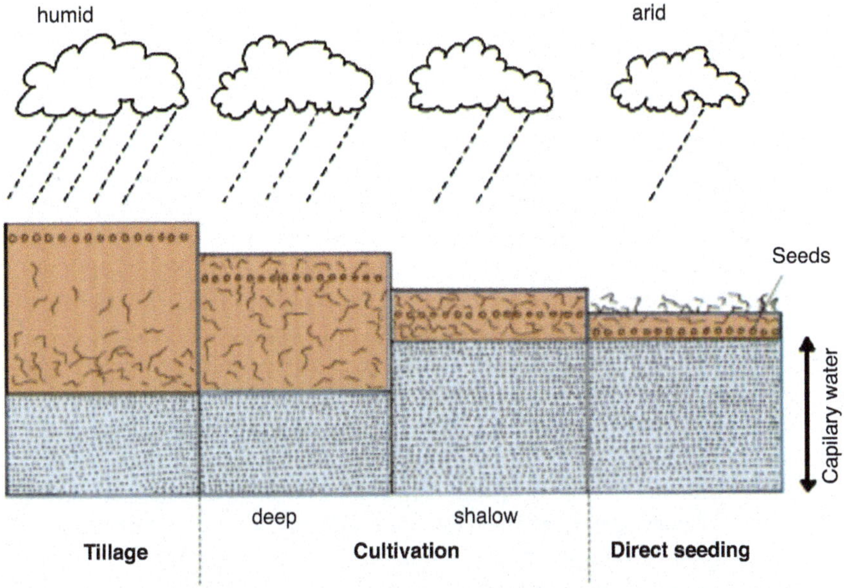

Fig. 2.31 Optimal tillage depending on the climate from humid to arid

spring. A lower soil temperature in the spring can then lead to a shortening of the growing season. Non-disturbance of the soil and the supply of postharvest residues to the soil surface bring suitable conditions for the development of soil biota. Restoration of soil biota and its functions takes some time. Laboratory experiments have shown that the positive effect of soil biota on C and N retention in the soil will only become apparent after about 1–2 years of application of no-till technologies (Toyota et al., 2013; Frouz et al., 2014).

Other shortcomings of no-till technologies are that no-till and till farming requires different mechanization, and it is a considerable investment for a farm to run both systems in parallel. Switching from one to the other is also costly and the yield is sometimes higher, sometimes lower. In colder conditions, a lower yield is common, probably due to higher moisture and the resulting shortening of the growing season. Conversely, in subtropics, where temperature is not an issue, greater water retention in the soil is a benefit of no-till technologies leading to higher yields. In addition to completely no-till technologies, there are various minimization technologies where either shallow cultivation or cultivation every few years. This is also related to the recommended rate of cultivation depending on climatic conditions (Fig. 2.31). See also Box 2.10 for major difference between temperate and tropical agriculture.

Finally, the last, so far rather theoretical way to improve and maintain soil quality, while achieving satisfactory agricultural production is to grow completely new crops such as perennial grasses bred to produce grain. Probably, the most successful attempt in this direction is kernza. Unfortunately, so far, the yields of this perennial cereal are relatively low to be competitive with conventional cereals. However, intensive research is underway in this area, which may bring significant improvements in the future (Box 2.9).

Box 2.7 Monocultures and Polycultures

In general, growing polycultures or intercropping is not really common in intensive agriculture focusing on large areas of high productive monocultures, which are much easier to cultivate, fertilize, harvest, etc., using heavy machinery. However, intercropping is common in subsistence agriculture with low use of machinery, low-inputs (high labor), and low-yield agriculture on small parcels. In such a situation, intercropping often obtains more yield per unit of land. Polycultures are also common in forage clops (Aponte et al., 2019). There are various ways to grow more than one crop in the same field during the same time (Fig. 2.32), sometimes called intercropping in broader sense (Brooker et al., 2015). These may include complete mixtures, as well as situations where small patch strips or rows of various crops are grown in close proximity. Some crops such as clover or alfalfa get established under cover crops such as oat or barely. Such undercropping can be used in some crops such as maize may bring additional environmental benefit such as reduction of erosion risk. Another way of getting two crops on the same field is to grow different crops in alternating rows or strips. Also, agroforestry represent other forms of intercropping (Fig. 2.32).

The reason why intercropping may bring better yield is that plants not only compete but also facilitate. Facilitation is an interaction where individual species change their environment in a way that is beneficial for them, but can be beneficial for neighboring plants. Neighboring plants may provide various benefits „exchange in benefits" that improve conditions for individual plants even

Fig. 2.32 Monocultures and polycultures. Freshly planted monoculture of potato in the Netherland (**a**), grooving lettuce in between rice paddy fields in Nepal (**b**), and polyculture in Papua New Guinea (**c**)

(continued)

Box 2.7 (continued)
further away. A classic example of polyculture widely distributed in subsistence farming in South America is the combination of maize, beans, and squash (Brooker et al., 2015). Squash provides cover in the early season reducing erosion, competing weeds and reduces transpiration. Beans and maize do the same later in the season and fix N, while maize may provide physical support for climbing bean varieties. Similarly, growing of grasses or cereals with legumes that fix N, and in some cases, it may grow more efficiently in combination with grasses than growing alone (Brooker et al., 2015). Another example is deep-rooting plants in dry areas: allowing better infiltration of water. These plants may even provide hydraulic lift of water which may spill in the surrounding soil and is utilized by the neighboring plants (Bayala & Prieto, 2020).

There has been extensive research about the yield of mixtures in comparison with single stands. A comprehensive review by Trenbath (1974) summarized an experiment with biomass productivity of 344 mixtures of grasses, legumes, and cereals, 39.8% were below, and 60.2% were above of mean of biomass of both components suggesting a slight positive effect of mixtures. Similar results were later confirmed by Jolliffe (1997) who used different calculation methods concluded that two species mixtures of various meadow plants are on average 12% or 13% more productive than monocultures. However, as mentioned above, positive and negative effects are common, and the outcome depends on specific mixture. Mixtures are however, widely used in forage production and may support a larger stability of yield than monocultures. Practitioners often use more species to establish forage mixture that should last longer. This finding is supported by the results of Jing et al. (2017) who showed that mixture complexity increased persistence and productivity in the long term. However, this result holds only in cut mixtures not a grazed one. However, this relates to growing of monocultures or mixtures of mostly forage plants which are harvested for total aboveground biomass (Aponte et al., 2019). Mixing plants grown for grain is less common in modern agriculture on arable land mainly for technological reason. Grain of different crops are often used for different purposes and using mixtures narrows its use to much smaller range of uses, preferably as animal feed. However, in grain production, many experiments were done with growing of mixture of various cultivar of the same crop. Also, here the effects vary and both positive and negative effects occur. In some crops, extensive testing of cultivar mixture do not show any substantial effects (Shorter & Frey, 1979), while in other, it may bring higher yield and higher yield stability (Hoersrna et al., 1985), namely, under stress conditions. Comprehensive review of Kiær et al. (2009) shows that for wheat and barley that relative mixing effect vary substantially from being negative (−30%) to positive (100%) with overall mean effect being slightly (2.7%) but significantly positive. Mixing effect increases as the mixed varieties differ in yield, disease resistance, and weed suppressiveness. Some

(continued)

Box 2.7 (continued)

studies also show that using cultivar mixtures may also support decomposer activity in soil and thus support better nutrient cycling (Chassé et al., 2019). It is noticeable from the examples described above that positive effect of agriculture occurs mainly when plant facilitation helps in acquisition of soil resources such as water or nutrient. This is consistent with the conclusion made in Sect. 1.4.2, that for plants, it is easier to coexist and partition resources when nutrients are limiting, than when nutrients are readily available and plants compete mainly for light. This is also consistent with the observation made by Brooker et al. (2015) that polycultures are mainly suitable for low-input agriculture.

Box 2.8 Wheat

We will demonstrate the main principles of cultivation of annual crops (monocultures and cohort) on the example of wheat. Many botanical species of wheat is grown but *Triticum aestivum* is the most common. Wheat yield depends on the number of spikes per unit area, the number of grains in the spike and the weight of one grain. Cultivation tries to optimize these parameters to achieve optimal number of spikes, grain, and grain weight. The number of spikes depends on the number of plants per unit area and the number of shoots that each plant produces. The first step that will affect future yield is the soil preparation and fertilization. As we mentioned, when fertilizing, we incorporate potassium and phosphorus fertilizers into the soil before sowing. We either do not fertilize with nitrogen in the autumn, or only a small part of the total dose. This is because the plants will grow only little by winter, and moreover, the natural decomposition of organic matter and nitrogen mineralization provide enough nitrogen for plant in autumn. The dose of fertilizer we use depends on the type of soil, its nutrient supply, the type of climate, and the yield we aim for. It can be imagined that for each climatic region and type of soil, or variety, there is a separate yield curve. In practice, however, farmers use tables that are organized the other way around. This means that depending on the natural conditions and the yield to be achieved, they show the required fertilization rate. Wheat is sown in the autumn, so its pre-crop must enable quality soil preparation so that at the depth where we sow the seed (Table 2.6), the soil is slender with capillarity above it. In central Europa, typical seeding density is 300–600 seeds per m^2. The number of seeds is the first step in influencing the future number of stalks. After overwintering, farmer check the vegetation and fertilize with nitrogen accordingly. If the vegetation is too sparse, farmers will try to support formation of stolons and new vegetatively formed plants (secondary tillers) this can be supported by nitrogen, or by mechanical stimuli, for example, by rolling. There are also phytohormone-based chemicals that

(continued)

Box 2.8 (continued)

support this vegetative reproduction. After adjusting the number of spike-bearing plants, another key period is when the plants begin to form stalks. At this point, the individual flowers in the spike begin to differentiate, and this helps determine the future number of seeds in the spike. Colder weather slows down this process. When the process is slower, more flowers are formed, that is, more grains are produced in the spike. Nitrogen fertilization during the peeling phase can also support a larger number of flowers, and thus grains in the spike. After flowering, the assimilates accumulate in the grain and this determines the weight of the individual grains. The vast majority of assimilates that accumulate in the grains are produced by the last leaf. Therefore, the breeding of wheat was largely concerned on the selection of plants with the largest possible area and the best possible geometry of the last leaf. At this stage, agrotechniques will also focus on making the last leaf work as long and as good as possible. This can be achieved by nitrogen fertilization, and protection of the last leaf against pests and fungal diseases. Subsequently, we monitor the maturity of the vegetation so that we can harvest at the appropriate time (Fig. 2.33).

With regard to the example of wheat, not only the total amount of N we apply is important but also the exact timing when we do that is important to

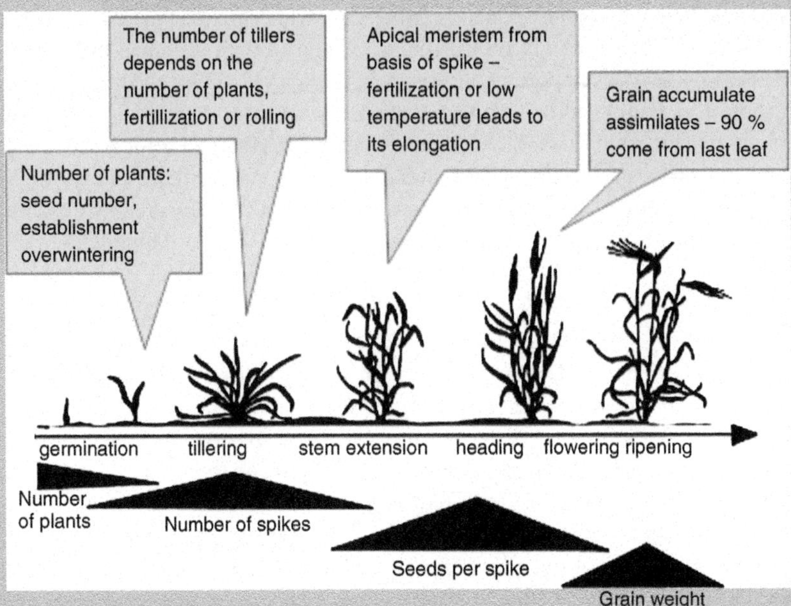

Fig. 2.33 Yield formation in wheat

(continued)

Box 2.8 (continued)

tune plant growth in a way that maximizes the yield. The same idea is also applicable to other operations such as pesticide application and also harvest which should happen in optimal time when grains are mature but not start to fall from the spike (which would increase harvest loses). In order to achieve this exact timing, farmers need that all plants in the field behave in the same way and in given time they are in the same phenological stage. Secondly, they need machinery that can perform the given operation in whole field and all fields where similar crop is grown during short period of time. To achieve that all plants will be in the same stage, farmers grow cohorts of monoculture of one species and usually one cultivar. We have already mentioned that cultural plants which propagate by seeds are selected to start germinating and emerge all at the same time (in contrary to wild plants). Farmers also prepare conditions in the field to achieve homogenous conditions that would allow very synchronized growth of all plants in whole field. We have already mentioned in chapter about mechanization that efficient use of machinery benefit from bigger field. At the same time, as we see later, the use of more powerful and heavy machinery has negative environmental effects. Also, machinery use forms more pressure on the synchronization of plant growth and development. This is particularly important as concerning harvest. If farmers harvest manually, it is possible to handpick only plants or fruits, etc., which are ready and leave other ones to mature but this is very difficult to achieve with available machinery. This also illustrates important difference between polyculture and monoculture (Table 2.6). In polyculture, the fact, that phenology and growth of an individual plant are shifted in time allows plants to share the available resource, namely, soil resources in low-input agriculture. However, monoculture allows high-input agriculture to maximize the resource (mainly nutrient) availability exactly in the stage of plant growth when plants need them to achieve high yield.

Box 2.9 Kernza and Other Attempts for Cereal-Like Perennial Crops

Virtually, all the cereals that play key role in mankind's diet are annual crops. There are a number of reasons for this fact. Annual crops use most of their energy to produce seeds, consequently, large part of the plant's biomass is concentrated in seeds. The time required to obtain the harvest is short when compared with the perennial plants. In contrast, perennial crops invest in long-term survival in one place. They build a powerful root system, invest to support of mycorrhizal fungi and other symbionts in the soil, fight pests and pathogens. All these factors cost energy (assimilates), which then cannot be used for seed formation. On the other hand, all these lead to perennial plants changing their environment so that they grow in long term just as well as in the beginning or even better. In contrast, most annual plants do not try to invest so much in their environment, so if we plant them twice in the same

(continued)

Box 2.9 (continued)

soil, the second generation usually grows worse than the first (phenomena called negative plant soil feedback). That is why in natural conditions, annual plants usually grow only on disturbed places. In arable land, we disturb the soil by regular soil cultivation. This increases the risk of erosion and loss of organic matter. We improve the condition by fertilization and other means which all cost energy and bring some negative environmental effect.

All these have led to the efforts of some experts to grow perennial cereals. This effort was most successful in kernza, which is a domesticated form of grass *Thinopyrum intermedium* which was bred by the Land Institute in Kansas, which seeks to breed perennial forms in other crops. Comparing kernza with wheat, it is evident that kernza has a much lower grain yield—about 150 kg ha^{-1} compared with 3 t ha^{-1} wheat grown in similar conditions. However, as a perennial crop, Kernza has a much larger root biomass. It protects the soil better against erosion and shows far less nitrate leaching compared with wheat (Culman et al., 2013), (Fig. 2.34). Kernza's low yield and

Fig. 2.34 Comparison of aboveground and belowground organs of wheat (**a**) and kernza (**b**)

(continued)

Box 2.9 (continued)

the fact that we use most of the cereals we produce for feed purposes leads some researchers to believe that instead of forcing perennial grasses to invest more in their seeds, we could harvest their entire biomass (as we do now for forage) and use it to isolate proteins and other components. The protein thus isolated could then be fed to pigs and other livestock, which do not normally feed on grass and similar forage crops and possibly directly to humans (Nørgaard et al., 2018, Santamaria-Fernandez et al., 2017). This would make it possible to replace a significant proportion of cereal crops with perennial grasses, clover or clover grass mixtures, which would significantly improve the provision of ecosystem services (erosion reduction, and nutrient loss, etc.) while providing a very efficient source of animal feed.

Box 2.10 Agricultural Production and Intensification in the Tropics

We already mentioned that in contrary to natural ecosystem, which are believed to have higher productivity in tropics, temperate agriculture reaches higher yield than tropical ones for most common crops (Haws et al., 1983) (see Table 2.7).

There are several reasons for this. Most of the tropical soils are old, moreover, in wet tropics, due to high rainfall and temperature, the soils are highly weathered and most of the nutrients are leached. This is particularly a question of P, but also basic cations and micronutrients. Soils are often rich in Fe, Al, and Mn, which may cause binding of P in unavailable forms. Moreover, dominant clay minerals have lower sorption. Soils in the dry tropics are affected by salinization and drought (Barrow, 1987). Temperate soils are generally more

Table 2.7 Comparison of the average yield of selected crops in tropical and temperate areas based on Haws et al. (1983)

Crop	Temperate t ha^{-1}	Tropical t ha^{-1}	%
Rice	4.1	1.9	48
Wheat	2.9	1.4	46
Maize	4.0	1.3	34
Sorghum	2.2	1.2	55
Soybean	1.6	1.0	64
Beans	1.0	0.6	59
Peanuts	1.7	1.0	62
Potato	18.1	8.7	48
Sweet potato	13.6	6.8	51
Cassava	11.8	9.1	77
Sugarcane	61.2	53.3	87

(continued)

Box 2.10 (continued)

favorable for agriculture although there are exceptions from this rule on both sides. Solar radiation is generally higher in the tropics, but during the vegetation season, the day is longer in a temperate climate. This causes a larger input of solar radiation during the vegetation season, also some crops that require longer days cannot be grown in the tropics. Additionally, solar radiation in tropics can be limited during the rainy season. A large number of crop pests, weeds, and diseases exist in the tropics which reduce the yield even further (Rosenzweig & Liverman, 1992). High temperatures also promote higher plant respiration which may cause loss of energy, in particular, high night temperatures may be important in this context. Considering all these, crop yield in tropics is about half of the yield common in temperate climate (Table 2.7). Besides these natural conditions, there are historical logistic economical and other reason which may reduce the yield gained even more.

When we displace natural vegetation by agricultural crops, we gain less yield in tropics than in temperate regions as was already explained. However, natural tropical ecosystems in average contain more C than temperate ones, which in fact mean that per unit of yield gained, we lose much more C in tropics than in temperate region (West et al., 2010). As mentioned earlier, the area suitable for crop production in tropics is relatively lower than that in temperate climate (arable land cover 20.4% temperate zone and 10.5% tropics). In tropics, relatively larger part of agricultural landscape is used as pastures. However, also the productivity of tropical pastures is lower than that of the temperate ones. Namely, the growth rate and milk production from cattle grazing solely on unsupplemented tropical pastures are usually lower than that found when grazing temperate pastures due to poorer feeding value of the tropical pastures (Minson, 1980). Besides deficiency in some micronutrients, such as Na and Zn, forage in tropics has lower content of water-soluble carbohydrates (which are instant source of energy) at 50–90 g kg^{-1} DM compared with the temperate region which is at 60–500 g kg^{-1} DM. Among others, this can be caused by higher temperature in tropics which trigger higher respiration rates and consequent loss of carbohydrates (Lowe et al., 2016).

As we see above, agriculture production is higher in temperate climate than in tropics, which partly corresponds with more suitable environmental, mainly soil condition, but also due to agriculture intensification. Most principles of agriculture intensification, such as fertilizer use, breeding varieties that dedicate larger part of their biomass in yield, development of heavy machinery that may help to replace manual work come from temperate climate where most of the developed countries are located. Application of these intensification measures in tropics is limited. These limitations are twofold: socioeconomic and environmental. Socioeconomic barriers result from less developed infrastructure, many farmers do not have enough capital for investment in more advanced technologies, which again have many reasons. Environmental barriers are among others in soil as already mentioned. For example, soils in tropics are

(continued)

Box 2.10 (continued)
often weathered, most of the nutrients are in the form of organic matter, and cultivation may cause fast decomposition of organic matter and nutrient loss. Fertilization in tropics is a more complex issue, as besides N, other elements, namely, P and micronutrients must be supplied. Many tropical soils fix P in unavailable form, which cause substantial loss of P added as fertilizer (Roy et al., 2016). All these made these agricultural practices more challenging in tropics. Many people start to believe that rather than trying to modify environment for plant needs, more suitable strategy particularly in tropics is to select and breed plants suitable for the environment.

Tropics are also likely to be more vulnerable to climate change. This is again due to both natural and socioeconomic constraints. Socioeconomic factors include larger dependency of economy on agriculture, higher level of poverty and lower level of infrastructure. As natural constraints are a concern, temperate C3 crops are likely to increase the production with increasing levels of CO_2 more than the tropical C4 plants. Tropical crops are closer to their high temperature limit of functioning and may thus be more affected by rising temperature, even if warming would be less pronounced in tropics. Moreover, pests and diseases, already abundant in wet tropics, may even become a more serious problem (Sanchez & Buol, 1975).

2.2.4 Permanent Grassland

Permanent grasslands resemble the middle stages of succession. In a substantial part of temperate climates, which would be naturally overgrown by forest, natural grasslands are rare. These are ecosystems above the tree line, in avalanche tracks, in waterlogged places or on shallow soils on rocky outcrops. Most other meadows or pastures are created and maintained by human activity, especially grazing or mowing. In the case of natural grassland ecosystems, their existence has mostly been maintained by occasional drought and the associated occurrence of fires and game grazing. Vinton et al. (1993) described the interactions of bison and fire in maintaining long-grass prairies. After a fire, the burned area is supplied with a pulse of nutrients. This leads to the rapid growth of grasses and herbs, which are preferred by bison. Bison intensively graze fast-growing species of grasses and herbs yet ignore species that leave large amounts of standing dead biomass. This leads to the rapid predominance of species leaving large amounts of dead biomass on bison-grazed areas.

Accumulation of dead biomass is a good fuel and will accelerate the occurrence of fires on the affected areas. However, the distribution of some natural grassland ecosystems is largely due to human activity. For example, it is believed that the extent of long-grass prairies in North America was significantly shifted eastward in the area that would be naturally occupied by deciduous forests far before European arrival. This was due to Native Americans who burned forests to increase the extent of the long-grass prairies and therefore bison and other large herbivores came into existence, which they subsequently hunted.

As we have already said, cultural meadows or pastures are maintained by grazing or mowing of vegetation. Mowing and grazing removes aboveground biomass and thus damages woody plants more than grasses and herbs. Woody plants grow slowly and must face the competition of grasses and herbs which grow faster. The advantage of woody plants is that their annual growth will be maintained until the next season, while grasses and herbs always start from the scratch. As a result, woody plants and herbs can overgrow through time. However, the periodic removal of aboveground biomass negates this advantage of woody plants and can thus ensure that meadows or pastures are not overgrown with woody vegetation. Mowing and grazing also suppress tall species and, conversely, may be beneficial for smaller species, in particular, to those species whose photosynthetic apparatus is located close to the ground (e.g., species with a ground leaf rosette). At the same time, grasses typically regrow better after grazing or moving.

When the top of the herb's is grazed, apical meristem is removed, and the herb has to regrow from the buds in lower ungrazed part of the plant. Grasses possess a so-called intercalary meristem, which is a meristem inserted between mature tissues. When the top part of the grass is removed, the grass can easily regrow from the intercalary meristem. This made grasses in general, better adapted for grazing or moving. However, there are other factors that may play some role. After mowing or grazing, the plant grows from the reserves which are stored in the lower part of the stem in the hypocotyl or in the roots. If we have a grass next to each other that will have storage at the bottom of the stem and in the hypocotyl, and a herb with supplies at the roots, mowing that removes the stems and part of the hypocotyl can support the herbs to regrow over the grass (Fig. 2.35).

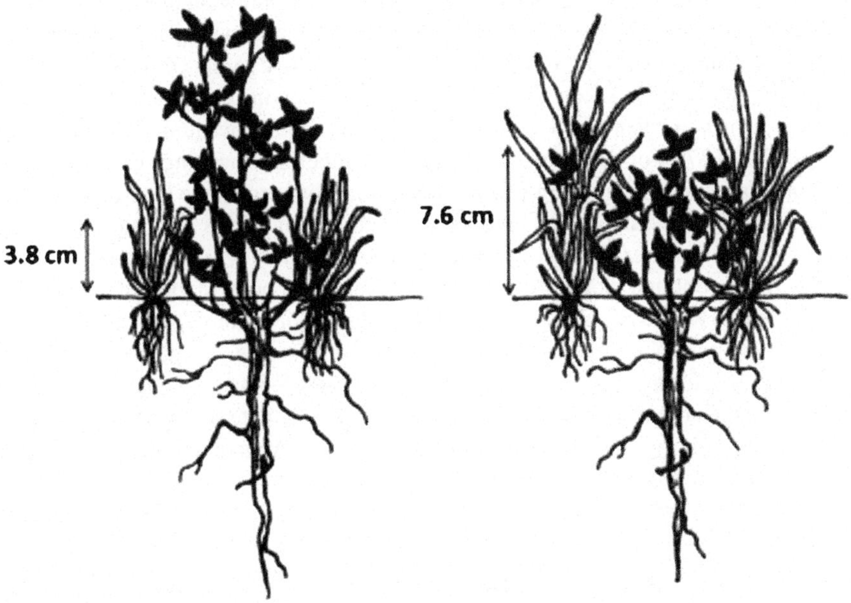

Fig. 2.35 Mowing low in a way that grass hypocotyl with reserves is removed, legumes may respond better as they are having reserves in root (left) and higher stalk after mowing supports grasses (right)

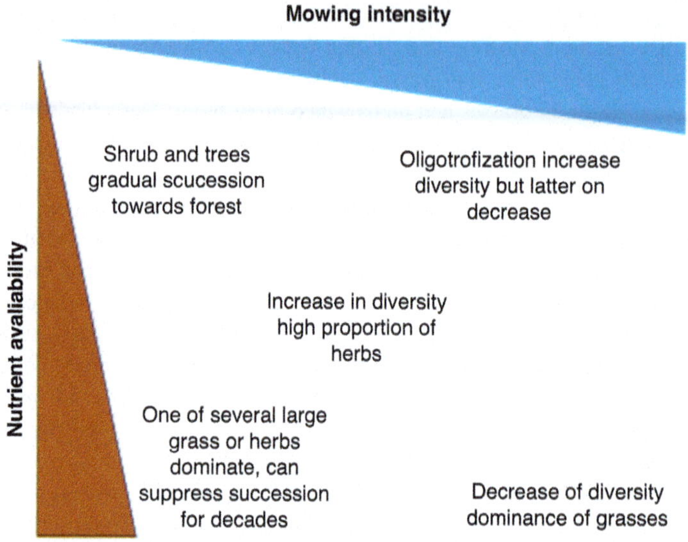

Fig. 2.36 Schematic representation of the effect of mowing and the amount of nutrients on the diversity and composition of the plant community

However, mowing and grazing change the environmental conditions in the long run. The grass that is mowed or grazed will produce more secondary tillers and the vegetation will thicken, especially if it has enough nutrients. The effect of mowing on diversity is complex and depends on the intensity of mowing and the amount of available nutrients (Fig. 2.36). On the other hand, less mowing leads to nutrient removal, which in the long run can also lead to a change in the composition of the plant community. The greatest nutrient removal by mowing and biomass removal is usually achieved when mowing happens just at the time of heading of the dominant grass (Fig. 2.37). For herbs, mowing happens just before flowering. This is also the optimal date for mowing in agricultural practice, as it ensures the maximum supply of nutrients for livestock. For production, fertilized meadows are usually mowed twice a year, however, for non-fertilized meadows, one mow is sufficient.

As for the optimal mowing time from a conservation point of view, it depends on the organism whose population we want to support. For example, for annual plants, the earliest possible mowing time is optimal. On the other hand, late mowing is advantageous for meadow orchid and to support insects that develop in flowers and plant seeds. Consequently, there is no one most suitable date for mowing to support a wide range of organisms, and the situation is optimal when a mosaic of differently mowed patches co-occur in the landscape.

There are also differences in the effect of grazing and mowing. Mowing causes larger removal of nutrients than grazing, where nutrients return in the form of feces. Also grazing typically produces larger spatial heterogeneity than mowing (Fig. 2.38) (more about differences between various species in Sect. 2.4.3.1).

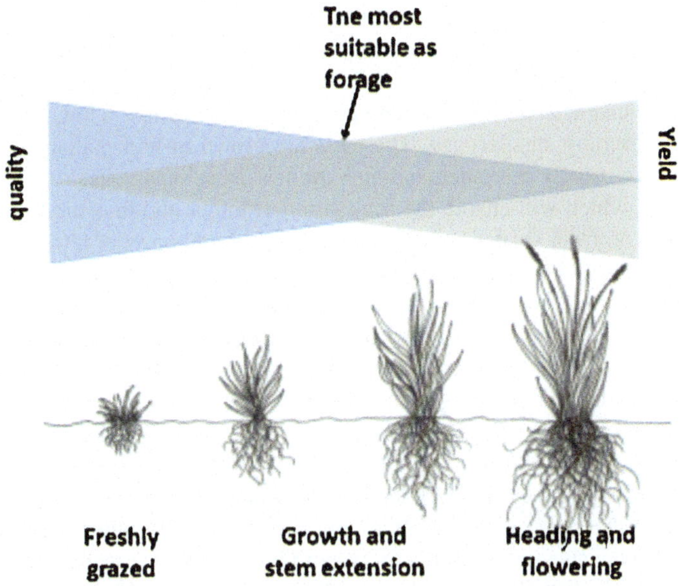

Fig. 2.37 Schematic representation of the relationship between grass yield and forage quality given the nitrogen content

Fig. 2.38 Comparison of mowed meadow (**a**) and grazed pasture (**b**) in Netherlands. Note higher spatial heterogeneity of vegetation in the pasture

2.2.5 Permanent Crops and Agroforestry

By permanent crops, we mean mainly orchards, vineyards, hop gardens and similar crops, which are based on semishrubs, shrubs, trees or vines, grown for a long time in one place. Permanent crops are a cultural analogue of the later stages of succession and generally, like meadows, have less environmental impact than field crops, especially when they are grassed under trees. All these crops can be grown more or less in an intensive manner. We will illustrate the differences between intensive and extensive cultivation taking an example of apple orchards. Intensively cultivated orchards are characterized by the cultivation of trees growing on low growing rootstock and reaching smaller size. The trees are grown in higher density (2000 ha trees) and are usually more demanding varieties. Trees on smaller rootstocks have fewer fruits, which will enable the acquisition of larger and thus more marketable fruits. Trees of lower shapes are characterized by a rapid onset of fertility (can bear fruit as early as 2–3 years), but they also have a short lifespan (10–20 years). They are usually grassed, but in the past, black fallow was often maintained in the rows between trees by cultivation or by using herbicides. When regenerating intensive orchards, a significant dose of nutrients is applied during renewal, both before planting and during the growth and production of orchards.

Extensive orchards tend to be on higher tree shapes, which take a longer time to achieve fertility (5–10 years) and this results in a longer lifespan (often inherited from generation to generation). Older, less-demanding varieties with a small to zero supply of fertilizers and other inputs are often grown with a smaller and often irregular yield. Extensive old orchards are important habitats for many species, especially insects, but also for other biota. Alaphilippe et al. (2016) showed that the overall environmental impact of orchards is mainly due to the way they are managed. Mainly, this is due to the use of N fertilizers and subsequently the application of pesticides. Erosion can also contribute to the environmental impact when black fallow is maintained.

In the past, extensive orchards were often used either for grazing or for the cultivation of field crops in the rows between trees. Therefore, they represented a form of agroforestry systems. Agroforestry has many forms, the most abundant in Europe being summarized by Mosquera-Losada et al. (2012). They include systems abundant in the Mediterranean today, where the cultivation of trees, whether forest or fruit, is combined with animal grazing (Silvopasture). It is a diverse group, which includes the above-mentioned grazed orchards on the one hand and grazed forests on the other. Another form of agroforestry is fruit or other trees, individually or systematically (e.g., in rows), planted in an agricultural landscape, often at the edges of fields or meadows. These stands served various purposes. They provided fruit, firewood, feed, and also protection against erosion. Strips of trees planted around watercourses or water sources also protect these water resources.

Another group of agroforestry crops are the planting of fast-growing, often nitrogen-fixing, trees, which in rotation improve soil quality and provide wood. Finally, there are systems for the simultaneous cultivation of trees and other crops. These can again be very variable, from forest areas used for the production or harvesting of naturally occurring crops for medical, ornamental or culinary use, to systems with widely planted trees, in between which, conventional agricultural crops are grown. An extensive study of agroforestry systems across several Western European countries has shown that the use of agroforestry systems reduces soil erosion and N leaching and promotes the accumulation of soil organic matter. This effect increases with the area on which agroforestry systems are applied and is higher when applied to intensively used land, where the effects of erosion and N losses by leaching were previously greatest (Palma et al., 2007).

2.2.6 Horticulture

Horticulture is a very broad area, including the cultivation of fruits, vegetables, herbs, medicinal and ornamental plants, but also the maintenance of gardens, parks, and other elements of landscape greenery. It is an activity that is performed both by professional and amateurs as a leisure activity. It might seem that the environmental impacts of hobby gardening are marginal. However, a study from the United States shows that ornamental lawns (private and public) cover 163,812 km^2, which is about 1.9% of the US territory. Irrigated lawns in the United States cover three times the area of irrigated corn (corn has the largest acreage among irrigated crops in the United States) (Milesia et al., 2005). The lawns are supplied with a large amount of water reaching up to 1900 mm of irrigation (or 1900 l m^{-2} $year^{-1}$) (Milesia et al., 2005) and are supplemented with a large amount of nutrients. For intensively mowed lawns, up to 200 kg N ha^{-1} $year^{-1}$ is recommended. Of course, the need for irrigation and further maintenance is greater in the less suitable conditions for growing lawns. They are especially extreme in arid areas of the USA. As we explained in the chapter on permanent grasslands, the combination of a high dose of fertilization and frequent mowing leads to a herb-free lawn with dense shoots of grass and low diversity. Paradoxically, we treat ornamental lawns as if they were a high-production system, with the only difference being production produces waste that is more or less difficult to dispose of. Interestingly, many of our lawn care practices have negative environmental impacts (nitrate leaching, biodiversity suppression, and high environmental cost of maintenance), despite the fact that large proportion of gardeners mention contact with nature as one of the reasons for gardening (Clayton, 2007).

Many gardening practices are driven by social conventions (Clayton, 2007). So how do we take care of the lawn, the whole garden or city park so that we can support the provision of ecosystem services and make this place a suitable habitat for other organisms? Let's start with the lawn. In areas where the lawn does not grow or grows with difficulty, we can be so radical and think about whether we really need to have a lawn and not to replace it with some other low vegetation, ideally based on local plants. If we want to have a lawn and if there are conditions for it, it is possible to keep it shortcut only on part of the garden and rest we can grow as a meadow, where we let the grass grow and meadow flowers bloom. Meadow flowers attract butterflies and other insects. You can also buy special seed mixtures that support the occurrence of butterflies and can also be supported by planting some shrubs. Flowerbeds or honey bushes can also be combined with a more combed part of the garden.

Another question is the choice of perennials and woody plants. Especially in areas where irrigation is required, it has been shown that native plants have lower nutrient and water consumption. It is also worth mentioning the fact that a large part

Fig. 2.39 Ornamental garden composed from native plant species (South Africa)

of the invasive plants has been introduced as ornamental plants. Although only a very small proportion of cultivated plants are re-wilded and invasive in a new environment (it is stated that 10% of imported species reach the stage of transient introduction, 10% of transiently introduced species domesticate and 10% of domesticated species act at the site of economic damage invasion—Williamson & Fitter, 1996), the large number of new species grown for ornamental purposes certainly increase the chances of further successful invasion. On the other hand, the motivation for gardening is also a social convention, so the effort to surround yourself only with native species is not entirely realistic and it depends on local communities. In some countries, however, such as South Africa, using native plants in gardening is very common and may people build their garden strictly from native plants (Fig. 2.39).

2.2.7 *Agricultural Mechanization*

As we already show dominance of annuals among crops we grow require frequent soil cultivation. This can be done in small spot by hoe or similar hand tools but plow allow to cultivate, (till) larger areas. The first plows were plow hooks or rakes. Unlike today's plows, which move the soil to one side and turn it, the first plows spread the soil to both sides. It could be equipped with a cutter, which cuts the soil and the rake then spreads it. The first plow hooks are depicted by Egyptian and the same rake principles were used in ancient Rome as well as for most of the Middle

Ages. During the eighteenth century, the first predecessors of modern plows appeared with an asymmetrical deflection plate, which turned the cut strip of soil in one direction. This created furrows. The first deflection plates were wooden, covered with a sheet of metal, and later all-metal.

The plow went through a number of small, but significant improvements in terms of functionality. One of them was the improvement of the European plows by the first colonists into the all-metal plow of John Deer, which made it possible to plow the long-grass prairies in the USA. Later, especially with the advent of steam engines and especially tractors, which provided greater traction, multiblade plows were created, which made it possible to plow more furrows at once. Another significant improvement was reversible plows. The classic plow turns the soil in one direction (to the right). So, if you drive in the opposite direction, your furrows are on the other side. Therefore, when plowing with a conventional plow, it was necessary to divide the field into an even number of belts so that the furrows in one belt had one direction and the other in the other, and at the edges of the belts they are either diverged or converged. Reversible plows have two rows of coulters plowing the soil in both directions. Thus, after plowing the furrows going to the left, the farmer can turn and when driving backwards, sweep the furrows to the right (assessed in the direction of travel), thus in the same direction as the previous furrows, to which they can follow smoothly. This, on the one hand, increases the work efficiency, and, on the other hand, enables plowing in one direction (e.g., against a slope on the whole plot), which as we will show, has great consequences, for example, in reducing erosion.

In addition to the coulters, disc harrows are also used for shallow plowing or stubble. It is a system of metal discs pulled obliquely in the direction of travel. The plates take up soil and turn it around, just like when you scoop a spoonful of butter. Rotavators are used when processing smaller areas. Rotavators use a set of rotating blades that separate the soil like a hoe, crushing and leveling it. Rotavators have a more intense effect on the disruption of soil aggregates and soil biota than the conventional plowing (Fig. 2.40).

Plowed land must be trimmed before sowing. This is done by very variable machinery for soil preparation. Its goal is to level the soil and create a thin loose layer, under which, is a layer of slightly compacted soil. The first tool used to level the furrows is the skid, it is a plate, a beam, and a traverse pulled behind the tractor, which pulls down the ridges of the furrows and levels the land. Other devices are harrows that often consist of a lattice structure on which the spikes are placed at the bottom. However, there are also harrows in the shape of cylinders covered with a spiral lattice structure around the perimeter, which roll on the soil surface when towing behind the tractor or the disk ones mentioned above. Harrows break lumps and level the surface. Stud harrows can also be used for other purposes, such as weeding. Combined toolsets are often used in soil preparation to prepare a compacted bed and a loose layer above it during a single run.

The seed drill was invented in 1701 by the Englishman Jethro Tull. His invention is considered to be the beginning of the agricultural revolution in England. It enabled a more even distribution of seeds compared with hand sowing. In addition, compared with manual sowing, where most of the seeds remained on the surface, the

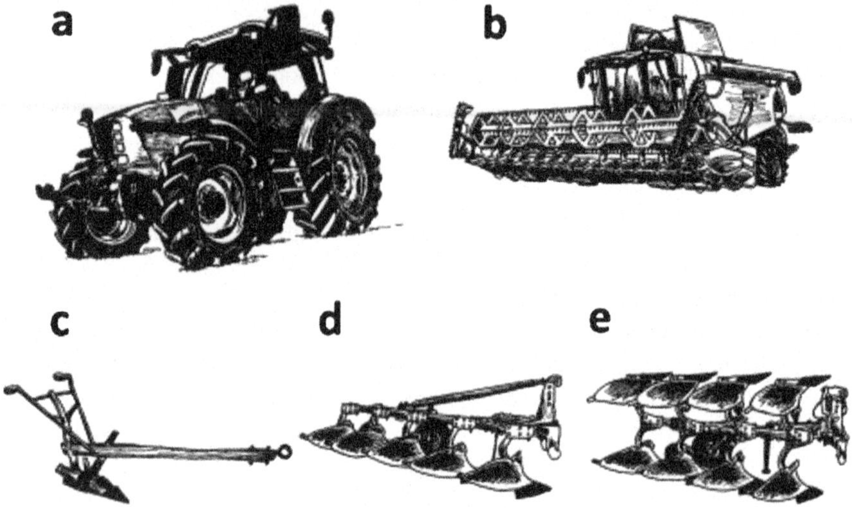

Fig. 2.40 Example of agricultural mechanization in the upper row of a tractor and a combine (**a**, **b**) in the lower development of a plow from rake (**c**) to modern plow (**d**, **e**)

seed drill incorporated them to a suitable depth, which increased germination. Gradually, a number of other types of seed drills were developed including precision seed drills, which lead to the exact placement of the seeds in a row.

The ancient Egyptians used sickles for harvest. These were first stone and later metal. The successor to the sickle was a scythe. In 1826, Scot Patrick Bell invented a cutting bar, working on the principle of fixed fingers, between which, trapezoidal blades move from side to side and cut vegetation like scissors. The cutting bar can be used to mow grain and grass. The grass then grows well thanks to a clean cut. However, the cutting deck requires even flat ground without stones. Therefore, lawn mowers, based on horizontally rotating blades, often used to mow grass, similar to most domestic lawn mowers. These can be pivotable and can be mounted on the edge of a vertically mounted or a horizontally rotating cylinder (drum). Such drum mowers mow at the edges and can thus mow tall grass, similar to string trimmers known from the home. Another group of mowers is based on a roller, which is carried parallel to the mowed surface and around the circumference of which chains or similar rotating segments. The cut of these mowers is the least clean or rather the most frayed, which results in the worst overgrowth. Biomass is also crushed and so they are often used in mulching. This type is widely used for mowing vegetation around roads. At harvest, the grain is first cut and then threshed. The first machines combining a cutting bar or a stick comb with a mobile thresher appeared in the second half of the nineteenth century. They were drawn by horses or other animals. The first self-propelled motor-driven harvester appeared in 1911 in the USA.

Tractors were firstly used for pulling the plow, but over the years they have become one of the most versatile agricultural machines. One of the things that catches the eye is the different sizes of the front and rear wheels. Larger rear wheels allow better force transmission and less opportunity to sink into soft soil, especially when carrying or towing tools, trailers, etc. Smaller front wheels, on the other hand, give the tractor a smaller turning radius and greater maneuverability. The development of tractors began in the second half of the nineteenth century, especially in the USA. They were first powered by a steam engine and later, from the beginning of the twentieth century, an internal combustion engine. At that time, designers in Europe focused on the development of the so-called motor plows. In other words, tractors integrated with a plow, however, later the concept of universal tractors capable of towing or carrying various types of agricultural machinery prevailed. Over the years, the performance of the delivered tractors has increased significantly. For example, John Deere models delivered in the early 1950s had an output of 17–51 hp. (approximately 13–38 kW). The models supplied today, have an output of 36–620 hp. (approx. 27–462 kW), while most of today's models have an output of over 100 hp. At the same time, however, the weight of these machines increased, which has implications for soil compaction.

As we have already said, the performance of tractors and other agricultural machinery is increasing. Larger and more powerful tractors made possible to connect machines with a larger working width, which could perform individual operations faster, the same applies for other machinery. In addition to work activities (plowing, sowing, spraying, and harvesting), the agricultural machine must somehow get to the field. Agricultural machinery is designed to have a large width with that, however, they cannot move on the road, for that they usually somehow fold, turn, etc. to their transport position. Upon arrival on the plot, the machine must be moved to the working position, adjusted, and in the same way in reverse order when moving to the next field, in addition to transport time as such. The smaller and more distant the fields, the more time these operations and moving between fields take from the working day. Even in the field itself, there are nonworking times such as turning or moving on field edges, in addition to the work operation. The proportion of time spent in this nonworking time depends on the shape and size of the field. It is better for larger fields to be in the shape of an elongated rectangle, where the machine moves along the longer side. This effect is also greater on machines with greater working width.

Reducing the share of nonworking hours, and thus increasing the efficiency with increasing field size, is not linear. It is largest in small fields, slows significantly in fields above 5–10 ha, and is practically zero in fields above 40 (Landers, 2000) (Fig. 2.41). However, such land size may be too large in terms of its impact on biodiversity or soil erosion. Part of the nonworking hours associated with crossings can be eliminated by having plots of land with the same crop, even of smaller dimensions, adjacent to each other. Then the crossing takes less time, and in addition, it is sometimes possible to move between plots with the machine in the working position, thus eliminating the time to rebuild the machine from transport to working position. An alternative is the complete or partial division of larger fields into smaller sections, for example, by appropriate anti-erosion measures (grass strips, hedgerows etc.), which under certain conditions, can also play an important role in promoting biodiversity. Rodríguez and Wiegand (2009) have shown that adjacent small fields up to 2.5 ha, or dividing a large field into such small sections, can be

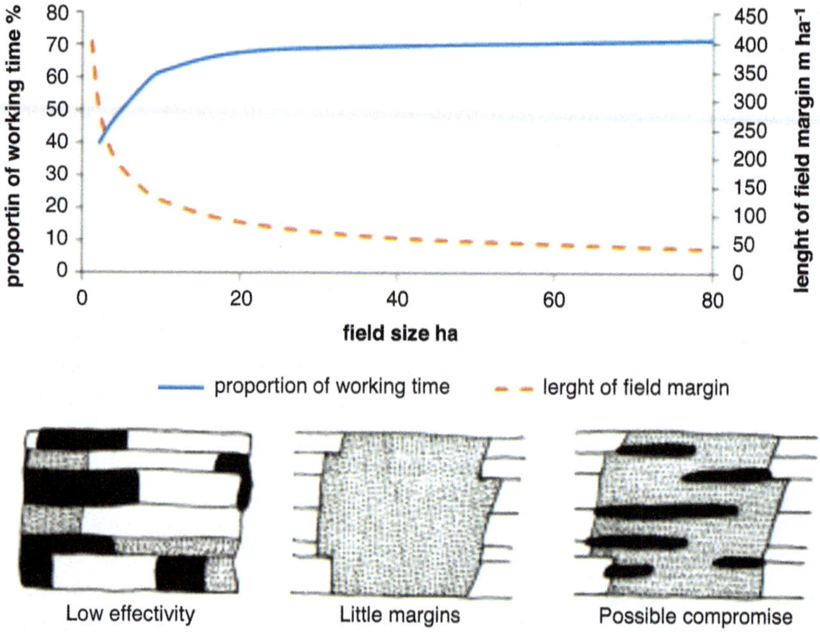

Fig. 2.41 Proportion of machine working time from the total machine time (the rest falls on non-working operations—adjustment, movement, rotation), according to Landers (2000). Ratio of edges to field area (as an approximate indicator of the number of stable habitats supporting biodiversity, erosion control, and other ecosystem services) for a hypothetical square field. Idealized maps in bottom show an examples of field with Low effectivity for machinery and large amount of edges (ecosystem function support) (left), filed with little margin and high machine efficiency (middle), and potential compromise between these two, which try to increase amount of edges and keep efficiency (right)

very effective in terms of the use of mechanization. There are also software tools available that enable to plan such field subdivisions in a way which optimizes the erosion protection and efficiency of machine movement (Kapička et al., 2017).

2.3 Crop Production and Ecosystems

2.3.1 *Impacts of Agriculture at a Landscape Level*

Agriculture affects many components of the environment, on various spatial scales with different temporal dynamics. The effects of agriculture also differ in the way in which agricultural production is increased. Figure 2.42 shows that in the EU and other developed countries, production increases mainly by increasing the yield per unit area. This leads to a more intense pressure on land use, increased doses of fertilizers and pesticides. In contrast, in developing countries, both intensification of

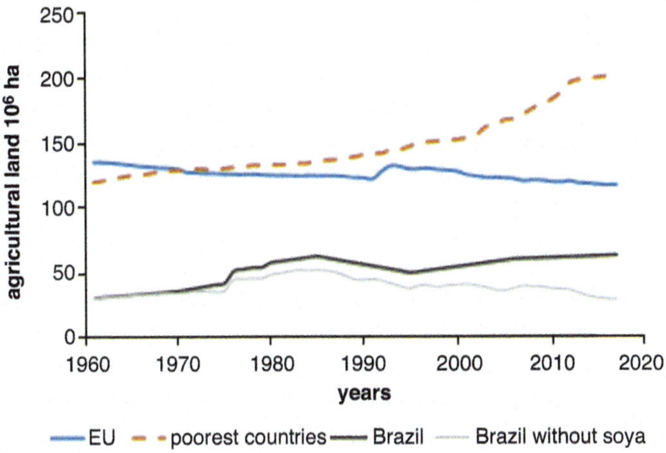

Fig. 2.42 Development of agricultural land area in the poorest countries of the world and in the EU countries, development of agricultural land in Brazil. The gray part shows the area of agricultural land devoted to soybean production. It can be seen that without soybean cultivation and exports, the area of agricultural land in Brazil would not increase. Data FAOstat

Table 2.8 Carbon yield and loss associated with the conversion of natural habitats to farmland in the tropics and temperate, West et al. (2010)

	Yield ha^{-1}	C loss connected with land conversion t C ha^{-1}
Tropics	1.7	120
Temperate	3.8	62

production and expansion of agricultural land at the expense of other land uses, especially natural habitats, contribute to increasing production. At the same time, in some tropical countries, agricultural land has been expanded to produce crops for export. For example, Fig. 2.42 shows that a significant part of the increase in agricultural land in Brazil is due to the increase in soybean cultivation, which is largely exported to other countries. This approach is all the more unfortunate because the conversion from natural habitats to agricultural land means a far greater loss of carbon and biodiversity than if such a conversion took place in the temperate zone. In addition, in the temperate zone, it would be possible to increase production by reusing the peripheral areas that we have left in the past, and thus the carbon cost of conversion would be much lower. In addition, the average yield of temperate soils is higher than that of the tropical soils, which makes such export of crops from the tropics to the temperate zone even more environmentally disadvantageous (Table 2.8), see also Box 2.10.

The overall impact of agricultural activities thus depends not only on the method of farming, but also on the size of the areas where this activity takes place and on the distribution of these areas in the landscape. It should be noted that many of the processes that create environmental problems have always been there. These are natural processes. Agriculture does not create them; it only modifies their intensity. If we look at erosion and leaching of nutrients, then traditionally the higher parts of the slope are exposed to greater or lesser erosion, depending on their shape. Erosion increases with the slope. On the contrary, in the lower parts of the slope, there is an accumulation of eroded material in the river floodplain. It can be formed into several river terraces reflecting periods of increased erosion in the recent geological past. In the upper parts of the slope, especially on steeper slopes, soil and nutrients are carried away, which then accumulate in the valleys or lower parts of the slopes. By respecting these processes, by appropriate placement of individual crops, we can reduce the negative impact of agriculture on erosion and nutrient removal. We will therefore try to place arable land where erosion happens the least. On the contrary, we will apply anti-erosion measures in the places most affected by erosion, or we will cover the land with high risk of erosion with forest or grassland.

Similarly, landscape layout affects biodiversity. For example, ponds close to each other are more populated by tree frogs (*Hylla arborea*) than isolated ones because it is easier for the frogs to migrate between them. Also, the occurrence of shrubs and high vegetation in the wider vicinity of the pond increases the chance that it will be occupied, because the presence of suitable shelters along the way again facilitates migration (Vos & Stumpel, 1995). In addition to the influence of location, individual influences acting at the level of individual fields are also reflected in the behavior of the entire landscape. Table 2.9 summarizes the relationship between the application of individual intensification measures at the level of individual fields and at the landscape level.

By applying certain phenomena in the whole catchment, these influences can be magnified and concentrated in certain places (e.g., fertilization of fields leads to eutrophication of waters). In addition, spatial homogenization (Fig. 2.43) and time synchronization makes landscape very sensitive to certain types of disturbances in certain parts of the year, which would normally have no or little effect. For example, heavy rains in intensely plowed areas in times when there is no vegetation in the fields can cause torrential floods and mudflow more easily than in a more diversified landscape (Fig. 2.43). Of course, landscape diversity affects species diversity.

In addition to the effect of landscape diversity, there is the effect of the field size itself (Fig. 2.43). This effect is not linear. Most studies show that increasing diversity with decreasing field size is the most pronounced in small fields of 1–3 ha (Konvička et al., 2016; Fahrig et al., 2015). For large fields, the difference between large and even larger fields (e.g. 10 vs. 15 ha) decreases and is practically lost in measurement errors. With a little simplification this is due to the proportion between field margins and field area. The area of the field determines the overall production, while field margins, edges, boundaries between fields, and also intermittent grass strips, provide support for diversity and also protection against erosion (a smaller field means that the boundaries will be closer and thus have a potentially shorter slope length; see the chapter on erosion). As the field size increases, the proportion of field perimeter (margins etc.) to field area decreases. This is because the circumference of the field increases linearly with the size of the edge, while the area increases with the square of the edge (Fig. 2.41). In other words, a smaller field has

Table 2.9 Individual intensification measures at the level of individual fields and their corresponding measures at the landscape level and their common environmental impacts, according to Tscharntke et al. (2005), modified

Intensification measure at field level	Intensification at landscape level	Environmental impacts
Specialization in a few varieties of a limited number of plants and simplification of crop rotation	In suitable conditions, most farmers specialize in similar crops which causes homogenization of the whole landscape	This reduces diversity at both local and landscape level, and possible refuges of species are more isolated
		Locally reducing the content of organic matter in the soil, on landscape level promoting erosion and reducing the ability to retain water in the landscape
Use of mechanization, soil compaction, use of larger, heavier and more powerful machines, creating conditions for their effective use by enlargement of field size	Increasing the size of fields, removing dividing lines and other barriers preventing the most efficient use of mechanization. This cause temporal and spatial landscape homogenization	Important local process is soil compaction and loss of organic matter, reduction of infiltration. Erosion is supported due to both the local processes mentioned above and increasing field size and reducing erosion barriers at landscape scale
Increasing the input of mineral fertilizers Increasing content of pesticides (herbicides, fungicides, insecticides, plant growth regulators, etc.)	Eutrophication of the landscape, increasing the chance of establishment and spreading of introduced and invasive species	Promoting water eutrophication Locally direct damage to invertebrate populations, especially some groups (pollinators, predators). This, together with other effects, contributes to the fragmentation of their populations. Support of invasive species and invasively spreading local expansion species and thus further homogenization of the landscape and suppression of diversity
Transfer of permanent and temporary grasses and clover to arable land, use of deep plowing instead of technologies with limited tillage	Increasing the share of arable land at landscape level. Due to the fact that these operations are synchronized, such a large part of the landscape is without vegetation for a significant part of the year	Local soil compaction and loss of organic matter, reduction of infiltration. Erosion support both through local processes and by increasing the size of fields and reducing erosion barriers. Reduction and fragmentation of habitats, especially during the winter
Increase runoff through drainage	Straightening of flows and acceleration of runoff, reduction of floodplains	Increasing water runoff from the landscape, increasing the risk of drought and floods. This is further supported by the homogenization of the landscape and the synchronization of water demand

Fig. 2.43 Example of different field sizes on the borders of the Czech Republic and Austria

Fig. 2.44 The concept of land sparing (**a**) and land sharing (**b**), which divides the landscape into a place for nature (dark) and for human (light). Figure (**c**) shows that both concepts depend, among other things, on the scale considered. Sharing plots in the upper left corner contain more sparing, and in the lower right corner more sharing

more edges per hectare than a larger field. For small fields with up to a maximum of 5 ha, the effect of resizing is significant. For large fields, they are certainly still changing, but these changes are too small to have any practical effect. This is bad news in order to achieve a compromise in size that would be optimal for both diversity and ecosystem functions as well as for efficient machinery use. As already mentioned, machinery uses may not be the only determinant of field size. It can be affected by political decision, various economic incentives, etc. We have already mentioned that extra-large farms from the communist era still cause very large field sizes in the Czech Republic (Fig. 2.2 and Fig. 2.43). So how to find a compromise between promoting diversity and providing other ecosystem services and efficient food production in such situations? One of the main reasons for enlarging fields was to increase the efficiency in the use of machinery. However, it depends not only on the size of the field but also on the aggregation of fields with the same crop. In other words, as Rodríguez and Wiegand (2009) have shown, even small fields around 2.5 ha can be effective in terms of the use of technology if they are close together, which makes work follow more smoothly. Here it must be said that from the point of view of erosion, it is not necessary to have, for example, five different fields (with different crops) next to each other, every 4 ha. It would certainly be better for diversity, but the division of one 20 ha field into 4 ha pieces using grass belts, infiltration belts or other similar means will have a similar effect. There will be one crop in this field and the farmer will manage it like one field. The loss of efficiency will then be minimal (see Fig. 2.41). Certainly, there is some (small) loss of acreage, but this may pay back on increase in ecosystem services provided.

Land sparing and land sharing. When we try to reconcile food provisioning and the impact on the environment, especially on biodiversity, we often talk about two extreme strategies called land sparing and land sharing. The land sparing strategy assumes that we will carry out the most intensive agriculture on a certain area of the landscape so that the production from this area will cover the food requirements of the population. This protects the rest of the landscape and leaves part of this saved area to nature, ideally the wilderness. In contrast, land sharing assumes that we will use the whole landscape, but we will use it more extensively, so that agricultural production and nature can coexist (Fig. 2.44). Both approaches are idealized, and conceptual scenarios depend, among other things, on the scale we are considering. For example, if small areas of intensive production and nature alternate, it is no longer entirely clear whether it is more about sparing or sharing. In the real world, the situation is even more complex. Areas are not only of different sizes, but also affected by different intensities, and the response of species to the increase in the intensity of land use is often not linear, etc.

Nevertheless, despite all that, these concepts are nice conceptual examples of how we should approach finding a balance between agricultural production and maintaining diversity and ecosystem services provision. Phalan et al. (2011) studied the distribution of different species of birds and trees in fragments of tropical native forests and in agricultural crops with different intensities. They found that more species were negatively affected by agriculture than being benefited from it. They concluded that land sparing is a more suitable alternative for biodiversity protection. In contrast, a number of authors (Pimentel et al., 1992; Tscharntke et al., 2005) pointed out that in the European landscape, more species live in habitats used by humans. Tscharntke et al. (2005) demonstrated that in Germany, around 25% of endangered species are located in protected areas, representing 2% of the country's total area, while the remaining 75% depend on areas managed by agriculture and forestry (covering 50% and 30% of the country area respectively). So, for the effective protection of species, it is certainly necessary to protect the existing remnants of the wilderness, but we cannot do so without the protection of species in the rest of the landscape. This rather implies land sharing. This is also consistent with the fact that 30% of the NATURA 2000 network of protected areas are areas with some agricultural management. The difference between these two different views may be in the historical development of the landscape. While nature in tropical developing countries is under great pressure from a growing population, this pressure has escalated relatively recently and there are still remnants of native wilderness large enough to preserve a significant proportion of species typical for native habitats and sensitive to agricultural land use. In contrast, nature in Europe has been under very intense pressure from agriculture for at least several centuries. The original wilderness was mostly completely eliminated or in the best case limited to very small fragments that do not allow the survival of species requiring, for example, large continuous areas of forest, and most of the species richness occurred in the extensively used agricultural landscape. For many of these species, both agriculture intensification and also abandoning traditional extensive landscape management is deteriorating. So, land sparing can be beneficial if remnants of original wilderness are still present, while sharing would be more important if the original wilderness is long gone and most of the species in the landscape depend on traditional (extensively used) cultural landscape.

Another consideration which is seldom accounted for is nutrient load. For example, in organic farming without mineral fertilizer input, yield reaches on average about 75% of the level of conventional farming which uses mineral fertilizers based on extensive meta-analysis of a Seufert et al. (2012). Let us see no input organic farming as land sharing. Then with the sparing option, we would need to increase the production in part of the field in order to save the rest for other purposes. This means we would need to apply conventional doses of N and other fertilizers to 75% of the land to reach the same total yield as in the share option and to „save" 25% of the area at the same time. If, for example, the saved part were a low input, species-rich meadow, then we could gain some biodiversity, but the overall input of N in the landscape would be much higher in the sparing than in the sharing option (75% vs. 0% or commercial dose). As we have mentioned already, some parts of the N may not be used by plants and is leached out of the field, which may potentially reduce biodiversity in other parts of the landscape, causing eutrophication problems in water bodies etc.

It is clear that the consideration of land sparing and land sharing represents a somewhat abstract situation. In a real landscape, we can have areas dedicated to nature, areas with intensive use for agriculture and areas where we will share the landscape with nature, and it will be a proportion of individual components. In real situations, when deciding on the proportion of these components, it is also necessary to take into account more the greater complexity of real systems (Fischer et al., 2014). The debate on sparing and sharing is often presented as a debate on food production and diversity. In fact, we grow much more than just food in production areas. There are biofuels, textile crops, animal feed. In the European context, only part of agricultural land is devoted to the cultivation of food intended for direct human consumption. There is therefore no guarantee that the area we will save by increasing food production will be dedicated to diversity when more biofuels, for example, can be grown on it. Production and diversity are not completely independent entities. By suppressing diversity below a certain threshold, we can jeopardize some of the biodiversity services that support production, such as pollination or pest control. Thus, there are combinations where an increase in diversity does not necessarily mean a decrease in production and vice versa. In addition, in the real world, we demand more ecosystem services from the landscape than just biodiversity, but also regulation of greenhouse gases, regulation of water flow, purification of water that could be used as drinking water, reduction of wind and water erosion, etc. The relationship between production and other ecosystem services is not simply linear; in certain cases, an increase in production may also support other services, while in other cases, it will be at their expense. Moreover, these relationships vary for individual types of services and finding the optimal value for all of these is difficult.

In addition, it is not only the intensity and field size, but also its distribution of patches in certain parameters in the landscape, which determines the connectivity of individual areas for different species of organisms or the flow of matter and energy between patches. Moreover, there is also connectivity of production through trade, and there is possibility of achieving higher production elsewhere (in another part of the country or another country), which can make a certain type of production economically unattractive. For example, the import of fodder from places where it can be produced more cheaply can create more pressure on the original ecosystems at the place of their production, and conversely, will lead to the abandonment of hay production in the foothill meadows. In both cases, this change in production will be accompanied by a loss of biodiversity. Finally, when we talk about the concept of land sparing and land sharing, the question of scale is also important. Sparing usually evokes the idea of larger homogeneous production blocks and large areas designated for biodiversity. On the other hand, sharing usually evokes the idea of the intertwining of production and diversity on a small scale (hedgerow, free greenery in the landscape, etc.). However, many authors discuss about sparing which refers to, for example, the uncultivated edge of the field, sown with a flowery meadow mixture to promote diversity. We can perceive such a measure as sparing at the level of one field, but at the level of the whole country, we will perceive it more as sharing.

2.3.2 *Water and Energy Flow*

Compared with natural ecosystems, their ability to retain water is often reduced in agricultural, especially arable land. This is due to the loss of organic matter and subsequent disruption of the soil structure, soil compaction (and thus loss of porosity) and, last but not least, drainage. For example, Frouz et al. (2010a) observed a decrease in field water capacity in drained meadows. One hectare of topsoil (top 30 cm) of drained meadow stored 240 m^3 less water per ha than an undrained seminatural meadow. The reduction in soil water-holding capacity, as well as the fact that arable land vegetation cover exists for only part of the year, caused arable land to have less overall evapotranspiration and overall greater runoff (e.g., Dias et al., 2015) compared with permanently vegetated areas. Higher runoff from heavy rainfall can cause local flash floods. The above-mentioned drainage accelerates runoff, both by accelerating the runoff from drained land and by indirectly increasing runoff from neighboring land. Moreover, in order to allow drainage to function, the capacity of watercourses and speed of water in them has to be increased, which increases landscape drainage even further (Just et al., 2005). This measure accelerates outflow from the land and may increase flood risk. In seminatural landscapes many small floods occur in small streams every year or several times a year. However, flood water spills into surrounding landscapes around small watercourses, is retained here and does not continue downstream, which reduces flood waves downstream. Through increasing the capacity and water speed of small streams, this brings water downstream faster, where water from larger areas is concentrated and may cause floods which would be more difficult to handle. Certainly, a more natural appearance of the landscape cannot be expected to protect us from all floods, but it can significantly reduce the peak wave locally. For example, Just et al. (2005) described a specific case where the revitalization of a 3 km long section of a stream reduced the height of the peak flood wave by about 20% during a flood caused by a 100 year flood (i.e. flood with in any given year occur with probability 1:100).

Changes in the flow of water are also reflected in the flow of energy in the landscape. As we have mentioned in the introductory section, when the sun's radiation hits the land surface, part of the energy is reflected, and another part is absorbed by the surface. The part absorbed by the surface is converted into the so-called sensible heat, which manifests itself as an increase in surface temperature and latent heat, that is, the heat used for the evapotranspiration of water. Evapotranspiration has two main components: water evaporated by soil (evaporation) and water evaporated by plants (transpiration). As we have said, vegetation cover is missing, or its cover is limited on arable land, for a large part of the year. This leads to a significant reduction in transpiration as mentioned above, and is reflected in the fact that most of the incident solar energy is used to heat the surfaces (turned to sensible heat). Fields with little or no vegetation are thus warmer than the surrounding landscape, and according to some studies, areas with a high proportion of tillage behave similarly to urban heat islands (Lal, 2017). At the landscape level, this phenomenon can be further enhanced by efforts to drain wetlands and waterlogged areas in order to increase the intensity of land use. When agricultural activities occupy a large part of the landscape, then these changes may affect local climate. For example, a study (Marshall et al., 2004) shows, following the example of Florida, that changes in land use have led to a significant reduction in small water cycles and a decrease in summer convective precipitation. But latent heat is not just about cooling. Condensation of water vapor, for example, at night, on the other hand, can slow down cooling. The above study, again in the example of Florida, showed that with the historical use of the landscape, especially with the greater proportion of wetlands in the landscape, the simulated effects of extreme frosts would be significantly mitigated. While there may not be general agreement in this regard and further detailed research is certainly needed, it can at least be said that there are studies suggesting that high areas of tillage and associated landscape drainage tend to strengthen the large water cycle at the expense of the small cycle, and increase the extremity of climate fluctuations.

2.3.3 *Nutrient Flow in the Landscape and the Cycle of Elements*

Nutrient turnover in natural ecosystems is almost closed. The main reservoirs of nutrients are minerals and organic matter in the soil. The nutrients are released slowly from both of these reservoirs, so that the released nutrients can be immediately consumed by the plants.

Early farmers supported the release of nutrients from these natural pools, namely, organic matter and biomass. They supported the mineralization of organic matter, for example, through cultivation. In slash-and-burn agriculture, the incineration of the biomass grown on the spot released the nutrients bound in it. However, this flush of nutrients is partly used by plants which are removed from the field and partly lost by leaching and other means. A decrease in nutrient availability causes a loss of fertility. In early farming, this was solved by the abandonment of the field so that the supply of nutrients could be restored, partly due to their release from less accessible mineral forms, and partly due to nitrogen fixation. The Norfolk crop rotation was revolutionary in this respect, where set-aside was replaced by the cultivation of N-fixing fodder such as clover, and the application of manure after root crops also returned the nutrients. In preindustrial, medieval agriculture, with a wheat yield of 1000 kg per hectare, approximately 21 kg N was taken. Due to the fact that there were other, for example, 3 ha of lands on which this N lost from wheat field, could be accumulated (pasture supporting livestock producing manure, fields with N_2 fixing plants, etc.), it was possible to compensate for these losses without the supply of industrial fertilizers. The efficiency of nitrogen utilization was high, but part of the nitrogen was obtained by its mobilization (mining) in the reserves of ecosystems, mainly in soil organic matter. The farmer mobilized N into available forms by accelerating the decomposition of organic matter, both by applying manure instead of litter, and by cultivation. This mobilization of N led to greater losses of N compared with natural ecosystems in the order of a few kg of N per ha, which needed to be replenished. However, before the introduction of industrial fertilizers, at the end of the Middle Ages and the beginning of the Modern Age, agriculture maximized the mobilization of nutrients available in the soil. At the same time, N was also removed from less productive parts of the landscape such as pastures or forests, which served as feed for livestock or litter sources. Both feed and litter contributed to the production of manure, which was then applied in the field to compensate for the loss from harvest and leaching. In doing so, N was gradually extracted from these extensively used ecosystems, which led to an overall depletion of the landscape by nutrients. This "nitrogen extraction from soils" then promoted the expansion of a number of oligotrophic habitats such as heaths (Fig. 2.45).

With the Industrial Revolution, N began to be supplemented from external sources. First by the extraction of Chilean saltpeter and subsequently, thanks to the discovery of industrial fixation of N from the atmosphere and in the form of industrial fertilizers. With the invention of industrial fertilizers during the last century, the supply of nutrients to agroecosystems and to the landscape as such has increased

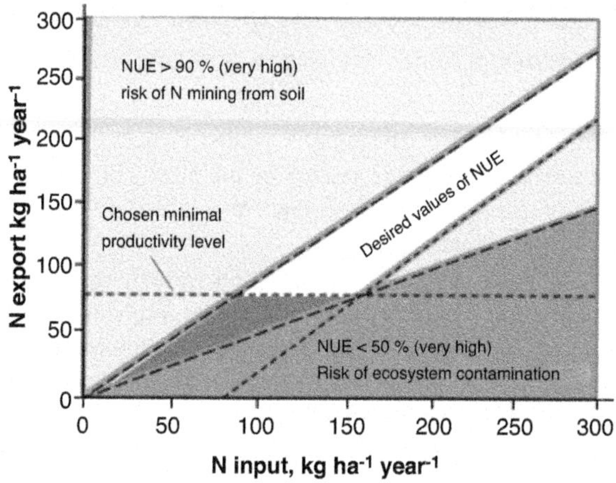

Fig. 2.45 Conceptual relationship between the efficiency of the use of nitrogen supplied to the ecosystem (NUE) and its impacts, with the designation of conceptual goals of nitrogen management according to EU Nitrogen Expert Panel (2015) modified. The wedge between the diagonals delimiting the NUE between 50% and 90% indicates the required efficiency values. Lower NUE values increase the N contamination of ecosystems, and higher ones increase the risk of N extraction from soils and the gradual oligotrophication of ecosystems. The horizontal line is an example of the chosen minimum level of productivity, the next diagonal represents the limit related to the maximum excess N, which do not significantly increase the ecosystem contamination

significantly. At the same time, at the level of individual fields, the importance of available nutrients bound in the available fraction has increased and the efficiency of nitrogen utilization has decreased. Nevertheless, despite this, nitrogen present in organic matter and microbial biomass and K and P in organic matter and minerals are still the main nitrogen reservoirs in agroecosystems. Let us first look at nitrogen, which is the most important nutrient in terms of increasing plant production. Nitrogen is released from the organic matter through microbial activity and is converted to ammonium and subsequently nitrate nitrogen under favorable conditions. The content of these available mineral forms of nitrogen peaks in the spring, when the microbial activity is high due to favorable humidity and temperature, and at the same time the uptake by plants, is low. During the year, due to nitrogen uptake by plants and a decrease in microbial activity due to unfavorable soil moisture, the content of easily available, mineral forms decreases. At that time, N is supplemented by fertilization. Following the harvest, if the temperature and humidity conditions are suitable, the content of mineral forms of N may again increase due to the decomposition of the postharvest residues. Mineral forms of nitrogen are very mobile in the soil and therefore N, which has not been taken up by plants or incorporated into microbial biomass, can be lost. The most significant losses of N occur by leaching, denitrification (especially for NO_3^-), and volatilization of ammonia. In addition, leaching of dissolved organic nitrogen can occur. Leaching is especially problematic for nitrates, which are about six times more mobile in the soil than ammonium ions. Nitrates can directly affect human health. Nitrates can be reduced to nitrites and may react with hemoglobin, especially in people with certain types of digestive problems, which prevents oxygen transfer. With increasing losses, the efficiency of utilization of the supplied nitrogen decreases and thus also the return on the investment decreases as well (Fig. 2.45).

Loss of nitrogen is connected with substantial environmental impacts which is mainly connected with the N effect on the eutrophication of aquatic ecosystems and also terrestrial ecosystems (both aspects will be discussed in detail in the following chapters). The issue of nitrate leaching into the environment and their reduction are subjects of the so-called nitrate directive (EU Council Directive 91/676/EEC). Leaching of nitrates is supported by the increase of total doses of N fertilizers, especially for doses at which the yield curve is close to saturation or is saturated (usually above the economic optimum). Besides the total dose, the application of a whole N dose at once or in a small number of applications that are inappropriately distributed during the year also supports nitrate leaching. Another factor supporting leaching is irrigation, especially sub irrigation. However, organic fertilizers can also significantly increase nitrate leaching. In this respect, long-term storage of manure in agricultural land and application of slurry in particular, during periods when plants have little chance of utilizing the released nitrogen is particularly undesirable. Leaching of nitrates can also reduce the cultivation of catch crops, cover crops, and a balanced fertilization with other nutrients. Leaching of nitrates from soils is also significantly affected by soil and environmental conditions. In addition to these farm-level factors, landscape-level factors are also important. Here, the nitrate load can be reduced by the presence of habitats, which can remove nitrates from the environment both by binding to biomass and by denitrification. These are mainly wetland habitats. In this context, river or stream floodplains have traditionally been of great importance (EU Nitrogen Expert Panel, 2015).

Denitrification is the process by which nitrates (NO_3) are converted to elemental nitrogen (N_2). From the point of view of optimizing nitrogen fertilization, it represents a loss of nitrogen; from an environmental point of view, it is not a problem in itself, because the resulting product does not represent any burden on the environment. However, the environmental risk associated with denitrification is the formation of nitrogen oxides NO_x as by-products of denitrification. In general, their formation in the soil promotes the availability of nitrates, organic matter, and increased soil moisture. Denitrification takes place optimally at higher temperatures (25–30 °C, below 10 °C it is significantly limited) and at pH 6–8, at pH below 5.5, it decreases significantly. Soil is a significant source of NO_x, but significantly smaller than burning fossil fuels. In Saxony, for example, soil has been estimated to account for about 8% of the total atmospheric NO_x emissions.

Volatilization of ammonia is another significant loss of nitrogen from the soil. Significant volatilization of ammonia occurs with inappropriate application of organic fertilizers. Ammonia volatilization occurs in alkaline soils and when the soil surface is dry. The main measures to reduce ammonia volatilization are related to the application of liquid organic fertilizers. Spray application over a large area of dry or frozen soil carries a high risk of ammonia volatilization. It is ideal to incorporate the fertilizer into the soil. Earthworms can also participate in this process in biologically well-revived soils (EU Nitrogen Expert Panel, 2015).

As we have shown, nitrogen fertilization significantly affects a number of processes at the level of individual ecosystems and the entire landscape. Globally, industrial nitrogen fixation is equal to the biological nitrogen fixation. In other words, by applying the Haber-Bosch process, we have practically doubled the entry of nitrogen into the ecosystems. Industrial fertilizers have the largest share in this increase. In addition, humans contribute to the increased supply of nitrogen to the ecosystem by burning fossil fuels and increasing the area of N-fixing plants (EU Nitrogen Expert Panel, 2015). As already mentioned and as will be described in detail in the chapter Impact on biodiversity, increased nitrogen supply leads to a decrease in biodiversity. In addition, the entry of nitrates into waters can support the emergence of eutrophication of freshwater as well as marine ecosystems, the occurrence of toxic cyanobacteria and anoxic conditions (detailed in the chapter Fisheries). In addition, nitrogen can cause acidification of soils and freshwater ecosystems.

Nitrogen generally increases the production of aboveground biomass. This, as we have already mentioned, has negative consequences for biodiversity. On the contrary, it helps absorb part of the anthropogenic increase in atmospheric CO_2 and thus reduces the impact of anthropogenic CO_2 production on radiative forcing and consequent increase in global planetary temperature (Hudson et al., 1994). Far more complicated is the effect of added nitrogen on carbon accumulation in the soil. Nitrogen supports the growth of aboveground biomass, but at the same time supports the decomposition of easily degradable parts of plant litter and slows down the decomposition of less degradable parts of soil organic matter. Due to the combination of these effects, the effect of nitrogen addition on soil carbon accumulation is different in different types of vegetation (Waldrop et al., 2004).

Another often overlooked consequence of the use of nitrogen fertilizers is the impact on the significant increase in the size of the human population over the past century. Erisman et al. (2008) believe that this increase was approximately half due to the development of industrial nitrogen fixation (Fig. 2.16). This statement is as daring as it may seem, however, corresponds well with a number of other observations. For example, Howarth (2008) has shown that more than 80% of the nitrogen found in human tissues today comes from the Haber-Bosch process. Without going into details, the population explosion of humanity is putting pressure on the further exploitation of natural resources with far-reaching environmental consequences. Some authors even link the increased nitrogen supply to the occurrence of some human and animal diseases (Johnson et al., 2010).

2.3.4 *Organic Matter, Erosion, and Other Effects on Soil*

Soil quality is one of the basic factors influencing crop production. Many soil properties are determined by factors that we cannot change (altitude or geological substrate). On the other hand, a number of other soil properties can be significantly influenced by land use and management options. One of the most important properties of soils is organic matter content, which in a number of ways directly or indirectly affects soil fertility, as described in Sect. 2.2.2.2. Soil cultivation usually significantly reduces the organic matter content. This is well documented by the example of long-grass prairies in the United States (Fig. 2.28). This is mainly due to the fact that plowing helps to disrupt soil aggregates, aerates the soil and thus contributes to the mineralization of soil organic matter. In the early days of agriculture, it was even desirable. At the beginning of the cultivation, there was plenty of organic matter that contained large amount of nutrients. By supporting its mineralization, these nutrients have been released and can be used for plant growth. However, in the long run, mineralization can lead to a significant loss of soil organic matter and a consequent deterioration of the soil's ability to retain water and nutrients. These negative effects are usually smaller where the humus (A) horizon was deep and the soil organic matter was bounded in aggregates and attached to mineral particles in some way, for example, in steppe and prairie soils (Fig. 2.46). On the contrary, the effects of cultivation are more pronounced in forest or peat soils. Drainage is also necessary in peat soils before intensive cultivation. Thus, the observed effect is the combined effect of cultivation and drainage effect. In the case of peat, organic matter losses can be considerable, but because peat can be quite deep, the loss may initially be manifested by a thinning of the peat layer, but not so much a decrease in the carbon content of the surface layer (Fig. 2.47). Conversely, where the peat layer was only shallow, the effects on the decrease in organic matter content are observed

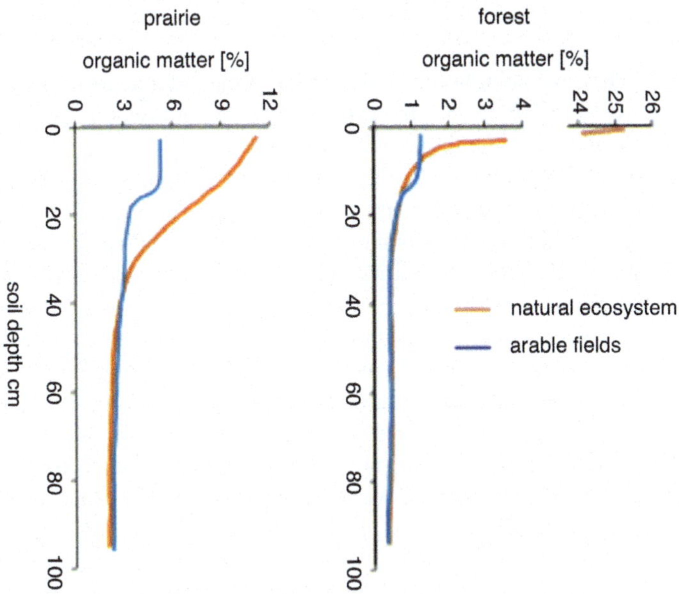

Fig. 2.46 Distribution of organic matter in forest soil and in soil of temperate grasslands (steppes and prairies and influence of plowing on C distribution in these soils according to Sanchez (2019), modified

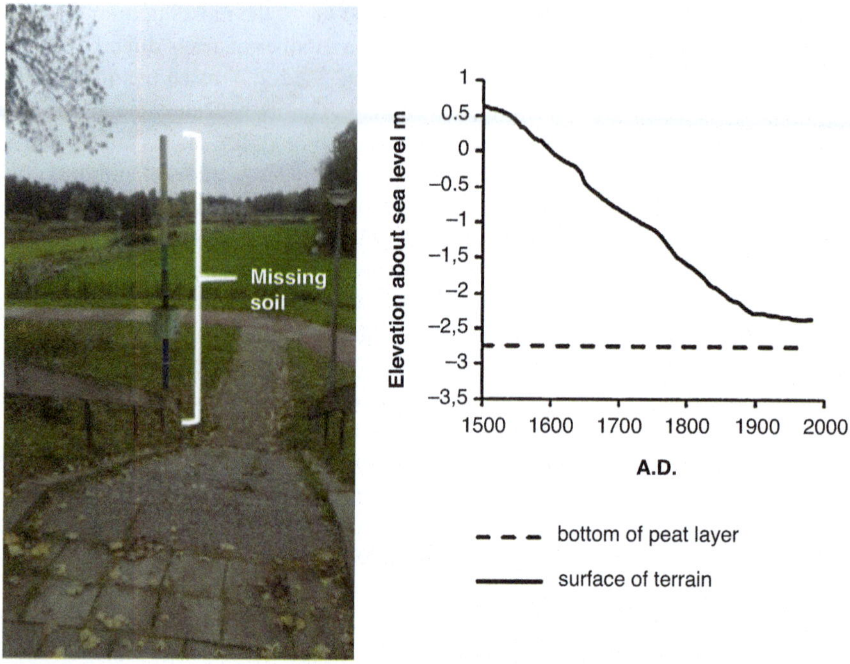

Fig. 2.47 Loss of organic matter due to peat drainage. The column on the left shows the loss of soil due to peat drainage around Amsterdam (photo courtesy of Jan Mourik), the graph on the right shows the time course of this decrease, according to van Mourik and Ligtendag (2015), modified

very quickly. For example, Frouz et al. (2010a) observed a decrease in the organic matter content from 19% to 5% in peat meadows around Senotín within 10 years from drainage and the beginning of cultivation.

It is also evident from Fig. 2.28 that the supply of organic matter in the form of organic fertilization can significantly slow down the decrease in the content of organic matter. In addition, the organic matter content stabilizes at values that are lower than that in the original prairies but higher than that in the variant without organic fertilization.

Another important factor influencing soil fertility is excessive soil compaction. We mentioned that the soil has a structure formed by soil aggregates. Inside the aggregates, the pores are mostly capillary, among them prevail the coarser non-capillary pores. The sum of all pores is called porosity, which changes during the year. After winter, soil often has a large proportion of macropores which is caused by cultivation, and volume changes due to the freezing and thawing of water. During the year, the soil gradually settles (compacts), the total volume of pores decreases, and the proportion of capillary pores increases. To some extent, this process can be beneficial for plants, as a larger proportion of capillary pores can improve the supply of water and the nutrients it contains. Some agrotechnical operations such as rolling even support this trend. However, extensive soil compaction has negative

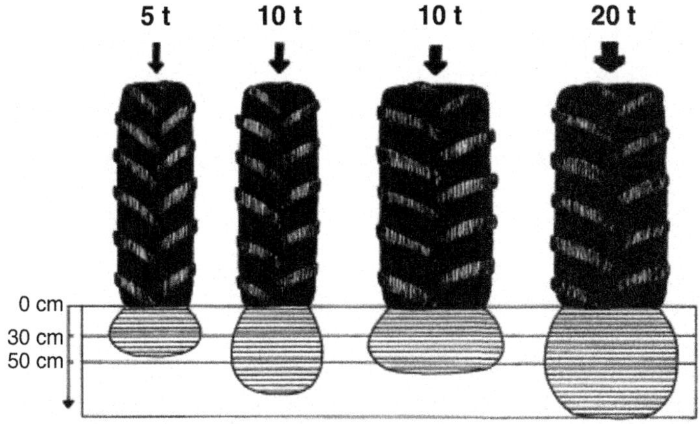

Fig. 2.48 Soil compaction increases with increasing ground pressure, and with the area on which the pressure acts. The first and second wheel have the same area, the same apply on comparison of third and fourth wheel, but in both cases wheel on the right had higher pressure per unit area and this cause larger soil compaction. The first and third wheel produce the same pressure per unit area, the same apply on comparison of third and fourth wheel, but in these pairs the right wheel have larger area and thus cause larger soil compaction

consequences for soil functioning and plant growth. The most common cause of soil compaction is the application of mechanical pressure on the soil by heavy machinery which compresses the soil. This reduces the volume of soil pores, increases the volumetric moisture of the soil and changes the soil structure to a sheet-like structure (Fig. 2.48).

This results in poorer penetration of water and roots through the soil, reduces the soil's ability to infiltrate water, and promotes anaerobic conditions. Due to the smaller pore diameter, pores are less available for roots and access to nutrients that are absorbed in the soil matrix decreases. This impairs plant growth and reduces yield. Arvidsoon and Hakansson (1991) estimated that only one pass of mechanization in beet harvesting could cause a loss of future yield of 3% if there were dry conditions in low-clay soil and up to 17% if clay soil is harvested in wet conditions. In general, dicotyledonous plants (herbs) are more sensitive to soil compaction than monocotyledonous plants (grasses). In the case of monocotyledonous plants, there is often an increase in yield with a slight compaction of the soil, but this is usually a degree of compaction corresponding to the extent of the natural variability of soil aggregation during the year (Fig. 2.49). However, it can be said that soil compaction, when the soil density exceeds 1.7 g cm^{-3}, generally has negative consequences on both the physical and biological properties of soils (Beylich et al., 2010), and in many cases, negative consequences occur even at lower compaction rates (Fig. 2.49).

The overall degree of compaction is determined by the pressure on the soil, the total area on which the pressure acts and the time for which the pressure acts. Furthermore, the degree of soil compaction is also affected by soil properties and other environmental conditions at the place and time when the pressure acts. The

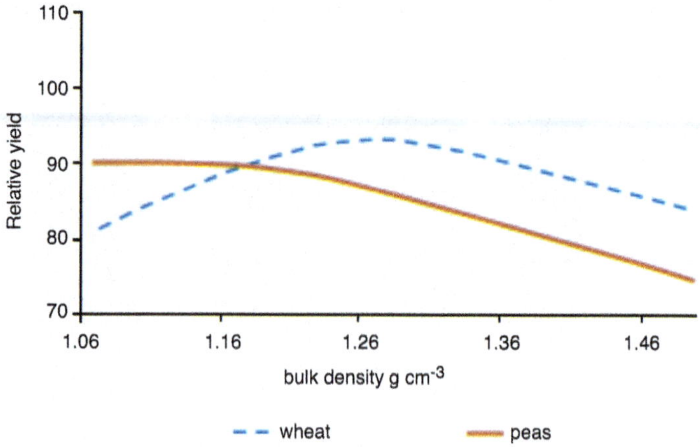

Fig. 2.49 Influence of soil compaction, here characterized as an increase in bulk density on the relative change in wheat and pea yield according to Beylich et al. (2010)

Table 2.10 Examples of static pressure of some objects on a substrate. This is the pressure of a standing object. There can be a significant increase in movement pressure, especially in humans, animals, and tracked vehicles. It should also be noted that this is the pressure at the point of contact with the pad and different objects exert a given pressure on different surfaces (e.g., the footprint of a horse is smaller than the footprint of a tank) according to Horn 2006 and company literature

Object	Pressure of the surface KPa
Humans with snowshoes	3,5
Humans	55
Caterpillar tractors	28–55
Tank T34	70
Tank Abrams M1	103
Horse	170
Wheal tractor	60–180
Offroad lorry	90–270
Harvester Komatsu 901XC	31–71

pressure depends on the weight of the object and the area of contact with the surface. The contact area of the vehicles is determined by the area of the wheels or belts that come into contact with the surface. For tires, this area increases in diameter and the width of the wheels. It is larger for underinflated tires, which deform more when in contact with the pad. Table 2.10 shows the pressure of various objects on the surface. This is the pressure of a standing object. In motion, shocks caused by the impact of the feet or irregularities associated with the transmission of pressure can lead to a two to threefold increase in pressure, especially in humans, animals and tracked vehicles. In addition to the pressure itself, the area on which the pressure acts is important. It is true that if we apply the same pressure on different areas, the pressure action will penetrate into the soil over a larger area and with greater intensity, the larger the area of action (Horn et al., 2006). The horse and the tractor have an order of magnitude similar pressure (per unit of area) on the surface, but the area on which the pressure acts is much larger for a tractor than for a horse. One passage of the horse therefore causes less compaction than the passage of the tractor. So, if we have two tractors, one heavier and the other lighter, applying the same pressure per unit area of soil, then the overall effect on soil compaction will be greater with a larger tractor, because the larger wheels touch the soil with a larger area. However, it will be better than if the heavier tractor had the same small wheels and thus much more pressure acts. The total affected area will then be also affected by the number of passes, that is, how much of the total area was trampled or run over and how many times. In addition, soil conditions are important. Clay soils are more sensitive to compaction than sandy soils. The organic matter content can reduce the sensitivity of the soil to compaction. Soils covered with sod are also more resistant to compaction than bare soils. Soils that have high porosity will be relatively more sensitive than soils that are already slightly compacted. Finally, the application of pressure on wet soil causes greater compaction than the application of pressure to dry soil.

This also includes possible measures to prevent greater soil compaction: use lighter machines with less pressure on the soil, reduce the number of passes over the land and, if possible, avoid wet travel. Another measure is to increase the content of organic matter in the soil. Unfortunately, as we said in the chapter on mechanization, the intensification of agricultural production leads to the use of ever larger machines. The compaction of the soil itself contributes to this. Compacted soils cause more resistance, so their processing, such as plowing, is more energy-intensive. So, we need a tractor with stronger traction, and such a machine is usually heavier, but this often puts more pressure on the ground, which can increase compaction and so on. Compaction is transmitted to the denser layers of the soil; compaction of the subsoil negatively affects the yield and may worsen the susceptibility to erosion (see below).

In addition to the action of mechanization, compaction is accelerated by the loss of organic matter. Also unbalanced fertilizers, especially with mineral forms of nitrogen, can promote the breakdown of soil aggregates and thus increase the sensitivity to soil compaction (Hati et al., 2008). Degradation of soil aggregates and subsequent migration of colloids of soil minerals in soil profile, which also cause compaction, can be triggered by chemical changes of soil solution. In general, the monovalent cations K and especially Na lead to the peptization of soil colloids, while the divalent cations Ca and Mg lead to their coagulation (Rengasamy & Marchuk, 2011). Peptization of mineral colloids (clays) occurs in saline soils but can be supported by irrigation with saline water or inappropriate fertilization. By depositing these colloids on the range of topsoil—subsoil, an intermediate layer can be formed that significantly affects the movement of water in the soil profile.

Finally, another negative effect on soil strongly affected by agricultural practice is erosion. Erosion is a process of material transport by water, wind, or ice. Erosion is a natural process that also occurs in natural ecosystems that are not affected by humans and play many crucial roles in ecosystem functioning. It is especially important in desert and similar ecosystems with sparse vegetation. However, in natural vegetation-covered soils, soil erosion is small and can be completely compensated by soil formation and a gradual increase in soil depth. In the following text, we will deal only with wind and water erosion on agricultural soils.

In temperate climate, wind erosion is usually less significant than water erosion. Wind erosion is typical for arid areas; however, it can also occur on dry lands in the temperate zone. Its occurrence is supported by the large size of the land, weak or no vegetation cover, the absence of obstacles to wind movement, such as forests, alleys, and windbreaks, etc. Water erosion is powered by the kinetic energy of water moving along the slope, which can carry soil particles downhill. The rate of soil movement by water erosion can be described by USLE (Universal soil loss equation), which well conceptualizes the main factors that affect erosion. So USLE can be written as:

$$G = R * K * L * S * C * P$$

where

G is the average long-term soil loss ($t.ha^{-1}.year^{-1}$), R is the rain erosion efficiency factor, expressed depending on the kinetic energy and intensity of erosively dangerous rains ($N.ha^{-1}$), K is the soil erodibility factor, expressed depending on the texture and structure of the topsoil, organic matter content and permeability of the soil profile ($t\ h\ MJ^{-1}\ cm^{-1}$), resp. after adjustment ($t.N^{-1}$), L is the slope length factor, expressing the influence of uninterrupted slope length on the amount of soil loss by erosion (dimensionless), S is the slope factor, expressing the effect of slope on the amount of soil loss by erosion (dimensionless—ratio of shear to shear on a unit plot of slope 9%), C is the factor of protective effect of vegetation, expressed depending on vegetation development and used agrotechniques (dimensionless—ratio of leaching to leaching on unit plot with permanent fallow), and P is the efficiency factor of anti-erosion measures (dimensionless—ratio of leaching to leaching on unit plot cultivated in direction slope of the plot).

So, the loss of soil by erosion depends on the intensity of the rain, which we cannot influence. It also depends on the slope. Thus, the steeper the slope, the greater the potential for erosion. The slope is difficult to change if we do not consider building terraces, which is very energy-demanding and costly. On steep slopes, risk of erosion will be very high, and therefore it would be preferable to use them as grassland or afforest them instead of using as arable land. Furthermore, erosion depends on the length of the uninterrupted slope, the longer the slope, the greater the erosion (Fig. 2.50). If we imagine two slopes with the same inclination, then a drop of water starting on a longer slope will simply be higher, will have more positional energy and can gain more kinetic energy during the way down the slope than a drop starting on a shorter slope. We can shorten the slope quite easily. We can put a barrier in the middle that stops the flow of water. The barrier can be vegetation with a large soil protection effect or, for example, an infiltration belt, a shallow ditch following the contour line, where the water stops and infiltrates to soil. The factor of soil erodibility is related to the formation of soil aggregates. Soil aggregates support the

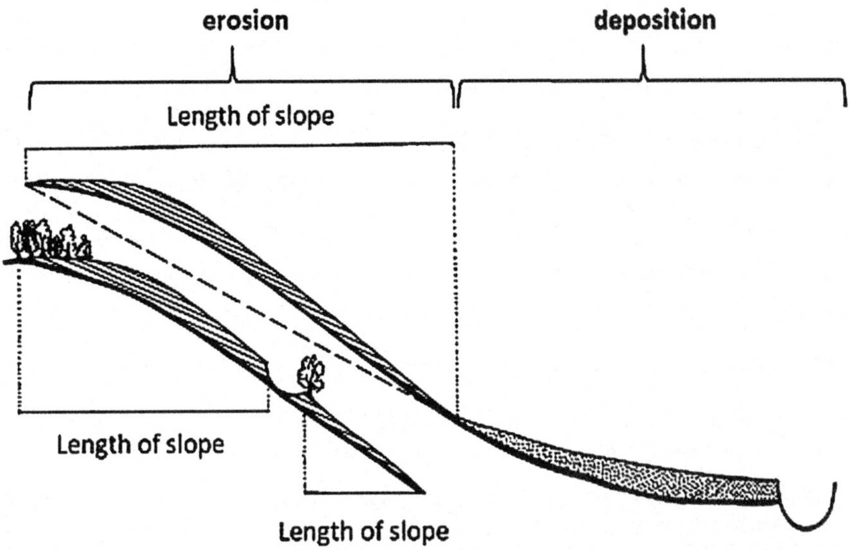

Fig. 2.50 Uninterrupted length of the slope on which erosion occurs. If the slopes were completely straight and smooth as an inclined plane, erosion would occur along its entire length. However, the slope is usually slightly convex or concave and therefore in some parts, the erosion is intense (light gray part), in others, on the contrary, the deposition of material occurs (dark gray part). These parts can be identified by the deviation from the uniform slope (dashed line). The length of the uninterrupted slope can be shortened by interrupting it with anti-erosion measures, vegetation that prevents erosion, etc.

resistance of soil to erosion because they are larger than the basic soil particles and are therefore more difficult to wash away. In addition, there are macropores between the soil aggregates that facilitate water infiltration. Water that infiltrates in instead of flowing on the surface does not have such an intense erosive effect. Conversely, factors that reduce water infiltration into the soil, such as compaction, promote erosion. Several cooperating processes contribute to the protective influence of vegetation. The leaves of the vegetation retain part of the rain (interception) or at least slow down the speed of the falling drops and thus they reduce the erosive effect of rain. For the same reason, soil covered with litter resists erosion better than soil without litter. Vegetation and litter on the soil surface increases the surface roughness, thereby slowing down the speed of running water and thus reducing its erosive action. In addition, the roots of the plants hold the soil together, making it less easy to wash away.

The effects of vegetation can also be classified as anti-erosion measures in certain circumstances. Others are the already mentioned infiltration belts, or various walls, boundaries and similar structures that interrupt the continuous flow of water.

From the above, it is clear that an area continuously covered with vegetation, such as a meadow or forest, is far less prone to erosion under otherwise similar conditions than bare soil without vegetation. Due to the fact that plowed fields are without vegetation for a large part of the year, they are more prone to erosion than meadows. The method of plowing also significantly affects erosion. Erosion is mostly supported by plowing along the slope, which during the vegetation period creates ideal conditions for the water movement downhill. If we plow along the contour line, it depends on where we dump the soil during plowing. If we tilt it down the slope, then in essence we help erosion to move the soil downhill. On the contrary, if we tilt it up, we move the soil uphill. It looks simply, but in order to do it so easily, we need a so-called reversible plow, that is, a plow that has two rows of coulters and we can always tilt the ground in the same direction. For a plow with one row of coulters, we always tilt to one side, which means that if we go in the opposite direction, we tilt the soil to the opposite side than when we arrived.

2.3.5 Crop Production and Biodiversity

Lyons et al. (2016) showed that with the course of Holocene there was a worldwide shift in the organization of communities of organisms, which is attributed to the advent of agriculture and human-made changes in the landscape in general. Agriculture occupies a significant part of the European landscape. Most non-forest habitats were affected by agricultural activity during the Middle Ages and the modern age. Moreover, many categories of non-forest habitats which are valuable for biodiversity, such as heathland, meadows, extensive pastures etc., are products of preindustrial agricultural use. Many of them are natural reserves but cannot sustain without management which in some extent mimic extensive agriculture uses. Almost a third of the sites of European network of protected sites NATURA 2000 are areas with some kind of agricultural management. However, it is mostly a traditional management that has developed in preindustrial agriculture. This shows that Europe's agricultural landscape is an important biodiversity hotspot, although for most species, it is a secondary habitat. Moreover, most of this diversity has been tied to preindustrial farming practices, which are now being replaced by other more intensive practices. The intensification of agriculture leads to a significant reduction in diversity of agricultural landscape. Unfortunately, we do not have really long-term data dating back a hundred or more years. Probably, the best example of this trend is the decline in diversity and abundance of wild birds (Reif & Vermousek, 2018) in recent decades (Fig. 2.51).

Donald et al. (2002) showed that the rate of decline in the number of birds is directly proportional to the intensity of agricultural production (Fig. 2.52). The preindustrial agricultural landscape differed from today's landscape in many respects. Note just a few major differences here. The first was at the level of biogeochemical cycles. Before the advent of industrial fertilizers and a significant increase in atmospheric nitrogen deposition from fossil fuel combustion, soils were far less supplied with available nutrients than today. At the same time, the increase in nutrient availability leads to an increase in the cover of a few aggressive plant species, which suppress other species by their competition, which leads to a loss of diversity (Soons et al., 2017), this reduces plant diversity which may affect higher trophic levels as

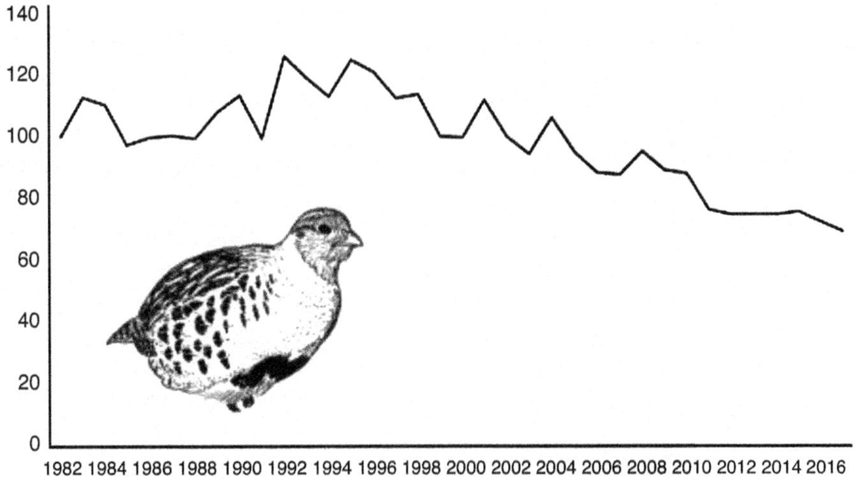

Fig. 2.51 Relative changes in the number of field birds (1982 is 100%) in the Czech Republic, Reif and Vermouzek (2018)

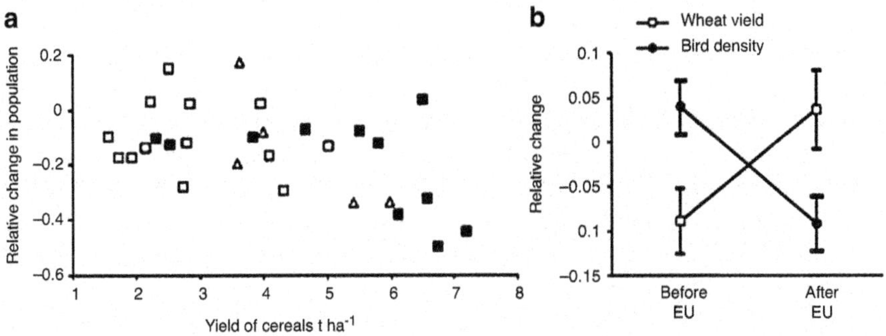

Fig. 2.52 Comparison of the relative change in the number of wild birds (between 1979 and 1990) for old and new EU members in relation to wheat yield (according to Donald et al., 2002) (**a**), change in the number of wild birds compared to 1982 for the Czech Republic before and after the EU accession and change in wheat yield in the same period (**b**), according to Reif and Vermousek (2018)

well (Wilson et al., 1999). In contrast, the preindustrial agricultural landscape was much more populated, and, among other things, it can be expected that there was a lot of pressure to take biomass in the form of feed and litter. This led to a reduction in the competition of growing plants for light, as well as to the removal of nutrients from the landscape. As is the case in many parts of Africa and the Middle East today, there may have been a risk of excessive biomass uptake, which could increase erosion. In comparison, in today's agricultural landscape, biomass uptake is declining, especially in marginal areas, partly due to declining numbers and changes in the composition of livestock, and partly due to the fact that part of the forage in the feed base has replaced imported feeds, for example, soya. Reduced interest in forage leads to the abandonment of meadows and pastures, which are gradually overgrown with shrubs and forests, which will have a negative effect on the decline of biodiversity.

The preindustrial agricultural landscape was also much more spatially diversified than it is today. There are a number of studies (Guerrero et al., 2011; Šálek et al., 2018) showing that a landscape with smaller fields, hosts more diversity and a greater number of birds, for example, than a landscape dominated by large fields (Fig. 2.43). It has many causes on different levels. Thanks to the ownership of smaller units, the fields were smaller and this was associated with a larger amount of woody vegetation in the open countryside. This non-forest greenery performed a number of functions in the landscape and provided habitat for many species of organisms. Over time, with the intensification of agriculture, the land was merged into larger units and large fields were created, on which it is possible to make better use of large mechanization. Due to the use of mechanization, the diversity of the landscape has also decreased at the level of individual fields of meadows compared to the historical preindustrial landscape. For example, when mowing of meadow is done by hand, it was often not possible to mow the whole meadow in one day, which creates a gradient of areas with differently developed vegetation. This mosaic enables to meet the demands of different plant and animal species and contributed to the diversity of the landscape. Today, we mow the entire meadow at once and thus support the species that are suitable for just one mowing date and thus is disadvantageous for all others. Subsidies for rural support under the EU's common agricultural policy contribute to this, requiring farmers to mow their meadows by a certain deadline for easier control. This makes the whole situation worse, because the mowing date is not the same within one meadow, but very similar in the whole landscape. Another similar trend is specialization within larger landscape units. Of course, even in the past, more specific crops, such as beets or potatoes were grown under conditions that were favorable. However, the farmer also grew other crops, which he needed as feed for his livestock or for his self-support, etc. At present, regional specialization is significantly increasing, which is reducing the number of species grown in a given part of the landscape, which again has negative consequences for biodiversity. Other factors such as soil compaction and chemicalization may have contributed to the reduction of species diversity, but their impact is far less quantified. Many studies however were done in pesticides and its negative effect on some insect species.

In the next part of this chapter, we will take a closer look at the relationship between agriculture, biodiversity, and the abundance of three important groups of organisms providing important functions in the agricultural landscape. These will be soil organisms, pollinators, and predators that may be involved in biological pest control.

Soil Organisms In natural ecosystems, together with plants, soil organisms play a significant role in the nutrient cycle, soil formation, and functioning. Even in agroecosystems, soil organisms play a key role in the transformation and mineralization of organic matter and the release of nutrients. Compared with the natural ecosystems, the role of soil biota in agroecosystems is somewhat suppressed. This is because, far more energy enters the detritus food chain in natural ecosystems in the form of dead organic matter. In agroecosystems, human appropriates a larger part of primary production and thus less is left for decomposers. On the contrary, the "work" that decomposers would do in terms of nutrient release and soil structure formation is replaced by agrotechnical operations at the cost of additional energy, which today is ultimately, to the vast majority, at the expense of fossil fuels. However, the effect of mechanization and decomposers on soil is not the same. For example, Frouz et al. (2014) showed that mechanical incorporation of litter into the soil leads in the long run to less accumulation of organic matter in the soil than when earthworms incorporate the same litter into the soil. In addition to a smaller input of organic matter (Miko & Storch, 2015), other factors also limit the activity of soil biota. Among the most important are tillage (van Capelle et al., 2012) and soil compaction (Beylich et al., 2010). There are a number of studies showing that increasing the intensity of agricultural production (more plowing, feed crop in crop rotation, more compaction, and lower input of organic matter into the soil, etc.) leads to a decrease in the abundance of dominant soil biota groups (Ponge 2003). This decrease in the dominant groups of soil biota then correlates with the decrease in key soil functions influencing soil fertility (de Vries et al., 2013).

Pollinators Approximately, 90% of the vascular plants depend on the pollination by animals, often by insects. If we look at agricultural culture, it is a little less. It is estimated that 85% of 264 agricultural crop species in Europe use insect pollinators (Aizen & Harder, 2009). In terms of yields it will be even less, cereals and root crops do not need insect pollinators. Even so, pollination is irreplaceable when growing fruit and many other crops. The most important pollinator in Europe is the honey bee and various wild bees and bumble bees. Intensive agriculture can affect pollination in several ways. It is a decrease in habitats suitable for nesting wild bees, which is associated with a decrease in landscape heterogeneity, a decrease in field margins and similar landscape structures. For example, coffee plantations within 1 km of the forest are reported to be better pollinated, which can increase yields by up to 20% (Foley et al., 2005). Another negative effect is the homogenization of the landscape, which synchronizes flowering and thus the occurrence of the main food sources in the landscape. This is reinforced by the distribution of subsidies within the common agricultural policy, where a certain date of meadow mowing reduces the food supply of wild-flowering plants, which negatively affects both wild bees and honey bees (Buri et al., 2014). Last but not least, the application of some pesticides may contribute to the reduction of bee numbers (Woodcock et al., 2017).

Predators Natural predators are an important component of integrated pest management. In many cases, the introduction of a specialized predator is difficult or not even available, and in addition, the introduced predator requires time to adapt to the new environment. In contrast, nonspecialized predators who are already in the environment can effectively capture the emerging pest. As with pollinators, natural predators suffer from simplification of the landscape mosaic and enlargement of fields, which leads to habitat loss (Marshall & Moonen, 2002), and as with pollinators, pesticide application can have a negative effect.

2.3.6 Crop Production and Global Climate Change

In the previous chapter, we explored how the agriculture can effect cycling of nutrients and energy in the landscape, which can effect ongoing climate change. On the other hand, the ongoing global climate change has a number of impacts on agriculture. Its overall impact is difficult to predict, both because of the complexity of the effects and because other processes, such as globalization, geopolitical change, plant and pest invasions are taking place in parallel. However, experiments with plant growth at higher CO_2 concentrations show that greater production of plant biomass can occur and plants can be more efficient in using water in higher CO_2 concentrations. However, this only applies if water and nutrients are not limiting. Therefore, an increase in production of irrigated crops is expected. However, the local availability of irrigation water remains an issue. For nonirrigated crops, due to the increase in temperature, greater variability in precipitation and changes in their distribution during the year, growth may be limited by a lack of water, which will result in reduced yields. At the same time, the increased temperature will lead to a shortening of the growing season and a shift in phenology. The overall impact on crop production can be both positive and negative. This can be reinforced by other influences. For example, raising the temperature will also speed up the development of insect pests and allow many of them to have more generations within a year. For example, Adams et al. (1990) assume for the USA, an overall slight decrease in the yield of the main crops, but with great variability within the individual US states, from a 50% increase to an equally large decrease in yield. In general, impacts are expected to be worse in tropical areas, while positive effects may prevail in colder climates.

These relatively small effects as described above will hold if the goals set by the international community can be achieved and the increase in global temperature is kept below 1.5 degrees compared with the preindustrial period. As global temperatures rise, yields will decline and their variability will increase. This can have consequences for food supplies. So far, we have discussed about the impacts at the country level. However, the question is how these fluctuations will be distributed globally. A small fluctuation in one country can currently be offset by imports from elsewhere, but if these fluctuations are global, this possibility is also jeopardized. For example, at present, the probability that in all major areas of maize cultivation in the world there will be such a fluctuation at the same time that world production will fall by more than 10% in a given year is very small. However, even when the temperature rises by 2 °C, this probability will be 7% for each individual year, which means that the probability that such significant fluctuations will occur once every 10 years will be relatively high. And with a global temperature increase of

4 °C, the probability that there will be such a significant drop in the global harvest worldwide each year will be 89% (Tigchelaar et al., 2018). In other words, it is very likely that major food shortages may become the rule rather than the exception.

2.4 Animal Production

2.4.1 Biological Basis of Animal Production

If we neglect animals that we keep just for fun, domestic animals provide us with either work (service dogs, draft animals, and racehorses, etc.) or biomass generated by secondary production (meat, milk, eggs, and wool, etc.). In the introductory chapters, we described the energy balance of animals (and other heterotrophs) as used in ecological studies. These principles were introduced into the field of ecology mainly through the International Geosphere-Biosphere Program (IGBP) in the 1980s. In parallel with this concept of animal energy budget, or even slightly ahead of it, the energy balance of livestock was developed by Max Kleiber (Kleiber, 1961). Thanks to this parallel development, a similar concept is used in the study of livestock feed utilization, but for historical reasons, it uses slightly different terminology and units.

We have already mentioned in the introductory chapters that food intake affects not only the amount of energy, but also the content of the necessary elements (ecological stoichiometry) and essential substances. If an animal feeds on a nitrogen-poor diet, it must eat more food than would correspond to its energy intake to meet its nitrogen needs. In ecological studies, ecological stoichiometry is usually assessed independently of energy. In animal nutrition, it is customary to correct for the main stoichiometric composition of the food. If the animal has an excess of nitrogen, it uses part of the nitrogen to replace its energy. To do this, however, he must deaminate the received amino acids and then get rid of excess nitrogen, which costs him extra energy. Metabolizable energy corrected for nitrogen balance is called balance metabolizable energy.

In agricultural practice, in addition to the energy parameters described above, the so-called conversion factor also called growth efficiency (ECI) or feed conversion ratio (FCR) is often used to simply express the efficiency of conversion of feed into production. Indicates how much food the animal must eat for a gain of 1 kg. The conversion factor takes into account the weight of the food and the animal at usual water content, it is not converted to dry matter, so the conversion factor of dry feed is usually lower than that of the feed containing a lot of water. It will therefore be smaller for hay than for fresh forage. It is specific to the type of feed and animal. For cattle, it is between 5 and 20, the lower limit applies to feed with a predominance of grain, the upper limit to feed with a predominance of grass. Similarly, in sheep, it can be 4–5 for young animals fed on grain and up to 40 for older animals fed lower quality forage. In pigs, the conversion factor is around 3–3.2, and in poultry, it can reach values of 1.7. Poultry is a prime example of the intensification of meat production. The conversion factor has decreased by more than 250% since the beginning of the industrial revolution (see also Box 2.11).

Box 2.11 Production of Broilers: An Example of Intensification of Animal Production

Table 2.11 shows a reduction in the time required to produce a broiler to less than half between 1935 and 1995, with a significant improvement in the conversion factor and market weight of the chickens. One of the basic preconditions for this development was putting broilers in cages (Boyd, 2001). Cages placed in the controlled environment of the halls not only limited the movement of animals and the associated energy losses, but also made it possible to regulate the animal environment in a very comprehensive way. Temperature regulation enabled the animals to save energy that they would otherwise spend on thermoregulation and subsequently to select for further breeding individuals who invest less in thermoregulation and even feather production. Daylight regulations regulate sleep time and thus again support the growth of broilers. However, caging the chickens was not without difficulty. From the beginning of the closing of chickens in cages suffer leg weakness due to the lack of sunlight and the related lack of vitamin D. At the beginning of the 1920s came research works that emphasized that the addition of vitamin D to the diet in the form of fish oil can eliminate this problem. In addition to vitamin D, another barrier to cage farming was the spread of infectious diseases. Therefore, as early as 1935, a plan was developed in the United States to control diseases in poultry farms under the auspices of the US Department of Agriculture. It was based on thorough testing, control and isolation of sick individuals in maternity farms and chicken coming from hatcheries. Subsequently, the fight against diseases was supplemented by vaccination. Absolute control of diseases is never possible, as new diseases, such as new virus mutations or more virulent forms of those already known diseases, are still emerging. Nevertheless, these efforts led to the control of diseases to the extent that they enabled profitable cage breeding. The benefits of caged chicken farming have been further enhanced by the electrification of rural areas. The various equipment of production halls, usually powered by electricity, has led to significant savings in human labor. In the United States, 250 man-hours were needed in 1940 to raise a thousand broilers. In 1950, it was only 48 man-hours, thanks to electrification. Intensive cage farming required a constant supply of freshly hatched chickens. This was made possible by the invention and the development of artificial incubation of eggs, again greatly enhanced by the development of electrification. Another important factor in the intensification of broiler farming was the development of complex feed

Table 2.11 Development of basic parameters of productivity of chickens broilers in the USA between 1935 and 1995—no data, according to Boyd (2001)

Parameter/year	1935	1945	1955	1965	1975	1985	1995
Market weight kg	1.3	1.4	1.4	1.6	1.7	1.9	2.1
Time needed for production—days	112	95	73	–	56	–	47
Conversion factor	4.4	3.8	2.9	2.5	2.1	2.0	1.9

(continued)

Box 2.11 (continued)
mixtures. The fact that chickens were popular laboratory animals and are associated, for example, with the development of knowledge about vitamins, certainly played a role in the development of complex feeds. The primary feed of the chickens was cereal grains. In addition to supplementing vitamins and minerals, it was necessary to strengthen the proportion of protein to increase meat production. After experimenting with a number of protein supplements of animal origin, soy came to the forefront of researchers as an ideal inexpensive source of protein. The development of the use of soybeans has been reinforced by the general increase of yields, which has liberated sufficient space for local soybean production. Use of soya even increased by massive import of soybeans from third countries, in particular, Brazil. Another important factor was the discovery of soy protein extraction, which enabled the preparation of very effective feed mixtures. The feed mixtures are carefully balanced with regard to the need for nutrients during the growth of the chicken. Mixtures with a high proportion of proteins are used at the beginning of fattening. As the size of the chicken increases, the proportion of weight gain to the chicken decreases, and the chicken must expend more energy to maintain its own metabolism. Therefore, the proportion of carbohydrates, which are a source of energy, increases in mixtures for older chickens. This will prevent energy losses that would relate to the need deaminated amino acid, and disposing of the excess of nitrogen, which may happen if the same protein-rich feed mixture would be used.

Another important step in improving the effectiveness of feeds mixtures has been the addition of antibiotics, which dates back to the 1950s in the United States. In addition, other supplements such as enzymes and substances that support the activity of the intestinal microflora are now part of feeds.

Another important factor was the breeding of poultry breeds suitable for broiler fattening. The development of new breeds of chickens and the application of the latest scientific knowledge in this direction date back to the beginnings of the application of genetics as a scientific discipline. A significant breakthrough was the effort to breed broad-breasted fat breeds of chickens in the United States in the 1950s (Chicken of Tomorrow) (Boyd 2001) (Fig. 2.53). Yet, another important feature of breeding effort was the maximum reduction of fattening time and strengthening of the initial growth rate. Although. in the USA, the use of antibiotics and other adjuvants is supported, they must not be used in broiler fattening in the EU.

It should be noted that, as has been suggested in several places, most of these innovations did not come by themselves, but they were the result of massive and intensive research that benefited, among other things, from poultry science establishing itself in many American universities very early. However, this great success in the mass production of chicken meat has taken its price in the form of environmental problems and deteriorating animal welfare, which we will discuss about in the next chapters.

(continued)

Box 2.11 (continued)

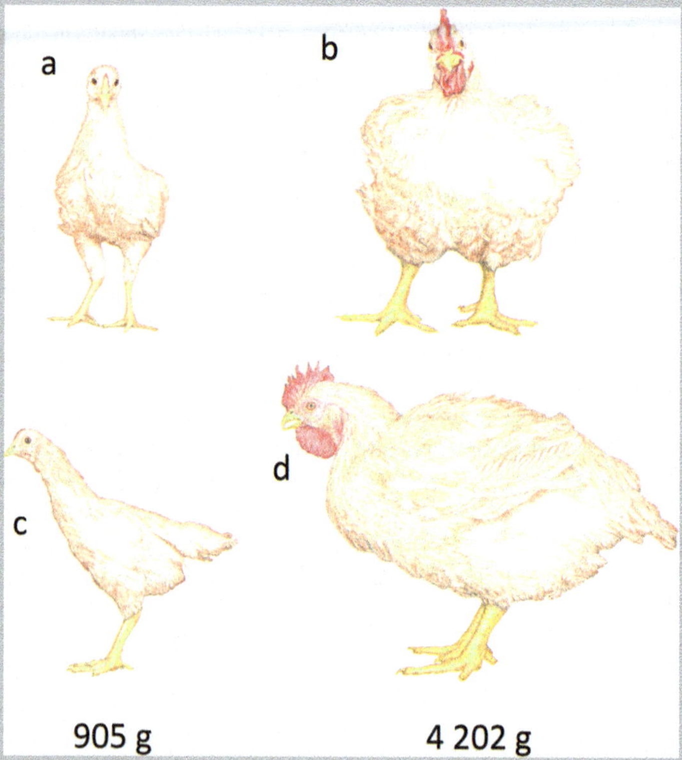

Fig. 2.53 Chicken (*Gallus gallus domesticus*) before and today. Example of changes in the morphology and body weight of broilers at the age of 56 days (**a**, **c**) representative of broilers bred in 1957; (**b**, **d**) example of broilers bred in 2005. (Adapted from Zuidhof et al., 2014)

2.4.2 The Main Species of Livestock and Their Origin

As with plants, only a small proportion of available species have been domesticated in animals. This is due to the fact that some species were more suitable for domestication than others. Herbivores were certainly more suitable for domestication than predators, and species less selective for food were more suitable among herbivores. Fast reproduction and a short reproductive cycle was also advantageous, which allowed the breeder to make a faster selection of suitable individuals. The ancestors of domesticated species were often highly variable species living in large area, which adapted well to different ecological conditions and often formed a number of geographical forms or subspecies. It was important that the domesticated species reproduced well in captivity. More suitable were calm, mild, nonaggressive species. Finally, those species that lived as wilds in herds with a hierarchical social structure

Table 2.12 Global biomass of humans domesticated and wild animals in 1900 and 2000 in t C 10^6 according to Smil (2011)

Year	Human population	Wild terrestrial animals	Domesticated animals
1900	13	10	35
2000	55	5	120

seem to be easier to domesticate. In addition to domesticated animals, we keep a number of wild species in captivity without being truly domesticated (Box 2.12). Domesticated animals may come from several independent domestication event, which then breed together such as cattle (McTavish et al., 2013) or may come from one domestication event in one place from where the domesticated forms spread over the world such as dog (Pang et al., 2009).

As with plants, domestication is a mutually beneficial relationship and provides the domesticated species with a greater degree of reproduction and subsequent distribution compared with the wild form. This agrees with the observation that in many widespread domesticated animals, their wild ancestors are either extinct or rare and endangered in nature. This can be very clearly demonstrated by the changes in the biomass of wild and domesticated animals in Table 2.12. It can be seen that the biomass of domesticated animals is an order of magnitude larger than that of wild animals, and this difference is constantly increasing.

Domestication brings a number of morphological and behavioral changes. Perhaps, most important is the reduction in aggression and less forceful response to external stimuli (numbness, one may say), this is accompanied by a reduction in the brain size. These changes in brain size are not reversible even with subsequent wildening. For example, in dingoes, it has been observed that their brains still have the proportions of a dog rather than a wolf. It is interesting that domestication is also manifested by external features, sometimes called domestication syndrome, characterized by lowered ears, twisted tail, white spots on the body, especially on the head. These morphological features often occur in animals selected for their tame. Domesticated animal not only become tame but many species, such as dog, and horses develop ability to understand the human gestures and speech (Miklosi & Soproni, 2006). In domesticated animals, there are often morphological changes related to the way they are used, and compared with wild forms, there is often a reduction in size.

Another change that occurs in some domesticated species is the loss of seasonality of reproduction, and reproduction takes place throughout the year. It is believed that there were several ways in which the animals were domesticated. One of them was a situation where animals became commensals close to humans, such as cats that hunted mice near grain warehouses. Another possibility was a situation where previously hunted prey was domesticated. Prairie, sheep, goats, or pigs are thought to have been domesticated in this way.

It is also noticeable that in many domesticated species such as dogs of cattle variation in body size or external morphology (Fig. 2.54), namely, variation between individual breeds, become substantially larger than variation in these morphological features in their wild ancestors (i.e., wolf or anchor). It is expected that this can be caused by the fact that initial population of domesticated animals was quite small (which may speed up propagation of genetic drift in the population) and more

Fig. 2.54 Buffalo and selected breeds of cattle, and domestic buffalo. (**a**) buffalo (*Bubalus arnee f. bubalis*), (**b**) zebu (*Bos primigenius indicus*), (**c**) Jersey cattle, (**d**) Charolais cattle, (**e**) Scottish Highland cattle, (**f**) Texas longhorn cattle (**c–f**, *Bos primigenius f taurus*)

Table 2.13 Production of the main commodities of animal production and stocks of animals that ensure this production. Worldwide data according to FAOstat for 2016

	pieces 10^6	Annual production per individual kg $year^{-1}$ $piece^{-1}$	Annual production weight 10^6 t
Pork meat	1478.2	79.9	118.1
Chicken meat	65847.4	1.6	107.1
Beef meat	302.0	218.4	66.0
Sheep and goat meat	1011.3	14.6	14.9
Other poultry meat	4388.3	4.8	13.1
Buffalo meat	26.2	146.1	3.8
Rabbit meat	980.8	1.5	1.4
Horse meat	4.8	154.2	0.7
Meat of camels and llamas	3.4	124.6	0.6
Game meat	0.7		2.0
Other meat	128.9		2.0
Cow's milk	273.8	2407.6	659.2
Buffalo milk	64.9	1709.4	111.0
Other milk	459.6	160.8	28.3
Eggs of hens	7671.2	9.6	73.9
Other eggs	212.0	32.4	6.9

importantly, the release of natural selection in domesticated forms may allow the survival of forms which otherwise cannot survive in wild, for example, due to lower metabolic efficiency. This leads to the accumulation of mutations, which are nonlethal, but it would be deleterious in wild. This could have facilitated increase in genetic diversity which could provide base upon which artificial selection has shaped modern breeds of some domestic species with large phenotypic variation (Björnerfeldt et al., 2006).

Table 2.13 shows the share of individual species of domestic animals in the total production of meat, milk and eggs. It is clear that five species of animals, cattle, pigs, chickens, sheep and goats, cover more than 90% of meat production. Of these five species, four were domesticated in the fertile crescent.

Cattle and the Like Zoologically, the domestic *Bos primigenius* was formed by the domestication of the wild form of the same species (Fig. 2.54) in the area of the Fertile Crescent about 7000 years ago. To the same genus belongs yak *Bos grunniens*, which was domesticated in the Pamirs and Tibet. Finally, we mention domestic buffalo (*Bubalus arnee* f. *bubalis*), which is one of the most important animals in the tropics. In terms of milk production, it is the world's most important milk-producing species after cow.

Sheep and Goats Sheep (*Ovis aries*) have been domesticated in the Fertile Crescent area, probably several times from mouflon (*Ovis orientalis*). The goat (*Capra aegagrus hircus*) was domesticated from the wild goat (*C. aegagrus*) also in the Fertile Crescent, about 10,000 years ago.

Deer Many deer are bred semiwildly in fields and today also in smaller enclosures, especially fallow deer and sika deer. However, the most important domesticated species of deer is the reindeer (*Rangifer tarandus*) which occurs wildly in the Arctic and has been (semi)domesticated by a number of tribes.

Camels and Similar Species The domesticated form of the camel (*Camelus ferus*) differs relatively little from the wild form, which was originally native to Central Asia and which still survives in remnants in northern China. The camel was domesticated by Mongol tribes about 4–5000 years ago. Lamas and vicuñas were domesticated in the Andes.

Horses and Donkeys The domesticated horse (*Equus ferus caballus*) is a subspecies of wild horses, which now included the extinct tarpan (*Equus ferus ferus*) and the endangered Przewalski's horse (*Equus ferus przewalskii*). The circumstances of the horse's domestication are still the subject of speculation. It is believed to have occurred in the area of present-day Kazakhstan. The donkey (*Equus asinus*) was domesticated in North Africa 5000–6000 years ago. Its ancestors are probably various subspecies of the African donkey (*Equus africanus*).

Pigs The domestic pig (*Sus scrofa domesticus*) bred in our country is a subspecies of the wild boar (*S. scrofa*), which was domesticated. The domestication occurred in the Middle East before 13,000 years ago, perhaps earlier.

Poultry The most important species of poultry farmed is the domestic fowl (*Gallus* f. *domestica*), which is a domesticated form of the red junglefowl (*Gallus gallus*) living in Southeast Asia (Fig. 2.55). Its domestication probably occurred in India and China more than 5000 years ago. The duck (*Anas platyrhynchos domesticus*) was also domesticated from a wild form of the same species about 4000 years ago in Southeast Asia. European domestic geese originated from the domestication of the greylag goose (*Anser anser*) while Chinese domestic geese are derived from the swan goose (*Anser cygnoid*). The time and place of domestication of geese are not entirely clear. It is believed that European geese were domesticated in the Eastern Mediterranean.

Small Mammals The domestic rabbit (*Oryctolagus cuniculus* f. *domesticus*) is a domesticated form of wild rabbit, with which it belongs to the same species. Its domestication occurred in southwestern Europe just before AD, and perhaps maybe even 1000 years ago in China.

We have already mentioned that most meat production and a large part of the world's farmed animals are animals that have been domesticated in the Fertile Crescent. Only domestic fowl domesticated in India were among them. As we can see from the above overview, many other animals have been domesticated in other parts of the world. Why did those livestock coming from the Fertile Crescent spread so much? It's probably a combination of several factors. In addition to their

Fig. 2.55 Common domestic animals, (**a**) pig (*Sus scrofa domesticus*), (**b**) fowl—male and female (*Gallus domesticus*), (**c**) horse (*Equus caballus f. Caballus*), (**d**) goat (*Capra aegagrus hircus*), (**e**) sheep (*Ovis aries*), and (**f**) rabbit (*Oryctolagus cuniculus f. Domesticus*)

adaptability to various environmental conditions, one of the other factors may be the fact that they were widespread in Europe at the beginning of the modern age during the period of European colonial raids. The animals were part of the equipment of almost all colonial expeditions, which spread them to many other parts of the world.

Within species, we distinguish breeds of livestock, which are a certain analogy of plant varieties. The breed is an artificially created unit include stock of animals of the same species, which have some characteristic features, which are heritable and usually having been developed by deliberate selection. There are many definitions for the breed, for example, regulation (EU) of the European Parliament and of the Council (2016/1012) state that "breed" means a population of animals sufficiently uniform to be considered to be distinct from other animals of the same species by one or more groups of breeders who have agreed to enter those animals in breeding books with details of their known ascendants for the purpose of reproducing their inherited characteristics by way of reproduction, exchange, and selection within the framework of a breeding program. Unlike plants, the emergence of a new breed is more of a collective work of a large number of breeders. Thus, in contrary to plants, there is no formalized assessment of what a sufficient difference is (according to the above-mentioned EU regulation is what one or more groups of breeders agree on) and even there is usually no exclusive copyright to the breed, as in the case of plant varieties. However, the method of keeping the breeding books (studbook or herd books) is formalized, which the designated authority entrusts to a selected subject. This de facto formalizes the breed. The studbook is used to register breeding individuals of the breed. The individual who is in the breeding book therefore belongs to the given breed and it is possible to compile a pedigree, etc. for him. (Examples are dogs with a pedigree). Individuals who are registered in the studbook and especially those who are intended for further reproduction also correspond to the description of the breed. On the other hand, there is a certain genetic drift within the breed.

Even before the start of targeted breeding efforts, there were regional breeds, which arose as a result of local conditions and intuitive selection of breeders. As with plants, new breeds are created by crossing and selecting suitable individuals. The selection of individuals is worked much more consistently here than with plants. This is also due to the fact that in animals, we are able to influence the mating of individuals. We therefore make sure that the individuals we use for further breeding have the best possible conditions to maintain or improve certain traits. As a result, these traits change over time within a single breed. To assess the suitability of a given individual for reproduction, the assessment of his traits associated with the expected performance, the performance of parents or siblings will help us. Artificial insemination even gives us the opportunity to use the performance of the individual himself. For example, we can collect semen from a bull during its life, from which we prepare a larger number of frozen insemination doses. We use some of them to fertilize several cows and then monitor his offspring. Subsequently, we will use the stored seed only from those individuals whose offspring are better than the average in the parameters determined by us. This procedure is due to the fact that we can store the seed for a long time, because at the time when we verify the quality of its offspring, the original bull is usually long after death.

Box 2.12 Are Elephants Domesticated?

The Indian elephant (*Elephas maximus*) is an endangered species listed in CITES Appendix 1. It is estimated that 37,000–48,000 wild elephants live in the wild (Lair, 1999). Their population is threatened not only by habitat loss but also by hunting. At the same time, Indian elephants have a long tradition of human exploitation dating back at least 4000 years, perhaps even further. Elephants are used in the military, logging, transportation of heavy loads and other activities, and currently in the tourism industry. Elephants are unique not only for their strength, but also for their longevity (40–65 years) and thus their long working life. It is estimated that about 16,000 elephants lived in captivity at the break of the century. The vast majority of them are elephants obtained by breeding wild elephants caught in the wild. This is because pregnancy in elephants lasts up to 22 months and then the female elephant takes care of its young for several years, during which time it is difficult or even impossible to use it for normal work activities. The pregnancy of a working elephant therefore means a long outage from the work process. An important part of the elephants that work today were obtained by hunting elephants in the wild. This hunt is a significant threat to the wild population. In many countries (Thailand and others), hunting of wild elephants has been banned. Mandatory registration of elephants has also been introduced and some countries, such as Nepal, have introduced a special program for breeding elephants in captivity (Fig. 2.56). However, in most countries where elephants occur, the number of newborn elephants in captivity is still small. In addition, elephants born in captivity usually have only one parent (mother) who lived in captivity. It is customary to have domestic females fertilized by wild males who actively seek out pregnant females (Lair, 1999). In principle, elephants have never been exposed to human-controlled breeding. Therefore, there have been virtually no domestication changes in the Indian elephant, and several previous experiments have shown that they can be released back into the wild very successfully. From this point of view, captive-bred elephants represent a kind of genetic reserve of the elephant population. Elephants are therefore a good example of wild animals, extensively kept in captivity, whose domestication in the true sense of the word is currently underway.

(continued)

Box 2.12 (continued)

Fig. 2.56 Elephants in Chitwan breeding station Nepal

2.4.3 Animal Husbandry

2.4.3.1 Pastoralism

Grazing, especially nomadic pastoralism, is one of the oldest ways of farming, although there is debate as to whether nomadic pastoralism preceded established agriculture or developed only as a specialization in the use of marginal nonirrigated, less fertile areas.

The intensity of grazing depends on the length of grazing and the number of animals. We can graze in different ways. During the year-round stay of animals on pasture (continuous grazing) we can count on production mesotrophic meadows with an intensity of about 0.5–1 LU ha $^{-1}$ (LU = livestock unit (also LSU) is the grazing equivalent of one adult dairy cow producing 3000 kg of milk annually, without additional concentrated foodstuffs, sheep or goat is about 0.1 LU, horse 0.8 LU). Rotation grazing is based on the division of pasture into a certain number of fenced sections—paddock, which are gradually grazed several times during the grazing period. Livestock rotate between padocks. During part of the season when paddock is grazed the density of livestock in paddock is much higher than during continuous grazing. The number of animals can also be calculated according to crop productivity and animal consumption. The yield of pasture can be estimated by harvesting areas with a known area. Daily consumption of green fodder by animals is as follows: the amount of saved green matter in adult sheep is in the range of 5–8 kg, sheep rams up to 10 kg, goat 5–9 kg, adult cattle up to 60 kg. The amount of unused biomass on production pastures is between 10% and 50%.

Different domestic animals graze in different ways. For the main livestock species most commonly used for grazing, these differences can be summarized as follows (Frost & Launchbaugh, 2003): Sheep prefer herbs > broadleaf grasses > narrowleaf grasses > sedges and bulrushes > shrubs. They are resistant to plant secondary metabolites. They bite the growth low above the ground, the grazed grass should not be high, optimally 10–15 cm. Goats—they are similarly selective as sheep, but up to 80% of the food can be woody plants (they are able to bite the higher leaves, for example, when standing on their hind legs, and they also bite the bark) and do not avoid throwing grasses either. When grazing, more mature vegetation focuses on the middle part of the stand. Compared with sheep, they prefer to graze higher above the ground. Like sheep, they are relatively resistant to plant secondary metabolites. Horses prefer narrow-leaved grasses over broad-leaved and herbs, bite the growth low. They need enough space to move. It divides the pasture into a place for grazing and excrement. During free grazing, a mosaic and saved and unsaved areas are created (Fig. 2.38), they create bare places by stepping on a dust bath. They need more time to graze than ruminants. Cows prefer narrow-leaved grasses over broad-leaved and herbs. Cows rather plucking the foradge (they pile it in their mouths with their tongue and plucking), they need rather higher vegetation—optimally 20–25 cm, grazing in vegetation with a height below 8 and over 30 cm is for problematic. They like to change the place of grazing. They graze more uniformly than goats.

By grazing, the animals obtain the necessary substances and energy, however, water must be accessible on the pasture and we supplement the animals with minerals. In more intensive forms of grazing, animals can be fed wit grain of complex feeds. If the animals are grazing for a long time, especially in the mountains or where adverse climatic conditions are common, there are shelters or screens on the pastures that provide protection for the animals in bad weather.

The most natural type of grazing, which most closely resembles the grazing of wild herbivores, is the year-round movement of a herd composed of several species of animals. This allows the herd to move naturally and create a wide gradient of variously grazed areas. This type of grazing also exerts more balanced pressure on different groups of plants, depending on the preferences of the individual species. The influence of animals on seed transport and the formation of spatial heterogeneity of vegetation will be more pronounced. Larger herbivores make the herd more resistant to predators. However, such an approach requires the completion of the necessary infrastructure for fencing (different species of animals respect different types of fences), shelter, watering, or feeding (different species have different requirements for mineral nutrition) and is only suitable for large areas. It is interesting from a conservation point of view, but due to the fact that there is a gradient of differently used places, the use of pasture is not very intensive. Seasonal grazing is similar. It uses the seasonal occurrence of grazing. In Europe, for example, it is used in the mountains, where higher pastures are used in summer and animals overwinter in valleys in winter. For smaller areas or if we want to increase the intensity of grazing, we usually graze only for a certain period of time in smaller patches most often fenced by electric fences.

According to the intensity, we distinguish several types of paddocks, where the size of the paddock and the time for which we offer grazing in this area decrease with increasing intensity. Pasture fencing can also potentially affect the movement of wildlife in the landscape, which can affect their populations (Harrington & Conover, 2006). There are a number of ways to modify fencing to reduce its impact on wildlife.

2.4.3.2 Keeping of Housed Animals

In more intensive farms or for some species, we keep animals in stables. Stables can vary from light to massive structures, depending on climatic conditions and the type of animals kept. The insulating properties of the material should ensure sufficient thermal comfort of the animals, at the same time, excessive condensation of water in the stable and drafts should be avoided, especially in winter. The animals can be housed either loosely or attached in individual stalls. In the stables with individual stalls, these are arranged in several rows provided on one side with a feeding aisle, through which feed is fed to the troughs and on the other side with a manure aisle used to remove manure. If dairy cows are housed in this way, they can be milked in stalls. In free-range housing, animals can move freely. It is fed from troughs, where the feed is replenished from bins (self-feeding). At present, various identification collars or other similar devices make it possible to individualize the feed for individual animals. If dairy cows are housed in this way, they are milked in milking parlors, where they typically move themselves.

Stalls or resting areas for animals may be provided with bedding. Winter straw is most often used as litter, however, in the past peat, forest litter or sawdust were used. The mixture of litter and feces forms manure after maturation. The bedding is changed either daily, then it usually consists only of straw, or once in a longer period of time (deep bedding). Deep bedding is formed by a combination of several materials, sawdust, straw, and for example, in pigs, it is replaced after breeding the entire animal rotation. Animals can also be housed without bedding. When raised without litter, sludge is produced, which is either flushed or collected in pits, from where it is distributed to meadows as fertilizer.

2.4.4 Feeding Livestock

We feed animals with various types of food. These include green forage, freshly mown or further cut green parts of plants. Forage can be either proteinaceous (various types of clover, alfalfa, clover grass mixtures, grassland, legume–cereal mixtures, grain or green legumes, and beetroot, etc.) or with a higher content of carbohydrates (corn, sunflower, etc.). Fresh forage is found available only in the growing season, and it is necessary to consume it immediately. Therefore, fodder is preserved in a number of ways. Probably, the oldest and easiest way to preserve green fodder is to dry and produce hay. Hay is produced by drying the mown meadow stand, or the clover stand, alfalfa, etc. Drying is usually performed by sunlight, spreading the mown vegetation on the ground and turning. Sometimes hay is dried on special dryers. Quality forage, such as young alfalfa, can be used for hot air drying, this is energy-intensive process, but allows them to produce high-quality feed locally. Another frequently used method of forage preservation is the production of silage. Silage is a process of biological preservation of forage, in which the forage is kept juicy without the access of air and is preserved with lactic acid formed as a result of acid fermentation. The process is reminiscent of pickling sauerkraut. To ensure suitable fermentation processes, fodder with a sufficient content of fermentable carbohydrates (maize), anaerobic conditions (sealing, air displacement, and sealing) and a suitable temperature (around 30 °C) are required. Artturi Ilmari Virtanen received the Nobel Prize (1945) for research in the field of forage conservation, especially silage. Haylage is mainly used to preserve wilted (partially dried—with a higher dry matter content) protein forage under anaerobic conditions. Several factors contribute to the preservation: physiological dryness of the material, anaerobic conditions, high concentrations of CO_2, which slow down microbial processes and partially also fermentation processes similar to silage can take place in haylage.

Fodder root crops (potatoes, fodder beets, fresh sugar beet pulp) are also used. An important component of feed is grain. In the past, feeds of animal origin (meat and bone meal, and fish meal, etc.) were used to supplement proteins in the feed ration. Today, their role is mostly replaced by soya or proteins extracted from it.

The feed ration is usually supplemented with a mixture of minerals supplementing important minerals, which are deficient in the plant diet of animals, especially calcium, phosphorus, sodium, chlorine, magnesium and trace elements iodine, iron, manganese, copper, cobalt, zinc, molybdenum and others. In addition, vitamins are part of complete feed mixtures.

From the specializations of agricultural production, it happens that breeders often do not obtain feed from their own production but buy finished feed mixtures from industrial producers. These may then contain a significant proportion of materials (e.g., soybeans) imported from third countries. The share of industrial feed mixtures is the largest in poultry and pigs (Roubalová, 2017). As already mentioned, soya is often imported, which may cause environmental problems in place of production. It may have negative effect on the environment also in the importing country. For example, when import of soya reduces use of local high-quality forage such as alfalfa or clover and thus reduces their production and hence proportion of these, soil improving crops, in crop rotation.

2.5 Animal Production and Its Effects on Organisms and Ecosystems

2.5.1 Impact on Habitats and Biodiversity

Table 2.12 shows the worldwide increase in livestock biomass, although as we have shown, locally the trend may be different (Table 2.14). This increase is related to the increase in human population and increase in the consumption of meat and others animal products. Meat consumption increases with the wealth of individual countries (Fig. 2.57).

Table 2.14 Meat production in kg per kg animal and year in developing and developed countries in 1980 and 2005 (Steinfeld et al., 2006)

	Beef		Pigs		Sheep & goats		Poultry	
	1980	2005	1980	2005	1980	2005	1980	2005
Developed countries	0,17	0,21	0,14	0,33	0,19	0,24	2,26	3,55
Developing countries	0,06	0,09	0,88	1,20	0,14	0,26	1,29	1,98

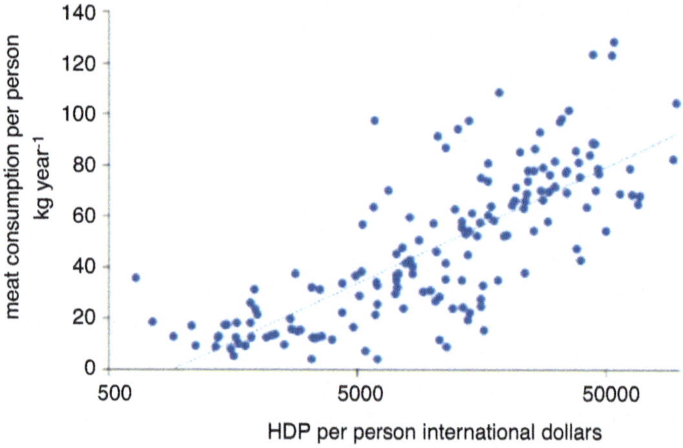

Fig. 2.57 Meat consumption in individual countries of the world in relation to GDP per capita, FAOSTAT

As we showed in the introductory chapters, the organism at a higher trophic level consumes more energy, because of energy losses that occur when energy is converted from lower trophic levels. Thus, meat consumption increases the impact of human population growth on resources. In other words, as meat consumption increases, the human population needs a larger and larger area to make a living. The growtth of meat production can be acieved by several ways. Fig. 2.58 shows a significant increase in the number of animals in least developed countries, while in the European Union (EU), for example, the overall numbers are rather stagnant or even slightly declining. Increase in the number of animals is taking place in underdeveloped countries, where extensification is to the detriment. Increasing the number of animals and using more land, which is obtained by converting natural habitats into pastures and fields, makes it possible to produce enough fodder.

Increasing production in developed countries is achieved by intensification of breeding and larger growth of farmed animals. Table 2.14 shows that the weight gain in kg of farmed animals increased more in developed than in developing countries, with the exception of sheep and goats. The impacts of increasing meat production on biodiversity thus differ in developing and developed countries (Table 2.15).

In developing countries, the species of animals that feed on pastures, that is, sheep, goats and cattle, are rising the most. Intensive grazing results in excessive vegetation removal, erosion and thus a reduction in the soil's ability to support

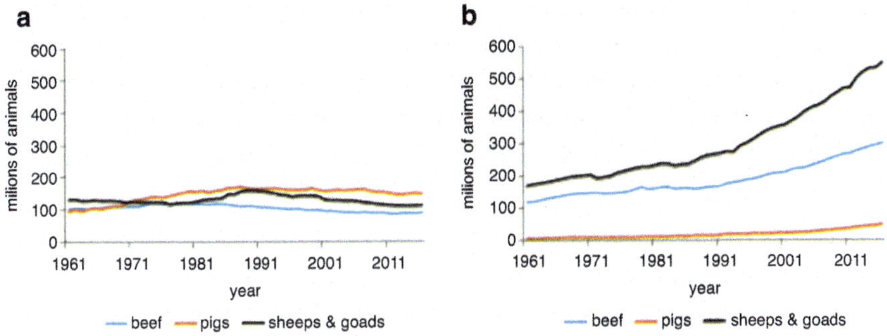

Fig. 2.58 Changes in the stocks of selected livestock in EU countries (**a**) and in the least developed countries in the world (**b**) FAOSTAT data

Table 2.15 Impacts of increasing meat production on ecosystems in developed and developing countries

	Developing countries	Developed countries
Changes in animal production	Lack of capital and infrastructure limits the intensification of farms. Increasing production is achieved by increasing the number of animals, especially grazing species	Intensification of livestock production leads to a higher concentration of stabled animals and often to a reduction in grazing
Impacts	Increasing the density of grazing animals is putting pressure on natural ecosystems and existing pastures. Intensive grazing leads to the loss of natural habitats, damage to vegetation, increased erosion, which starts a spiral of degradation of soils and entire ecosystems	By reducing grazing, extensive pastures disappear, and the demand for forage decreases, which leads to the abandonment of meadows, especially in marginal areas. This destroys valuable, species-rich habitats of extensive meadows and pastures, to which a large part of the diversity of the cultural landscape has been tied

vegetation, which is then even more easily subject to excessive grazing. Excessive grazing can reduce production, increase erosion, cause loss of soil organic matter and lead to land degradation (Belnap & Gillette, 1998) especially if some other environmental factors such as water shortage limits vegetation growth. This leads to the degradation of habitats and thus to a decline in wildlife populations. The pressure of domestic animals is most pronounced on islands, where, for example, goat populations cause significant changes in vegetation, reduce biomass, species cover, lead to erosion and soil loss, which in turn has led to the extinction of many species (Gizicki et al., 2017). In contrast, adequate grazing supports the growth of the root system and the entry of organic matter into the soil, which can help counteract erosion (Hubbard et al., 2004), support biodiversity, etc. Restricting grazing in developed countries therefore also has negative consequences for biodiversity and the functioning of ecosystems.

This problem would seem to be mainly a matter for developing countries, but the intensification of crop production requires sufficient feed. It is no exaggeration to say that most of the consequences of intensive crop production, which have been described above, are indirectly linked to animal production, respectively with meat consumption, because with lower meat consumption, we would not need a large part of crop production. In addition, quality feeds require a suitable source of protein, which is most often soya.

Soybeans can either be grown locally or are more often imported from middle-income tropical and subtropical countries such as Brazil (more than 30% of Brazilian soybean production is exported to the EU). Developed countries use these imports to export the problem of habitat degradation associated with obtaining the necessary feed to other countries. At the same time, the intensification of livestock production leads to less interest in green forage, especially in marginal areas, and this leads to the abandonment of meadows and pastures. As we have shown in the chapters on meadows and pastures, the greatest diversity is at the optimal (medium) density of herbivores. When the pressure of the herbivores drops or rises significantly, the diversity decreases. Therefore, this decrease in herbivore pressure (abandoning meadows and pastures) due to intensification leads to a further decrease in biodiversity.

In addition to the above-mentioned impact on habitats, animal husbandry but also animals kept as pets can have other mostly negative consequences for wildlife. We will give only a few examples here. Domesticated animals can affect wild populations through genetic erosion. We have mentioned above that during domestication, a number of changes occur that facilitate the coexistence of animals with humans, but also decrease their chances of survival in the wild. Hybridization of a domesticated and wild populations can lead to changes in the wild population, which then may deteriorate or even extinct or live as commensal alongside the human population. An example is the crossing of a rock pigeon, the ancestor of a domestic pigeon, with wild individuals of the domestic form. Most wild pigeons today carry the genes of domestic pigeons. Breeds of homing pigeons are thought to have contributed the most to this hybridization (Stringham et al., 2012). Similarly, most wild populations of the red junglefowl are now contaminated with domestic hen genes (Peterson & Brisbin, 1998).

Predation of domesticated animals in wild populations can also be significant. Man keeps either as pets or as fur animals a number of small predators. These can then become important predators of small animals. The North American mink, which is widely bred in Europe and over time, part of the population escaped into the wild, where it multiplied. The mink occupies a similar ecological niche as the otter and thus both competes with it and increases the damage to pond management, which is also often attributed to otters, which further aggravates the social debate on its protection. Another worth mentioning example is cat predation. Cats have been imported to most countries of the world and are the cause of the extinction of island populations of 33 species of small vertebrates. In the United States, domestic and feral cats kill 1.3–4.0 billion birds per year and 6.3–22.3 billion small mammals, making them the most significant anthropogenic cause of small bird and mammal mortality in the USA (Loss et al., 2013).

Other impacts on biodiversity and the functioning of ecosystems are related to the use of veterinary medicines. In the breeding of domestic animals, a number of drugs are used, both against diseases and preventively to reduce the occurrence of parasites. "Deworming" preparations against helminths in the digestive tract are often used, which subsequently become part of the excrement. Some of these preparations may be toxic to insects and may thus affect the subsequent colonization of feces by insects (Floate et al., 2016). In the case of excrement produced on pastures, this will affect not only the communities of coprophagous insects but also the rate of fecal decomposition and nutrient recycling (Lumaret & Kadiri, 1995). For example, dichlorvos, used to deworm horses, was very toxic to insect populations, especially beetles living in horse manure, which significantly slowed its decomposition (Lumaret, 1986). It is also assumed that the application of this supplement, together with the boom in tourist horseback riding, significantly damaged the population of coprophilous beetles on the Mediterranean sand dunes in France at the turn of the century.

The influence of veterinary antibiotics, which, together with antibiotics used in human medicine, enter the environment is also significant. It is estimated that the amount of antibiotics excreted by domesticated animals is thousands of tons per year worldwide. These antibiotics remain in the environment for various lengths of time, especially in soil, the persistence of some antibiotics can be considerable, leading to their accumulation. Increased concentrations of antibiotics also lead to the selection of strains of soil microorganisms resistant to one or more antibiotics (Kemper, 2008).

2.5.2 Impact on Biogeochemical Cycles, Pollution, and Global Change

As shown in Table 2.12, the biomass of domestic animals is an order of magnitude higher than the biomass of wild animals, but it is also several times higher than the biomass of the human population. The animals thus constitute important consumers of plant production, and thus are also an important player in biogeochemical cycles of key biogenic elements, especially N. The annual production of manure and urine from livestock contains considerable amount of nitrogen. This is also because only a small part of ingested nitrogen is used by livestock, on average about 10% and the rest (90%) is left in excrements and urine. Nitrogen efficiency varies between animal species, with the highest in poultry (34%) and the lowest in cattle (5%). Globally, nitrogen in the excrement and urine of livestock represents 102 MT N year^{-1} (Vayssières & Rufino, 2012). Amount of nitrogen in excreta and urine is substantially higher than the amount of N in meat produced by domesticated animals (c. 12 MT N year^{-1}). Only a small part of the nitrogen contained in the urine and excreta of farm animals is taken by plants. The greater part of nitrogen either goes into the atmosphere (about 42 MT per year) by volatilization of ammonia and by the production of NO_x or denitrification, or is leached from the soil, particularly in the form nitrates (37 MT annually). For comparison, the volume this unused nitrogen contained in the urine and excreta is an order of magnitude comparable to the volume of N, which is annually fixed from the atmosphere by Haber-Bosch synthesis (Vayssières & Rufino, 2012). Animal production is responsible for 65% of anthropogenic production of NO_x and 64% of anthropogenic ammonia production (Steinfeld et al., 2006).

These losses of nitrogen take place in three main steps: (1) during the collection and processing manure, or sludge, (2) during application of manure on the land or during deposition of manure and urine to pasture by grazing animals, and finally (3) due to loss by leaching or washing away from the soil. During the production of the manure in stables and its subsequent processing, in particular, N loses ammonia volatilization. In average, about 20% of the nitrogen contained in the urines is lost at this stage. The smallest losses by ammonia volatilization occur in this phase occur in sheep (5–10%), while the largest losses in poultry and pigs (36%). After the application of manure or dung deposition on grassland again, the most important loss happens by the volatilization of ammonia. Again, there is considerable variability, on average, about 20% of the nitrogen contained in the manure is lost. Losses are generally greater in the tropics than in the temperate and in monogastric animals than in the ruminants. N losses after application are the largest source of NO_x, which is a significant greenhouse gas. The observed losses range from 0.1% to 4.6% N in manure piles, on average, 0.6% is calculated when applying manure and 2% when depositing excrement on pastures. After application to the soil, leaching and runoff are a significant form of N loss. These losses are very variable and involve nitrogen from added manure and mineral fertilizers, as well as mineralization of soil organic matter. It depends on the amount and form of N entering the soil, timing, soil type, and climate, etc. It is estimated that in clay soils, 10–40% of the N added to the soil can be leached, while in sandy soils, it is 25–80% (Vayssières & Rufino, 2012).

Nitrogen losses can be reduced by a number of measures at different levels. Higher efficiency of nitrogen fertilizer utilization and thus lower content of excreted N can be achieved by better optimization of N supply with regard to the growth needs of animals, and then the regular removal of manure. Storage losses can be limited by covering manure and especially slurry tanks. In terms of application, the greatest volatilization of ammonia occurs when manure or slurry is sprayed on the soil surface. This can be limited by using hoses for application on the soil surface or incorporation into the soil (Vayssières & Rufino, 2012).

The reduction of NO_x production is possible by longer storage of manure or by manure processing in anaerobic digestate for methane production, because then less carbon remains in the applied material. However, this process increases the volatilization of ammonia and, with prolonged storage, the production of methane is increased. The efficiency of the use of nitrogen fertilizers, mineral and organic can be achieved by better synchronization of the application of fertilizers with the growth needs of plants (Vayssières & Rufino, 2012).

On the other hand, as shown in the chapter on fertilization, with proper treatment and use, livestock manure can be an important source of nitrogen, and a higher share

Table 2.16 Soybean yield in the tropics and temperate and carbon loss required to convert one ha of (semi) natural habitats to soybean fields, according to West et al. (2010)

	Yield from ha $^{-1}$	Loss C associated with a change in land use t C ha $^{-1}$
Tropics	1,7	120
Temperate zone	3,8	62

of organic fertilizers in the total nitrogen dose increases the nitrogen efficiency to some extent.

Another significant environmental impact of livestock farming is methane production. Animal husbandry, manure production, and other related activities are responsible for 37% of the anthropogenic methane production (Steinfeld et al., 2006).

In addition to optimizing the handling of animal excrement in the production and use of livestock manure (Sect. 2.2.2.3) (in addition to reducing meat consumption), the way to reduce these impacts of livestock production is to find other sources of animal protein, see Box 2.13 and of course reduction of food waste (Box 2.14).

As we have already pointed out in the section on biodiversity, the production of nitrogen feeds is often associated with the import of soybeans from tropical countries where soybeans are grown at the expense of natural habitats. As already mentioned, this has negative consequences on biodiversity and also on biogeochemical cycles. The transformation of new areas in the tropics leads to the loss of a large amount of carbon contained in the biomass. This loss is disproportionately larger than if these areas expanded in the temperate. However in temperate yone there are no or very little free areas for such actions (West et al., 2010). This discrepancy is enhanced by the fact that yields in the tropics are smaller than that in the temperate zone (Table 2.16).

Box 2.13 Insects as a Food Source

Raising and catching insects was and is a significant constituent of human food. In terms of carbohydrates, honey is probably the most important insect product. Honey is obtained partly through beekeeping, but significant amount of honey, especially in tropical areas, is obtained from wild bees. The nests of these bees are typically completely destroyed during honey harvest.

Insects and some other invertebrates are also important sources of protein. Ramos-Elorduy (2008) mentions more than a thousand species of edible insects. Terrestrial insects are most often consumed, include Orthoptera and mainly insect larvae from the orders of butterflies (Lepidoptera), beetles (Coleoptera). Insect biomass offer a high-quality source not only of proteins, but also of energy of trace elements and other substances (FAO, 2008; DeFoliart, 1994). Insects also offer a number of new taste variations (Ramos-Elorduy, 2008). However, it can be expected that, at least in the near future, the focus will be on the use of insects rather than on the production of proteins that will become part of various food supplements or as an additive in food

(continued)

Box 2.13 (continued)

production. The advantage of insects over warm-blooded animals is fast growth and high production efficiency. The conversion factor for different types of food ranges from 2 to 12 (Ramos-Elorduy, 2008). At the same time, the ability to process less valuable forms of food, which would often be unusable as other animal feed, is important.

Box 2.14 Wasting Food

We discard about a third of the food we produce without being used. Table 2.17 depicts a comparison of the amount of food thrown away in different parts of the world. The total amount is surprisingly very similar in both developed and developing countries, but in developing countries, most of this loss is on the side of suppliers, while in developed countries, it is on the side of consumers (so in developed countries we throw away what we have already purchased). The main cause of losses in developing countries is the lack of infrastructure associated with the transport, processing and storage of food. On the contrary, in developed countries, on the producer side, it is an effort to perfect the appearance of products, which leads to the elimination of visually imperfect products that are otherwise completely edible. Furthermore, the effort to offer the customer a wide range of products (e.g., usually more than a hundred different assortments of fruits and vegetables) increases the chance that some of these products will not be sold. In addition, in developed countries, the customer spends a relatively smaller portion of his income on food than in developing countries (Fig. 2.59). This can lead to less concern for possible economic loss in the form of uneaten food.

Another reason is the labeling of food with an indication of the time of its recommended consumption. For perishable foods, the use by date, month, day

Table 2.17 Food losses per person in different parts of the world in kg per person. (According to FAO, 2011)

Region	Total	At suppliers	At consumers
Europe	280	190	90
North America	295	185	110
Developed Asian states	240	160	80
Sub-Saharan Africa	160	155	5
North Africa	215	180	35
South and southeast Asia	125	110	15
Latin America	225	200	25

(continued)

Box 2.14 (continued)

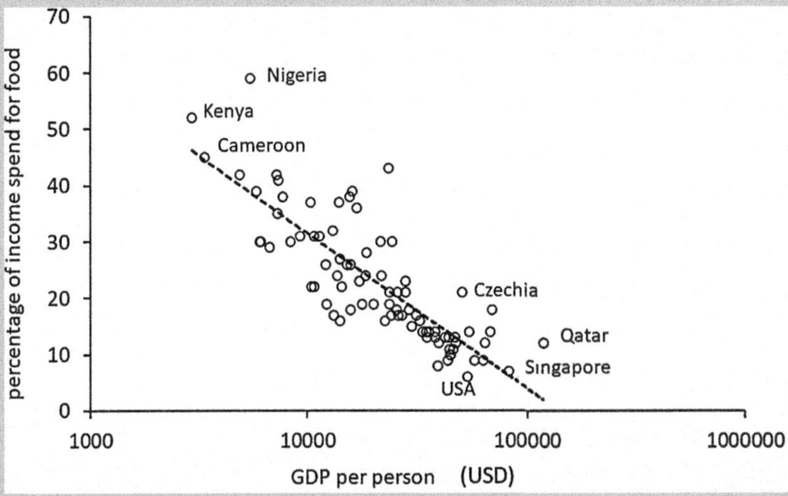

Fig. 2.59 Percentage of income spent on food in relation to GDP per capita in individual countries of the world

and year shall be indicated. After the expiration date, foods labeled in this way may not be brought to the market, although many experts point out that some of them are still edible. Durable foods are then marked with the best before date. Foods with an expired best before date can be sold if they are specially labeled and are harmless to health, and the seller assumes responsibility for this. However, there is usually less risk for the seller to get rid of such foods. Of course, individual consumers also buy more food than they can consume, and the relatively low price of food supports this. In addition, of course, food intake is much higher in developed countries than in developing countries, while energy expenditure is about the same in both groups of countries. These are foods that we do not throw away, but we somehow eat in addition to what we need.

2.5.3 *Living Conditions of Domestic Animals*

As we already mentioned, from an evolutionary point of view, domestication can be assumed as beneficial for domestic animals. It has led to an expansion of their populations to such an extent that there are usually far more domestic animals today than their closest relatives living in the wild. However, domestic animals have lost many properties important for their survival in nature. On the contrary, the animals acquired a number of skills important for living with humans. An example is the ability of dogs to understand human gestures in some respects better than apes (although there is still debate about the extent to which domestication is responsible for this phenomenon, Miklosi & Soproni, 2006). Regardless of how domestic animals understand us, it is an undeniable fact that the adaptation of animals for our needs has been beneficial to humans, and hence the moral obligation to provide animals with dignified living conditions. We discuss about animal welfare. It means ensuring sufficient adequate food and drinking water, suitable accommodation, and living condition for the animal, whether it is a stable or a temporary. The accommodation should be large enough, with an appropriate heat and light regime, an appropriate microclimate (humidity, air flow, etc.) and with an environment otherwise suitable (allowing natural behavior) for the species. It is necessary to ensure regular maintenance and cleaning of the quarters, veterinary care, regular monitoring of the animal's health and prevention of diseases and injuries. Furthermore, it is necessary not to expose animals to stress and to give the animal the opportunity to realize its natural behavior. It must be said here that although domestic animals have adapted to life with humans, they have often retained, in some modified form, certain patterns of behavior inherited from their wild ancestors. In many cases, this behavior does not serve its original purpose today. Dogs mark their territory, hens tend to perch as a remnant of spending the night on branches, hen rummaging in residual search of food, etc. Lack of opportunity to perform this natural behavior, and or environment with a lack of stimuli which promote such behavior, can negatively affect the psychical status of animals. Similarly, disturbing influences, noise, disturbance of the light regime, inappropriate and frequent handling, some preventive veterinary interventions, etc., can cause stress in animals. Absolute elimination of all disturbances, stress elimination, etc. is probably not technically possible (we have not achieved this in humans either) and a certain degree of stress is natural in the lives of humans and animals. In addition, some methods of intensification of animal husbandry suppress their manifestations of their natural behavior. The question is what is still acceptable and determining the extent of what is no longer acceptable. In most countries, these minimum limits to what is already considered animal cruelty are set out in Law (e.g., in EU country, it follows Council Directive 93/119/EC). The law regulates not only the living conditions of farm animals but also, for example, the living conditions of laboratory animals. Obtaining certain livestock products (meat, fur) or performing certain experiments on laboratory animals requires the killing of the animals. In addition to setting framework rules for animal welfare, the law also regulates the manner and circumstances of the killing of these animals. Setting rules is a matter of measure, and the rate set by law is a minimum that protects animals from cruelty. We can say that there is a social consensus that what goes beyond the law is unacceptable. On the other hand, it cannot be said that everything that meets the law automatically ensures maximum animal well-being. Therefore, many breeders, for example, in organic farming, apply higher standards of animal welfare. This often entails higher costs, which are then reflected in the higher price of the products thus obtained.

2.6 Organic Farming and Other Approaches to Environmentally Friendly Agriculture

2.6.1 Organic Farming and Its Impact on the Environment and Human Health

An organic product is a raw material, animal, feed, seed or propagating material that has been produced in organic farming. (Council Regulation (EC) No 834/2007 of 28 June 2007 on organic production. It will enter the EU from 1 January 2021, the new Regulation No. 2018/848 on organic production and labeling of organic products enters into force and will thus replace the regulations mentioned above.)

A similar definition applies to organic food. In other words, it is an agricultural product that is produced according to certain standards set by the law for individual commodities, by an subject (farm) that is registered and controlled for organic farming. If a farmer farms "organically" (for example, observes the principles of integrated pest management) but does not want to be registered due to increased control, for example, he cannot label his production as organic according to the provisions of the above regulations. As we have already stated, in addition to registration, the product in question must be manufactured in accordance with certain rules. These vary from commodity to commodity, but in general, organic production is characterized by a ban on the use of chemical-synthetic pesticides and rapidly dissolving mineral fertilizers by excluding GMOs and taking into account livestock welfare, natural cycles, and dependence.

The impact of organic farming on health is often discussed. In terms of hygienic production standards, organic food is as safe as conventional food. Extensive independent tests have shown that the incidence of bacterial contamination in organic foods is not statistically significantly higher than in conventional foods (Bourn & Prescott, 2002; Mukherjee et al., 2004). Organic foods have a number of positives. People who regularly consume organic food have fewer allergic problems, less problems with overweight and obesity. However, this may be related to the fact that this group of people lives generally a healthier life, from which the effect of organic food is difficult to separate (Mie et al., 2017). Epidemiological studies that try this have not yet shown that organic food consumption improves the health of people who consume it (Bourn & Prescott, 2002; Mie et al., 2017). This is despite the fact that there is evidence of the negative effects of the background occurrence of pesticides on children's development (Mie et al., 2017). At the same time, there are individual case studies pointing to higher residual pesticide contents in some conventional foods compared with analogous foods grown organically (e.g., Baker et al., 2002; Rekha et al., 2006). The nutritional differences between organic and conventional foods are small (Bourn & Prescott, 2002; Mie et al., 2017), for example, higher content of phenolic substances in some vegetables and fruits, higher proportion of omega-3 in the meat of animals kept in organic production, etc. (Mie et al., 2017). For many commodities, better sensory values (aroma, taste, etc.) are reported for organic foods, but even this difference is not consistent across all commodities (Bourn & Prescott, 2002). In summary, it can be said that organic food is certainly safe and in a number of parameters shows higher contents of substances considered beneficial and lower concentrations of pesticides and other substances considered harmful. However, the effect is not large enough to be captured in epidemiological studies.

As for the impact of organic food on the environment, it is very important if we look at the problem in terms of the impact on the unit of area used or production achieved. Land managed under organic farming has a significantly higher organic matter content, larger topsoil depth, higher polysaccharide content, and less soil erosion than commonly cultivated land (Reganold et al., 1987), also has lower nitrogen losses per unit area and support more species and larger populations in a number of groups of organisms (Mader et al., 2002; Hole et al., 2005; Tuomisto et al., 2012). It can therefore be said that organic farming shows better environmental parameters per unit area.

The production of individual commodities per unit area is generally lower in organic farming than in conventional farming (Goklany, 2002; Seufert et al., 2012). This difference is about 25% for average conditions but is highly variable and dependent on environmental and crop conditions. In general, under favorable conditions, the differences between conventional and organic farming are smaller than under less favorable conditions. In terms of individual commodities, the smallest differences are in fruit, while the largest are in vegetables. Regarding crop types, small differences are in the cultivation of clover and permanent crops, while they are larger in the case of annual crops (Seufert et al., 2012).

Life cycle analysis shows that organic products have lower energy consumption per unit of production for most plant products. The decrease in energy demand is large for organically grown wheat and rape, on the contrary, more energy is needed for tomatoes (Williams et al., 2006). The overall lower energy consumption is due to the nonuse of nitrogen fertilizers and pesticides, production of which is energy-intensive (Williams et al., 2006). Also, for organic livestock products, energy consumption per unit of production is generally lower than for conventional agriculture, with the exception of chicken and eggs (Williams et al., 2006). This is due to the extremely high intensity of production of these commodities in conventional agriculture and the higher production per unit of inputs. In addition, some of the environmental impacts of product processing may be greater, because production is generally small and many of the impacts associated with building and operating infrastructure are relatively significant, because they are budgeted for smaller production (Tuomisto et al., 2012). The slightly lower production of organic farming is also the reason why, in terms of per unit of production, organic food has a higher loss of nitrogen, and the associated greater eutrophication and acidification potential was reported in some studies (Williams et al., 2006; Tuomisto et al., 2012,). On the other hand, it should be borne in mind that, for example, the effect of nitrogen on diversity is not linear, and so simply dividing the eutrophication potential by the production is not entirely appropriate. On the contrary, a small increase in N may even increase the diversity (as explained in Sect. 1.5). Similarly, it is with eutrophication. If the local concentration does not exceed a certain limit, nitrogen is used by the given ecosystem and there is no further displacement and thus no eutrophication of other parts of the landscape. However, this certainly only applies to situation where the eutrophication potential has actually spread over the landscape, for example, during extensive grazing. In addition, many of the positive effects associated with carbon accumulation and water retention may have off-farm impacts and may not be fully covered in life cycle studies. No study has addressed the interaction of commercial and organic farming areas. For example, unexplored remain the question whether nitrogen flowing from conventional areas cannot be recycled in organic farming areas with a low supply of nitrogen. In addition, many species require some extensive use to promote diversity. This is evidenced by the fact that a third of the NATURA network has some agricultural management.

Overall, organic farming leads to the promotion of biodiversity and the maintenance or restoration of soil fertility at the cost of somewhat lower yields. This may not be such a problem in European conditions, where overproduction generally predominates. In addition, the biodiversity of the cultural landscape has been well adapted to extensive land use and essentially cannot exist without it.

2.6.2 Alternative Approaches to Environmentally Friendly Agriculture

Organic farming has provided considerable inspiration for a number of other directions that it seeks to reduce the environmental impact of agricultural production (Matson et al., 1997). There is a group of farmers who are interested in increasing the environmental friendliness of their production, but for various reasons, they are not registered organic farmers, for example, because they are bothered by the administration associated with registration or want to focus on achieving some specific environmental goals which are measurable, for example, by promoting diversity, instead of complying with general organic farming rules. An example is the Hope farm described in Box 2.2 on the socioeconomic factors affecting agriculture. These producers are not certified anywhere, they are not organic farmers, and their products cannot be labeled as organic production. However, there is an interesting shift in approach to the problem. While organic farming legislation seeks to follow some well-defined practices that are expected to lead to more environmentally friendly production, Hope and other similar producers seek to be competitive with conventional farmers while achieving some specific measurable environmental goal (e.g., improving the population of wild birds). Therefore, the exact procedure is not prescribed, but some indicator is monitored showing that the production is really environmentally friendly. Of course, the monitoring of this improvement is done by the producer himself, so there could certainly be dishonest people who would just declare something like that. On the other hand, monitoring these indicators on a large scale for each farm would certainly be technically and organizationally difficult. From this point of view, compliance with the rules is certainly easier to monitor. In the future, it might be interesting for the development of environmentally friendly agriculture to combine a somewhat looser list of restrictions with specific, easily measurable indicators of the impact of agriculture on the environment. Remote-sensing methods might offer suitable indicators in this regard (Zaks & Kucharik, 2011).

In addition to efforts to promote biodiversity, there are a number of research directions and related applications that seek to achieve the same or better production than commercial approaches with lower nutrient inputs and environmental impacts, especially in reducing greenhouse gas emissions (Chen et al., 2014). These procedures are based on better optimization of nutrient inputs depending on the given conditions. This includes the large-scale movement of conservation farming (Gonzalez-Sanchez et al., 2015).

2.7 Alternatives for the Development of Future Agriculture

It must be said at the outset that today's agriculture is very diverse, it represents the direction of different approaches. In the same way, different approaches and procedures will certainly intersect in the future, and it can be expected that future developments will also be a mixture of different and often diverse concepts. The following text should be understood in this regard.

Continuation of the Current Development (Business as Usual) Agriculture can certainly develop for some time as before, with the parameters of the individual tools used gradually improving. Environmental problems associated with agricultural production will be solved primarily through technical solutions, which, however, in the long run may lead to their deepening. New varieties and agrochemicals will be developed. The trend, especially in countries where high doses of fertilization are applied and can probably no longer contribute to a further increase in yield on their own, is the addition of "friendly" microorganisms. An increasing share of GMOs and modern mutagenesis methods (such as CRISPR and other similar techniques) can also be expected, especially in developing countries. Gradually, the share of human labor will decrease even more. There will be even greater globalization of the market. Developed countries will import a significant part of feed or cash crops, such as cut flowers, vegetables, fruits, herbs, from third countries with cheap labor and suitable climatic conditions. This can lead to the "export" of the environmental problems associated with this production and to the promotion of large-scale export production at the expense of small growers in these countries. Overall, intensification will lead to increasing discontinuities in material and energy flows (nitrogen is fixed on soybean plantations in third world countries will be removed, will be part of food chains in developed countries and will be replaced by nitrogen from mineral fertilizers). Reducing the share of human labor, specialization and the associated higher transport costs will lead to an even greater reduction in the energy efficiency of agriculture as a whole (Box 2.15). All these are already happening, but

Box 2.15 Development of Energy Efficiency of the Agricultural Landscape and Agriculture as a Whole

In Sect. 2.2.2.1, we dealt with the amount of additional energy required to grow a crop in one field and the energy balance of this in another field. We have shown that the amount of energy needed to grow the same amount of product has multiplied with the intensification of agricultural production. This energy return can be measured as the share of energy in the final product divided by the amount of total energy input. We call this EROI (Energy Return on Investment). However, the energy efficiency of agriculture as a whole is not the sum of the energy of individual crops grown. Additional costs are required, for example, we need to bring product grown to it to the point of consumption. Doing that the energy value of the product will not increase, but the amount of energy invested will increase. The more different parts of the landscape specialize in individual crops, the greater the transport costs and the overall energy intensity. Specialization also leads to the decoupling of traditional material flows within farms. Traditional agriculture was based on the recycling of biomass (BR—Biomass return) in one place, usually within one farm. The straw was used as animal litter, and together with the feces, it became manure and it returned to the field. With the specialization of agricultural production, there was a separation of plant and animal production. In animals, production began to dominate animal species that can be farmed with great intensity, such as pigs and poultry. Feeding for these animals is imported in the form of a prefabricated compound feed, the individual components of which have often been imported from great distances, thus increasing energy requirements. Animal feces often do not return to agricultural production, but in certain circumstances, they become waste that needs to be disposed of at the cost of further energy losses. It is difficult to capture all these flows, but there are studies that are at least partially trying to do so. Parcerisas and Dupras (2018) describe the changes that occur in the energy return of Canadian agriculture. It is evident how the EROI decreased with increasing intensification and specialization. It is worth mentioning that at the same time, the ratio of biomass recycling to invested energy decreased. In other words, instead of recycling biomass from agricultural production, the energy of pesticides, machine wear, and fuel consumption, that is, energy that comes from fossil fuels in one way or another, is increasingly contributing to the energy invested. Similar results have other researchers, for example, Marco et al. (2018), who studied the historical changes in the energy performance of Catalan agriculture (Fig. 2.60). A similar trend is shown by a comparison of different types of agriculture (Table 2.18).

It can be noted that during the development of agriculture, as the share of human labor decreases and it is replaced by the work of animals and subsequently by machinery, energy efficiency decreases. This was also evident in the examples given in Sect. 2.2.5.1. Lough (1999) attributes this to the great

(continued)

Box 2.15 (continued)

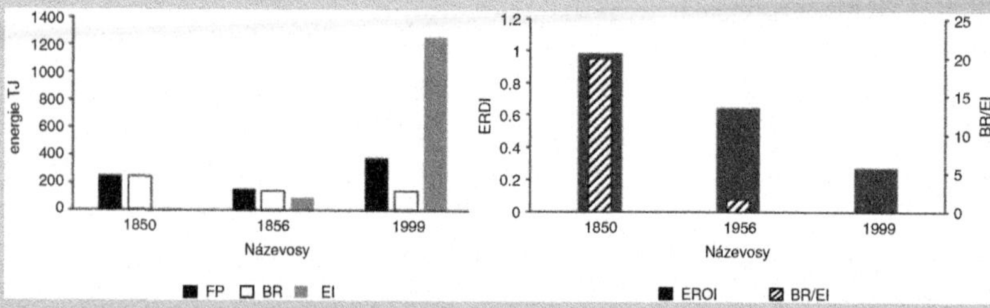

Fig. 2.60 Changes in the energy metabolism of farms in Catalonia from 1860 to 1999. FP, final production taken by humans; BR, recycled biomass; EI, external energy inputs (additional energy); EROI (Energy Return on Investment) index describing the energy return of input energy. According to Marco et al. (2018)

Table 2.18 Return on energy input EROI estimated for different agricultural production methods according to Lough (1999)

Method of production	EROI
Pre-agriculture communities	50
Gardeners, swidden shifting agriculture	40
Self-supporting agriculture	12.4
Hunters and gardeners	10
Western preindustrial agriculture	1–5
US agriculture 1970	0.11

efficiency of human labor in agriculture. On the other hand, increasing the input of additional energy (and reducing energy efficiency but increasing the effectivity of human labor) has enabled the development of modern society, that is, the fact that one person can feed more and more people and they can then specialize in various other activities. At the same time, people are devoting less and less of their energy to providing food and can devote the time saved to other work or entertainment. At the same time, as shown in Fig. 2.20, the share of biomass recycling for production is decreasing, which can be understood as a degree of system closure. In other words, such systems are more and more dependent on external energy inputs, which in today's economy come from fossil fuels. At the same time, as shown in Box 1.3, reducing biomass recycling has negative consequences for biodiversity and the ability of ecosystems to provide ecosystem services. Some authors (e.g., Lough, 1999) then conclude that such a low level of efficiency leads to habitat wear and tear faster than they can regenerate.

this trend will deepen. The continuation of business as usual therefore faces the very difficult problem of ensuring an increase in production while not increasing or, if possible, reducing the carbon intensity of production.

Continuation of Intensification: Robotization—Precision Agriculture Further intensification of agriculture presupposes a reduction in the share of human labor. This is to be achieved by a larger share of robotics, where most of the work would be done by autonomous vehicles and drones. The use of robots will enable the acquisition and processing of a large amount of information. This will allow greater use of precision agriculture. We have said before that a farmer uses information about soil properties, the amount of nutrients in the soil and the expected yield when planning fertilization. This information is now available at the level of individual plots. However, these properties are very variable even within individual fields. The development of robotics and a number of other techniques, such as remote-sensing, allows to obtain and process a large amount of information about the properties of individual plots. Robotization in individual agricultural operations will then enable the use of this detailed spatial information on soil properties on individual plots, for example, for fertilizer dosing in a spatially differentiated manner per square meter of land so as to make the most of the yield potential of each individual site. This information can also include information on the current state of plants, etc. Similarly, access to individual livestock, etc. can be optimized. The question is how this trend will develop in relation to the size of the plots and other effects of the current intensification. Some analysts expect robotization to replace unpopular manual labor, such as manual harvesting, unification, and more. And this could support the development of small farms specializing in production, which today is limited by labor shortages. This is especially probable for crops, where the determining parameter of profit is quality rather than yield (herbs, fruits, vegetables, and spices). Here, robotics could also lead to an increase in the proportion of small farms specializing in the usual crop types, especially, if it is supported by the growing interest of consumers in local products. On the other hand, robotization can be very well, perhaps easier, applied in combination with the existing large-scale production technologies. Currently, a number of workplaces are testing the use of more robotic units, such as more tractors with implements, which will move together with the control tractor with the operator. For these units, work efficiency increases with the number of units deployed and with the size of the field (Zhang & Noguchi, 2017). It can even be said that in order to maximize the increase in efficiency compared with using one tractor, we need a larger (longer) field the more units we add. In addition, from the manufacturer's point of view, it may be easier to implement new technologies into a larger and thus more expensive machine, which, thanks to its higher price, also provides a larger margin that covers the cost of innovation. From this point of view, it might seem that expectations that the implementation of robots has reversed the current trend of enlarging land may be too optimistic, but this is not excluded. However, the precise application of agrochemicals can lead to a slight reduction in their overuse or at least stop the growing trend of overuse.

Agriculture as a Lifestyle (Hobby Farming) and Community Agriculture In contrast to this trend of reducing the number of people in agriculture is the emerging trend of developing agriculture as a lifestyle. A growing proportion of people buy small farms, either rent land or maintain farming as part of their lifestyle. In the UK, these hobby farmers are the dominant owners of small farms with an area of less than 16 ha. These are often people who are not economically dependent on the profit from farming, they are mainly concerned with doing this activity right according to their ideas. The concretization of what is the right approach can be different, but it is often a proclaimed effort to respect nature and the effort to exist in connection with it. However, the question is to what extent this trend may be significant in the future both from the point of view of food production and from the point of view of the impact of agriculture on the environment.

Another way of farming focused on growing local products is community farming. Despite the fact that this type of agriculture does not necessarily mean environmentally friendly agriculture, this aspect is often emphasized. The principle of community - supported agriculture (CSA) is that the community of users subscribes to the production of a given farm, which is planned to provide a diversity of local products, which it then delivers to customers at regular intervals. The farmer obtains capital to grow his production and has a secure sale of the production, and the customer has a regular supply of local products produced in a way that he can at least in some extend control (he knows that the produce is local and can learn how the produce is grown).

Urban Gardening, Vertical Agriculture The share of the urban population is constantly growing and thus a large part of the population is not in contact with the food production process, which has long been part of our culture. There are a number of concepts dealing with the use of urban areas for the production of food, especially vegetables. In addition to commercial opportunities and more meaningful land use, the cultivation of fresh vegetables in cities has a number of social and cultural impacts. In this context, it is important to be aware of multiple aspects of food that we, as a part of culture, have lost or started to lose. Technically, urban agriculture has a number of forms, which are, among other things, influenced by the part of the municipality where it is operated. On the outskirts of cities in less densely built-up areas, there may be classic greenhouses or vegetable gardens, while in the city center, there are various roof or vertical gardens. Vertical farming is based on growing crops in multiple layers on top of each other in order to maximize the production per unit area. Often in a controlled environment aimed at optimizing plant growth and using soil-free plant cultivation techniques, such as hydroponics, aquaponics, and aeroponics. The advantages of vertical farms are the high yield per unit area and the production of fresh food independent of the season at the point of sale. Its disadvantage is the high energy intensity (Birkby, 2016).

Organic Farming and Food Production in More Environmentally Friendly Ways A separate chapter has been devoted to organic production, and the share of people who prefer products produced in the mode of organic production or in other nature-friendly ways is certainly growing. As stated in the chapter on organic farming, organic farming, as understood by the current legislation, is based on compliance with a set of standards which, based on our current knowledge, can be expected to be environmentally friendly. In addition, there are a number of producers who seek to modify existing commercial agricultural practices in order to achieve specific environmental objectives (such as the presence of certain indicator species on their land) while remaining competitive. In addition to organic farming, it can therefore be expected in the future to develop other directions of agriculture aimed at achieving specific environmental standards. Here, it is important to highlight the recent growing popularity of products of local origin. The culinary uses of apples imported from the other side of the globe and apples grown locally will be very similar, but their environmental footprint can vary considerably. In the same way, the popularity of buying food from small producers has been growing recently, who, only because they are small farmers and have smaller plots, have a number of environmental benefits, but most of them at least declare environmentally friendly production as part of their business strategy. It is possible that the increase in this popularity "food with story" will lead to a gradual increase in the volume of this sector. As explained in the chapter on the socioeconomic drivers of agriculture, cooperation with small producers is more difficult for large trade organizations, which largely determine the market, and are therefore often avoided. Conversely, the Internet and other platforms could help to create a direct link between small producers and consumers. This also applies to other producers who, for some reason, have not registered as organic farmers, but are striving for more environmentally friendly production. Establishing mutually beneficial and comfortable relationships for the customer and producers can be a way to strengthen environmentally friendly management in the landscape. After all, this is already happening in part. An interesting issue is the use of GMO technologies, especially modern methods of mutagenesis (such as CRISPR and other similar techniques) in organic farming or in other areas of environmentally friendly agriculture (Lehrman, 1994).

Technologies Based on New Dominant Crops and Still Minor Technologies Many of the problems associated with agriculture stem from the cultivation of the soil and the cultivation of annual crops grow as even aged cohorts. There are therefore efforts to build future agriculture to grow perennials. One of the best-known efforts is the efforts of the Land Institute in Kansas, USA, to breed perennial cereals based on botanical grass species other than current annual cereals. Probably, the most significant success is the breeding of perennial kernza cereals. So far, it provides only small revenues. In addition to the effort to domesticate perennial plants, there is also an effort to make existing annual crops permanent crops by gradual crossing. This is certainly an interesting project, but so far it encounters a small grain yield of these plants.

Another way is for researchers to look at the possibility of very intensive production of existing varieties of perennial grasses, which, after some processing, could be used to produce feed suitable for monogastric animals, pigs, and poultry. More in the "Kernza" box (Box 2.11).

The development of completely new technologies of food production is also on the scene, for example, based on microscopic algae. Microscopic algae are a rich source of biomass with very interesting nutritional values. They have been considered the food of the future for several decades. So far, however, their commercial production is far from sufficient to become a crop that will contribute to the EU food security. This is due to a number of obstacles, both natural (especially climate) and a number of financial and regulatory barriers (Vigani et al., 2015).

Optimization of Individual Ecosystem Services Change in agricultural subsidy policy, change in consumer behavior. An important component of the income of many farms in the EU is subsidies, which farmers receive to ensure efficient food production and at the same time, ensure that their management not only brings food production but also keeps the landscape and ecosystem in a state capable of providing quality ecosystem services. For this purpose, the farmer fulfills a set rules that may contribute to reasonably sustainable farming (cross compliance), more in Box 2.2. It sounds simple and nice, but so far, as the examples in Box 2.2 show, the volume of ecosystem services obtained by CAP payments policy does not fully correspond to the level of payments entering the common agricultural policy (representing tens of percent of the total EU budget). If truly meaningful measures can be recommended to promote diversity and the provision of ecosystem services, it is hoped that with considerable money from the European taxpayers, we would receive significant improvements in ecosystem services (and thus greater protection against drought, cheaper drinking water treatment, etc.). Another problem is the fact that the consumer has virtually no real opportunity to assess the environmental impact of the product he is buying. There is no general system for assessing the environmental impact of individual products, the result of which could be easily communicated to the customer in the form of one or a few graphical scales covering the main components of the environment, or a simple table similar to that of the dietary properties of the product deciding which product to buy. The indicator could also include the average value of the impact in a given product class, which would make it easier to decide whether it is better to buy another product in the same category or to reduce and replace that category in consumption with something else. Certainly, it would not oblige the customer to buy only the most environmentally friendly products if he does not like them or they are too expensive, but it would be an additional information entering his choice. Of course, even today, the customer can avoid some products that he doubts are not environmentally friendly, such as monitoring whether they contain palm oil or another ingredient, but it is laborious, often difficult to detect, requires additional knowledge. There are already procedures in the academic sphere such as life cycle analysis that could form the basis of such a labeling system. However, the introduction of such a system would certainly not be entirely free and would bring a number of initial organizational difficulties. Retail chains, as described above, are often described as causing great pressure on the environment in an effort to reduce costs and thus the price of products supplied. Some retail chains are aware of the shift in customer priorities and are trying to offer products that meet certain standards of environmentally friendly product or avoid a certain negative impact on the environment. An example is the efforts of food producers and retailers associated in the Sustainable Good Forum to stop deforestation in connection with the production of certain commodities through a systematic effort to remove from their supply chain products that required further deforestation.

Chapter 3
Forestry and Hunting

3.1 Forest and Biological Bases of Forestry

3.1.1 Definition of Forest and Its Function

We usually understand a forest as an ecosystem or land cover in which trees occur in high density (in a canopy). Mostly, it is a rather long-lived stand, for example, plantations of fast-growing woody plants, which are harvested after a few years, and we would not usually call a forest (but this is also a question of definition). There are many definitions of forest; the FAO definition (FAO, 2010) states that by forest we mean land with a tree cover of more than 10% on areas larger than 0.5 ha. Trees should reach a minimum height of 5 m in adulthood in situ. Young natural stands and all plantations established for forestry purposes which have not yet reached a crown density of 10% or a tree height of 5 m are included in forests, as well as areas that usually form part of a forest area that are temporarily free of canopy due to human intervention or natural causes, but where they are expected to return to the forest. Not every group of trees forms a forest. This is reflected in our experience that the forest creates specific properties, such as the microclimate and soil conditions, which only apply when the group of trees is large enough. The fact that trees grouped in the forest bring qualitatively new properties is reflected in a number of common phrases such as "he couldn't see the forest for the trees". The forest creates a specific environment that hosts a number of other organisms and gives rise to a complex network of ecological interactions (example in Box 3.1). We see that the definition of something as common as a forest can be complex and very diverse and several hundred definitions of a forest can be found in the literature.

The word "forest" probably comes from the Latin word *forestare*, which means forbidden or protected area and was originally used to describe a royal game reserve. It was therefore an area that is not only necessarily covered by woody vegetation but also included pastures, shrubs, and also a woodland, which served to protect the

© The Author(s), under exclusive license to Springer Nature Switzerland AG 2022

J. Frouz, J. Frouzová, *Applied Ecology*, https://doi.org/10.1007/978-3-030-83225-4_3

game that was subsequently hunted by the nobility; in other words, it was the exclusive hunting ground of the ruler. The etymology of word "woodland," in contrast, has clear connection to the growing of trees although paradoxically the woodland is rather used as a term for forest with a low density of trees. The origin of commonly used words for the forest is therefore related to both the occurrence of trees and the fact that this area was used for the protection and hunting of game. This is also one of the reasons why we connected hunting with forestry, although hunting does occur in many types of ecosystems.

As mentioned above, the forest has a number of functions from the point of view of human society; we can divide them into production, protective and regulatory functions, and cultural and social functions. We usually understand the production function as the production of wood. Wood is indeed the most important product of the forest (Box 3.2). However, the forest produces a number of other products, such as game, berries, mushrooms, various spices, herbs, vegetables, medicinal plants, honey, resin, maple syrup, and the like (Schulp et al., 2014). In addition to production functions, important are regulatory and protective functions: support of biodiversity, carbon sequestration in biomass and soil and thus mitigation of global climate change, regulation of water regime in the landscape, contribution to water self-purification, protection against both water and wind erosion, microclimate control, sanitation functions, reduction of pollution in the form of dust, noise reduction, protection against floods and other catastrophic phenomena, etc. An example of the protective function of a forest illustrating the diversity of protective functions may be to mitigate the impact of a tsunami in places with a well-developed mangrove forest. Equally important are the functions of social, recreational, artistic and spiritual inspiration, etc. By their value, nonmarket ecosystem services often significantly exceed market services (Costanza et al., 1997).

Human's use of the forest and forest perception as an economic entity has evolved over time. Prehistoric people used the forest as a habitat for game that they could hunt and also as a source of other products (honey, fruits, herbs, etc.) that they could obtain in the forest, which included wood, but with far less importance than today. Preagricultural societies had already started burning the forest, creating non-forest habitats or a mosaic comprising anthropogenically created areas of sparse forest, shrubs, and grassland habitats in order to increase the occurrence of game, which they could then hunt.

Emerging agriculture promoted deforestation not only in order to acquire agricultural land but also in order to obtain wood usable for house or shipbuilding, or as fuel, which led to significant deforestation, for example, in Roman times (Grove Rackham, 2003) in many parts of Mediterranean.

During the Middle Ages, in Europe, the forest was largely the property of the nobility and served as a hunting ground. Significant use was also made of the forest as a source of other nonwoody compounds such as rakes of litter, resin, or forest pasture. The forest was certainly also a key source of construction and firewood. For the intensive production of firewood, the original forests were transformed into coppice forest, where the fuel was obtained by cutting off large branches or multiple stems that regenerate from the stump. After cutting, stumps regenerate and form shrubby vegetation. In this coppice forest, individual trees were sometimes left as a source of construction wood. Construction timber was also obtained by selective logging in otherwise little-altered natural forests. Coppice forests were particularly common in western Europe (McGrath et al., 2015). Coppicing was often accompanied by forest pasture in silvo-pastoral systems in which temporary cropping, husbandry, hunting, gathering, etc., formed very open landscape where woody and grassland components were mixed (McGrath et al., 2015). In the Middle Ages there existed, side by side, both little-affected unmanaged forest (pristine forest) and various forms of intensively managed forest, which often had much lower tree density and or height than the forest today (high-stand forest). An example of such a forest would be Montado, still preserved in southern Portugal, which combines three types of use, namely sparse trees used for cork and charcoal production, forest pastures, and individual fields (Kabrda & Bičík, 2010). The proportion of unmanaged forest decreased over time and also from east to west across Europe (Fig. 3.1). Between the seventeenth and the first half of the nineteenth centuries, population growth drove increased demand for fuel wood which increased the share of forest in coppice management with a simultaneous decrease in forest cover. After the 1830s, an increasing substitution of coal for fuelwood resulted in coppice forests losing importance and decreasing in area.

As already mentioned, during the modern age there was also great pressure on the forest, both due to the expansion of agricultural land and due to the growing consumption of wood by the emerging industry. All these led to deforestation of the landscape. In Central Europe, deforestation of the landscape culminated in the first half of the eighteenth century during the period of Maria Theresia (Kabrda & Bičík, 2010). Also around the 1850s there was a peak in the use of the forest for litter raking (McGrath et al., 2015). This led to increasing effort of reforestation and afforestation and an increase in the cover of more densely stocked stands. This was often accompanied by a shift from broadleaf to conifers (McGrath et al., 2015).

With the growing demand for wood, owners became increasingly aware of the significant income potential of forests. Consequently, forest management became a subject of study by economists. See, for example, Viitala (2016) for a historical view. Among the many pioneers of these efforts is Hans Carl von Carlowitz and his book *Sylvicultura oeconomica*, first published in 1713 and also available in modern reprint (Hamberger & von Carlowitz, 2013). These efforts have led to systematic afforestation and reforestation. Various economic schools have been established, the aim of which was to optimize forest management in order to achieve the highest possible yield. Here one of the most striking differences of forest, as compared to other natural capital used by mankind, is the long production cycle. In boreal forestry, the time needed from tree seedling to harvestable tree is often close to, or even over, a century. Although several conceptual schools developed more or less in parallel, one of the first was the school that prioritized highest net yield from the forest. The subject of economic considerations here is the forest as a whole (all components of the ecosystem including soil and trees). The school's philosophy is based on the fact that the first forest manager took over the forest that was there and takes profit from it every year, especially by selective logging of individual trees or small

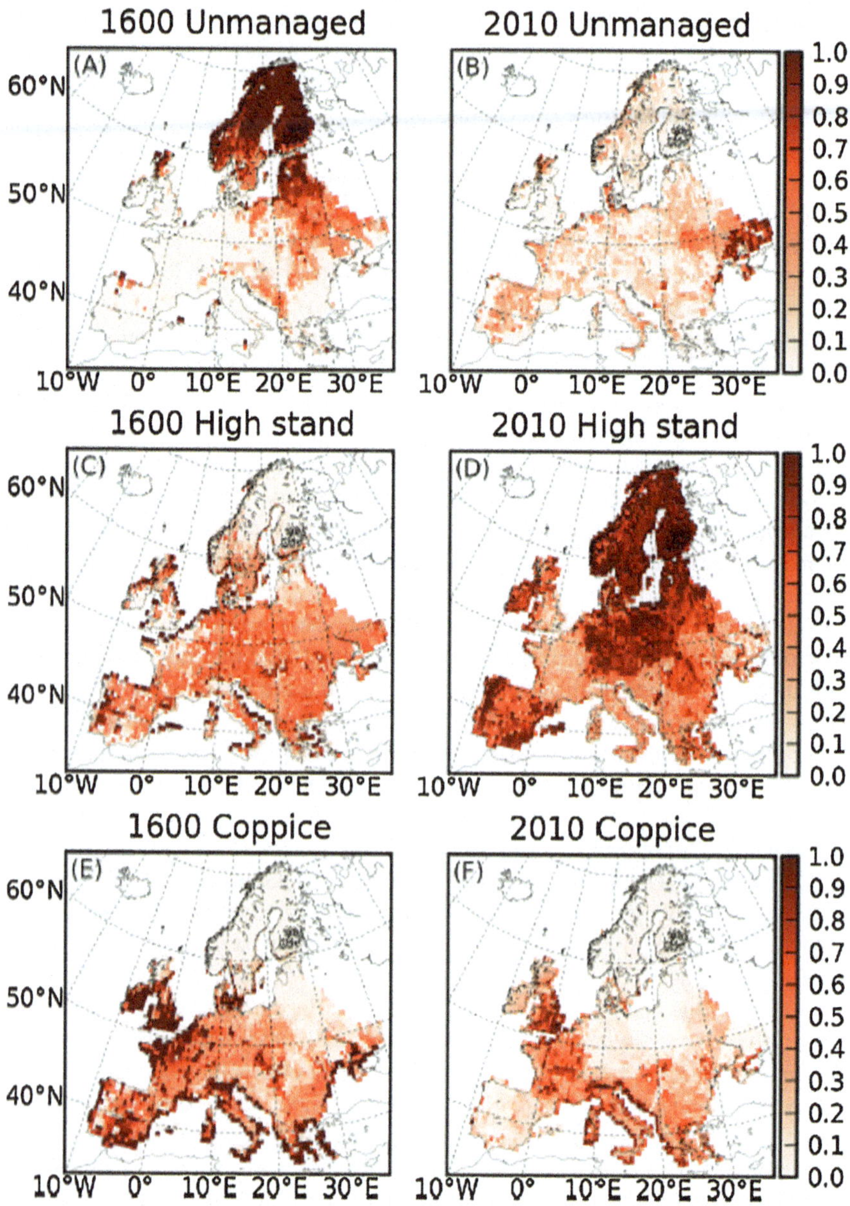

Fig. 3.1 Distribution of management strategies over Europe. (**a**) Unmanaged forests in 1600 and (**b**) in 2010. (**c**) High-stand forests (close to cultural forest as we know it today) in 1600 and (**d**) in 2010. (**e**) Coppice management in 1600 and (**f**) in 2010. The color bar shows the fraction of the forest managed by a given strategy. Pixels along the coastlines often contain few forests, frequently resulting in the entire pixel being under a single management strategy. (Reprinted from McGrath et al. (2015) with permission under creative common license (Creative Commons Attribution 4.0 Licenses, Biogeosciences journal))

Box 3.1 Wood Ants and Forest

Wood ants are conspicuous and well-known insects of Euroasian temperate forests. We will use this example to demonstrate the complexity of the relationship between the forest and individual species that inhabit it. Euroasian wood ants belong to the genus *Formica*, subgenus *Formica* s. str. In the Eurasian forests, representatives of this subgenus form characteristic conical nests made of needles, which can sometimes be quite large. The tallest nests can exceed a height of 2 m, and the nests with the largest volume can have over 10 m^3. It is interesting that the subgenus *Formica* s. str. Also occurs in North America, but they do not form similar nests there; instead, large nests are formed by another subgenus of the same genus. Some European species have only one queen, which establishes the nest itself with the help of other species of ants, and the nests of these species typically occur solitarily in the forests. In other species, there are several dozen queens in the nest, and their nests are largely propagated by division (budding), and disintegration of the mother colonies. By budding, large colonies of related nests are formed; their inhabitants are not aggressive with each other and can even cooperate. Such colonies can include hundreds of nests and cover an area of tens of hectares (Fig. 3.2). We have little idea of how many such anthills occurred in the original untouched forests. In many European fragments of pristine forests, there

Fig. 3.2 Wood ants subgenera. *Formica* and wood ant nests: winged queen, male, and worker are shown (from left to right) in inserted circles on the left

(continued)

Box 3.1 (continued)

are few or no forest ants. The situation is somewhat different in Scandinavia, where forests are often not so dense, but even here forest ants prefer the forest edges (Stockan and Robinson, 2016). Ants significantly affect the character of the forest in which they live. The exclusion of ants' access from part of the forest undergrowth led to significant changes in the composition of vegetation (Frouz et al., 2016a). During the year, ants collect tens to hundreds of kilograms of building material, tens of kilograms of insects and hundreds of kilograms of honeydew and bring all this to the anthill. In addition, during the growing season, ants maintain a relatively constant temperature in the anthills, that is higher than the ambient temperature, and some anthills may be moist inside. This leads to accelerated microbial processes and faster mineralization of organic matter and release of nutrients. The anthill thus becomes a place with a high concentration of nutrients. For example, it has been calculated that about a third of the available phosphorus in the soil of spruce stands with anthills and is concentrated just below the anthills. Thus, it could be assumed that trees near the anthill will grow faster than trees farther from the anthill because they will be better supplied with nutrients. Indeed, trees growing within 1 m of the anthill have been shown to grow faster than trees 5–50 m from the anthill. At the same time, however, it turned out that trees growing even further away, in places where ants no longer existed, grew even faster. This unexpected result is due to the fact that a substantial part of the energy and nutrients that come from anthills come from honeydew produced by aphids. Ants support aphid populations and honeydew production. At the same time, ants actively spread and protect aphids, so the collection of assimilates is much greater where there are ants. The trees near the anthill are thus exploited by aphids to a greater extent. The amount of honeydew taken away is several hundred kilograms per nest, which means a large loss of assimilates for trees, which then grow more slowly than if they had not been exploited in this way. However, the trees can offset this partly by the greater availability of nutrients, so it grows faster than the tree at intermediate distances. So, it might seem that ants are actually bad for trees, but ants also eat a lot of insect herbivores. In some types of forests, such as in the north of Finland, it has been observed that during the calamities of some pests (specifically in Finland it was the autumn moth), green islands of surviving trees form, in the otherwise completely eaten forest, around the anthills. Thus, it seems that in at least some situations, ants can protect trees from damage caused by calamitous pests. From this point of view, a small slowdown in growth due to the consumption of honeydew would actually be quite a tolerable tax. This small example shows how complex the relationships of different species of organisms in a community are and how different conclusions can be drawn if we consider them on different spatiotemporal scales (Frouz et al., 2016a).

Box 3.2 Wood Characterization, Production Trade, and Use

Wood is a complex tissue that gives woody plants enough mechanical strength so they can raise their photosynthetic apparatus higher, and as a result, more efficiently compete for light with ground vegetation (grass and herbs). The wood thus performs several key functions: it provides mechanical strength that holds the whole structure in place and ensures connectivity that allows the transport of water and nutrients from roots to foliage. Chemically speaking, the major component of wood is lignocellulose consisting of cellulose, hemicelluloses, and lignin. Cellulose is a beta(1–4)-linked chain of glucose molecules; individual chains are bounded by hydrogen bounds which contribute to the resistance of crystalline cellulose to degradation. Hemicelluloses are polysaccharides consisting of various 5- and 6-carbon sugars such as xylose. Lignin, consisting of closely linked phenolic polymers, is largely responsible for the stiffness of the wood and also provides hydrophobic surfaces. Although there is substantial variation, wood is roughly comprised of 45% cellulose, 25% hemicellulose, and 25% lignin (Sjostrom, 1993).

Cellulose forms microfibrils which combine with hemicellulose and lignin to form macrofibrils which are located in cell walls and support its mechanical stability. Macrofibrils are located in layers within cell walls, so organization of these determines the properties of wood. This differs substantially between angiosperms (broadleaves) and gymnosperms (conifers) (Fig. 3.3). Wood of angiosperms is called hardwood. Its wood has vessels that allow the rapid

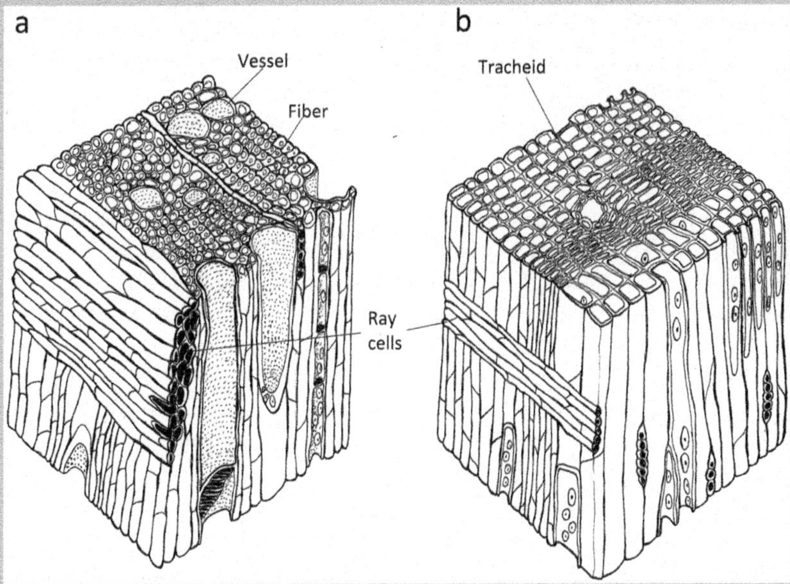

Fig. 3.3 Schematic of hardwood (**a**) and softwood (**b**) structure

(continued)

Box 3.2 (continued)

vertical transport of water; under a microscope, on perpendicular transection, these vessels form pores. Pores can be distributed regularly through the wood (most common in tropical trees), or concentrated in rings (most common in temperate trees), with larger pores being produced early in the growing season. Wood fibers are long cells with strong cell walls that support mechanical stability of the wood. In cross section, these are too small to be distinguished by naked eyes; instead, their aggregations can be distinguished as colored areas, forming wood's endgrain. Finally, rays are structures running perpendicular to other fibers and connecting cambium, sapwood, and pith. These consist mostly of parenchyma, living cells that serve as storage and in the production of secondary compounds. Gymnosperm wood is termed softwood, although it is not necessarily less strong or dense than that of hardwoods. In softwood, tracheids transport water and contribute to mechanical stability. Tracheids account for over 90% of softwood cells so parenchyma are much less frequent in many species, and occur in rays that are much smaller than those of hardwood, usually only one cell wide. Under a microscope, softwoods have no visible pores. The softwood may contain resin channels, and those may look like pores at first glance, but they are not represented by a cell, but rather a cavity in between cells. Although the term "hardwood" implies that it may be harder than softwood this is not always true; instead, one can say that hardwood is more variable in density and mechanical strength.

As already mentioned, the wood offers mechanical support, connects root and foliage, and allows transport of water and nutrients, but wood also faces many threats such as mechanical damage by wind or snow, embolism due to freezing or drought, and attack by insects and fungi. In previous chapters, the leaf energy spectrum was described, with plants having faster growth generally having thinner leaves, a greater specific leaf area and shorter leaf lifespan, and providing litter that decomposes faster than that of slower growing species. Despite some mixed results, there is some indication that wood density may be a good proxy for "wood energy spectrum." Wood density correlates with many other traits and shows a positive correlation with mechanical resistance (modulus of rupture MOR) (Fig. 3.4). Wood density is negatively correlated with leaf area; trees with lower wood density (which is cheaper to build) also tend to have thinner leaves, shorter leaf life span, and faster growth, while trees with higher wood density (which is more expensive to build) have slower growth and larger survival rate, a typical indicator of conservative growth strategy (Chave et al., 2009).

In terms of production and trade the amount of wood is most often expressed in cubic meters (this will be also used here). Cubic meters are sometimes also called solid meters (m^3), which is a cubic meter completely filled with wood, and it can be imagined as a cube (1 × 1 × 1 m) composed of closely stacked wooden blocks. In addition to solid meters, the term stacked or loose cubic meters may be used, which are approximately 0.7–0.4 m^3 respectively.

(continued)

Box 3.2 (continued)

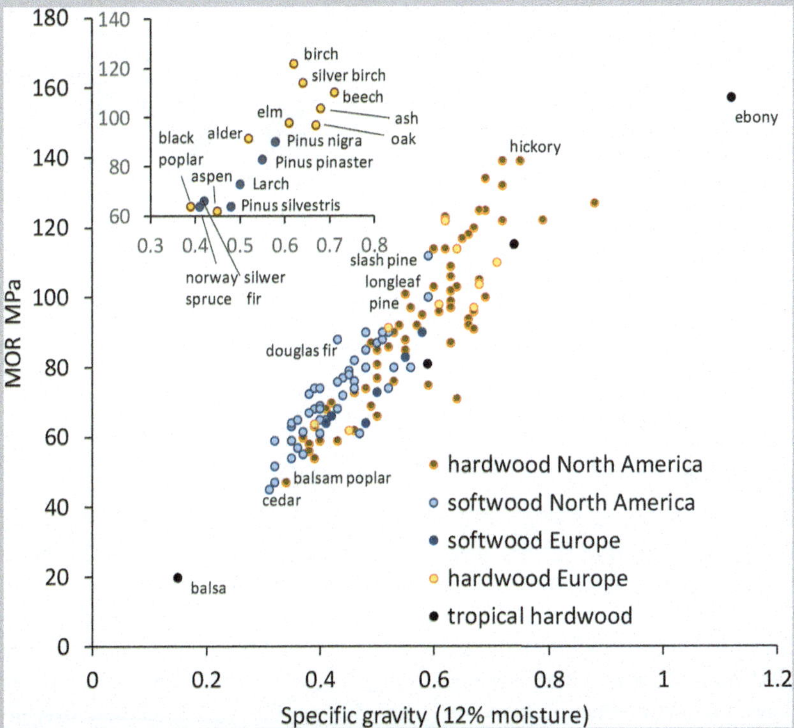

Fig. 3.4 Relationship between wood specific gravity and mechanical strength expressed as modulus of rupture

The trend in worldwide wood production is shown in Fig. 3.5. One can see that wood production in 1800 was a bit higher than hundred years after, but since then wood production has been increasing steadily. In 2017 the most important wood producers, in terms of round wood volume, were the USA, India, China, Brazil, Russia, Canada, Indonesia, Ethiopia, Congo, and Nigeria. These 10 countries produce over 55% of world wood products. Based on FAO statistics, less than 5% of that volume is subject to export and import in the international market. The 10 largest exporters of wood worldwide were Russia, New Zealand, USA, Canada, Czechia, Australia, France, Germany, Norway, and Papua New Guinea. The largest importers of wood were China, Austria, Germany, Sweden, Republic of Korea, Finland, and India (data for 2017, based on FAOSTAT). Here it should be mentioned that there is quite a significant amount of illegal trade of wood. In 2007–2009, the UNEP report estimated value of world wood trade was US $327 billion (Nellemann & INTERPOL, 2012) and estimated that illegal logging accounted for 10–30% total logging (worth US $30–100 billions). Some estimates even say that when laundering of illegal wood is included, the numbers may reach 20–50%. Most of that illegal

(continued)

Box 3.2 (continued)

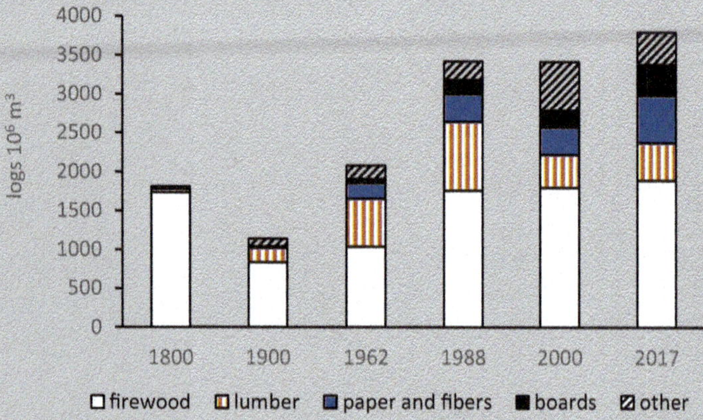

Fig. 3.5 Worldwide timber production and main uses of the FAO

logging is happening in tropical countries (Nellemann & INTERPOL, 2012). Large amounts of money are involved, and this illegal operation may be highly organized, at least in some cases. Illegal logging has many forms, it may include logging of rare wood in protected areas, using of conflict zones for illegal logging, illegal logging in remote areas or various form of logging on various permit frauds. Some illegal logging is more focused on obtaining land for cash crop production than on wood itself. In addition, the ways to launder illegally logged wood are very variable, from mixing illegally and legally harvested wood to various form of falsification or fraud. The most important buyers of illegally logged wood are China, USA, EU, and Japan which together trade about 80% of illegally logged wood based on the Nellemann and INTERPOL (2012) estimate. Until recently, once the wood has been exported outside the country of origin, it has been difficult to prove that it is coming from illegal logging. Many developed countries adopted legal schemes to combat illegal logging, which change this practice. Those include several regulations of the EU (EU Forest Law Enforcement, Governance and Trade Action Plan (FLEGT) 2003 and EU timber regulations 2010), the Green Purchasing Law (Japan, 2006), the Lacey Act (US 2008), and Australian Illegal Logging Prohibition Act (2012). Although these legal acts represent an important step in reducing illegal wood trade, they have some limitations. They have variable philosophy and requirements. Some of them are voluntary for exporting countries (FLEGT). More importantly, they are not synchronized and are enforced in different ways in different countries. There are no standardized requirements for documentation that will prove the timber origin in international trade, and there is limited exchange of information between law-enforced agencies who control the trade. There is, for example, cooperation between EU member states, but other international networking is limited. All these make illegal logging and laundering of illegal timber easier.

(continued)

Box 3.2 (continued)

As can be seen from Fig. 3.5, worldwide, more than half of the timber harvested is consumed as fuel, either directly or as charcoal. At the beginning of the nineteenth century, it was more than three-quarters of the harvested wood, but gradually the share of firewood decreased. At present, a large part of firewood is burned in developing countries, when cooking in open fireplaces. The efficiency of such fireplaces is around 10%. For example, about 4 kg of wood is used to cook 450 g of beans. The efficiency of the stove can be over 60% and for modern wood boilers over 90%. Different types of wood differ in their heating value (Table 3.1). However, these differences are significant especially when comparing the volume of wood; if we convert the heating value to the weight of wood, the differences are less pronounced. The heating value is strongly influenced by the moisture content of the wood, so in operating conditions the dryness of the wood is very important for the efficiency of combustion. It is the water balance that distinguishes the heating value from the calorific value heat, which we became acquainted with in the introductory chapters. Calorific value is the heat that is released by the perfect combustion of a given material, while it is assumed that any released water condenses and therefore the latent heat that served to evaporate water is not included. On the other hand, when measuring the heating value, sometimes also called net calorific value, it is assumed that the released water leaves in the form of water vapor, and therefore the heating value is lower than the calorific value. Therefore, if we calculate the efficiency of boilers in relation to the heating value of fuel, there may be boilers with greater than 100% efficiency – condensing boilers, where we ensure the condensation of the

Table 3.1 Selected properties of different tree species that commonly grow in Europe

	Heating value GJ m^3	Density m^3	Hardness	Durability years[a]	Culinary use
Spruce	7.4	470	Very soft	6–10	
Pine	8	520	Very soft	6–15	
Larch	9.3	600	Soft	10–15	
Fir	8	720	Soft	6–10	
Douglas fir	10.2	480	Soft	10–15	
Birch	10.5	600	Soft	2–6	
Alder	8.3	520	Soft	2–6	Good for smoking
Poplar	6.5	450	Very soft		
Ash	11	720	Hard	2–6	
Beech	10.8	720	Hard	2–6	Wooden dishes
Oak	10.9	760	Hard	15–20	Baric barrels
Hornbeam	10.8	820	Hard	2–6	
Apple	10.4	750	Hard		Good for smoking

[a]Durability when in contact with soil or water

(continued)

Box 3.2 (continued)

released water vapor in the boiler. Calorific value varies between tree species and ranges from 14.7 to 21.1 $MJ.kg^{-1}$. The difference is mainly due to the ratio of lignin, which has a calorific value of 25.5 $MJ.kg^{-1}$, and cellulose, which has a calorific value of 18.8 $MJ.kg^{-1}$. The calorific value is also affected by the resin content. The calorific value of pure wood mass of fast-growing poplars is 19.5 for poplar biomass with a natural bark content of 20–20.4 ($MJ.kg^{-1}$). Combustion of wood, although considered carbon neutral because we release CO_2 previously fixed from the atmosphere, poses environmental risks. Wood burning is a significant source of NO_x and PM_{10}, especially compared to natural gas (Nussbaumer, 2003).

A significant proportion of harvested wood is processed as lumber, which is then used to build structures, manufacture furniture, and other wood products. Hardwood tree wood is mostly used for the production of furniture, tools, floors, stairs, and decorative items. Hardwood trees provide very variable wood, and it is usually more expensive, so it is less used in construction. Historically, a number of items were made of multiple types of wood: the spinning wheel was composed of seven types of wood, the wooden part of the plow had four types, a rake of two to three types, etc. Filková et al. (2015) identified wood of 18 local tree species used in various products in Czechia in the nineteenth and first half of the twentieth century. However, it should be noted that the most frequently used wood here was beech (*Fagus sylvatica*) and spruce (*Picea abies*). However, at least four other species occurred with a frequency greater than 5% (fir, birch, elm, and oak). At the same time, logs of smaller diameters were very often used for the production of objects. The handle on the rake was created by processing the smaller diameter trunk, not by turning from a lumber cut from the whole trunk. The variability of the properties of different types and assortments of wood was used. Today, most items are made of one type of wood supplemented with plastic, where the properties of wood would not fit.

When making structures, not only mechanical properties, weight, and strength but also resistance to decomposition are important. Wood, as a biological material, is subject to decomposition which is initiated mainly by fungi. Wood generally does not decompose at humidity below 20%. When in contact with water or soil, the resistance of different types of wood is very variable (Table 3.1). It is interesting that the resistance to decomposition is not determined by the hardness of the wood. Hornbeam (*Carpinus betulus*), which is a hard wood, decomposes very easily, while softer Douglas fir is much more durable. Durability can be increased by a number of technical procedures. Softwood is more often used for construction because it is usually cheaper and more easily accessible (coniferous lumber makes up 80% of all lumber). Conifers usually have straight logs, which provide long straight construction segments, and the wood is usually lighter (carries less of its weight) and easy to work with. Of course, there are exceptions. Hardwood was used for construction where

(continued)

Box 3.2 (continued)

there was no other wood locally, or on special occasions. Such special occasions were, for example, the trusses of the Notre Dame temple. Many experts now point out that after its fire, it will be difficult to obtain the same amount of oak (*Quercus* spp.) trunks for its reconstruction, simply because similar trees are not commonly found in forests today. Entirely woody structures made of solid wood are usually smaller buildings. These are, at maximum, 5–6 storeys; one of the largest is a 6-storey wooden castle in the city of Himeji in Japan. At present, in addition to solid wood, variously treated wood, wooden boards, CLT panels, glued beams and various composite materials using wood, such as carbon fiber reinforced beams and similar materials, are used for construction (Ramage et al., 2017). It is assumed that these structural elements could be used for the construction of even larger buildings. There are already several greater than 10-storey buildings based on wooden construction (Tall wood building in Vancouver, HoHo in Vienna, etc., Fig. 3.6). The use of wood in the construction of buildings is also interesting from the point of view of carbon sequestration, because the carbon built into the wood of trees remains bound in the structure of buildings and thus prolongs its turnover time. By using 1 m^3 of wood, we bind about 1 t CO_2 for an average of 20 years. Considering all necessary processing if wood replaces concrete, in construction then the net emission reduction is 0.3 t CO_2, when replacing steel, 1.2 t CO_2 per 1 m^3 of wood (Ramage et al., 2017).

A significant amount of wood is used for the production of paper and pulp. Both hardwood and softwood can be used for paper production; however, softwood is more advantageous. In contrast to lumber, there are no such demands on the diameter of logs, and various materials such as branches or wood gained during thinning can be used for the production of paper.

Fig. 3.6 HoHo in Vienna, an example of large wooden buildings: general view (**a**), construction (**b, c**). (Picture courtesy of Kerbler Holding GmbH)

groups of trees, and invests the costs in forest management. The aim of management is therefore the long-term maintenance of the forest unit's reserves in order to maximize growth, and thus, annual profits. You may notice that the school's philosophy is very similar to the theory of maximum sustainable yield introduced earlier. The aim is to optimize harvest in a way that tree harvest is as profitable as possible; for example, because harvested trees are at the peak of their economic value, but their removal also reduces competition and triggers establishment and growth of young trees, which maximizes future profit. The fact that trees of various ages are present, even on small areas of forest, allows for an estimate of the number of trees that are harvestable not only now but also in future. The fact that harvest happens in this forest patch regularly over short time intervals allows one to directly compare profit from wood gained each year to associated cost spent that year, and because forest in maintained in similar age structure, the costs are also similar from year to year.

After 1860, the so-called school of net revenue began more popular. A significant conceptual difference is the separate perception of forest soil (land) and forest cover. The land is seen as an investment facility providing conditions for forest cultivation and the stand is then the product of this cultivation. It is assumed that the forest manager has taken over the land on which he grows the forest. This makes it possible, among other things, to compare the economic return on forestry with other uses of the landscape. The forester has been using the land for a long time and gradually invests in forest productivity. It is important to realize that in economics, the price of money changes over time. The money one can spend today has more value than the one available at some point in future. After all, everyone who borrows money knows this. The bank gets paid in the form of interest for lending us money (i.e., for the fact that we could spend the money before we made it). The amount of interest (discount rate) determines the price of money and thus the overall costs. Thus, costs increase with the time of forest cultivation, not only because we invest more work in it and spend associated costs over a longer period of time but also earlier because investments get more expensive due to interest rates. At the same time, the value of the forest stand grows over time and thus the yield that we obtain by harvesting it. The relationship between these two quantities is described by Faustman's formula named after German (Hessian) forester Martin Faustmann (although similar ideas appeared several times before him: Viitala, 2013, 2016). Using Faustman's formula, we can optimize the time when the profit (difference between revenues and costs) is the largest. The school of net revenue opened the possibility of planned, economically profitable, clear-cut forest cultivation (Viitala, 2013, 2016; Matějíček & Zádrapa, 2014). In other words, the growing of even aged cohorts of trees is in a way similar to cultivation of cereals on arable land. At the same time, it indirectly stimulated the development of other optimization steps, such as the optimization of stand density so that maximum growth is achieved with the best possible technological properties of wood, etc. Such optimization is easier for monocultures than for mixed stands, and the school of net forest revenue has thus led indirectly to the development of forestry based on the cultivation of even-aged monocultures of trees. The cultivation of such stands has enabled better use of mechanization and greater work efficiency. On the other hand, the optimization of the stand in terms of profitability has led to a reduction in its resilience in terms of resistance to various adverse biotic and abiotic factors (pests, diseases, bad weather). This indirectly supported development of many disciplines of forest science such as development of new machinery and optimization of machinery use or forest protection against various abiotic and biotic stress. Growing of highly productive even aged forests also lead to reduction in the provision of some other ecosystem services, or at least to larger fluctuation in service provisioning over time. Efforts are currently being made to include other forest services in the forest economy, including ecosystem services (Masiero et al., 2019), but, in practice, this approach faces a number of problems similar to those described earlier in the chapter on ecosystem services and payments for these services.

3.1.2 Types of Forest

The forest covers 31% of the earth's land (over 4 billion ha). Five countries, Russia, Brazil, Canada, USA, and China, have on their territory more than half the world's forest area. Globally, many of the world's biomes are predominantly covered by forest. From the north, it is the taiga, the deciduous temperate forests, then the shrubby Mediterranean vegetation, and finally the deciduous and evergreen tropical forest. However, the area of the forest is constantly changing due to both anthropogenic and natural causes (Box 3.3).

The taiga and other boreo-alpine coniferous forests are located in colder climatic conditions (between 50th and 60th parallels), especially in the northern hemisphere, as there is very little land in the southern hemisphere in suitable climatic areas. Similar forest types are also found in warmer areas at higher altitudes – the closer to the equator the higher. Annual precipitation ranges from 300 to 900 mm, and most of it comes in the form of snow. The average temperature is 0–5 °C with large differences between summer (up to 20 °C) and winter (up to −40 °C). The growing season is short (100–130 days); however, the length of the day in the growing season is significantly longer than 12 hours. The transition between summer and winter is very sudden. Typical vegetation is an evergreen coniferous forest dominated by spruces, other abundant genera are pines (*Pinus* spp.), firs (*Abies* spp.), larch (*Larix* spp.), with a few deciduous trees, mainly birch (*Betula* spp.) and aspen (*Populus* spp.).

Temperate forests, as the name suggests, are located in a temperate zone (between 25th and 50th parallel). It is characterized by an average annual temperature of 5–20 °C with significant annual temperature fluctuations being as low as −30 °C and as high as 30 °C, and an average annual rainfall of 750–1500 mm. It is characterized by alternating four seasons. Most trees shed leaves in winter. Vegetation consists of deciduous trees such as, oaks, beeches, and maples (*Acer* spp.), often with coniferous admixtures.

Tropical forests are located in the tropical zone; temperatures range between 20 and 30 °C; and the average annual temperature is in a similar range, which shows that seasonal temperature fluctuations are very small. The annual total precipitation

Box 3.3 Development of Forest Area in the Past

If we are interested in long-term changes in the forest, we will get a very different picture depending on how far back we look. In this section we will look at forest development at three very different time scales, first briefly describing forest development during the last geological epoch, then focusing on human influence on forest area over the last few thousand years and especially on recent changes in this and the last century.

We will first look briefly at the evolution of forest ecosystems in the recent geological past the Pleistocene and the Holocene; four ice ages (donau, günz, mindel, riss and würm) alternated during the Pleistocene, among which were interglacial periods, whose alternation was triggered by different solar energy input due to small changes in the ecliptic described by Milankovich cycles (see Box 1.8). Glacials lasted hundreds of thousands of years, and interglacials for tens of thousands of years. During the glacial, up to 30% of the earth's surface was covered with ice. The climate was drier, and ocean levels dropped. A large part of the forests was replaced by steppe and forest-steppe vegetation. For example, during the last glacial maximum (about 18,000 years ago), the Amazon rainforest reduced its area significantly below the area given by the current anthropogenic deforestation and disintegrated into a number of forest islands isolated within forest-steppe vegetation (Wang et al., 2017). On the contrary, the interglacials were wetter and warmer; some of the previous interglacials were warmer than ours. The development of vegetation during the interglacial was influenced by two main factors, climate change and soil development. We will demonstrate the impact of climate on our interglacial. In terms of the effect of nutrients, glaciers have removed large parts of the soil and exposed the subsoil or the underlying geological substrate. In addition, a large number of these exposed areas were further eroded by both water and wind or covered by sediments formed by deposition of this eroded material. This access to fresh geological substrate resulted in the relative abundance of nutrients that plants obtain from the soil, especially phosphorus (P), at the beginning of each interglacial. This led to a sharp increase in vegetation followed by gradual depletion of nutrients and subsequent gradual loss of vegetation associated with an increase in the occurrence of conifers using ectomycorrhiza to obtain less accessible forms of P (Kuneš et al., 2011). The first period after the end of the last ice age is called the preboral (8300–7700 years BC). Average temperatures were about 5 °C lower than today, and the soils were strongly underdeveloped. Gradually, open forests developed with birch and Scots pine (*Pinus sylvestris*), larch, stone pine (*Pinus pinea*), and mountain pine (*Pinus uncinata*). Gradually, hazel (*Corylus* spp.), oak, and elm (*Ulmus* spp.) spread, and once the boreal began (7700–6000 years BC), open pine forests with hazel undergrowth developed and spruce spread in the mountains. Compared to previous drier periods, there was a significant humidification of the climate, and in the Atlantic (6000–4000

(continued)

Box 3.3 (continued)

BC) there was the development of extensive deciduous forests dominated by oak, lime, elm, and maple. The upper elevational limit of the forest was about 400 m higher than today. Subsequently, in the Epiatlantic (4000–1250 years BC), the climate gradually became drier and colder, and the upper limit of the forest receded. Human's influence was manifested by the spread of steppe vegetation. During the Epiatlantic, there was a sharp spread of beech and fir. Beech survived the ice age in a number of refuges in the Mediterranean; however, while some of the refuges expanded only slightly, others laid the foundation for the current European population. It is also interesting that the expansion of beech in our interglacial period did not occur analogously in the previous interglacial, where beech was ecologically replaced by elm. The influence of humans was even more evident in the sub-boreal (1250–750 BC), when the climate was warmer and drier than today. This was followed by cooling and some humidification of the climate in the sub-Atlantic (750 BC to 600 AD), which was accompanied by a massive development of beech and fir. In the subrecent (from 600 AD) there was a gradual decrease in humidity and drying of the landscape, the forest receded due to agriculture and the upper edge of the forest decreased due to grazing.

Human has been influencing the size of forests for a very long time. While some authors believe that the greater influence of humans outside the developed centers in Europe did not manifest itself until after 1750, other authors believe that temperate forests in Eurasia and other parts of the world were reduced by humans by 20% of the area as early as 3000 BC. A similar rate of exploitation was achieved around 1000 AD (Ellis et al., 2013). The pre-industrial impact of humans was probably greater than it might seem today, but ecosystems have largely been able to regenerate. For example, Kukla et al. (2019) estimate that most of the rainforest in Papua New Guinea has been used by indigenous people to form temporary gardens, at least once since the onset of agriculture. But even the more pronounced traces of great civilizations could be largely erased by natural processes. An example is the large agglomeration of Angkor Wat in Cambodia, which has been largely swallowed up by the forest. A number of recent studies also show that the Amazonian region, in the pre-Columbian era, was home to an advanced civilization (de Souza et al., 2018) which created a number of fortified settlements and other structures, and most likely significantly deforested and fragmented local forests.

In pre-industrial times, there were quite clear relationships between density of human population and deforestation. This resulted in gradual decrease of forest cover interrupted by two major events where forest cover increased: around 600 AD, after the breakdown of Roman empire during transition to the Middle Ages, and at 1400 AD, shortly after the Black Death (plague epidemic). The relationship between log of population density and deforestation was generally sigmoid, being more gradual in ancient times and becoming more

(continued)

Box 3.3 (continued)

abrupt in pre-industrial times. This may reflect the development of agricultural technologies that allow for the support of more people on the same amount of land, and at the same time, technological advances to clear forest fast occurred when the inflection point was met. Later on, with ongoing industrialization and urbanization, the agricultural productivity increased sharply which generated less pressure on forest cover, that then stabilized and in some cases even increased (Kaplan et al., 2009).

Regarding the recent development of forest cover, the data are inconclusive. FAOSTAT reports a decrease in forest area of 3.1% between 1990 and 2016. In contrast, Peg-Song et al. (2018), who monitored the development of forest cover between 1982 and 2016, showed that the overall rate of forest cover has increased over 35 years. Respectively, there was a decrease in forest in tropical areas, and in contrast, there was an increase in forest area in colder areas (Fig. 3.9) resulting in a total increase in forest area over this period of +7.1% (Peg-Song et al., 2018). In general, all mountainous areas increased in forest cover, while in arid and semiarid areas, this cover decreased. It should be remembered that we are talking about an area covered with forest, and we would get a different picture by comparing different metrics, for example, biomass. It is also probable that the loss of diversity, resulting from the loss of forest in tropical areas, is not offset by the increase in forest cover in temperate areas.

If we talk about the deforestation of tropical forests, the most significant wave of deforestation took place from the 1960s to the 1990s of the last century, when, according to some estimates, up to half of the tropical forests

Fig. 3.9 Changes in forest area in recent decades, black: deforestation horizontal, hatching: stagnant forest area, gray: increase in forest area, dots: degrading areas in arid areas

(continued)

Box 3.3 (continued)

disappeared. In terms of ecosystem type and geographical location, deforestation in the Amazon rainforest in highly medialized. In Southeast Asia there was only slightly less absolute forest loss; however, the baseline was smaller so relative loss is much bigger, based on FAOSTAT data. However, the FAO in this statistic only includes an absolute change in forest area. Secondary forest or tree plantations are considered a forest for this purpose. Including the transformation of rainforest into plantations in Southeast Asia, the proportion of pristine forest lost would be even higher. This indicates that deforestation and the loss of native forests are an even bigger problem in Southeast Asia than in South America. Similarly, in addition to the tropical rainforest, tropical deciduous forest and dry tropical forests may be even more endangered (Table 3.4). For example, the Mata Atlantic, a deciduous tropical forest along the Atlantic coast of Brazil, has been deforested by more than 85%. Similarly, the Cerrado, dry forest savannah which once covered 20% of Brazil, decreased in area by 53% between 2001 and 2017. In conclusion, most deforestation is happening in the tropics, and tropical forests in Southeast Asia are more at risk of deforestation than South America, and dry and deciduous forests more than the evergreen rainforest (Table 3.2).

The main cause of deforestation is considered to be the development of cattle breeding, especially in South America, and the development of soybean cultivation, not only in South America but also in North America. Other important drivers of deforestation are the development of oil palm plantations, especially in Southeast Asia, and the consumption of wood for charcoal production, especially in Africa, and to a lesser extent also in India and Thailand. In all areas, there is an increase in population and thus an increasing area needed for agriculture. Globalization and the increase in world timber trade can also contribute to the growth of deforestation. The engine of this acceleration is China, which is the world's largest importer of wood and the largest exporter of wood products.

As mentioned above, the main wave of deforestation in Europe took place during the Middle Ages, and today the forest cover is being restored.

Table 3.2 Annual average percentage of deforestation in tropical forests in different tropical regions and different forest types between 1980 and 1990. (FAO)

	All tropical forest	Evergreen rainforest	Deciduous tropical forest	Dry tropical forest
Africa	0.7	0.5	0.9	0.7
Asia	1.2	1.1	1.5	1.1
South America	0.8	0.4	1.0	1.3
World total	0.8	0.6	1.0	0.9

is 1250–6000 mm. The two seasons are the dry season (summer) and the rainy season (winter). While at the equator the drought period is usually humid, and the rainy season is extremely humid; further from the equator, the total amount of precipitation decreases and at a precipitation below 2500 mm the drier period is already noticeable and the trees drop leaves for 6–8 weeks. The growing season is the entire year, or almost the whole year, but the time of daylight ranges between 11 and 13 hours a day. The vegetation consists of deciduous or evergreen hardwood trees, of which there are a large number of species, usually dozens of species per hectare.

Natural and cultural forest. A significant part of natural forests have been more or less altered by human activities. There are many definitions of naturalness, and similar terms are often used with different meaning. Buchwald (2005) proposed complex terminology using 14 degrees of naturalness which include 10 natural levels and four plantation levels. In general, natural forest is here defined as "forest stands predominantly composed of self-sown native trees." This is forest that consists of native trees and has structure and species composition close to that of natural forest, and follows closely to natural forest dynamics. Often it is mentioned that natural forest includes old trees and dead wood. FAO uses the term "primary forest," which is defined as naturally regenerated forest of native species, where there are no clearly visible indications of human activities and the ecological processes are not significantly disturbed. Globally, the area of primary forests is estimated to be 36% of all forests (FAO, 2010). The largest share of natural forests in the total forest area occurs in South America, the smallest in Africa (Figs. 3.7 and 3.8). The country with

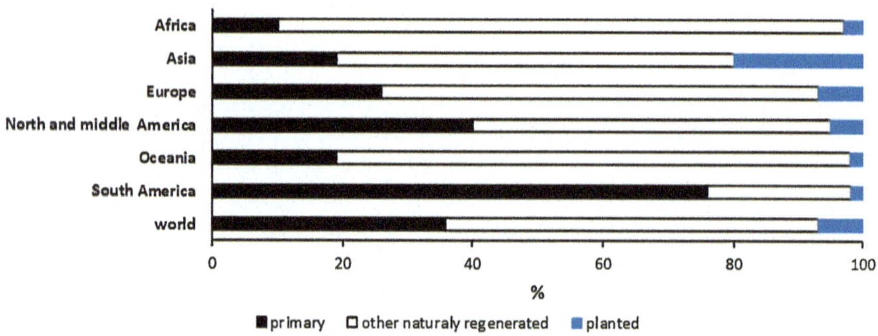

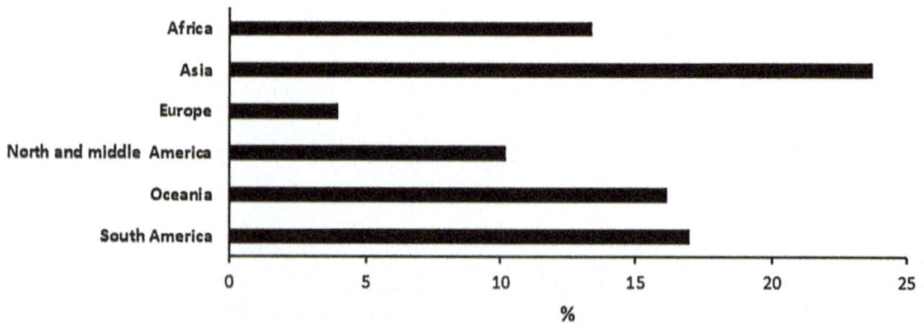

Fig. 3.7 Share of primary forests (black), other naturally regenerated forest (shaded), and plantations (white) (above), and share of protected areas from the total forest area in individual continents (bottom). (FAO (2010))

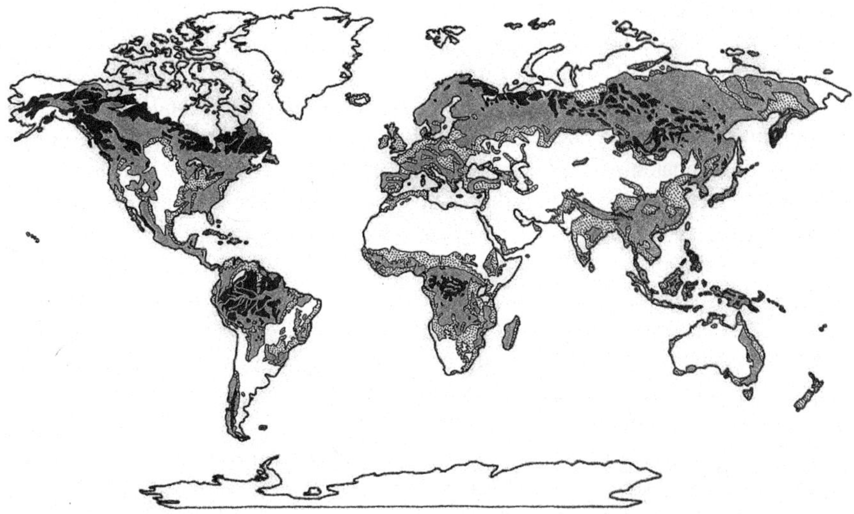

Fig. 3.8 Map of the distribution of natural (black) and cultural (gray) forests. (FAO)

the largest area of natural forest is Brazil, followed by Russia and Canada. In Europe, native natural forests occur mainly in Northern Europe, elsewhere the remnants of native natural forests form only small fragments (Box 3.3). Natural forest is a forest with a natural species composition but a changed structure because it was influenced, for example, by selective logging.

3.1.3 Growth and Development of Natural Forests

Like an individual tree, the whole forest has its life cycle. We distinguish between the so-called small- and large-scale forest cycles. The small-scale forest cycle, or forest gap dynamics (Yamamoto, 2000), consists of three main stages: the optimum stage, breakdown stage, and growth stage. The forest in the optimum stage is visually similar to the cultural forest, where trees of the largest diameter classes significantly predominate. The number of individual trees per unit area is the smallest of all parts of the cycle. The trees are in a continuous canopy, and the intensity of light falling on the ground is small. However, the trees that occur here are very different in age, despite their similar size, due to the fact that individual trees have a significantly longer lifespan than the time of their intensive growth. This means that in addition to young trees at the end of their intense growth period, there are also much older trees from the previous generation. At the end of this stage, the oldest trees begin to die. The breakdown stage begins. Death or injury of old trees form gap in a canopy; in these gaps more light comes to the soil surface which triggers growth of young seedlings in the opening and the regeneration begins. The number of older generation trees is decreasing; the number and growth of the new generation in the gap opened by death of old trees is increasing. The distribution of trees is very patchy and irregular. The stock (biomass) of wood of the dying generation is not fully replaced by the growth of young trees, and therefore, the total stock of wood is declining. Dead wood accumulates on the soil surface. The share of new trees is increasing. The patch enters the growth stage. Younger trees, which are in the phase of intensive growth, grow much faster than older trees that remain from previous generations after the breakdown phase. The canopy is gradually closed, the thickness, height, and spatial differentiation of the stand reaches its maximum. From the point of view of the older tree generation, trees which were old already in optimum stage, if these individuals are still present, it is a phase of survival. As the younger trees grow faster than the old ones, the size of old and new trees become similar and patches gradually approach the optimum phase.

It should be noted that not all stands follow simultaneously such that the entire extensive forest stand would be at one stage of its development. On the contrary, the forest is divided into smaller patches which are in individual phases of the cycle, so that in natural forests the whole stand consists of a mosaic of these patches. The size of these patches varies between different types of forest; in deciduous forests it can be 300–700 m^2, in spruce, 0.5 ha or even more. However, in real natural forests, it can happen that the stands in some phase of the forest cycle pass very quickly or may be almost completely skipped (Král et al., 2018), so the final spatiotemporal pattern can be even more complicated, as we indicated in Fig. 3.10.

The large-scale cycle of the forest begins after a more significant and large-scale disturbance, when a larger area of forest is affected. The first stage in the large cycle consists of invasions of pioneer trees (such as birch, willow (*Salix* spp.), aspen). These are light-demanding tree species with a large production of seeds that are carried by the wind, which allows them to spread quickly. These trees grow fast,

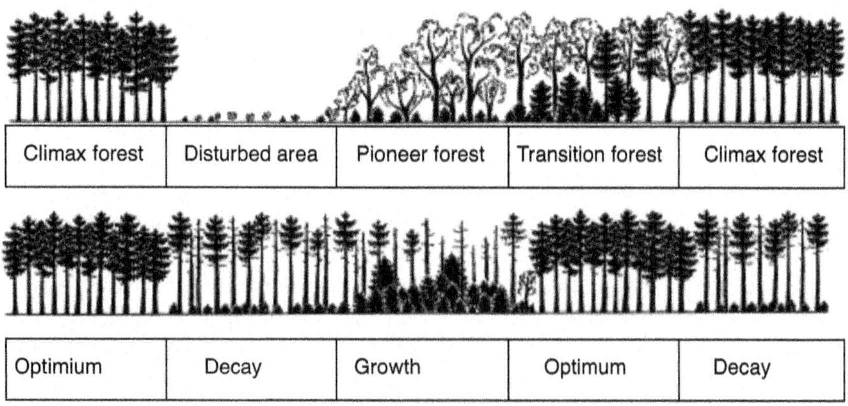

Fig. 3.10 Large-scale (top) and small-scale (bottom) forest cycles. The small-scale cycle of the forest consists of the stage of optimum, decay, and growth; however, in their regular rotation, various deviations can occur, when the forest "skips" some phase. (Král et al., 2018)

especially in youth, but have a lower competitiveness. In the undergrowth of these pioneers, suitable conditions are created for the establishment of more demanding woody plants, usually (semi-)shade tolerant (fir, beech, spruce, maple, etc.). These trees are better competitors, tolerate shading, but do not tolerate climatic fluctuations and do not grow well in open areas. They grow more slowly in their youth, but gradually, these trees will outcompete the pioneer trees of the first generation and a climax forest will be established.

3.1.4 The Main Species of Forest Trees

Unlike crops used in agriculture, where few species dominate in terms of production and area used for their cultivation, the use of tree species by human is much more diversified. Most forests are based on local species native to the area, despite fact that some species have been successfully introduced in new locations outside their original geographical range. When speaking about trees as source of wood, both conifers and broadleaves are important. Technically, stems of some monocots, namely bamboo and palms, are also used as wood but in comparatively smaller amounts so most of the non-conifer wood comes from dicotyledons (dicots). Dicots are also called broadleaves in contrast to conifers, which often have needles. Dicot trees are also often called hardwoods, and conifers softwood, although some hardwood may have much softer wood than most of the conifers. Looking at the proportion of conifers and dicots in wood production (Table 3.3) worldwide, non-conifers greatly dominate in fuel production which is the dominant wood use worldwide. Conifers are more important as a source of sawlogs and pulpwood. Fuel is the major use of wood in Africa and South America, and here dicotyledons also play a major role in firewood production. In both South America and Africa, dicots also play an

Table 3.3 Contribution of conifer and non-conifer (mostly Dicotiledonae) in production of wood for certain uses worldwide and in certain region in 2016, wood quantity in millions of m^3. (Based on FAOSTAT data)

	World	Africa	South America	North America	Europe
Wood fuel, conifer	213.8	18.0	8.7	12.6	58.2
Wood fuel, non-conifer	1714.7	668.4	171.5	51.6	112.2
Sawlogs and veneer logs, conifer	711.4	7.9	49.0	252.7	311.7
Sawlogs and veneer logs, non-conifer	375.5	27.8	40.1	50.2	48.2
Pulpwood, round and split, coniferous	355.6	3.3	31.3	148.5	144.1
Pulpwood, round and split, non-conifer	327.6	7.4	94.8	62.6	62.7
Other industrial roundwood, coniferous	31.4	0.7	2.9	4.1	13.1
Other industrial roundwood, non-conifer	118.2	27.8	9.2	11.1	13.3

important role in the production of wood for other uses. In North America and Europe, firewood is less important, although here dicotyledons play a major role in its production. Sawlogs and pulpwood are major wood uses here, and both of these categories are greatly dominated by conifer wood.

The high usage of conifer wood in Europe substantially changed the forest composition compared to the natural tree composition. To conclude what the composition of the natural forest would have been is tricky because the development of the forest composition changed significantly during our interglacial period and human affected forests for a long period of time. Despite that, we have a reconstruction of natural vegetation which is based on many proxies that describe how the European vegetation would look if there had been no human influence (Fig. 3.11). Based on this reconstruction we can roughly say that Scandinavia and Northeast Europe was covered by coniferous forests, mostly dominated by spruce, while most of the Central and Western Europe would have been naturally covered by broadleaf forests (except in mountainous areas) dominated by beech and oak (Bohn et al., 2003). Table 3.4 describes the proportion of major tree genera in European forests that roughly follows a decreasing proportion of conifer and increasing role of broadleaves from north to south. However, in many central and western European counties, the proportion of conifers is substantially higher than the expected natural occurrence mainly due to planting of spruce and pine. These trees were native in

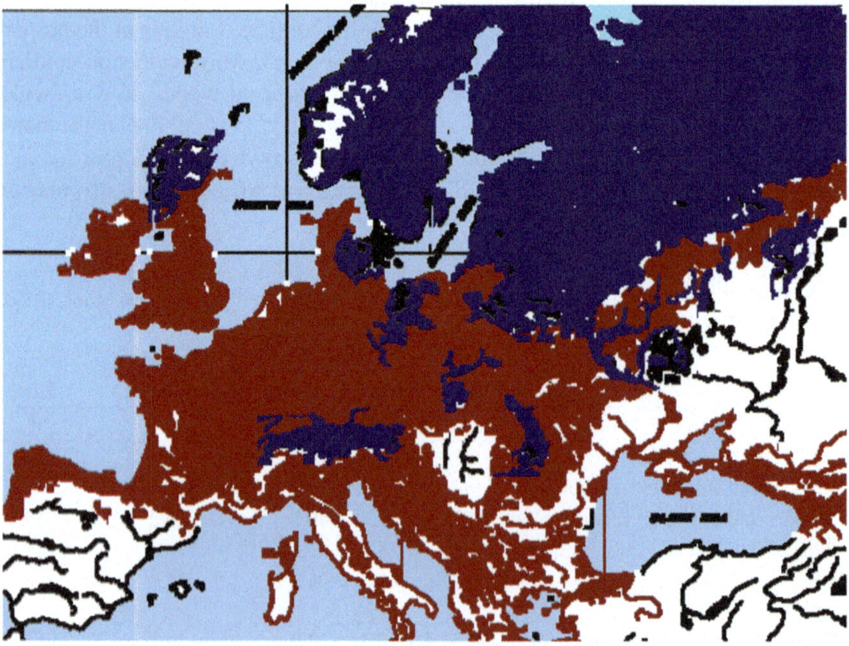

Fig. 3.11 Simplified map of potential natural forest vegetation in Europe; blue represents coniferous forests or mixed forest dominated by conifers, brown represents broadleaf forest or mixed forest dominated by broadleaf species. (Based on Bohn et al., 2003)

Table 3.4 Percentage of conifers and broadleaves and percentage of selected tree genera within forested area in selected European countries based on national forest inventories. Mixed stands were either counted as that of the dominant tree species, or divided between species if this distinction was not possible

	Beech (*Fagus*)	Oak (*Quercus*)	Broadleaf	Spruce (*Picea*)	Pine (*Pinus*)	Conifer
Finland	0	0	10[a]	25	64	90
Sweden	0	0	11	40	27	86[b]
Poland	8	7	31	7	58	69
Germany	16	11	44	26	23	56
Czechia	11	8	42	43	10	58
Slovakia	28	10	69	21	5	31
Austria	11	2	27	59	5	73
France	10	27	65[c]	8	12	35
Bulgaria	47	12	70	5	15	30
Italy	12	34	80	7	7	20

(Sources: Šmelko et al., 2006; INFC, 2009; Milewski, 2017; Foglar-Deinhardstein et al., 2015; IGN, 2019; SLU, 2020; https://bwi.info, http://nil.uhul.cz/, https://www.icp-forests.org/pdf/NatRepBulgaria2005.pdf, http://statdb.luke.fi/)

[a]The most important broadleaf in Finland is birch (*Betula* spp.)

[b]Conifer/broadleaf mixtures accounted as half to broadleaf and half to conifer

[c]The second most important broadleaf genera in France is chestnut (*Castaneus*)

their respective counties but their natural range was usually restricted to certain habitats (such as mountains or poor soils). They were often artificially planted extensively outside their ecological niche as we can see from large proportion of conifers in some European countries where natural vegetation would have been dominated by broadleaves (Table 3.4). We can also see that spruce, pine, beech, and oak cumulatively form majority of trees in most European counties with one of these being the most dominant genera in most of the countries.

Forest composition in North America or Japan, in terms of dominant genera, is more diversified (Table 3.5). In North America there is gradual decline in conifers from north to south; the highest broadleaf proportion and the lowest conifer proportion is in the northern USA, where *Acer* is the most important genera. Within the USA, a higher proportion of conifers can be seen in the south and west. In contrast to Canada, where spruce dominates, in the southern USA pine (*Pinus*) is the most important genera. In the western USA *Pinus* is also dominant but more conifer genera take an important share.

In the following paragraphs, we will describe a selection of European and North American tree species, and we also describe species that were extensively introduced outside their original distribution range and mention a few examples of the most iconic tropical forest trees. From the list below, and also from Tables 3.4 and 3.5, it may become apparent that there are fewer genera of native trees in Europe than North America or the temperate zone of Asia. This discrepancy has attracted the attention of researchers for long time. Reid (1935) came up with a hypothesis that the east-west orientation of major mountains in Europe, not only the Alps but

Table 3.5 Percentage of conifers and broadleaves and percentage of selected tree genera in Japan, Canada, and USA based on national forest inventories. In Japan and Canada, data are based on proportion of area of pure and mixed stands to total forest area. Mixed stands are either counted as the dominant tree species or divided half and half if this distinction was not possible. In the USA the percentage is based on the proportion of basal area of given genera from the sum of basal area of all trees recorded in the inventory in that given area. This approach was used for the USA due to high proportion of mixed forests. North and south in USA refer to northeast and southeast of USA

		Japan	Canada Boreal	Canada south of Boreal	USA north	USA south	USA west
Spruce	*Picea*	3	65	16	4		5
Pine	*Pinus*	4	7	16	8	31	23
Fir	*Abies*	5	3	14	5		13
Douglas fir	*Pseudotsuga*		6				22
Cedar	*Cryptomeria*	19		4	1		
Cypress	*Chamaecyparis*	10					
Conifer		50	79	66	25	37	87
Poplar	*Populus*		12	9	5	3	3
Maple	*Acer*			10	21	5	1
Birch	*Betula*		3	8	5		
Beech	*Fagus*	4					
Oak	*Quercus*	10			17	18	5
Broadleaf		44	15	30	75	63	13

(Sources: Forestry agency Japan, 2019, https://nfi.nfis.org/en/, https://www.fia.fs.fed.us. If a species was not mentioned it was missing or has a proportion below 1%, or was included in a conglomerate we cannot separate)

also Caucasus and Pyrenees, caused obstacles for tree migration southwards when they retreated in front of a progressing glacier. Later, Huntley (1993) showed that rather than migration obstacles, the geography of Europe caused a much smaller area that was left for refugia during glaciation maximum in Europe than in North America and Far East. It was recently concluded that interglacial climate may be more important to these differences than we earlier thought, but the differences in glaciation history may also be important, namely for the difference between Europe and the Far East. In Europe the refugia, in addition to being small, they were also fragmented which led to the formation of specific populations and or subspecies. This bottleneck during the glacial periods also resulted in populations from some refugia that were more successful. This is particularly the case for beech where population of beech that came from certain refugia now dominate contemporary Europe, and made beech a species that is now much more successful than it was in previous interglacials (see also Box 3.3).

In the following paragraph, we will describe a selection of European and North American tree species (Figs. 3.12, 3.13, and 3.14), and species that were extensively introduced outside their original distribution range will be described here, finaly we will mention also few examples of the most iconic tropical forest trees. For a more comprehensive overview, the reader can consult specialized monographies which

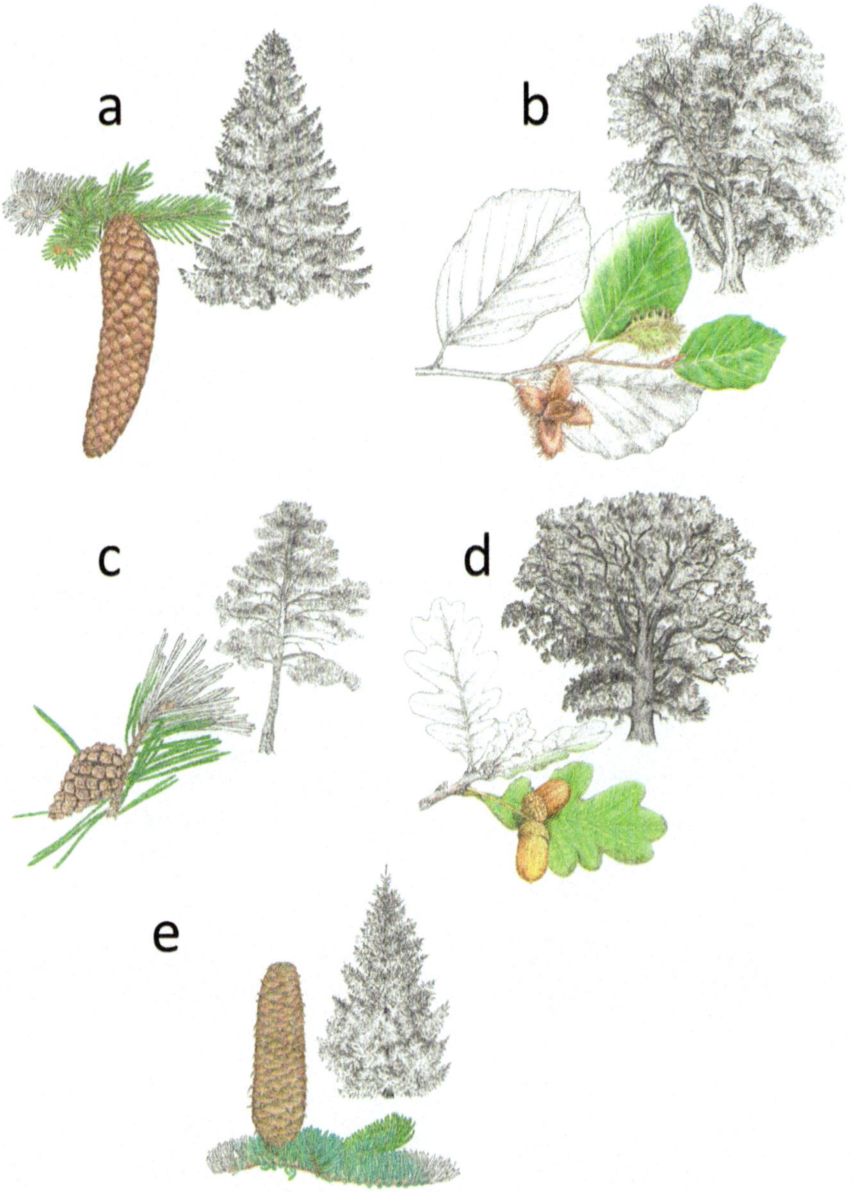

Fig. 3.12 Common European forest trees I, (**a**) Norway spruce (*Picea abies*), (**b**) Beech (Fagus sylvatica) (**c**) Scots pine (*Pinus sylvestris*), (**d**) English oak (*Quercus robur*), (**e**) Silver fir (*Abies alba*)

Fig. 3.13 Common European forest trees II, (**a**) ash (*Fraxinus excelsior*), (**b**) elm (*Ulmus* sp.), (**c**) larch (*Larix decidua*), (**d**) aspen poplar (*Populus tremula*), (**e**) white birch (*Betula pendula*), (**f**) black alder (*Alnus glutinosa*)

were also consulted for this brief list (Brockman, 1968; San-Miguel-Ayanz et al., 2016; Russell, 2017).

Spruce, in particular Norway spruce (*Picea abies*), is one of the most common species in European forests (Fig. 3.12). It grows up to 60 m tall, is an evergreen tree

Fig. 3.14 Common North American forest trees, (**a**) Douglas fir (*Pseudotsuga menziesii*), (**b**) pecan hickory (*Carya illinoinensis*), (**c**) sugar maple (*Acer saccharum*), sand pine (*Pinus clausa*), slash pine (*Pinus elliottii*), pole pine (*Pinus concorta*)

with needles individually arranged on branches, and has overhanging cones. The largest (in terms of volume) spruce can reach a height above 60 m, but spruces in common cultural forests are about 30 m tall. Spruce is a light-loving tree, but it tolerates shade well. The rooting is shallow, it is not demanding of nutrients but it needs enough water, and it also tolerates wetting and peat soils. It provides soft solid lightwood for use in construction and furniture. It is native throughout Europe, mainly in the north in Central and Southern Europe in the mountains. However, in many European counties it has been extensively planted outside its natural distribution range not only in mountains but also in lowlands. Another interesting European spruce is Serbian spruce (*Picea omorika*) which is endemic to the Drina River Valley in the mountains between Serbia and Bosnia and Herzegovina. Although it has a very small natural range, it is a common ornamental tree in Europe and North America. Among North American spruces we mention black spruce (*Picea mariana*), also called bog or swamp spruce, that is an abundant conifer occurring in most of the Canada up to northern limit of trees in North America. White spruce (*Picea glauca*) has a similar geographic distribution. Among other American species is blue spruce (*Picea pungens*), native to Colorado but often used as an ornamental tree.

Among European pines, Scots pine (*P. silvestris*) is an abundant European species. It is a very light-loving, pioneering tree. It grows well on poorer, sandy soils, and is very undemanding with respect to the quality of the substrate. Other species of pines are the mountain pine (*P. mugo*), which occurs as a shrub growing above the upper forest border, and stone pine (*P. cembraus*) growing in the mountains of Europe, especially in the Alps and the Carpathians. Black pine (*P. nigra*) comes from Southern Europe. Many pine species are important in North America, particularly noteworthy is loblolly pine (*P. taeda*), the most oft-planted species in the southern USA. Longleaf pine (*P. palustris*) is native to the southeastern USA, and it is also important cultural symbol of southern states. It is highly adapted to periodic natural fire. Longleaf and loblolly pines are the most planted pines in the eastern USA. Also important, especially in the western USA, are ponderosa pine (*P. ponderosa*) and lodgepole pine (*P. concorta*). Several pine species have been extensively planted on forest plantations outside their natural range. The most noteworthy in this context is Monterey pine (*P. radiata*). This species is native and occurs as several subspecies in a very limited range on the Monterey peninsula (California, USA) and vicinity. However, it has been extensively planted outside this natural range, not only in USA but also in many other countries such as South Africa, Chile, Spain, Australia, and New Zealand (Richardson, 1998). Interestingly, although the species has become very abundant outside its natural range, its native population suffers from pine pitch canker caused by *Fusarium circinatum*, a fungal disease introduced from the southeastern US.

European firs are represented by silver fir (*Abies alba*) native to the mountains of Europe. It is a stout tree up to 65 m tall. In adulthood it has a cylindrical crown. It also has a more cylindrical trunk than spruce. It prefers deeper soils, is not demanding of nutrients, but needs more soil moisture. It tolerates shading well and has slow growth. In Europe it is native, occurring naturally in the hills. The wood has similar uses to that of spruce, but is water resistant. It is sensitive to game grazing and air pollution.

Douglas fir (*Pseudotsuga menziesii*) is a massive tree (it grows to a height of 65 m, the largest individuals are said to be over 120 m, the trunk is 1.5–2 m in diameter, the largest individuals are over 4 m). It is native to the west coast of North America. It is the most common tree by volume in USA with nearly twice the volume of the next most common species, loblolly pine. It produces valuable wood with versatile uses, for it is resistant to the effects of the weather, and so is also used for outdoor structures. It was introduced in many European countries and also to South Africa, Australia, and New Zealand (Da Ronch et al., 2016).

Larch (*Larix decidua*) is a light-loving deciduous conifer native to open mountain forests. It grows to a height of 50 m, has small cones, and needles grow in tufts. The wood is soft but harder than other conifers, and it is flexible and hardy. Several other larch species are found in mountainous regions of the northern hemisphere, while others are found throughout the boreal forests of North America (*L. laricina*) and Asia (*L. siberica*, *L. gmelinii*, and *L. Cajanderi*).

Japanese cedar, or sugi (*Cryptomeria japonica*), is endemic to Japan although extensively grown in China as well where it has naturalized. It is large tree (70 m tall) and produces lightweight but strong wood that is resistant to decay. Japanese cypress (*Chamaecyparis obtusa*) is a cypress native to Japan and East Asia, and is widely cultivated in Japan and elsewhere. It is a slow growing species reaching up to 35 m tall. It produces valuable timber with very versatile uses from temple construction to table tennis paddles. The major pillars of Himeji castle are made from Japanese cypress. Several species of the genera *Tsuga* are abundant in forests of East Asia and North America. The word "Tsuga" comes from the Japanese word for *Tsuga sieboldii*, also called the southern Japanese hemlock. *Tsuga canadensis* can be mentioned among the American species. Last but not least, we mention the Sequoias; the specimens of *Sequoia sempervirens* native to northern California and southern Oregon are the tallest measured trees (116 m) but not the heaviest; this record belongs to a specimen of giant sequoia (*Sequoiadendron giganteum*) growing naturally in Sierra Nevada, with an estimated weight of 6.2×10^3 t.

European beech (*F. silvatica*) is a stout, up to 40 m tall, deciduous tree with gray bark. It tolerates strong shading and has the ability to strongly shield the soil. It is moderately demanding of the amount of nutrients in the soil and of soil moisture. It is native to most of Europe, providing medium-hard, easily chipped wood. Several other species of the genus are important, among them we mention Japanese beech (*F. crenata*) and American beech (*F. grandifolia*).

The hornbeam (*C. betulus*) is a monoecious deciduous tree growing to a height of 25 m. It tolerates shade well and shades well, and likes deeper, moist soils. It is native throughout Europe, and it tolerates game grazing, pruning, and cutting well, so it is often part of hedges, and it provides hardwood. There are several original species of oaks in Europe, including summer oak (*Q. robur*) and winter oak (*Q. petraea*), and Austrian oak (*Q. cerris*) which are massive trees growing to a height of about 30–35 m. They are light-loving woody plants, prefer warmer habitats, and occur in lowlands and hills. Oak provides hard, heavy wood, resistant to water. *Quercus suber*, commonly called the cork oak, is a medium-sized, evergreen oak tree native to Southwest Europe and Northwest Africa. It is the primary source of cork. There are several other evergreen oaks in Mediterranean including holm oak (*Q. ilex*) and shrubby kermes oak (*Q. coccifera*). *Quercus alba*, or white oak, is one of the most important hardwoods of eastern and central North America. The name comes from the color of finished wood which is used for furniture, barrels, and other uses. Red oak (*Q. rubra*) is another species native to North America, and has been introduced in many places in Europe.

Maple or *Acer* is an important dominant of many North American forests. Sugar maple (*Acer saccharum*) is an important tree of northern USA and southern Canada, and is the source of maple syrup. Red maple (*Acer rubrum*) is one of the most widespread trees in eastern North America, and is tolerant of shade and of a wide range of soil types. Notable European species include Norway maple (*Acer platanoides*) which is native to eastern and central Europe and western Asia.

Hickory (*Carya*) is a genus of temperate trees with pinnate leaves, and hard wood that is valuable for flooring. Many species are common in North America, and some produce edible nuts such as the commercially cultivated pecan (*Carya illinoinensis*). Sweetgum (*Liquidambar styraciflua*) is a light-loving deciduous tree native to warm temperate areas of eastern North America, and is common in the southeastern USA where it prefers areas with wet soils.

European ash (*Fraxinus excelsior*) is a tree growing to a height of 30–40 m. It is light loving in youth but tolerates shade. It grows from lowlands to foothills and low mountain locations in humid waterlogged places. *Fraxinus americana* is common in eastern and central North America. Elm is a genus of woody plants characterized by asymmetric leaves. American elm (*U. americana*) is native to eastern North America, from Nova Scotia west to Florida, but was decimated by Dutch elm disease, a fungal pathogen, and has disappeared from most of its range in recent decades although other elm species are still common. There are several European species (*U. laevis*, *U. minor*, *U. glabra*). In the past, elm in some parts of Europe was also strongly decimated by Dutch elm disease. Elms provide flexible wood, and it was used to make bows.

Alders are fast-growing woody plants (Fig. 3.13), often found growing in wetter places. They are light loving, but young plants can adapt to canopy cover. They are demanding of soil moisture and do not like poor soils. By means of symbiosis with the actinomycete *Frankia*, it fixes atmospheric nitrogen (N). Black alder (*Alnus glutinosa*) and gray alder (*Alnus incana*) are common European species, and black alder was planted and has become naturalized in northeastern North America.

Birches are pioneer light-loving species of fast-growing hardwood often have white bark. Silver birch (*Betula pendula*) is characterized by white bark, and grows to a height of 20–25 m. It is widespread from southern Europe to the alpine foothills, while in Finland and Norway it reaches the northern border of the forest. It is a light-loving tree, very undemanding of the site, and is one of the first colonizers of abandoned areas. In the far north at the treeline, downy birch or mountain birch (*B. pubesces*) is common in Northern Europe. Among many North American species is the paper birch (*B. papyrifera*).

Aspen (*Populus tremuloides*) is one of the most common hardwood species of northern part of North America. It is characterized by vegetative reproduction, forming clones that cover large areas of forest that are, in fact, one individual. The largest clone has an estimated weight even greater than the largest specimens of giant sequoia. Similarly, the European aspen (*Populus tremula*) is another pioneer species, growing up to 40 m on disturbed areas and forest edges where it occurs from lowlands to mountains. Many other poplar species occur in North America including the eastern cottonwood (*Populus deltoides*), native from the Canadian prairies to eastern and southwestern USA. Many species of poplar have been grown in forest plantations including eastern cottonwood and black cottonwood (*Populus trichocarpa*). In Europe, white poplar (*Populus alba*) and many hybrids are used in plantations.

Eucalyptus is a large genus of trees native to Australia. Many species have been planted around the world, and some have become naturalized around the globe. Due to their fast growth and adaptability to a wide range of conditions, *Eucalypts* are some of the most commonly grown plantation trees in many tropical and warm temperate regions of the world. They are harvested for biomass, fiber, timber, and essential oils, while other species are often planted as ornamental and shade trees.

There are more than 6000 species of tropical trees so we cannot mention comprehensively even the most important ones. We mention only a few iconic species and or species which would be later important for better understanding the examples presented in this book.

Mahogany is wood from tree genus *Swetia*. The wood is a dark brown or reddish color with a fine texture, hard, stable, but comparatively easy to work with and can be polished to a smooth and shiny surface. Due to these properties, it represents one of the most valuable woods for making furniture, boats, musical instruments, and many other items. Originally, the term "true mahagony" was used for *Swietenia mahagoni*; but now it is also for other species of the genus, namely *S. macrophylla* which occurs in Central and South America. *Swietenia macrophylla* was the most commonly exported wood originating from selective logging in the Amazonian rainforest. Since 2001, *S. macrophylla* has been listed in CITES II (CITES is an international convention that regulates trade of endangered organisms and products made from them). This stops official export of mahogany, but there is a belief that it boosts illegal logging and trade (Chimeli & Soares, 2017).

The term "mahogany" is often used also for other species of family Meliaceae. Among them the most noteworthy is an African tree of genus *Khaya*, called African mahogany. However, many other trees producing similar wood are often called mahogany. For example, some genera of *Shorea* from the family Dipterocarpaceae produces wood that is sometimes called Philippine mahogany.

Ebony is another iconic wood that is mostly black or with some texture; it is very heavy (it sinks in water), can be polished to a mirror finish, and is used for carving and various decorative objects. It is produced from heartwood of several species of genus *Diospyros*. This genus has a wide distribution of many species. The trees of this genus produce fruits which are bitter when immature and sweet when ripen; *Diospyros kaki* produces a popular subtropical fruit originating from China but now grown in many subtropical countries.

Teak is another durable tropical wood, resistant to water and termites due to its natural production of teak oil, that is often used in construction, boats, furniture, etc. True teak is produced by *Tectona grandis,* a topical hardwood from the family Lamiaceae. It is native to the mixed hardwood forests of southern and southeastern Asia. Two species of the genus *Milicia*, *M. excelsa* and *M. Regia*, in the family Moraceae produce wood known as African teak.

Dipterocarpaceae is a family of tropical trees containing more than 900 species occurring in most tropical areas, but the largest diversity occurs in Borneo. Many of the species are large, emerging trees reaching up to 70 m which provide valuable wood, and are a source of plywood, essential oils, resins, and other products. Especially in Borneo, they are often the subject of overcutting.

The balsa tree, *Ochroma pyramidale*, is a fast-growing tree in the family Malvaceae that produces a very light wood called balsa. It is native to Central and South America but is introduced in many other countries. It is a pioneer species that occupies clear-cuts and similar patches. Most of the commercial balsa now comes from plantations.

3.2 Forest Cultivation, Clear-Cut, and Continuous Forest

As we explained in the introduction, to produce yield, we can work either with even-aged cohorts, which we then harvest at one time, or populations from which we remove suitable individuals. In forestry, the analogy of these practices is the clear-cut and continual cover forestry (CCF) (Fig. 3.15). In German literature term Dauerwald is used for forest based on similar principes as CCF. The clear-cut forest cycle starts with the clearing of a certain part of the forest, the establishment of a more or less even-aged forest, and its thinning and subsequent logging. In CCF, the principal activity is selective logging. At more or less regular intervals, once every few years, we select the largest trees that have already passed the phase of intensive growth and harvest them (Pommerening & Murphy, 2004; Kuuluvainen et al., 2012). At the same time, we will harvest trees that have little prospect of further growth, but their removal

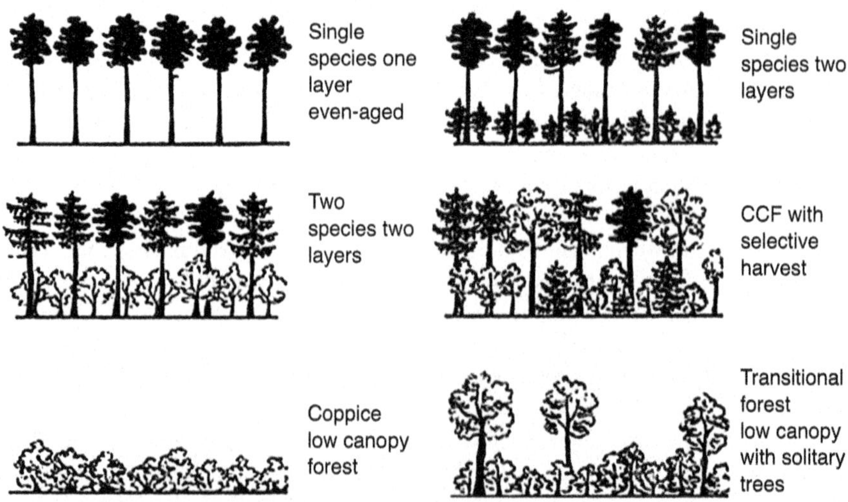

Fig. 3.15 Different clear-cut and continuous cover forestry methods; the first two rows are high forest

will promote the growth of other trees. The volume of such felled trees represents only a small part of the trees present and thus forests continue over time. In the gaps generated by tree felling, young seedlings establish and as they grow their number decreases by gradual self-thinning. The age structure of the CCF forest has a pyramid shape, with a smaller number of large individuals suitable for harvest and a gradually increasing stock of younger individuals who continuously replenish these removed individuals. It is often said that the CCF forest has a similar age structure to the natural forest. This is only partially true. Certainly, its age structure is more complex than that of the clear-cut forest and in this respect is more similar to the natural forest. However, in the natural forest the age structure and the representation of individual strata change significantly during a forest cycle; on the contrary, in the CCF forest the age and size structure is kept more or less constant, and similar to the natural forest at the beginning of the growth phase. This is because we constantly remove trees which have finished their intensive growth, in contrast to a natural forest where these trees can survive for a considerable time and affect age, spatial, and height structure, by both its very presence and through competition with other trees.

Although CCF can technically be operated in a single species monoculture, many forest organizations operating with CCF synonymize this approach with nature-friendly forestry that also seeks to achieve more natural forest compositions, which often means a mixed forest with multiple tree species. It is expected that this forest should provide a more stable provisioning of regulatory ecosystem services, such as erosion protection and water regime regulation. Due to the greater diversity of the tree layer and the more complex age structure, it is expected that the CCF forest will also be better able to withstand stress factors such as drought, gradation of forest pests, and so on. There are few empirical studies comparing the two large-scale approaches, but it can be expected that increasing the share of CCF may contribute to increasing diversity. On the other hand, in these forests, light-loving tree species are less common, and also some groups of animals and plants tied to sunny habitats of clearings and forest edges may be reduced. It can therefore be expected that the highest diversity will be achieved by combining both approaches rather than by applying only one of them. As a matter of fact, Dauerwald approach did not relay on selection of individual trees only but combine this with selection of small groups of trees. The CCF forestry relies, to a large extent, on the natural regeneration and self-thinning of the stand, which reduces the labor requirements. On the contrary, selective harvest, which is associated with CCF is usually more laborious. The CCF is also much more demanding on the knowledge of cultivation interventions and their application. It provides less favorable conditions for the application of mechanization and is therefore more demanding on human labor. A simple transition from clear-cut to CCF forest is usually accompanied by an overall increase in production (Vacek & Podrázský, 2006). However, the success of CCF depends on the harvested trees being continuously replaced by growing individuals, and especially when moving from an even-aged forest to a CCF, there may not be enough trees to grow to harvestable size until the next harvest. This gap in the age structure can then be reflected in the overall decline in yield. Another factor that complicates the comparison of CCF and clear-cut management is the fact that this transition is often associated with a change in the species composition of the stand. Some tree species have greater production potential than others, and the question is in which direction we will move on this gradient. However, CCF brings greater stability of production. First, this is because harvest does not take place once in 100 or more years, but in a way that once every 10 years we cut down only the largest individuals representing, let us say, 10% of the forest. Another reason for the expected greater stability is the assumption that the stand will be less affected by gradation of forest pests or climatic extremes, due to species and age variability, and thus its production will fluctuate less over time. In addition, at least in theory, CCF management provides the potential for better adaptation to market requirements. In the clear-cut forest, we plant a culture of one type of tree and grow it with the aim of a certain expected application. However, this growth may take 100 years or more, and it is not quite easy to predict what the demand for different wood assortments will be at that time. In the case of CCF forest, which would be composed of several species of trees, we can still influence the proportion of individual tree species or assortments somewhat better in the medium term.

The difference between clear-cut forest and CCF is not as sharp as it might seem from the previous text. Both approaches have a number of modifications. Certain intermediate strategies between them include the situation where the stand consists of small, more or less homogeneous patches. The subject of selective harvest is not individual trees, but small groups of trees or combination of individual selection with selection of small patches. The clear-cut forest can consist of several stands of different ages which were harvested at different times. Another category is represented by the coppice forest which, after harvest, regrows from the stumps. The coppice can be combined with leaving some large solitary trees, which makes spatial and age structure of the stand even more complex. Another technique that largely blurs the differences between clear-cut and CCF is the variable retention harvest. It combines clear-cut on part of the stand with selective logging with leaving part of the stands (min. 15%) intact so that there may be overgrown trees, dead wood, etc. The system is used by large forest owners, for example, in Canada. Its aim is to extract large areas at once and thus make the most effective use of available mechanization, while leaving habitats for biodiversity and supporting the ecosystem functions of the forest. After the extraction, there is a natural regeneration in the excavated places, which is of different speeds, and leads to a mosaic occurrence of different habitats.

3.2.1 Reforestation

Reforestation can be natural, artificial or combined (Table 3.6), combining the two previous methods. The basis of artificial reforestation is the planting of pre-grown seedlings, while natural regeneration is based on the sowing by locally occurring trees and the subsequent use of spontaneously established seedlings for further forest cultivation. Pre-grown seedlings are usually grown in forest nurseries from seeds from recognized stands. The seeds are allowed to germinate, then they are usually transplanted into loose soil (free-seeded seedlings) in a bed or in containers. Bare-root seedlings are most often used for forest regeneration, and planted seedlings are

Table 3.6 Comparison of the main characteristics of natural and artificial reforestation

Natural regrowth	Artificial reforestation
The species composition of the renewed stand is derived from the species composition of the existing stand. The new stand contains not only species, but also genotypes, that have already grown on the site. During seedling establishment, individuals adapted to local conditions are selected	By tree planting or seeding, we can also introduce species, or genotypes, that do not occur in the stand or in the vicinity
Natural regeneration depends on the production of seeds by the mother stand or on the occurrence of seed years, which do not come every year. Some trees may also produce new stems by vegetative spreading from root and stumps. Coppice forest regenerate mainly by this vegetative production of new stems from stumps	Thanks to the collection of tree seeds over a large area, a seed storage system and a seedling nursery system, one can plant new seedlings every year
The vegetation created by natural regeneration is spatially heterogeneous; it may be interesting in terms of habitats, but it may complicate thinning and other interventions	The growth created by artificial regeneration is usually very regular
Natural regeneration can be applied only in combination with some types of harvesting, generally only in a situation where there are trees in the area that are the source of seeds, or there are enough seedlings before harvesting	Tree planting or seeding can be used after, and in combination with, any type of tree harvest including large clear-cuts and even on non-forest land
It is usually cheaper	It has a higher cost

most often 1–2 (pine, oak, beech) or 3–4 years (spruce) old. For some tree species, we can obtain seedlings (clones) vegetatively (poplars, willows). Alternatively, trees can be reforested by seeding. The techniques used in artificial reforestation can be used also for afforestation of non-forest areas.

Forest regrowth requires the release of competition from the large trees in the stand, usually by tree harvest. Therefore, the method of forest regeneration is closely connected with the way we will harvest a forest. At the same time, we have a wide range of different options to increase light availability to the midstory and the forest floor: we can lighten the stand only a little, for example, by selective logging, or fell almost all trees and leave only selected individuals as a seed source, or clear-cut the whole area. On this gradient, the amount of light that reaches the soil surface increases, but at the same time other properties change: the decomposition of litter accelerates, nutrients are released, and growth of grass and herbs in the understory increases. Microclimates change as well. The more we open the stand, the more microclimate of the forest stand is disturbed and the soil surface is exposed to greater daily and seasonal fluctuations in temperature and humidity. These changes may be problem for some species, namely climax species such as European fir or beech. These species not only tolerate shading well, but excessive illumination also would not be desirable as they will not grow better in full light and conversely will suffer from fluctuations in the microclimate, competition from other plants, etc. The degree of shade also determines the shape of the future tree (more in Sect. 3.2.2). On the contrary, when restoring light-loving species such as pine or oak, it is advantageous to open vegetation more. In this respect, spruce is an interesting species as it tolerates a certain amount of shading in its youth, but it can also be planted on clear-cuts that are completely open.

In addition to planting seedlings and self-seeding, we can also restore vegetation by sowing seeds of forest trees. Seeding (seeds) can be used to supplement the natural regeneration of the stand, to introduce species that are missing in the stand.

In many European countries and in the USA, game grazing is a significant problem, and both seedlings planted and grown from seeds, especially deciduous and fir seedlings, may need to be protected from the game by fences.

When regenerating the forest we also, to a large extent, influence its species composition. This brings us to the question of growing monocultures or mixed forests. The native forests in the temperate zone often included a mixture of species. On the other hand, it is true that even a number of natural forests, especially spruce and beech, created almost pure monocultures under certain conditions. Extensive analysis by Vilà et al. (2013) show that across European forest stands, mixed forests have an average about 24% higher productivity than monocultures. These authors also conclude that wood production was enhanced with increasing tree species richness in almost all forest types. In a more detailed analysis of German forests, Jucker et al. (2015) show that tree diversity has the greatest potential to positively influence forest productivity at either end of the latitudinal gradient, where adverse climatic conditions limit productivity and lead to the development of less densely packed stands.

If we create a mixed forest by natural regeneration, we can influence its composition by the species composition and distribution of adult trees, which will be the source of seeds, but this is often limited by what trees we have from the previous generation. When planting a mixed stand, a planting pattern of various regularity and complexity can be applied.

Use of natural regeneration and tree planting varies substantially among individual countries (Fig. 3.16).

3.2.2 Thinning

The aim of thinning is to maintain the optimal density of the forest stand, at which we both maximize the production of wood and achieve the optimal properties of the produced wood. For example, for wood produced for lumber it means ensuring that the resulting log would be as straight as possible, have a large and ideally a cylindrical diameter, and have the smallest possible proportion of knots. For firewood or pulpwood these properties are not so critical, but selling these products may not bring so much profit.

As we said in the introductory chapters, as the number of individuals increases, their competition increases, and as a result, individuals grow less. Thus, the total growth, and similarly the total biomass, of individuals firstly increases with the increasing number of individuals, because the total growth of two individuals is greater than that of one individual and at low densities the effect of competition is small. Later, however, this increase in total biomass begins to decline or stagnate,

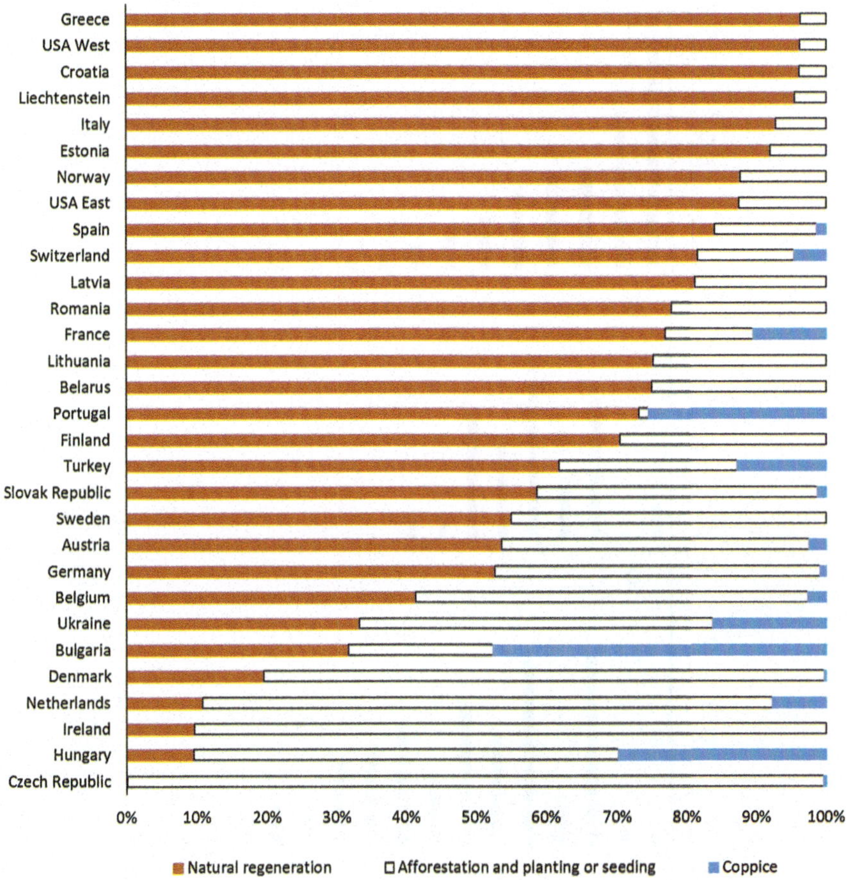

Fig. 3.16 Application of natural regeneration planting and coppice in selected European countries and the USA. (Based on data from Forest Europe (2015) and Oswald and Smith (2014))

because the influence of mutual competition prevails. The degree of competition depends on the size of the individuals. Young individuals are small and therefore their mutual competitive pressure is smaller, on the contrary, as individuals grow, their mutual competitive pressure increases and the same area can support fewer older than younger individuals. This is also common in natural populations, and it is also the principle of self-dilution in natural forests. In the forest gap created by the fall of a large tree a dense carpet of small seedlings appears, but as they grow their number decreases.

From this point of view, it would actually be enough to rely on self-thinning. We can do this, but winning competition through self-dilution is not so easy, so in a forest where there have been no thinning interventions, there will be more trees of smaller diameters and fewer trees of larger (commercially interesting) diameters (Fig. 3.17). In CCF forest which we harvest by selective logging it may not be so

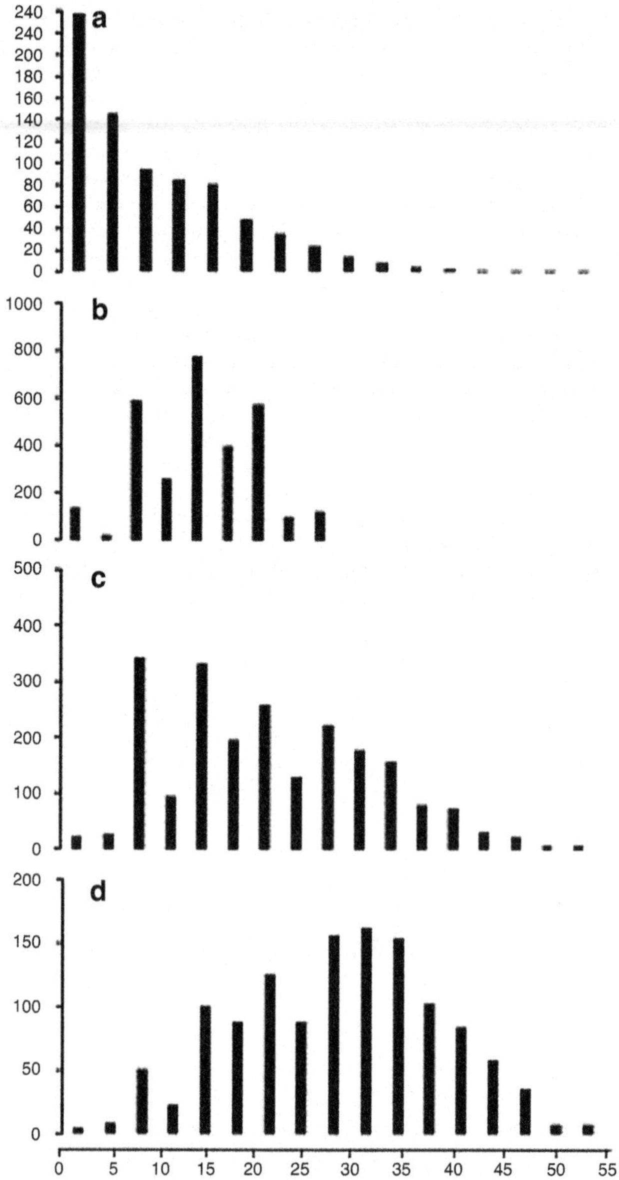

Fig. 3.17 Size structure of selected forest (**a**) and clear-cut forest of different ages (**b**–**d**)

critical. In such case we would select only the largest trees and leave the others for next time. Logging will also open the space for other trees and allow them to grow. So in case CCF we would rather focuss on opening enoufh space to maximize overal forest growth. However, in the case of a clear-cut forest, increasing the share of larger diameter trunks in the cohort is very important for the overall economic result.

What if we planted trees at the same spacing that we would expect to find in a forest that is ready for harvest? On the one hand, we can intuitively expect that we need to have a few extra trees in order to have some in reserve, in case some seedlings do not survive, etc. However, more importantly, planting trees far apart would lead to the development of stand that does not have the required technological properties. If we look at solitary growing tree and a tree that grows in a continuous canopy, we see that their appearance is different (Fig. 3.18). For conifers, we can observe that for solitary trees, the branches reach to the ground and the trunk is somewhat more conical in shape, while for trees growing in a canopy, the branches are only at the top of the crown and the trunk is somewhat more cylindrical. Not only in hardwood trees but also in pines, we can observe that the trunks of solitary trees are more often more curved and branch into several large branches instead of one central trunk, while one straight trunk is typical of trees growing in a canopy (Fig. 3.18). These properties of trees subsequently significantly affect the use of wood. As we already mentioned, for lumber, a straight cylindrical trunk with wood free of large knots is ideal. Branching of the trunk, and its curvature or twisted growth, limits the possibility of cutting the necessary lumber profiles from it. Similarly, in the case of a fairly conical trunk, there will be a larger proportion of edges. The number of knots in the wood is related to the number of branches; if the trunk has no branches at the bottom, it will be without knots, or at least there will be fewer of them and they will be small. Conversely, trunks with large branches all the way to the ground will have a very high number of large knots (see Fig. 3.18). For this reason, we try to keep the trees constantly in a continuous

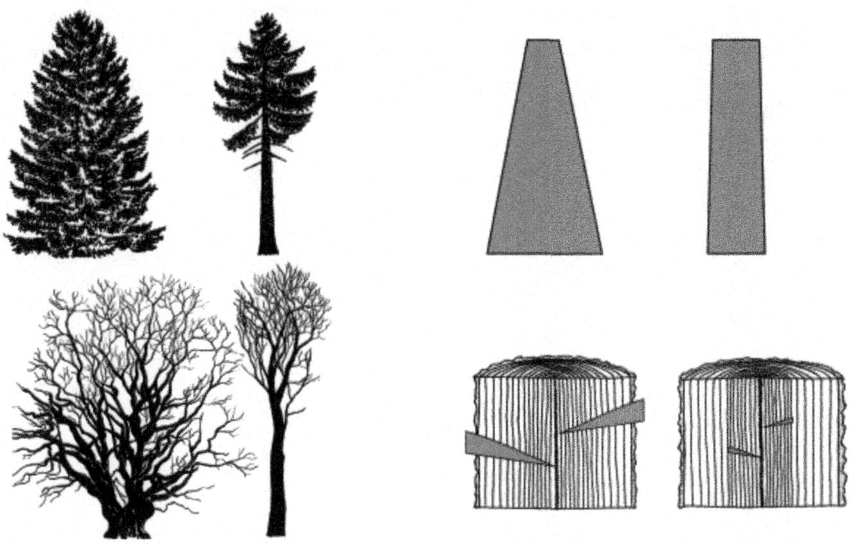

Fig. 3.18 Comparison of different aspects of a solitary tree (left of each pair) and a tree that has grown in a closed canopy (right). A tree that has grown solitary has a conical trunk with branches all the way down and can twist and branch more, while a tree in a canopy has a more cylindrical trunk, is more straight, has fewer branches and therefore fewer knots at the bottom of the trunk. Examples are chosen to magnify the difference and make them more visual

canopy. The canopy and competition of similarly large individuals may not have the same effect as shading given by higher tree layers, and as we have said, some tree species may grow better with some shading (see also Sect. 3.2.1). Ensuring optimal conditions to ensure optimal growth and quality of wood can be complex. For example, larch is a type of tree that requires it to overgrow other trees and not be shaded, on the other hand, we want it to have a crown only at the top. Ensuring optimal conditions is possible by growing it as an admixture in another stand and ensuring sufficient light by thinning. If we have a stand with a predominance of larch, it is possible to plant a fast-growing, shade-tolerant deciduous tree, such as linden, under the larch. This will shade the lower part of the larch and ensure the absence of branches, while at the same time it will also provide some production after the larch removal. At the same time, it provides soil cover, and the litter of deciduous trees may have beneficial effect on soil, etc.

Thinning is therefore aimed at ensuring the optimal number of trees per area in an optimal, that is, usually regular, spatial arrangement. Optimization of tree density to reach maximum production and wood quality is one of the most important problems in forest science (Skovsgaard & Vancay, 2008; Pretzsch, 2009). In the introductory chapters, we discussed how an increasing number of individuals increases the overall biomass production until a certain point; with further increases in the number of individuals, the competition prevails, and biomass increase will stagnate. In trees, this relationship is bit more complicated, and tree competition changes with tree age. Younger trees are smaller and need fewer resources than larger, older trees, so the same area can support fewer larger trees and more, smaller ones. There are several ways to conceptualize this (Allen et al., 2020), but Langsaeter's Curve is very popular. It relates forest stand growth or increase in biomass to existing biomass. In forestry, we use tree volume instead of biomass, and we relate tree volume increase to existing volume of trees (Allen et al., 2020) (Fig. 3.19). In a young stand, the volume of wood is small and so there is a small increase in wood volume, but as the tree grows, the volume of wood increases and the annual wood production increases as well. At a certain point however, the competition starts to have an

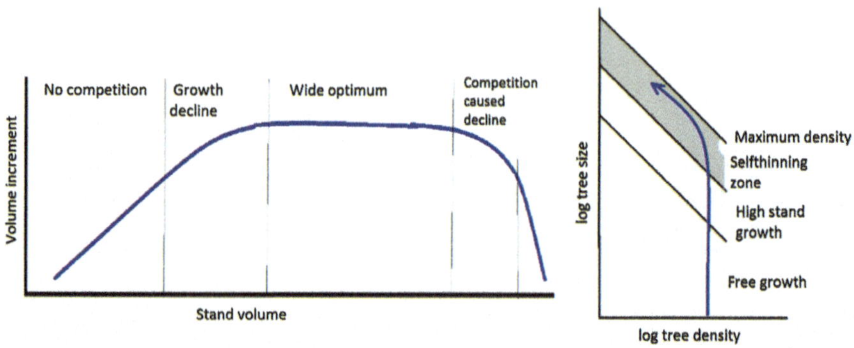

Fig. 3.19 Langsaeter's curve (left) depicting the relationship between forest stand volume increment and existing wood volume, and Reineke line (right) showing the relationship between logarithm of tree size and logarithm of tree density

increasingly negative impact on the growth and eventually flatten the curve as the tree growth slows due to competition. Competition may cause not only the individual trees to grow less, but also (particularly at higher densities) some trees die and number of trees is reduced in a process called self-thinning. Both of these processes can cause the volume increase to slow, then at very high levels of wood volume, to decrease. The self-thinning effect can be visualized by projecting the logarithm of tree size against logarithm of maximal tree density (Fig. 3.19). This line represents the boundary of tree numbers at a certain size that can be supported by a certain area of forest (with certain site properties), limited by self-thinning. Reineke (1933) shows that this relationship results in a line with a constant slope for a large variety of sites and tree species (Fig. 3.19). Other authors use a similar chart where the position of the x- and y-axes is switched, but these have a very similar meaning, as again there is a fixed-slope line that delineates the maximal number of individuals of certain size that a certain area of forest can support (Skovsgaard & Vancay, 2008). One of the most important pioneers of this approach in Europe was Assmann (1961). Although now we know that slope may vary, the assumption of a fixed slope is widely used in the construction of growth models (Skovsgaard & Vancay, 2008). Assuming the slope is constant, we can conveniently express differences between sites by the height of the line. Sites that are more productive can support a larger number of trees of the same diameter than can poor sites, or when we keep tree number constant, they can reach larger a diameter on a more productive site. The height of the line can be characterized by the site density index (SDI). This relationship has become a basic concept in forest science allowing for the prediction of optimal tree density of even aged forest stands. It allows production of various tables and other tools that can indicate to forest managers the optimal density of trees, of a certain size, that should be reached to maximize overall wood production.

However, the difference in site quality translates to a difference in production in a nonlinear manner, as site conditions affect not only tree density and DBH but also tree height, which again changes with tree age in more less predictable, but site specific manner (Skovsgaard & Vancay, 2008). So, based on an evaluation of stand height at certain age, one can estimate tree height development with tree age. Taller trees can potentially support a greater total volume of wood production. Here, we should mention that these optimizations were largely done for even-aged monocultures and (together with Faustmann formula mentioned before) have become a base for industrial-scale clear-cut forestry, which is based on growing even-aged monocultures. Moreover, some coniferous trees such as spruce or pine that produce valuable, fast-growing lumber have received great attention in this effort (Daniels & Burkhart, 1988; Allen et al., 2020), which goes hand in hand with their popularity in clear-cut even-aged forestry.

The wood obtained during thinning can be used, for example, for the production of paper or for energy purposes. Carrying out intensive thinning at long intervals is undesirable for several reasons. If the stand grows too dense for longer than desired, the trees compete too much, they are slimmer than desired, and this negatively affects not only production but also their stability and resistance to wind, heavy snow, ice, etc. In addition, a tree that has been shaded for a long time has adapted to the situation, and if it is suddenly released, it is a shock for it which can lead not only to a slowdown in growth, but also to the death of the tree. Some species, fir trees, for example, are very sensitive to a sudden surplus of light.

As already mentioned, in CCF we rely mostly on self-thinning and the creation of gaps by felling individual trees. There is great variability in conditions for natural regeneration inside each single gap formed as a result of selective logging. Some parts of it are more illuminated, others more shaded, etc. In addition, the seedling density is different based on the position relative to the seed source tree. All this gives individual seedlings very different chances to win in competition with surrounding seedlings. Here, a factor which is once advantageous may become disadvantageous in other conditions, or at a different stage of development. This greater variability and stochasticity of conditions gives a better chance that among self-seeded seedlings the potential winner(s) of the competition may appear sooner than when the conditions are very homogeneous.

3.2.3 Forest Harvest

Forest harvesting is usually closely associated with its regeneration; by clearing the stand we not only obtain wood, but also at the same time open space for new trees. The logging method is then closely related to the method of restoration and further management of the forest. In order to maximize the yield from the forest, it is advantageous to carry out the logging when the trees have just passed the phase of intensive growth. We measure tree growth as a change in selected dendrometric variables, such as a change in the tree DBH. This change during the life of the tree can be well described by the sigmoidal curve. The tree grows slowly at first because its total biomass is small and so the growth may be small. Later, as a tree ages, its growth accelerates exponentially until it reaches its maximum. Later on, as the tree ages even more, the growth begins to slow down until it theoretically comes to a complete stop, although in practice this asymptote may almost never be reached. How can we recognize that the intensive growth phase has just passed? There are two ways how we can measure tree growth, or increment, as some dendrometric parameters such as DBH. The first one is mean annual increment. This is basically the size of a given parameter, such as DBH, divided by the age of tree or tree stand. Let us say a tree is 20 cm in diameter (DBH) and is 40 years old; the mean annual increment (MAI) will be 20 divided by 40, which is 0.5 cm year^{-1}. The other way is to measure tree diameter repeatedly, then the growth can be expressed as diameter increase divided by the length of interval, to give the periodic annual increment. Let us assume that the tree mentioned in the previous example has a DBH of 18 cm 2 years ago. Its periodic annual increment (PAI) will be 20 minus 18, divided by 2, which is 1 cm year$^{-1.}$ By analyzing the growth curve, we can see the inflection point of the sigmoidal curve where it begins to approach the asymptote (i.e. the moment when the tree is already past the fastest growth period), and this is the time when the

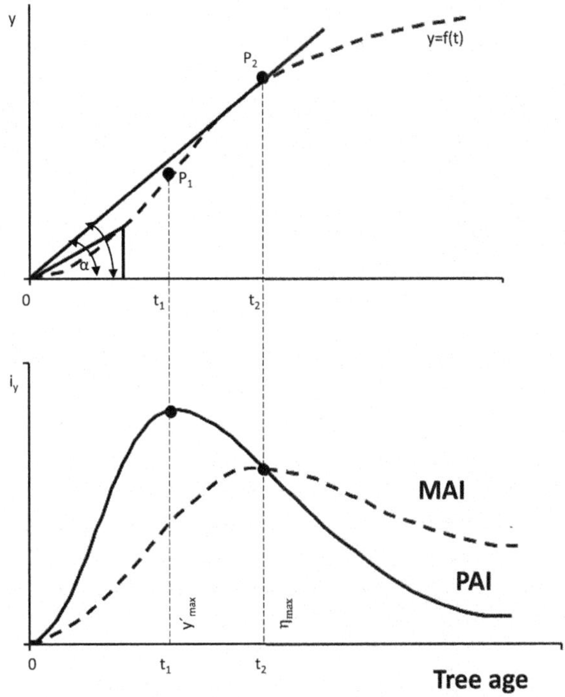

Fig. 3.20 Determination of the end of intensive growth based on the relationship between the growth function and a comparison of PAI and MAI. The top chart shows the size of the tree, the lower chart shows tree increment (both PAI and MAI in the same scale)

MAI and PAI is equal. Therefore, if we regularly measure individual trees in the forest and know their age, we can determine exactly when the individual trees, or the entire forest stand, reached this inflection point (Fig. 3.20) and are suitable for logging. For common practice, this procedure is a bit complicated and therefore the correlation of this point with a certain size, or age, of the tree is used. The growth dynamics in even-aged and uneven-aged forests differ somewhat. In an even-aged clear-cut forest the growth increases very quickly, but it reaches a maximum and starts to decrease sooner than in an uneven-aged forest. In uneven-aged forest the initial growth is slower but it remains high for a longer time and decreases only slowly. Therefore, the trees in the selectively logged forest can reach larger diameters when the maximum growth is completed, in comparison to similar even-aged sites.

We carry out selective logging in an uneven-aged forest. We do this at regular intervals approximately once every 5–10 years, when we harvest all suitable trees. Tree selection for logging is based on it having reached the target thickness. However, we also remove damaged trees including trees that suffer damage due to harvest operations, and also those trees that have been suppressed for many years and will rarely increase growth rates to the same level as trees that had not been suppressed. Consequently, selection of harvested trees should consider also the further potential growth of residual trees and other issues, such as how the extraction of a given tree will affect the growth of the surrounding trees, or how much this particular tree will be marketable at the moment, compared to its potential in the next logging period, etc. As previously mentioned, tree extraction will create gaps that will be filled by natural regeneration. This periodic harvesting is necessary to maintain the character of the uneven-aged CCF forest; without it, the height of the trees in the stand would equalize and, as a result, the tree growth and regeneration would stagnate. Selective harvest requires good site accessibility by roads, so harvested logs can be extracted. Because of risk of damaging remaining trees it requires extra skilled workers and limits the use of mechanization which cause lower work effectivity.

In an even-aged forest, we basically use the same approach to determine site maturity but we harvest whole stand at once, or over several periods coming at comparatively short intervals. The second option (shelterwood) is particularly useful if we want to restore a stand of shade-loving trees, either by their natural regeneration or by planting under the existing stand. We can harvest a portion of the trees to lighten the stand and enable better growth of planted seedlings, or possible natural regeneration. If we want to use the natural regeneration of light-loving tree species, we can leave a suitable generation of trees in small but sufficient numbers (tens per ha) on the felled area, wait until these trees produce enough seeds and natural regeneration is established, then go back and harvest the remaining trees. Another alternative method or harvest consists of felling a narrow strip of forest at the edge of the parent stand, which then is a source of seeds for natural regeneration. After this belt is restored by natural regeneration, we can continue toward the interior of the forest and take next strip. Of course, nothing prevents us from planting this area with seedlings from the nursery. Finally, we can harvest a larger piece of forest and create a clearing, which we usually plant with nursery seedlings, and we call this method of logging "clear-cut." Clear-cutting offers the greatest possibilities for using mechanization (more Box 3.4) and is the most efficient from the point of view of work efficiency, while selective logging is the most laborious, requiring skilled workers, etc., but on the other hand allows the easiest natural regeneration. A certain compromise between efficient use of mechanization and use of natural regeneration is variable retention harvest (Franklin et al., 1997), which is applied by larger forest owners, for example, in Canada. Here we combine different types of logging in a larger area in a planned systematic manner, leaving a small part of the area completely untouched, for example, for protection of a watercourse or for other important landscape element, a portion we extract only partly, leaving seeding trees, and other parts we clear-cut. This will create a more spatially heterogeneous landscape structure, support natural regeneration, and at the same time we will concentrate logging operations in one place which enables us to optimize the use of mechanization (Fig. 3.21).

So far we have talked about logging, which was planned. In addition to that, incidental harvests, sometimes called salvage logging, can be carried out to mitigate the action of some stress factors (wind, insect outbreaks, drought, etc.). Its purpose is, on the one hand, an effort to use and economically recover wood mass damaged by a given harmful factor, and at the same time is an effort to reduce other secondary damage (more in Sect. 3.2.4). More about the significance of such disasters for ecosystems is shown in Box 3.5.

Fig. 3.21 Different methods of forest harvest

Besides forest harvest related to forest regeneration, there is also clearing of forest associated with deforestation and crop change, for example, to pastures, oil palm plantations, etc., which is a major cause of deforestation in the tropics. In this case we usually do not care too much about the previous growth and development of the stand, the purpose is to clear the area, and wood is a secondary consideration. In forest stands, especially in the tropics, selective logging is used to harvest only the most valuable types of wood. A typical example is logging of mahagony in the Amazon or Dipterocarps in Borneo. Asner et al. (2005) reported that selective logging in Amazon caused loss averaging 23 m^3 of wood ha^{-1}, which is mostly not reported in deforestation statistics. This represents carbon (C) loss of about 8 t ha^{-1} in roundwood and up to 50 t ha^{-1} including associated fine and coarse debris. Similar, or even higher, C losses are reported by Saner et al. (2012) from selective Dipterocarp logging in Borneo where about 36 t C ha^{-1} is removed and about 50 t C ha^{-1} represents downed woody debris.

The relationship of those gaps and, especially establishment of valuable tree species, is more complicated, than what is described above for temperate forests. Despite the fact that during this harvesting only a small amount of wood is taken from the total stock, the structure of the stand is disturbed not only by removing the harvested trees but also by damaging the surrounding trees. The tropical forest stand is spatially complex and multistorey, and trees are connected by lianas, so the felling of trees often leads to damage to other trees. It is often necessary to cut down other surrounding trees, etc. Further damage is caused by transport of extracted logs. This is partly due to the fact that it is necessary to build a number of approach roads into the closed forest, which opened the forest even further. All these lead to the fact that selective harvesting in the tropics can be far more destructive than in the temperate zone. If such extraction, and in particular, the handling of extracted logs, is carefully planned with a view to minimizing such damage, it is called reduced impact logging (RIL). Besides this extensive damage, seed source and seedling establishment differ from the way it was described for temperate zone. In a tropical forest, there is generally a better chance of successful development where there are seedlings that have established further from the mother tree (see also Sect. 3.4.1). After selective extraction or natural death of the tree in the tropical forest a stand gap will be created, which will be used by the trees already usually present in the stand. While in temperate forests there will usually be seedlings or small trees of the same species in the undergrowth of such a tree, small trees of other species will predominate in the tropics. The Janzen-Connell hypothesis deals with the explanation of this phenomenon. It assumes that pathogens and pests accumulate under the mature tree, and these restrict the growth of seedlings of the same species. In contrast, the number of seeds, similar to temperate species, decreases with distance from the parent tree. This hypothesis also explains the great species richness of trees in tropical forests (Clark & Clark, 1984). Seed dispersal in the forest, done mostly by animals, is thus critical for successful regeneration. If dispersal is affected, for example, due to fragmentation, the whole system can be negatively affected.

Box 3.4 Forest Mechanization

Probably the most typical piece of forest mechanization is a chain saw. The first chainsaws were handheld, used for cutting bones, and appeared in the early nineteenth century. The first portable chainsaw usable in forestry then appeared in the early twentieth century (1918). The chainsaw is a basic tool for felling and processing the tree. However, the felled tree must be pulled out and brought closer to a suitable place from where it can be taken out of the forest. In the past, the tree was branched and debarked on the spot and allowed to dry and cut on the spot. This reduced friction and weight and made it easier to pull out, usually with the help of horses, ideally in winter. This method is the most gentle on the forest, but it is the least effective. Gradually, pull-out tractors of various capacities were used. With increasing traction it was possible to drag the whole trunk and then cut it elsewhere, or even move the whole felled tree with branches. At the point where the entire trunk was felled, further processing does not have to be done manually, but by a harvester, a self-propelled machine that can be used for felling and processing whole tree, which will significantly increase work efficiency. Of course, more force is needed to pull out the whole trunk than to pull out the individual partially dried and debarked pieces. The traction force depends not only on the properties of the trunk but also on the character of the surface (better on snow, worse in the understorey vegetation, etc.). Similarly, as we have shown, with agricultural mechanization, with increasing traction, the weight of the tractors increases which increases pressure on the ground and thus the compaction of the soil. As a rule, there are fewer travel passes in forest land than in arable field; travels are often concentrated in one path and there can be dozens of them. Harvesters can have an adverse effect on soil compaction, as their

(continued)

Box 3.4 (continued)

weight ranges from 4 to 20 t. In addition, a loaded front-end loader tractor can weigh over 20 t. Forest soils naturally have a lower bulk density (higher porosity), so the consequences of compaction can persist for a long time. As with field soils, compaction caused by travel worsens with increasing soil moisture; on the contrary, damage can be reduced by covering the soil surface with branches, and it is also less on frozen soil. It turns out that the first crossings have a very significant effect on soil compaction; this results in reduced infiltration, increased surface runoff and risk of erosion, impaired soil biological activity, and reduced growth of tree seedlings (Cambi et al., 2015). The time required for the regeneration of forest soil after heavy machinery crossings is difficult to determine precisely, but there are studies showing that the observable effect of mechanization passage on some soil parameters can persist for many years and decades (Cambi et al., 2015). Dragging a larger number of trunks or unbranched trunks increases the risk of damage to understory vegetation and standing trees. The trees can also be retrieved using cable skidders. The cable skidders are used in the western USA and Canada, and also in Europe. Cable skidders are suitable for retrieving wood in otherwise inaccessible terrain; their advantage is that they exert less pressure on the soil surface than front-end loader tractors. The cable skidder can again be advantageously combined with a processor.

Box 3.5 Disasters and Their Significance in Nature

In general, we tend to perceive large natural disturbances as large windbreaks, forest fires, and outbreaks of forest pests as disasters that somehow deviate from the normal course of nature. It is also often believed that these large disturbances were caused by human influence and did not occur in preindustrial times. However, there are a number of studies showing a relatively regular course of such disturbances. For example, regular oscillations of the larch weevil (*Zeiraphera diniana*) over the last 1200 years have been documented in the Alps (Esper et al., 2007). Monitoring of forest fires in Ontario (Canada) showed a regular occurrence of forest fires with a periodicity of 260 years (early Holocene) to 120 years (late Holocene). The increase in fires in the late Holocene is associated with a colder climate and the presence of more flammable species (Hart, 2016). A number of records of large windstorms can be found in historical chronicles. In many places it is difficult to carry out such studies because suitable archives are not available, but we can assume that the occurrence of such large disturbances with a period of once every hundreds of years could be very common even in a landscape unaffected by human. However, human has certainly influenced their frequency and intensity. How would the long-term absence of such disturbances affect

(continued)

Box 3.5 (continued)
ecosystems? This is clearly shown by a study of the islands on the Swedish lakes Hornavan and Uddjaure. On these lakes there are small uninhabited islands covered with forest. Dating of organic soils on these islands has shown that the smaller the island, the older the ecosystem on it. This is due to the fact that the main disturbance factor here is forest fire arising from lightning strikes. If lightning ignites the vegetation, the whole island usually burns. Larger islands are a larger target and are therefore ignited more often by lightning than small islands, leading to the aforementioned negative relationship between the age of the ecosystem on the island and the size of the island (Wardle et al., 2003). If we now look at the islands of different ages (different sizes), it can be seen that biomass and tree production is higher on the young islands than on the old ones. This is an example of a more general phenomenon called retrogression. Retrogressive ecosystems have lower biomass and production, especially of the. This is due to the fact that key biogenic elements, especially phosphorus, have gradually been leached from the soils of these ecosystems. In addition, the remaining phosphorus is bound in forms that are not readily accessible by most plants. In addition, due to soil development, unweathered rocks are too deep to serve as a source of phosphorus for plant roots (Peltzer et al., 2010). Fires or similar disasters lead to the death of large amounts of biomass and thus allow the recycling of phosphorus and other biogenic substances. Even major disasters such as glaciers or lava flows can then expose minerals from the newly emerging geological substrate. In addition to this effect on biogeochemistry, periodic disturbances allow species that use such vacant areas to be maintained in the landscape. A study of the succession of animals after a forest fire in North America shows that many animals prefer some of the early successional stages of forest development to climax forests (Oliver et al., 1998). Based on that, we can conclude that despite fact that natural disasters certainly interfere with our economic activities, they occurred in natural ecosystems which were adapted to them, and in many aspects they were essential for normal functioning of ecosystems.

3.2.4 *Forest Protection*

The aim of forest protection is to improve forest resilience and protect forests against various adverse factors which may be both natural and anthropogenic, namely factors that damage or reduce forest production, usually the production of wood. Of course natural forest ecosystems have always had similar disturbances, often large, natural ecosystems are adapted to these phenomena, and it can even be said that without these disturbances, ecosystems would not be able to function normally. These harmful phenomena, or natural disasters, were important for the natural course of biogeochemical cycles (Box 3.5); many species of light-loving forest trees and herbs and many species of forest animals were bound by their existence to habitats that created such disturbances. In many protected areas, therefore, these disturbances are respected as part of the natural course of natural processes. However, in commercial forests, these processes can hamper our economic interests. Factors damaging the forest can be divided into abiotic (wind, drought, snow and ice, fire, flooding, etc.); some of these can be anthropogenic in origin (acid rain, ozone and other pollution) and some biotic (damage by game, insect pests, fungal diseases, etc.).

3.2.4.1 Abiotic Factors

Wind The wind causes both the upheaval of trees, when the tree is uprooted, or fractures of the trunk or crown of the tree, which the tree may subsequently regenerate, sometimes with the formation of multiple trunks that are not suitable for lumber production. The wind effect can be simply the result of the wind's force, but very often its effect is dynamic, as it acts repeatedly (in gusts); it can vibrate the tree and the subsequent resonance then amplifies the effect of individual gusts. Strong winds also cause the formation of reaction wood in the stem, and this causes timber to twist or bend when it dries, greatly reducing its value. The occurrence of strong winds is relatively high; in the last 30 years there have been over 20 hurricanes (winds with speeds above 118 km h^{-1}) in Central Europe. Consequently, assuming time from planting to harvest of clear-cut forest to be around 100 years, chances that it will face such winds at least once are high. The probability of wind damage is increased by the terrain configuration, as a location on the windfacing slopes increases the effects of the wind; on the contrary, the leeward position and the rugged terrain in the direction of the incoming winds reduces the effects of the wind. Woody plants with a dense, high-set crown, or with brittle wood are more easily damaged (spruce, poplars, birches, aspens). Taller vegetation is more often damaged and species with shallow roots suffer more from upheavals, particularly on wet or waterlogged soils. Trees infested with rot are also very vulnerable. Spruce monocultures are particularly endangered, especially in places where harvest has exposed the edge of the stand. Collateral to wind damage, such damaged trees are further attacked by other pests such as bark beetles. Wind also causes many species to slow or cease growth for a period of time. Protection against the wind is mainly in the spatial structure of the vegetation, namely in formation of buffer strips which slow down the wind speed. Buffers are usually densely branched edges of the stand, formed by different species of trees, situated perpendicular to the direction of the winds, and composed of shrubs and trees of different heights which form a dense continuous edge of the stand gradually rising from the ground to the treetops. It should have a high proportion of deciduous trees (min. 30%). Its goal is not only to slow down the flow of wind, but it also helps protect against other harmful factors. Other buffers to slow down the wind flow can be then built inside the stand, such as those formed by strips of resistant woody plants (e.g. beech, maple). These vegetative buffers should not be an impermeable barrier, which would only raise the wind to a height that would then cause damage when it falls again. The goal is to create a semipermeable wall that takes some of its energy from the flowing air.

Snow and Ice Snow and ice damage the trees mechanically due to the considerable weight of snow or ice that is retained on the branches. The branches, or whole trees, can break or bend. Conifers and trees that retain leaves over the winter are more damaged, as are dense, slender trees. The possibilities for protection lie in the right thinning so that there is no overcrowded and slender vegetation.

Drought Drought damages all types of trees, for when there is a lack of water, the plants cannot transpire, the formation of wood stops, and if the drought lasts for a long time, the tree suffers irreversible damage or even dies. An important factor that influences susceptibility to drought is competition for water. In denser stands, where more trees compete for the same amount of soil water, the effect of drought is more pronounced than where competition is less severe (Gleason et al., 2017). Reducing the density of trees by thinning can thus be one of the measures that increase the resistance of forest stands to drought. Not all trees are equally sensitive to drought; spruce is very sensitive. It starts to have problems when the water balance during the growing season is zero (water balance is calculated here as the sum of precipitation during the growing season – potential evapotranspiration of grassland + soil water supply at the beginning of the growing season). Beech is more resistant (tolerates a deficit of up to 25 mm), oak and Douglas fir are even more resistant (tolerates a deficit of up to 150 mm), and the most resistant is pine (up to a 180-mm deficit) (Albert et al., 2017). The situation is even more complicated in mixed stands; for example, it has been shown that spruce and oak in mixture behave similarly as in monocultures in terms of resistance to water stress, but the resistance of beech in the mixture increases, especially in a mixture with oak (Pretzsch et al., 2013). The situation is even more complicated because it depends on all the trees in the mixture and, in addition, the behavior of the same species in the same mixture may be different in different climatic regions. It is expected that ongoing climate change may shift areas suitable for conifers to more maritime climates as well as to higher in the mountains, and further north. On the contrary, many areas in Central Europe which are now extensively planted with conifers will become more suitable for broadleaf species (Fig. 3.22).

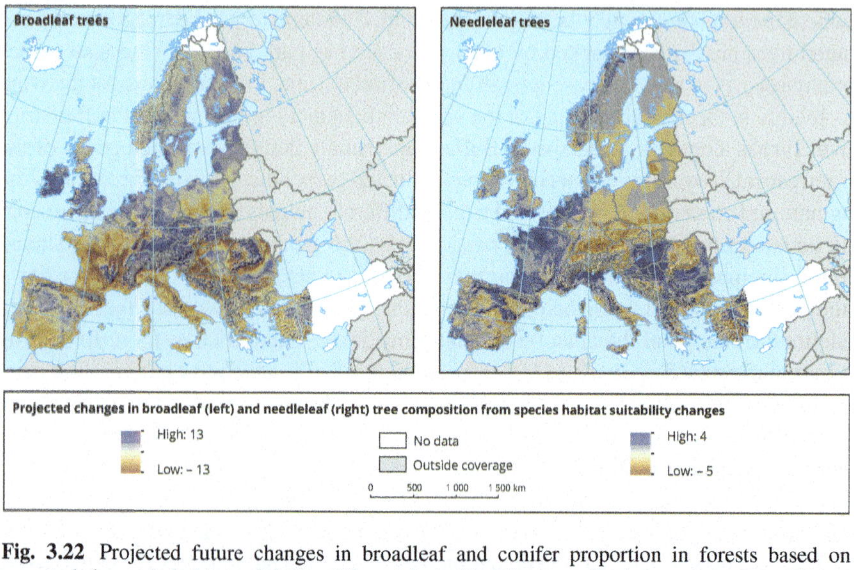

Fig. 3.22 Projected future changes in broadleaf and conifer proportion in forests based on expected change in habitat suitability. (Copyright holder: European Environment Agency (EEA))

Atmospheric Emissions and Ozone Atmospheric emissions, especially sulfur and nitrogen emissions, are the cause of forest dieback in many places. In Europe, especially the so-called black triangle between the Czech Republic, Poland, and the former East Germany, was famous for massive forest dieback. The effect of this acid rains on the trees is complex, but damage is typically the result of nutrients leaching out of foliage and soils. Acid rain could have a direct effect on the photosynthetic organs of plants, especially in the mountains, where there was a high incidence of fog. In addition, when the soil pH drops below 4.5, the solubility of aluminum increases sharply, which is then toxic to plant roots. As part of the fight against acidification, a significant part of endangered forest stands has been limed. The effects of this intervention are very complex. Increasing the supply of magnesium to air-damaged trees reduced the symptoms of yellowing and improved their health and growth (Vacek et al., 2009, 2019). The soils of mountain forests are characterized by slow decomposition and a thick organic horizon with low pH (mor type of humus). Under the influence of acid deposition, the mineralization of organic matter may be further slowed down (Mulder et al., 2001). As the pH increases due to liming there may be a sudden increase in the mineralization of accumulated organic matter associated with nitrification (Corre et al., 2003). Nitrates released in this way can release other basic cations from the soil and thus, paradoxically, further acidify the soil and at the same time eutrophize surface waters (Armbruster et al., 2000; Nohrstedt, 2001). At the same time, however, liming has not been found to affect wood production (Ingerslev, 1997; Sikström, 2001; Hindar et al., 2003). Despite the fact that coal-fired power plants were mostly desulfurized during the 1990s and the death of forests due to acid deposition practically disappeared, the overall rate of defoliation remained high in many European countries. One reason may be that while sulfur deposition has decreased significantly, nitrogen deposition has not (Wright et al., 2001; Hettelingh et al., 2005). In addition, there has been an increase in the concentration of ground-level ozone, which can also damage the photosynthetic apparatus of plants. It is manifested as interveinal damage to the upper part of the leaf, especially in older and more exposed leaves. Beech is especially sensitive to ozone.

Fire Fire is a natural phenomenon occurring during most of the geological history of terrestrial plants (Scott & Glasspool, 2006). Fire is an important factor influencing natural forests in many countries (see Box 3.6). Fire has very variable effects on the forest depending on fire severity and return interval, and the adaptation of trees to fire. Fire may affect only the ground layer if it is of low intensity, or it may reach higher strata of forest vegetation until eventually reaching the canopy. Fire that reaches the canopy is called a crown fire, and it is usually severe fire with high burning temperatures and severe effects on the forest. On the other hand, many forest species are adapted on ground fire. The occurrence of fire depends on the availability of fuel, its flammability, and climatic conditions. Heat waves, dry weather, etc., can promote fire. Prevention against fire can be the removal of fuel by prescribed burning and minimization of ignition sources.

3.2.4.2 Biotic Factors - Pests

Insects Forest trees are attacked by a number of insect species that damage various parts of the tree. There are hundreds of tree/pest combinations however only some are economically important. Analysis in the USA shows that there are more (about two-thirds) pest/tree combinations in broadleaves (angiosperms) than in conifers (gymnosperms); however, severity of pest attack on conifers is greater than on non-conifers (Potter et al., 2019). The same authors also show that exotic pests are also a larger problem in conifers. Consequently, damage from disease is greater in the western USA where there is a larger prevalence of conifers and forest communities are simpler. Also, in Europe, large-scale outbreaks can cause significant damage to trees grown for commercial timber production. Besides fungi, bark beetles, namely those such as *Ips typographus* that damage conifers are most often mentioned among the most important forest pests (BIO intelligence service, 2011). In both the USA and EU, invasive pest species are a serious problem.

Among economically important insects is the spruce bark beetle *I. typographus*, to which we dedicate a separate box (Box 3.7). Several bark beetles of genera *Dendroctonus*, such as mountain bark beetle *D. ponderosae*, are important pests of pines and other conifers in North America. Among leaf pests, we should mention the gypsy moth (*Lymantria dispar*), which is a polyphagous species that feeds on the leaves of many tree species and can cause severe damage when overgrown. It is native to Europe but introduced to North America. A relative, also polyphagous, is *L. monacha*, which can cause calamity to spruce stands. Another pest that can cause calamity to oaks is *Tortrix viridiana*. Disaster can also be caused by the autumn moth (*Operophtera brumata*), and its attacks are frequent but short-lived. A noticeable example of a newly emerging pest is the horse chestnut weevil (*Cameraria ohridella*), whose caterpillars overwinter in chestnut leaves. Interestingly, this species was only discovered and described in the 1980s in Northern Macedonia, and by the end of the last century had spread throughout Europe, where it began to cause devastating deaths of horse chestnuts. It is a textbook example of a suddenly appearing pest.

Fungi Woody plants are infested with a number of fungi, of which we will mention root disease caused by *Armillaria* species. *Armillaria* are among the largest known organisms in the world, comparable to some poplar clones mentioned above, at least in terms of the area it covers. One colony in Michigan covers an area of 37 ha and is estimated to be 2500 years old.

Damage by Game Grazing Game can be divided according to their way of eating, into browsers who bite young shoots (such as deer and moose) and grazers who feed on a volume of lower quality food such as grass, but many other species are generalists and move between these strategies according to food supply. Browsers, in particular, are problematic concerning damage to young trees. In many European countries the game grazing is one of the most severe problems causing damage to young seedlings (BIO intelligence service, 2011).

Box 3.6 Fire as a Factor Shaping Natural Ecosystems in the USA

Fire significantly affects the functioning of ecosystems (Senici et al., 2010, 2013, 2015). Different ecosystems create conditions for fire to various degrees, and vegetation is an important factor in determining fire frequency and severity (Fig. 3.23). Many natural ecosystems are dependent on fire. If we look at Florida's natural or nearly natural ecosystems, out of a total of 81 habitat types that are differentiated by conservation mapping, 18 of them depend on the periodic presence of fires. The ecological impacts of a fire are mainly influenced by two factors, the frequency of the fire and its intensity. The occurrence of fire is largely influenced by habitat factors such as the groundwater level and its fluctuations, soil texture, slope steepness and aspect etc. Other major factors influencing the occurrence of fire are the accumulation of fuel, its flammability, continuity and quantity. Therefore, there is a certain trade-off between the frequency and intensity of the fire. Where large amounts of highly flammable fuel, such as dry grass, dry palm leaves, and so on, are produced, fuel is rapidly accumulated and allows fire to spread easily, which results in frequent fires. However, the intensity of such fires is low. On the other hand, if the fuel is made of less flammable material such as shoots, small twigs, etc., fire starts and spreads less easily, but during that time more fuel accumulates, and when it does burn it produces more heat so the effects of the fire are then far more severe. An example is the comparison of flat forests dominated by mature *Pinus eliota*, *P. serotina*, and *P. palustris* pines, in the undergrowth of

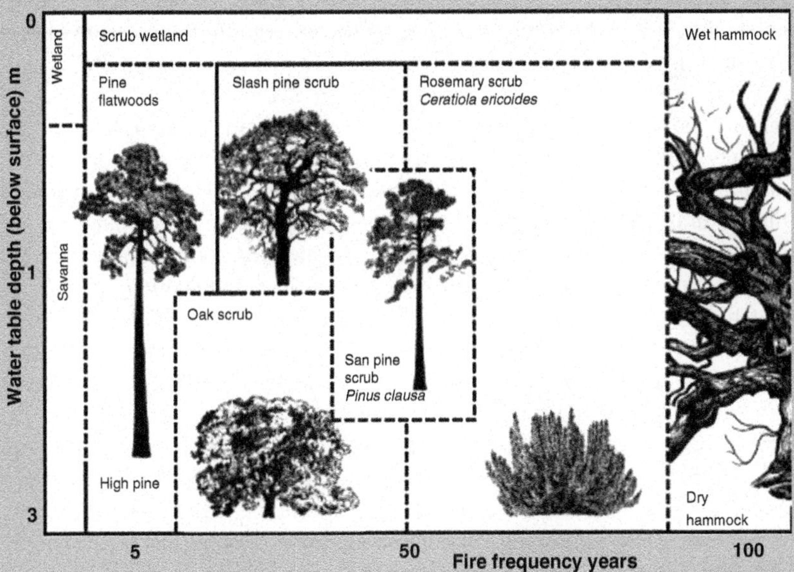

Fig. 3.23 Formation of different types of ecosystems as a result of different groundwater levels in m and the intensity and frequency of fires in Florida, USA

(continued)

Box 3.6 (continued)
which grow a number of grasses and a small palms (*Serenoa repens*), which produce large amounts of dry, highly flammable leaves, which cause frequent fires. Fires are common once every 2–5 years, but are not very intense and vegetation grows quickly from surviving ground rhizomes of palm, and grass roots in the soil. *Pinus palustris* seedlings have a well-insulated meristem that protects the growing tip, and even young trees have a thick, insulating bark to protect the stem. In contrast, areas dominated by *Pinus clausa* and several shrubby evergreen oaks often lack a continuous herbaceous layer and accumulate fuel more slowly. Fuels consists of small twigs and needles that do not carry fire easily, and so the fire occurs less often (once every 30–50 years), but when the fire occurs, it is very intense. *Pinus clausa* is adapted by having cones that do not open until after the fire, and the seedlings germinate on bare, fire-burnt soil. It is a short-lived species so the seedlings will grow until the next fire at which time the next generation of cones is ready.

Nowacky and Abrams (2008) show that in the USA, European colonization resulted in landscape changes that led to a reduction in the frequency of fires. This was due to the suppression of fires, the subsequent thickening of vegetation, and development of shade-loving species which produce litter that decomposes quickly, thus leaving less flammable material on the soil surface. In the USA, a law was passed in 1934 ordering the extinguishing of all forest fires by 10:00 a.m. the day after the fire was reported. The longer the time after the law came into force, the more difficult it was to comply; a phenomenon called the Wildfire Paradox. The more thorough the suppression of fire, the more their intensity increases, because more fuel accumulates. If this happens on large continuous areas, conditions are created for the occurrence of mega-fires, which are difficult to manage. At the same time, a study of the ecology of fires in the USA has shown that many species are dependent on the natural occurrence of fire. The Wilderness Act of 1964 thus recommended that natural occurrences and effects of natural processes, including fire, be encouraged in protected areas. In many cultivated forests preventive burning is used every 2–5 years, which prevents fuels from accumulating.

Changing the natural dynamics of fires can also have devastating effects. In Florida, most fires naturally occurred due to lightning, especially in summer, which is traditionally humid. These fires are usually of low intensity, burning ground vegetation and dry litter; however, most of the organic matter in the soil along with the seeds and vegetative organs of plants survive, as do trees adapted to low-intensity fires. In addition, the fire released a considerable amount of nutrients and led to a slight increase in the pH of otherwise acidic forest soil, thus improving nutrient availability beyond that provided by decomposing organic matter alone, and opening room for intensive growth of many plant species adapted and stimulated by low-intensity fire. Nowadays, fires occur also in the winter due to the greater influence of winter tourists

(continued)

Box 3.6 (continued)

visiting Florida. In winter there is less rainfall, dry periods are usually much more intense, and winds are stronger. Vegetation is often damaged, overlying organic horizons with a bank of seeds and vegetative organs of plants are burned, and hydrophobic layers are formed which subsequently leads to ash erosion during rains and nutrient removal. Ecosystem restoration in such areas is then very slow.

It is difficult to assess the role of fires in the natural dynamics of forests in Europe. As we have shown in previous chapters, most forests in Europe are heavily influenced by humans, and many past forest use practices (e.g. litter removal) have significantly reduced the risk of fires. However, paleolimnological studies show that in the past, especially in the Neolithic, fire and controlled forest burning have played a key role in shaping the European landscape (Tinner et al., 2005). In the future, however, it is necessary to take into account the increasingly frequent occurrence of fires in the landscape and the increase in the size and intensity of fires, which cannot be prevented only by fighting individual outbreaks. It will be necessary to change our understanding of fire in the landscape, which is currently understood more as a consequence of carelessness and human failure, which could be completely avoided. We should begin to perceive fire as a logical consequence of fuel accumulation and other factors supporting fire, and as a necessity for fire-dependent species. There will be a need to focus more on reducing fuel accumulation and the continuity of areas with high fuel accumulation, the appropriate choice of landscape mosaic and, if necessary, prescribed burning.

Box 3.7 Bark Beetle

The *Ips typographus*, a 3.5- to 5-mm long brown beetle with yellow hairs (Fig. 3.24), lives on spruce and occasionally on other conifers. The swarming takes place in April and June. It primarily attacks weakened and uprooted trees. At lower (endemic) population densities the beetle represent only little treat for healthy trees. But if, due to drought, wind damage, etc. more weakened trees occur, the bark beetle population may multiply and reach a higher (epidemic) population density. At epidemic densities, bark beetles begin to attack healthy trees. The beetle prefers trees 60–150 years old (trees under 40 are rarely attacked), particularly those on the sunny forest edges. Because it is difficult for one beetle to overcome a tree's defenses, a synchronized attack by multiple individuals is needed to defeat the tree. Males therefore secrete an aggregation pheromone to attract other males, but if the density of the secreted pheromone is high, it acts as a repulsive, dispersive signal. As a result, males that have arrived attracted by the aggregation pheromone will not settle on the tree if the population is already too dense. Males who have not attacked a

(continued)

Box 3.7 (continued)

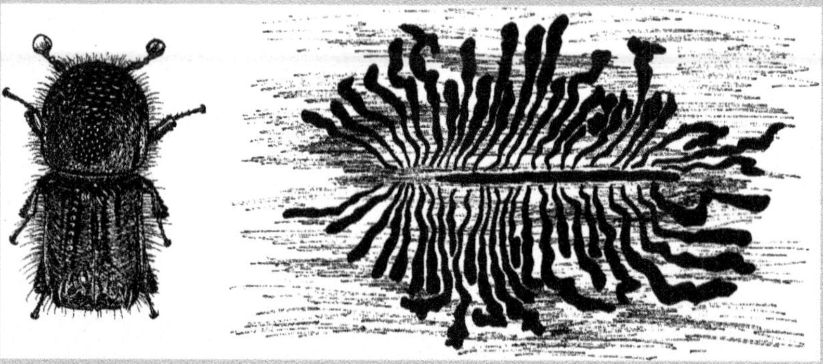

Fig. 3.24 Bark beetle (*Ips typographus*), adult beetle, and damage under bark

given tree can attack trees in the immediate vicinity. This creates groups of infested trees in an otherwise healthy forest. In the same way, individuals who have hatched from an infested tree will attack trees in the vicinity (up to 400–500 m). It is also clear that the spruce monoculture supports the spread of bark beetles better than does a mixed and diverse forest. Males then build the wedding chamber in the bark and attract females, males fertilize one to three females, which then excavate corridors in the phloem into which they lay about 60 eggs. Laying eggs takes about 7–10 days. After 6–18 days, larvae hatch from the eggs and build vertical feeding corridors. Their development takes 7–50 days followed by 8 days as pupae, depending on the temperature conditions. Older literature states that there is one, and occasionally two generations a year, but today in Central Europe two generations are relatively common, and there may be three. This is due to an increase in temperature, which leads to accelerated development. The bark beetle overwinters either as a pupa under the bark, or as an adult beetle in the litter. The bark beetle transmits the fungi *Ophiostoma* and *Ceratocistis*, which in fact can kill the tree, fungi are beneficial for the beetle as they help weaken the tree's defenses and also serve as food for larvae.

Bark beetle outbreaks have been observed in Central Europe since the nineteenth century, and before World War II, they occurred mainly in mountainous areas above 800 m. The first great outbreak at lower elevations was observed at the turn of the 1940s and 1950s. The bark beetle is a major pest throughout Europe, with the exception of the Mediterranean and the British Isles. In the last three decades, bark beetle outbreaks have been described, during which tens of millions of cubic meters of wood were harvested from Scandinavia, the Baltic countries, and Belarus. Long-term waves of drought, which weaken the trees, contribute to the occurrence of these catastrophic outbreaks. In addition, rising temperatures lead to an increase in the number of generations that bark beetles can have, which exponentially increases the size of the population.

3.3 Lignicultures and Plantations of Fast-Growing Woody Plants

Tree species with a short rotation are grown on both forest and agricultural land, as part of agroforestry systems. Especially in the tropics, the cultivation of eucalyptus or pine trees with a short rotation is very widespread. In Europe, poplars and fewer willows are grown in this way, but some other woody plants such as alder and black walnut have also been tested. Growing on agricultural land with a high content of nutrients, and cultivating the soil, results in very high growth rates. Lignicultures can be used to produce timber that can be used as lumber, but often they are used as a source of pulp or to produce energy biomass. The spatial structure and the structure of the community are even simpler than in a commercial forest. The area is planted not only with the same species, but usually also with the same, vegetatively propagated clone. In ligniculture, thinning may occur, and trees are planted more closely than the desired final distance to reduce branching in the lower layers of the stand, which reduces knot formation. However, often the thinning is omitted. Energy trees are planted in a denser stands and are harvested after a period of 3–7 years, after which they regrow by sprouting from the stumps, which will be the basis for the next harvest. In this way, several harvests are carried out without the need for further planting. This is basically an analogy of coppicing. The result is biomass, which is used as an energy, most often for direct combustion, but it can also be used for the production of liquid fuels such as ethanol, or construction materials, boards, pulp, and so on.

The assessment of the environmental impacts of growing plantations with short rotation times depends to a large extent on the land use which was converted to plantation. The conversion of the original forest into eucalyptus and pine plantations certainly leads to the loss of natural habitats, and the loss of biodiversity and many ecological functions such as carbon storage (Fig. 3.25). On the other hand, growing energy trees on degraded agricultural land both in the tropics and in the temperate zone can bring a number of benefits such as protection against erosion and the accumulation of soil organic matter, as compared to the original habitat, and is a better habitat for many organisms. Specifically, eucalyptus plantations are often planted to slow down land degradation, for example, to stop erosion. However, planting eucalyptus also has a negative effect; the intensive use of nutrients can reduce nutrient availability in the soil, which can reduce growth of other plants. Allelochemicals released by eucalyptus reduce growth of understorey plants even further (Bayle, 2019). Eucalyptus plantations have been shown to have lower diversity of birds and understorey vegetation compared to native forest (Goded et al., 2019). It has also been shown that eucalyptus plantations reduce total, as well as microbially driven decomposition of litter in streams running through the plantation, as compared to natural vegetation (Ferreira et al., 2019). Even in its native range, a commercial plantation harbors lower insect diversity than that of native eucalyptus forest (Cunningham et al., 2005).

Fig. 3.25 Stands of exotic tree plantations in South Africa (**a**) *Eucalyptus camaldulensis*, (**b**) *Pinus radiata*

3.4 Forest Management and Ecosystems

3.4.1 Forest Management and Biodiversity

Deforestation and fragmentation of the forest is often mentioned as one of the major problems of tropical forests. Birds, which are one of the well-studied groups, show that the effect of fragmentation can vary between tropical and temperate forests. In general, forest fragmentation, the division of the forest into smaller islands, leads to the loss of species specialized for forest interiors, and in turn supports species specialized for forest edges in both the tropical and temperate zones. However, in the tropics, species specializing in forest interiors contribute more to the total number of species, while in the temperate regions, species specializing in forest edges contribute more (Báldi, 1996). This leads to the result that forest fragmentation in the tropics causes a more significant decline in species than in temperate regions, where in some cases it can lead to an overall increase in the number of species. In addition, in the tropics, many species specializing in forest interiors are unable to overcome even a small (several tens of meters wide) gap in a continuous forest. In addition, bird species specializing in forest interiors often play a key role in the distribution of forest tree seeds. This can be crucial for maintaining the diversity of forest trees and the forest itself. One of the hypotheses (Janzen–Connell hypothesis) explaining the great diversity of trees in the tropical forest assumes that the incidence of seeds decreases with distance from the parent tree, and at the same time, seeds that germinate near the parent tree have little chance of survival due to accumulation of specialized pathogens and herbivores. On the other hand, seeds that germinate as far away from the mother tree as possible have a better chance of survival. That is, in a place where there will probably be no other tree of the same species nearby (Connell, 1970; Janzen, 1970). Therefore, organisms that transmit seeds over long distances, such as those specialized for the interior of the forest, or some mammals, are important for maintaining the diversity of the tropical forest. These organisms may be very sensitive to forest fragmentation (Sekercioglu & Robinson, 2011). Experiments with fragmentation of the Amazon rainforest (Laurance et al., 2018) have shown that in small fragments these seed carriers disappear, and the fragments are not even visited by carriers from the surrounding environment. This stops the supply of seeds from distant areas and local seeds are, in turn, restricted by pathogens and herbivores as described above. This leads to the complete breakdown of the forest vegetation in small fragments. In contrast, in temperate ecosystems, forest fragmentation effects are related to increases in forest edge area. Especially in situations where the cultural forest is fragmented, this leads to an increase in the area of habitat with greater spatial heterogeneity, providing habitats for many species. An increase in the number of these species can then offset the decrease in the number of species specialized for the interior of the forest. Forest edges often host early and mid-successional species and often have a larger proportion of invasive species (Laurance et al., 2007), so forest fragmentation and similar disturbances can help to spread these.

Due to the effects of various, globally widespread forest management practices on biodiversity, a meta-analysis was carried out by Chaudhary et al. (2016). Based on the analysis of a number of published studies, the authors ranked the main forest management methods according to their impact on biodiversity as follows: variable retention harvest > gentle selective logging > conventional selective logging > grazing > agroforestry > fast-growing lumber plantations > energy tree plantations. This is a worldwide study, and therefore the comparison is necessarily very rough. As will be shown in detail in Box 3.8, under certain specific conditions, coppice forests (low forest) can be important habitats for many species and have considerable potential for conservation. In general, however, it can be said that farming methods leading to the simplification of the spatial structure of the stand also reduce diversity.

Another procedure that significantly affects biodiversity is the retention of individual old trees and dead wood. Leaving old and declining trees in the harvested stand can serve as a habitat for a wide range of bird and insect species (especially species that live in tree cavities) and can reduce the loss of biodiversity that results from clear-cutting. In Scandinavia and some other countries, this is part of the standard harvest practices (Gustafsson et al., 2010; Vitkova et al., 2018). The size of the trees left and their location in the stand are important for the impact on biodiversity. In general, it is preferable to leave larger trees to support biodiversity. Very small trees do not form a large enough habitat. Sunlit trees are particularly important for many specialized saproxylic insect species. Similarly, leaving dead wood can promote the diversity of many other types of organisms such as bryophytes and fungi. In addition, in natural forest settings, dead wood creates an important micro habitat for the establishment of tree seedlings. In general, leaving more trees brings greater benefits in terms of biodiversity (Gustafsson et al., 2010); on the other hand, it means loss of production. Another practical risk of leaving fresh dead wood is the potential spread of some pests, especially root pests (Vitkova et al., 2018). Advantageously, existing dead wood, trees of unsuitable dimensions, damaged trees, or trees in hard-to-reach places can be preserved, as can tall stumps of trees having an unsuitable shape at the base, etc. (Vitkova et al., 2018).

As described above, the forest represents an important reservoir of carbon and other biogenic elements, and their dynamics is influenced by the type of forest trees and the way the forest is grown. Forests have great potential to contribute to the restoration of ecosystems that have been significantly damaged, for example, by mineral extraction (Box 3.9). The forest management cycle influences the biogeochemistry of the forest, especially logging which removes a large amount of biomass and at nutrients. In addition, harvest results in a significant increase in solar radiation at ground level which causes a change in the microclimate and accelerates decomposition. This effect is especially significant where most of the litter accumulates on the soil surface, that is, in stands dominated by conifers or other trees that produce litter with a high C:N ratio. The increase in decomposition, supported by the decomposition of residues, especially needles and small branches with a high content of nutrients that remain after harvest, causes the release of nitrates and basic cations. Due to the extraction of forest biomass, their uptake by roots is limited and nutrient outflow into watercourses increases (Kreutzweiser et al., 2008).

Box 3.8 Some Forms of Historical Use of Cultural Forests Interesting for Nature Conservation

In addition to the remnants of the original forests, or forests where the effort is to restore the forest character by minimizing or eliminating human intervention, there are practices of forest use which were widespread in the past and created a substitute habitat for conservationally interesting species of plants and animals. The first example of such use is the coppice, which may also be called the low forest. Coppices are a very old form of forest use in Europe, which mainly covered the need to supply firewood. In Britain, for example, coppices have been known since the Neolithic. This type of forest is characterized by vegetative propagation by means of stump or root sprouts and a short rotation time. Alder, oak, hornbeam, maple, ash, elm, willow, and poplar are especially recommended for growing coppice forests. Rotations can be 25–40 years (oak, beech, hornbeam), but it can be less for willow twigs (5 years). The production per one rotation is smaller than in high forest; but as the rotation is short total production of woody biomass may be equal to a high forest or even higher (when compared over the period of high forest rotation); but the wood is jagged, curved, and especially suitable as firewood. Logging is usually carried out in such a way that a certain part of the forest, corresponding to the need for

(continued)

Box 3.8 (continued)
firewood, is cleared every year. Harvest should take place during the dormant period, and the remaining stump should be as low as possible. In the mixed forest, in addition to stump and root sprouts there are also solitery full sized tall trees, which can be eventually the subject of selective logging with the rotation time of these trees being the same as in the normal high forest in the area. As can be seen in Fig. 3.1, this practice was common in Western and Central Europe at the end of Middle Ages, but recently the coppice proportion has decreased (McGrath et al., 2015). In addition to conservation use, a low forest is less sensitive to drought and thus can be suitable for soil protection while at the same time serve for the production of biomass.

The conservation expectation from the low forest is that it will be an open forest with high light availability that will create a suitable habitat for many species of open forests such as the longhorn beetle (*Cerambyx cerdo*). The brightness of the forest is due to the lower height of the stand, as well as the practically constant presence of mixed and overgrown parts of the stand. Another interesting consequence of coppice management is the gradual reduction of the supply of nutrients thanks to the continuous harvest and removal of wood, which has a larger proportion of bark and young branches, that is, parts richer in nutrients. This may not cause the same nutrient shortage as in preindustral times. This is due to much higher nutrient imput in contemporary landscape. Moreover, in history, the low forest has often been combined with forest grazing, raking of litter, and other uses that have supported nutrient removal and greater openness.

Forest grazing was a traditional way of using the forest, which, depending on the intensity of grazing, led to the creation of an open forest or forest-steppe. Grazing suppresses fast-growing species. The effect of grazing on woody plants is often more pronounced than on grasses and herbs, because after removing the treetop, the tree sprouts and establishes a new treetop which is often grazed again. It is a matter of debate to what extent forest grazing could be a mere replacement for the grazing of the original forest megafauna, that became extinct at the beginning of this interglacial (Vera, 2000), and to what extent it is a significant human intervention accompanying other human activities that support the semi-open (forest-steppe) landscape. Either way, human activity has significantly contributed to the emergence of an open landscape, and many species specialized on that landscape are now subject to protection in many European countries. The classical theory describing the coexistence of trees and open areas assumes that unpalatable species (e.g. thorny) were left on the pasture, and solitary trees could have grown in their shelter. However, this theory only works in the presence of large, nonselective herbivores such as horses or cattle, and probably their ancestors. This theory also explains only the presence of trees on the pasture, not the origin of the pasture as such. Samojlik and Kuijper (2013) believe that grazing alone can

(continued)

Box 3.8 (continued)
lead to a change in forest species composition, for example, to support grazing-tolerant species such as hornbeam, but does not lead to the formation of open, forest-free areas within the forest. Additional disturbance, most likely fire, would be required to create such an open pasture. Fire may have been anthropogenic, such as burning that was done by indigenous communities in North America and Australia. But fire frequency would also increase with a greater cover of grasses, which are promoted by grazing.

Grazing of domestic animals, or the replacement of game grazing by domestic pasture animals, also affected the forest species composition. Most of today's domestic animals, such as cows, sheep, and goats are grazers, preferring to graze continuous areas of grass and herbs. When grazing in the forest, these species prefer deciduous tree seedlings rather than conifers. In contrast, roe deer (*Capreolus capreolus*), currently one of the most abundant forest game in Central Europe, is a browser specializing in the selective grazing of young woody shoots. Therefore, the expansion of forest pasture for domestic animals (which suppressed beech more) and the simultaneous suppression of game grazing (which had suppressed fir) led to the expansion of European fir at the expense of beech (Vrška et al., 2009). This trend was further promoted by the raking of litter (see below). On the contrary, high levels of wildlife are today a major obstacle to the greater spread of fir in many European countries.

In the mountains, historic pasture also pushed the upper treeline to lower elevations by strengthening grass competition. Pasture also leads to the redistribution of nutrients, creating places rich in nutrients and places from which nutrients are removed by grazing. In mountainous locations on the forest border, nutrients accumulate in the lower positions of the pastures, and conversely nutrients are removed from the higher positions. Similar to grazing in agriculture land, intensive forest grazing can inhibit vegetation regeneration, increase erosion, and lead to soil loss and gradual degradation of the ecosystem (this has led to legal restrictions on these techniques). In Central Europe, for example, namely in the Austrian Hungarian Empire, forest grazing was banned by a decree of Maria Theresa in the 1700s.

Raking of litter was a common technique for most of the Middle Ages and the beginning of the modern age. Litter served for housing livestock in case of lack of straw, which, on the contrary, could be used otherwise as feed, building material, etc. Raking litter has led to significant nutrient removal from the forest ecosystem. For example, Hofmeister et al. (2008) calculated that nutrient removal by long-term raking in the Šumava region caused as significant a loss of basic cations as did their leaching due to acidification in the 1980s. Raking litter also locally supported species that prospered on the exposed mineral soil, as compared to species that grew well even on a thick layer of litter. An example is the promotion of fir at the expense of beech, because fir seedlings grow

(continued)

Box 3.8 (continued)

better when in contact with mineral soil and have difficulty establishing in thick layer of litter, whereas beech, whose seeds are larger and carry more storage substances, had a greater chance to penetrate the sediment and fermentation layer to reach the mineral horizons (Vrška et al., 2009). Again, because extensive litter raking may have negative effects on the forest, it was banned in many countries. In today's agriculture there is no economic reason either for forest grazing or litter raking. This is due to the decline in the number of livestock and the decline in grazing, in general, and partly also by litter-free housing and a lower consumption of straw as a raw material for the production of structures and other objects. Due to the fact that this technique, in some situations such as lowland forests, supports the occurrence of some rare and endangered species, today this procedure is experimentally applied as a management intervention in some protected areas.

The techniques described above (coppice forest, pasture, litter raking) describe some historical forms of forest use, to which a number of species of fauna and flora have been secondarily adapted. For many of these species, their original habitats have disappeared and the species have been preserved only in these anthropogenically formed secondary habitats. These forms of forest use, or their imitation, thus represent an important management element for the protection of these species. In contrast, in other ecosystems, species and communities linked to natural processes and natural disturbance regimes have still been preserved, and therefore nonintervention is an important element in the protection of these forests. This shows us how diverse (from massive interventions that mimic historical forms of use, through a nature-friendly CCF forest, to nonintervention) management interventions and measures in the forests of protected areas can be, depending on what the subject of protection is. It is therefore very difficult to lay down any general rules for growing forests in protected areas. Different protected areas are established to protect different species or phenomena (objects of protection). However, some general principles, which are similar in most areas, were established for management within the systems of protected areas NATURA 2000. These principles are summarized by Vacek and Podrázský (2006) as follows: (1) not to increase the share of geographically non-native trees and other woody plants, (2) to modify the species composition in favor of climax woody plants through educated interventions during thinning and regeneration (unless this is in conflict with the subject of protection, which in some specific cases are tied to earlier successional stages), (3) to emphasize the natural regeneration of vegetation, and (4) to promote a more complex spatial structure of vegetation, which can be achieved by a more diverse species composition.

Box 3.9 Use of Forest Stands for the Restoration of Disturbed Areas

The forest has ideal properties to be used to restore heavily disturbed ecosystems such as post minning dumps, disturbed post-industrial sites, soils heavily degraded by erosion, and similar areas. Growing forests in such areas can bring a rapid reduction in erosion and the restoration of a number of other ecosystem services, including timber production, which is, however, often not a priority in this situation. Forest restoration is one of the most common approaches to restoring post-mining sites (Macdonald et al., 2015). During reforestation, the site is usually leveled and trees are subsequently planted. In some cases, topsoil is also stratified and herbaceous species are sown (Frouz et al., 2001; 2013a). However, in many places of the dumps, there was a spontaneous growth of forest vegetation in areas that were not treated in any way after overburden dumping (Frouz et al., 2008). A number of studies point to the great importance of these areas as habitats for rare and protected species. The first key question in remediating any disturbance is whether it is really necessary to remediate such disturbance. Here we must consider the fact that some disturbances are part of the natural processes. For example, avalanches are a typical example of a process that plays a key role in the functioning of many of the mountain ecosystems that depend on this disturbance, and efforts to rehabilitate these areas would, if successful, lead to the eradication of many of these ecosystems. However, in the case of post-mining heaps, the development of new ecosystems is certainly desirable. However, in many cases, natural processes can account for the rapid development of forest ecosystems. It depends mainly on the quality of the substrate and the climate (Fig. 3.26). Ecosystem restoration is generally difficult on heavily acidified substrates with a high content of heavy metals, or on areas with a very alkaline pH. Such areas with extreme properties can occur on dumps, for example, due to the oxidation of pyrite, or on sludge ponds where material that has been leached with extremely acidic (or basic solvents) is deposited. In that case, it is necessary to somehow separate the phytotoxic layer from the layer that will serve as the medium for plant growth. This is usually done by application of some clay layer, over which some topsoil or topsoil substituting mixture of various materials is applied. In some cases, topsoil or topsoil substitutes are applied even on nontoxic overburden. Although this usually leads to the improvement of some soil parameters (higher content of organic matter, higher nutrient content), it is often at the cost of greater soil compaction. Therefore, pros and cons of this measure should be carefully balanced.

Soil compaction is one of the main obstacles to reforestation. In areas which are not formed by extreme substrates, as described above, and are not compacted in any way, and, if possible, a loose and rough surface created by dumping is preserved, then the establishment of pioneer woody vegetation is generally relatively easy (Frouz et al., 2018). In Central Europe, the growth of such woody plants is usually formed by a combination of willow, birch, and

(continued)

Box 3.9 (continued)

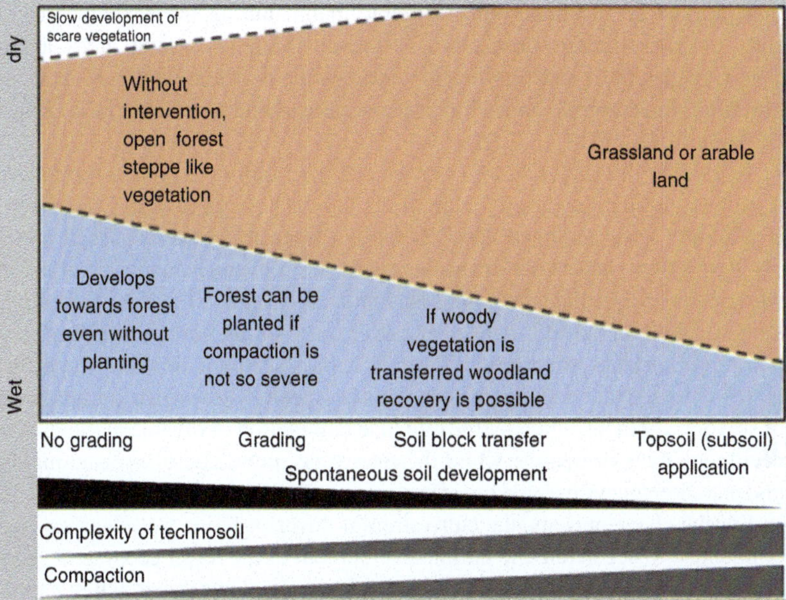

Fig. 3.26 Influence of climate and substrate modifications on vegetation development and possibilities of forest regeneration after mining based on Frouz (2021)

aspen which can reach the full involvement of crowns in 20-year-old sites, and by 30–40 years sites often have biomass production that is similar or greater than in areas reclaimed by tree planting (Frouz et al., 2015a). In areas compacted by leveling or topsoil application, spontaneous forest development is generally worse although planted seedlings usually achieve a full canopy by 10 years of age. However, in some cases, focus on achieving closed canopy as soon as possible may not result in the best long-term outcome, and studies in Europe as well as Eastern USA show that long-term woody production can be comparable or even higher in unreclaimed regrowth than on reclaimed forest plantation (Frouz et al., 2015a; Frouz & Franklin, 2014).

One of the main goals of forest reclamation is the renewal of soils and the associated renewal of the soil supply of biogenic elements. It should be emphasized that the supply of biogenic elements in the soil is generally much higher than the amount of nutrients supplied by fertilization, so restoring the soil pool of biogenic elements, especially nitrogen, is key to plant growth. Nitrogen is stored in the ecosystem mainly in the form of soil organic matter. In addition, the accumulation of organic matter in the soil affects the formation of the soil structure, which again has a positive effect on the infiltration of water into the soil and the retention of water in the soil. It should be noted

(continued)

Box 3.9 (continued)
that the rate of accumulation of organic matter during the development of soils on post-mining substrates is relatively high (in young areas about 2.4 t C ha per year) and significantly exceeds the typical rate of carbon accumulation, for example, under forest plantations on former arable land (Frouz & Vindušková, 2018).

Regarding the choice of tree species and their influence on soil formation, it can be said that the development of the humus horizon of soils (A horizon) is generally faster the lower the C:N ratio of litter. It is therefore faster under deciduous trees, especially in species fixing N. Conversely, under conifers, in which the C:N of foliage and litter is significantly higher, there is an accumulation of litter on the soil surface, and over time, strong horizons of overlying humus (O horizons) are formed. Overall, carbon accumulation under conifers is slower than under deciduous trees. This is especially true in the initial stages of soil development, and later the rate of accumulation of organic matter under deciduous trees decreases, while the rate of accumulation of organic matter under conifers remains the same or even increases (Vindušková & Frouz, 2013).

For the above reasons, the cultivation of nitrogen-fixing trees is often recommended to accelerate the initial development of the forest ecosystem. This is certainly a good idea if nitrogen-growing woody plants are grown in a mixture with other woody plants. Nitrogen fixers serve to boost initial N accumulation and may be species that predominate during the growth of the first generation of trees. In the long term, nitrogen-fixing plants remaining throughout the first rotation no longer have such an effect, because in older areas with N-fixing woody plants production decreases, thus soil formation and accumulation of soil organic matter slows down. In addition, excess nitrogen in the soil promotes grass competition and may adversely affect ectomycorrhizal fungi, which negatively affects the development of a possible subsequent generation of climax trees (Frouz et al., 2015b).

The fact that a forest has an extraordinary ability to restore ecosystem functions, even in heavily disturbed areas and often in an economically acceptable way, sometimes leads to the assumption that any kind of afforestation of any area may represent universal solution that will instantly improve ecological functions in targeted area. However, this may often not be true, especially when trying to afforest otherwise non-forest ecosystems such as peat bogs, savannas, and so on. Such an approach often does not lead to the desired effect.

For example, afforestation of savannas, often using exotic species, can significantly impair rather than improve the functioning of ecosystems (Bayle, 2019). This is especially true in the world's grassy biomes, where massive planting of trees is especially harmful (Veldman et al., 2015). Tree planting in savannas typically reduces biodiversity with a particularly strong impact on endemic species; it decreases water discharge which may cause water scarcity to local populations; and it shifts carbon storage from belowground to

Box 3.9 (continued)
aboveground which increases the sensitivity of the carbon pool to fire and causes carbon loss when wood is eventually harvested and used as fuel (Veldman et al., 2015). Interestingly, the deforestation of rainforest receives much public and media attention but disappearance of natural savannas due to inappropriate plantation and woody vegetation encroachment receives almost no attention, despite severe ecological consequences. Afforestation is therefore an appropriate tool for restoring some ecosystems, especially in the disturbed areas in forest biomes, and if the restored forest is based on species that are native to the given area. However, planting any trees anywhere, especially planting exotic trees in non-forest biomes, may not bring the expected ecosystem benefits and may likely lead to deteriorating ecosystem services.

3.4.2 The Importance of the Forest for the Water Movement in the Landscape

Let us first look at the behavior of water in a forest ecosystem. Rainwater falling on the forest may first be retained by the canopy. This phenomenon is called interception. In temperate forests this is roughly 25–30% of the precipitation, occasionally even more. The amount of interception depends on the architecture of the trees and their foliage (trees with denser crowns retain more water) and the amount of precipitation. The greater the precipitation, the smaller the proportion of water retained by the interception. This is due to the fact that the tree canopy has a certain capacity to retain water, and when it is saturated, rain is no longer retained. In tropical forests, the interception is usually smaller. Some of the water that is retained by interception evaporates, while some flows down the trunk and may be absorbed by trees or epiphytes on them. Water that has fallen through the crowns (throughfall) or flowed down the trunk (stemflow) may become part of the surface runoff, or may soak into the soil (infiltrate). A portion of infiltrated water may evaporate, part is retained by the soil and can be used by plant roots, and the remainder may become surface or subsurface runoff (Fig. 3.27).

Forest evaporation thus consists of evaporation from the soil, transpiration by understorey vegetation, transpiration by trees, and evaporation related to interception. The last two components are usually larger in the forest than in non-forest areas, while the opposite is true for the first two. Overall however, the evapotranspiration of the forest stand is generally greater than the evapotranspiration of grasslands (Fig. 3.28) under similar conditions (Zhang et al., 2001). This difference increases with increasing precipitation, and it becomes significant at a precipitation rate of around 1000 mm per year (Fig. 3.28)).

Greater transpiration of the forest stand has a number of consequences. The temperature in the forest is more balanced, and during the day it is colder than the agricultural field due to the greater amount of energy that is consumed for

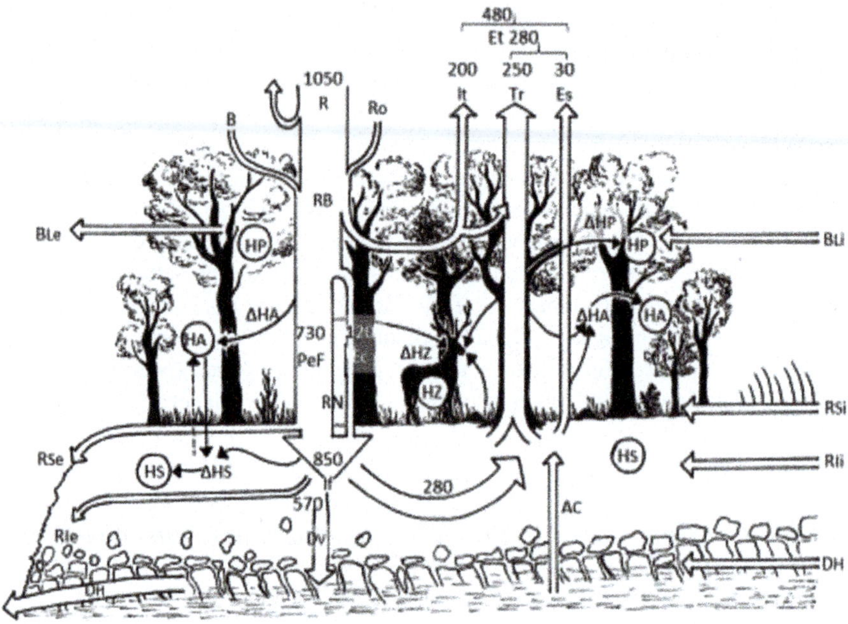

Fig. 3.27 Forest water balance, flows in mm per year

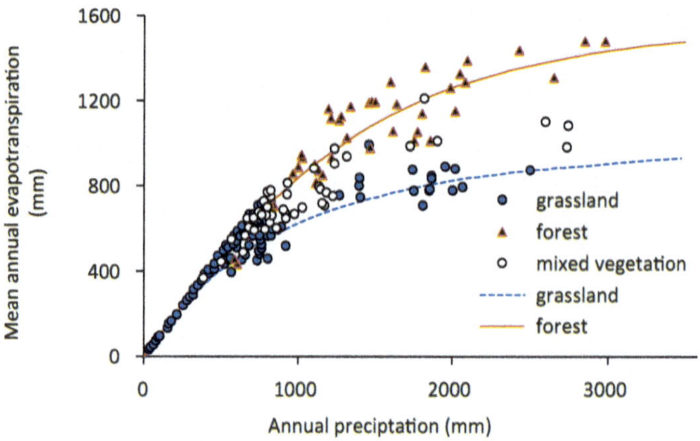

Fig. 3.28 Comparison of evapotranspiration from forested and grassed river basins at different rainfall intensities according to Zhang et al. (2001) (modified)

3.4.3 Forest and Biogeochemical Cycles

Forests are an important global reservoir that affects the global carbon cycle. Forests globally contain about 1146 Gt of carbon, of which about 37% is at high latitudes (50–75°), 14% at medium latitudes, and 49% at low latitudes (25–0°). Globally, evaporation. On the contrary, at night, due to the higher humidity, water vapor can condense more easily and at the same time the latent heat (previously consumed for evaporation) is released; thus, at night the cooling effect is reduced. Greater transpiration, and thus more water vapor in air over large forest stands, promotes local precipitation and a short water cycle. The forest can also support a short water cycle by releasing aerosol particles into the atmosphere, which can promote water condensation and thus promote the formation of raindrops.

Contrary to some earlier ideas, forest catchments often have less runoff than similar non-forest catchments (Calder et al., 2007). However, runoff depends not only on vegetation cover but also on the slope of the terrain and the ability of soils to retain water. In the case of very flat catchments, deforestation may lead to wetting of the terrain rather than increasing drought. In contrast, on sloping terrain, deforestation often leads to an increase in runoff. Greater runoff, especially surface runoff, leads to greater erosion. Forest lands generally have a very good ability to protect the soil against erosion, and this is usually lower than for most other land uses. This is due to the above-mentioned reduction of surface runoff, as well as interception, which will reduce the impact of precipitation on the soil surface, spread it over a longer period and reduce the erosive energy of rain. In addition, the surface of forest soil is usually covered with either vegetation or litter and is often intensely rooted, which increases its mechanical resistance to water. Forest soil also tends to be more porous and has greater infiltration due to the activity of roots and soil fauna. Measurements in Serbia have shown that forest soil under a deciduous forest has 2.5 times greater infiltration than meadow soil, and 6 times greater infiltration than arable land (Gajic et al., 2008). Cejpek et al. (2017), who compared different forest stands (alder, oak, spruce and pine) on dumps, found that all forest stands, except oaks, showed a higher rate of infiltration than meadow stands. The fact that forest areas have a greater infiltration rate than agricultural areas is confirmed by other authors (Gray, 1973; Wood, 1977; Bharati et al., 2002). A higher rate of infiltration again reduces surface runoff and thus the erosive action of water. A number of authors (Bharati et al., 2002; Seobi et al., 2005; Gajic et al., 2008) have shown that forest soils have a higher porosity and water field capacity and consequently are able to retain more water than soils of meadows and arable land. Under similar conditions, deciduous trees appear to produce soils with greater water retention capacity than soils under conifers (Fig. 3.29); however, the overall effect on runoff is a combination of other factors (interception, infiltration). It is expected that the occurrence of local floods decreases in catchments with a high forest cover. However, it cannot be expected that forests could stop floods resulting from long-term extreme rainfall.

Forest cover also improves the quality of runoff water. The reduced risk of erosion is associated with a lower incidence of suspended solids and thus turbidity. A higher rate of infiltration and thus a higher proportion of subsurface runoff can reduce the content of various pollutants and nutrients, which can either bind to the soil organic matter or be used by plant roots. In addition to continuous forest stands, riparian vegetation can also play an important role in water purification.

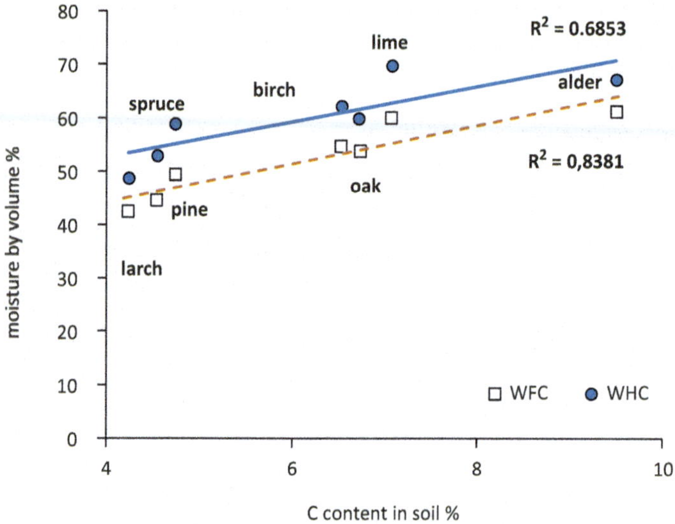

Fig. 3.29 Water absorption (WHC) and field water capacity (WFC) under stands of different tree species, depending on the carbon content in the soil below. (Frouz, 2018b, modified)

about two-thirds of carbon is in forest soils and peat deposits (Dixon et al., 1994), predominantly in large carbon stocks of boreal forest soils (Table 3.7). It is therefore not surprising that deforestation and dewatering of peatlands are among the most significant land use changes, promoting CO_2 emissions into the atmosphere (Harris et al., 2012).

Of course the carbon stock in soil and vegetation given in Table 3.7 is the average for large areas within which there is great variability between different tree types. Figure 3.30 shows the carbon stock in soil for different tree species grown in similar conditions (so-called common garden experiment) on dumps in the Czech Republic (Frouz et al., 2013b) and on forest soils in Denmark (Vesterdal et al., 2008). It can be seen that conifers, especially spruce, tend to accumulate more carbon on the soil surface, while deciduous trees accumulate more carbon in the mineral soil. On young soils, where the forest grows in its first generation, this difference is very significant, while on soils that have been covered with forest for several previous generations of the forest the differences in carbon content of the mineral soil are much smaller.

For the carbon stock in the soil, it is assumed that the soil is generally saturated with carbon and its content in the soil does not change much. For wood, it is more complex, and the average long-term carbon stock is calculated as the average amount of carbon in the tree biomass that increases during tree growth, as well as the carbon contained in wood products (more Box 3.2).

The global carbon stock in vegetation (especially trees) and soils has gradually increased over the last century with increasing concentrations of atmospheric CO_2. This absorption of part of the anthropogenically produced CO_2 into biomass is one of the reasons why the impacts of CO_2 on increasing the greenhouse effect are

Table 3.7 Carbon stocks (t ha^{-1}) in above-ground biomass and soil in forests at different latitudes

		Vegetation	Soil
Boreal forest	Canada	83	282
50–74°	Alaska	28	484
	Russia	39	212
	Average[a]	64	343
Temperate	USA	62	108
Forest	Europe	32	90
25–50°	China	114	136
	Australia	45	83
	Average[a]	57	96
Tropical forest	Asia	132–174	139
0–25°	Africa	99	120
	America	130	120
	Average[a]	121	123

[a]The average represents the average weighted by the area of forests in individual areas

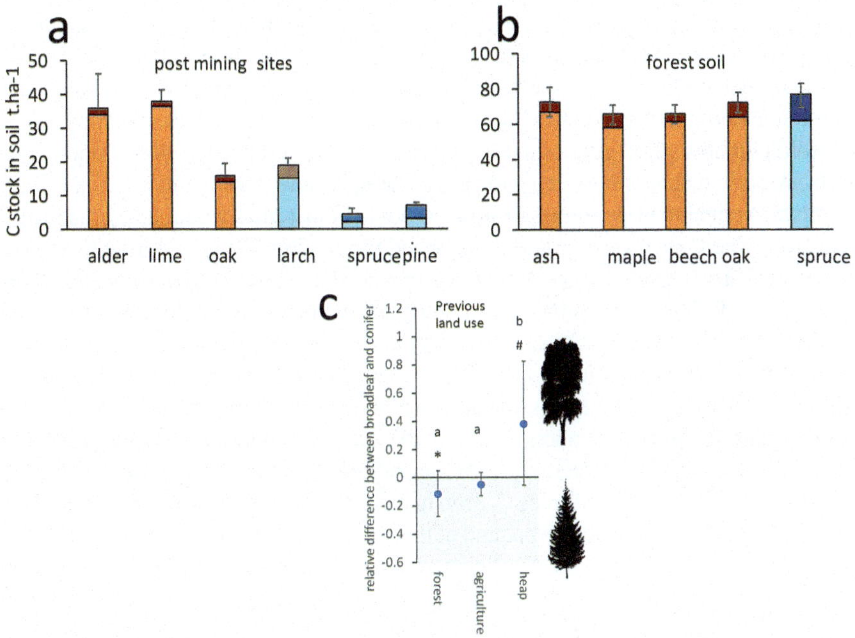

Fig. 3.30 Carbon stock in biomass and soil under different tree species, in an experiment where these were planted close to each other in one area on dumps around Sokolov (left), and on forest soil in Denmark (right). Conifers are blue, deciduous broadleaves orange, deciduous conifers blue and orange, carbon in mineral soil is in lighter color while C in forest floor is darker, based on Frouz et al. (2013b) and Vesterdal et al. (2008). Very right figure show proportion of C under broadleaf and conifer planted on soils with various previous land use, based on Frouz and Hublová (2021)

somewhat smaller than would correspond to their production. Janssens et al. (2003) estimate that European ecosystems thus absorb 7–12% of European anthropogenic CO_2 production. This is partly due to the increase in CO_2 concentration itself, because it is known that the rate of photosynthesis increases with increasing CO_2 concentration. In the long term, however, growth increases due to increasing CO_2 that may not be sustained as the trees become nitrogen limited (Warren et al., 2015). Another important factor that accelerates tree growth is nitrogen deposition. Due to the production of mineral fertilizers and the burning of fossil fuels, human society produces significant amounts of nitrogen compounds that can enter soils as atmospheric deposition (Thomas et al., 2010).

So far, we have dealt with the size of the carbon pools at a more or less steady state. However, it is also interesting to study how much carbon we can accumulate after afforestation. Vindušková and Frouz (2013) monitored the development of carbon accumulation in post-mining sites reclaimed by forest plantation. It turns out that afforestation of such substrates can achieve significant carbon sequestration. Soils in such ecosystems accumulate an average of 2.4 t C ha^{-1} $year^{-1}$. The initial rate of sequestration is higher in deciduous trees and decreases with age, whereas in conifers the initial rate is lower than in deciduous trees, but remains higher for a longer time. Interestingly, the rate of carbon sequestration in these heavily disturbed soils is much greater than, for example, in afforested agricultural land. This meta-analysis also showed that the effect of individual tree species on soil carbon accumulation is climate dependent. Conifers show the highest rate of carbon accumulation in cold climates while grasses are the highest in warm climates, and deciduous forests are somewhere in between and are less temperature dependent.

If we wanted to grow forests to store carbon, our priorities would be a little different from when our goal is to maximize wood production. We would have to respect the fact that a significant part of the carbon in the forest ecosystem accumulates underground. This accumulation is controlled by mechanisms that are not fully understood, but a significant part of this accumulation of C is related to the accumulation of peat. It is known from a number of peat forests that their drainage leads not only to increased wood production but also to the decomposition of peat and thus the loss of carbon (Hommeltenberg et al., 2014). Drainage will allow better growth of trees; this will support greater transpiration and thus a further drop in water level, and thus greater development of trees. A drop in the water table will cause carbon loss due to accelerated peat mineralization, and at the same time increases carbon storage due to production of new biomass. The balance of these two processes may vary but often it may cause overall loss of soil carbon storage (Simola et al., 2012). Peat forest drainage is often associated with boreal climates, but tropical swamp forests are also an important sink of carbon and their drainage may cause significantly release of CO_2 due to mineralization of peat and greater sensitivity to fire (Page et al., 2002).

In localities where peat and similar organic horizons do not form, the resulting carbon sequestration in the ecosystem is the result of a complex interaction between soil carbon accumulation, wood production, and its subsequent use (Nabuurs & Schelhaas, 2002).

3.4.4 Forest and Global Change

As already mentioned, forests retain large amounts of carbon and thus remove CO_2 from the atmosphere, which can lead to a reduction in the atmospheric concentration of CO_2 and thus reduce the intensity of the greenhouse effect and slow down the warming of the planet. It is estimated that about a third of the planned reduction in CO_2 production and release to the atmosphere, that states have promised to combat climate change, should be achieved by applying nature-friendly solutions (Griscom et al., 2017), among which forests play an important role (Pan et al., 2011). As already mentioned, peatlands have a similar effect; so in the decision as to whether to support forest growth (e.g. drainage) or, rather, to support peat accumulation, then usually, purely in terms of carbon accumulation, peat is a better solution. However, in addition to this discrepancy between forest and peat, there are a number of other mechanisms that can make the relationship between forest and global climate change even more complicated and, as a result, make the net effect of forest cultivation far more variable. Thus, a certain type of forest under certain conditions may not lead to a slowdown in global warming, but rather the opposite.

The problem is that the relationship between forests and global warming is more complicated and is due to the joint action of several mechanisms. The first of these is the already described removal of CO_2 from the atmosphere, the process of photosynthesis, and its long-term "storage" in the form of carbon in biomass and soils, which in turn leads to reduced atmospheric CO_2 concentration and thus to the reduction in greenhouse effect. Another factor is the change in reflectivity, or albedo. Forests are usually darker than many other natural surfaces; this reduces albedo and leads to more solar energy intake and thus heating. This can be particularly important, for example, in boreal forests, where the albedo of the forest is much lower than the albedo of snow, and the spread of boreal forests can significantly increase the total amount of sunlight absorbed (especially in winter) and thus contribute to warming (Betts, 2000). On the other hand, the forest evaporates a considerable amount of water, as described in Sect. 3.4.2, thus significantly cooling the local microclimate. If large areas are present, this increased evaporation can promote condensation and a short water cycle (Sheil, 2018). In addition, forests contribute to the production and transformation of other gasses and volatile substances, which can significantly affect the greenhouse effect both positively and negatively (Unger, 2014). These are mainly substances with short- to medium-term persistence in the atmosphere, especially tropospheric ozone, aerosols, and methane. Methane is an important greenhouse gas that enhances the greenhouse effect more strongly than does CO_2, but has a shorter residence time in the atmosphere (in the horizon of 100 years it is about 25 times more efficient a greenhouse gas than CO_2, but its concentration is an order of magnitude lower than CO_2). A number of studies today show that trees, as well as other plants, can transport and emit methane produced in the soil, but it is not known what the overall impact of this phenomenon is on the methane cycle, for example, whether it can significantly reduce methane oxidation in soil surface layers (Pangala et al., 2017). Tropospheric ozone contributes to the greenhouse effect in two opposing ways. Ozone directly promotes the greenhouse effect as a greenhouse gas. At the same time, however, the decomposition of ozone due to solar radiation produces OH radicals and these contribute to the oxidation of methane. In addition, forests produce a number of volatile substances such as isoprene. Isoprene and other volatile substances both help to create tropospheric ozone, which tends to increase the greenhouse effect, and can also promote the formation of aerosols that, in turn, reduce the greenhouse effect. As can be seen, the impact of volatile substances is complex, but it is believed that the impact on the greenhouse effect is somewhat enhancing (Unger, 2014). Certainly, the direct influence of trees on atmospheric chemistry deserves further systematic research.

An equally important question is how the forest responds to climate change, and what adaptation measures can be taken to better cope with the change. Climate change brings very complex changes to forest growth, at the same time, as already mentioned, the forest is a major player in the mitigation of climate change. One of the important aspects of adaptation to climate change is therefore the question of the social value of individual ecosystem services, in other words, redefining what we, as a society, want from the forest, and what tools should be used to achieve that.

If we look in more detail at the change in conditions that climate change brings, we can expect greater unevenness in the supply of precipitation. With a similar total amount, more precipitation will fall outside the growing season. At the same time, there will be a gradual warming that will prolong the growing season, and it will start earlier. This will lead to a faster depletion of the soil water supply and increase the threat of drought. This will be exacerbated by the more frequent occurrence of extreme weather events, windstorms, tornadoes, heat waves, and heavy rain. The combination of drought and these climatic extremities will, in turn, increase the risk of fire, increase outbreaks of pests, increase the incidence of fungal diseases, or promote the outbreaks of rodents. An increased occurrence of ground-level ozone can also be expected. In addition to these phenomena that are primarily related to increasing greenhouse gas concentrations, climate change is accompanied by a number of other anthropogenic changes that are not necessarily primarily related to the greenhouse effect, but simply occur simultaneously. These include an increasing supply of nitrogen to ecosystems due to the use of industrial fertilizers, but especially due to the burning of fossil fuels. Nitrogen, on the one hand, supports the growth of plants, but on the other hand it causes acidification of soils and subsequent loss of basic cations, and leads to the suppression of ectomycorrhizal fungi which endangers tree species that use this type of mycorrhiza. Greater globalization of the world increases the chances of the spread of invasive species. In addition, we have done a number of things in the past, in terms of overall land use, that can affect our landscape's sensitivity to climate change.

The adaptation strategy includes both a change in the species composition and a change in the way the forest is grown. Figure 3.22 shows expected changes in dominant tree species across Europe as a result of global change. As mentioned in Sect. 3.4.2, in several European countries conifers have spread largely in sites which are outside of their ecological niche. Among them, spruce is the most sensitive to water shortages. On the contrary, broadleaves that would naturally occur in these areas such as beech and, especially, oak are much more resistant. To put it in a more general way, the tree selection should follow closely the expected changes in climatic conditions. It is expected that the way the forest is grown should also change. As a promising adaptation measure, it is proposed to convert forests into CCF. Abandonment of clear-cut even-aged forestry is more gentle on the soil as it reduces the loss of soil organic matter in the clearings. Species, age, and spatial diversity should increase the resilience of vegetation to stress factors. The maximum use of natural regeneration will then support the selection of individuals that can better withstand drought and other adverse factors. As mentioned in Sect. 3.4.2, many authors suggest a reduction in stand density. A frequently discussed issue is the effort to cultivate other exotic species that are expected to be more resilient to ongoing climate change. A greater use of coppice systems is also recommended as a suitable adaptation to climate change. Coppice polycormons can rely on extensive root systems that make the root:shoot ratio much higher than in normal trees, and this can help the polycormon to forage for water over a large area. At the same time, individual stems of the polycormon can redistribute assimilated water, and this may help them to adapt to spatiotemporal fluctuations in insolation and water supply.

The use of exotic species generally carries a number of risks, and so it would be better to avoid their extensive use. The situation is somewhat different for domesticated species, which have been grown for several generations, are currently grown on a large area, and with which we have extensive experience. All this is certainly not a reason to expand their cultivation massively, but on the other hand, it certainly deserves a serious discussion of all possible benefits and risks. See Box 3.10.

Box 3.10 Douglas Fir in Europe

Douglas fir is among the most controversial trees, and many European foresters see it as a substitute for spruce. On the other hand, it is an introduced species with a potential impact on local biodiversity. Many environmental organizations are opposed to its dissemination, which may affect its market preference (Spiecker et al., 2019). Douglas fir was introduced to a number of European countries in the first half of the nineteenth century. It is grown in a number of European countries, both alone and in a mixture with domestic woody plants, in both clear-cut and CCF forests. At present, there is an extensive professional discussion about further cultivation of Douglas fir in many European countries (Spiecker et al., 2019; Remeš et al., 2020).

Douglas fir provides wood with very good technological properties, and under European conditions it has a very good production potential, better than that of spruce. It affects soil chemistry and the abiotic environment similar to native tree species, especially spruce. Many forest organisms are able to live together. However, a reduction in the diversity of fungal communities and a reduction in the number of arthropods, especially during the winter which has possible negative consequences for birds have been observed. It has been

(continued)

Box 3.10 (continued)
shown that Douglas fir can negatively affect the abundance of some endangered species. It is generally considered that the ecological impacts of Douglas fir are less severe than those of other exotic trees, such as pines in South Africa or *Robinia pseudoacacia* in Europe (Schmid et al., 2014). Douglas fir has not yet been the subject of extensive pest outbreaks in Europe, but they have significant pests in their home range, and the possible introduction of these could cause significant problems. Like mature trees, it is very resistant to small- and medium-intensity fires due to their thick bark. They have highly flammable needles, and although they can survive without fire for a long time, in their homeland, regeneration is associated with fire because they do not regenerate in the shade of the forest as well as other species such as western hemlock (*Tsuga heterophylla*), with which they often grow. Douglas fir is gradually replaced by hemlock. After intense fires, the Douglas fir will be significantly restored.

3.5 Game and Hunting

In this section, we will deal mainly with hunting, and we will only touch on the care for game marginally, in connection with its potential environmental impacts. Game hunting, along with fishing and insect collection, has been one of the main sources of animal protein throughout human development. Hunting was performed also by many hominid species that were the ancestors of modern humans. In addition, the hunting of animals, especially large mammals, was one of the major social events, largely shaping the culture of the original human societies. This cultural and social role was retained by game hunting even after the domestication of animals, which became a more important source of animal protein than game. This role is also evident by the origin of the word "forest," which refers to an exclusive area for hunting rather than an area overgrown with trees (see Sect. 3.1.1). Gradually, in many societies, game hunting has become a social and cultural issue rather than a livelihood activity, and certain types of hunting such as hunting certain attractive prey, or hunting in a certain way (e.g. falconry) has become, and to some extent still is, a way to manifest social status. However, this does not mean that hunting game becomes an economically marginal activity. On the contrary, many people are willing to pay a substantial amount of money for hunting and the opportunity to hunt. For attractive game trophies the fees range from tens to over hundred of thousands of dollars apiece, but even hunting more common prey can be a significant source of income for many landowners and national economies. For example, in 2017 alone, the US Fish and Wildlife Service collected over US $853 million in fees for hunting permits and other related permits. It is estimated that Americans spend US $3.5 billion on various federal, state, and local fees connected to hunting; moreover they spend another US $11.3 billion on hunting expeditions and another US $40 billion on equipment, clothing, and other related purchases.

3.5.1 Most Important Game Species

Similar to trees, the number of species that has been hunted by human is tremendous and it is determined not only by availability but also by social convention. In this overview, we will review some important or interesting game species of Europe and North America in order to facilitate orientation in the following text (Figs. 3.31 and 3.32). Densities of most important game species in Europe are summarized in Table 3.8.

Roe deer (*Capreolus capreolus*) is one of the most abundant representatives of game in Europe. This representative of the deer family has adapted very well to life in the cultural landscape. Other deer are the European red deer (*Cervus elaphus*) and the spotted fallow deer (*Dama dama*), which characterized by its spotted coloration. The fallow deer comes from the Mediterranean and Asia Minor, and was first mentioned in Central Europe in the fifteenth century. Two subspecies of red deer (*Cervus nippon*), originated in eastern Asia, were introduced into Europe since the turn of the nineteenth and twentieth centuries. A common North American deer, occurring from southern Canada to northern Brazil, is the white-tailed deer (*Odocoileus virginianus*), which has been recently introduced to Europe. Related to the white-tail is the mule deer (*Odocoileus hemionus*) which has a much more restricted range, occurring only in western North America. A noteworthy game deer species in North America is the elk or wapiti (*Cervus canadensis*), which is one of the largest deer occurring in several subspecies in North America and Eastern Asia. Another large species is the moose (*Alces alces*), the largest species of family Cervidae; males have characteristic broad, palmate antlers. Moose have broad distribution form fenoscandinavia across Russia to North America.

Another species of game is one of the probable ancestors of the sheep, mouflon (*Ovis musimon*), native to Corsica and Sardinia but introduced elsewhere in Europe. The wild boar (*Sus scrofa*) is another species of game that has a close relative, the pig, among domestic animals. This is one of the species that is very well adapted to life in the cultural landscape, and their numbers have grown significantly in recent years. On the contrary, the European hare (*Lepus europaeus*), which belonged to a very abundant species in the past, is now gradually declining. Among North American hares are the snowshoe hare (*Lepus americanus*), which is the principal prey of lynx and with which it shows apparent cycles in population dynamics. The black-tailed jackrabbit (*Lepus californicus*) is a large hare that lives in desert scrubland and prairies of Western America, where they often penetrate into the cities. The wild rabbit (*Oryctolagus cuniculus*) came to Western and Central Europe from the Mediterranean during the Middle Ages. The cottontail rabbit is one of the most widespread rabbits in the USA; it is a complex of about eight species of the genus *Sylvilagus*. Among other small mammals

Fig. 3.31 Examples of main game species, (**a**) red deer (*Cervus elaphus*), (**b**) spotted fallow deer (*Dana dana*), (**c**) wild boar (*Sus scrofa*), (**d**) roe deer (*Capreolus capreolus*), (**e**) hare (*Lepus europaeus*) and (**f**) pheasant (*Phasianus colchicus*)

Fig. 3.32 Examples of main game species II, (**a**) white-tailed deer (*Odocoileus virginianus*), (**b**) elk (wapiti) (*Cervus canadensis*), (**c**) moose (*Alces alces*)

Table 3.8 Densities of the most important game species in various European countries (individual per 1000 ha of land area). For the empty cells, no data were available. (Based on Schulp et al., 2014)

Country	Red deer	Roe deer	Wild boar	European hare	Pheasant
Austria	18.2	9.1			
Czech Republic	3.24	38.2	7.4	42.3	36.7
Estonia	0.4	12.9	5.5.	5.7	63.6
France	1.8	21.8	16.0	18.2	
Germany	4.7	86.0	26.0	10.3	5.73
Hungary	7.7	34.7	10.7	11.6	44.9
Italy	1.7	10.6	20.4		
Netherlands	0.9	14.7	0.6	147.6	17.1
Poland	4.8	22.6	6.9	17.4	13.5
Slovakia	8.3	17.7	6.1	42.2	39.6
Spain	5.5		6		0.1
United Kingdom	14.6	20.6			32.9

worth mentioning are the muskrat (*Ondatra zibethica*) and the beaver (*Castor canadensis*) originating from North America. Muskrat was introduced at the beginning of the twentieth century in Europe. We avoid covering predatory species in this short list.

Probably the most important game bird in Europe is the common pheasant (*Phasianus colchicus*), a pheasant originating from Asia where it occurred from Northeastern Turkey to China, documented in the Central Europe from 1330 when it was bred in captivity, then later (from the nineteenth century) found in the wild. In addition to the pheasant, partridge (*Perdix perdix*) and quail (*Coturnix coturnix*) were important hunting birds in the past, but their numbers are significantly declining, despite the fact that they are protected throughout the year. The interesting thing about quail is that they are probably the only landfowl bird that migrates regularly. European quail overwinters in North Africa and the Sahel. In the past, migratory birds, especially in the Mediterranean, have been intensively hunted. Among North American upland game birds are the largest representative, the wild turkey (*Meleagris gallopavo*).

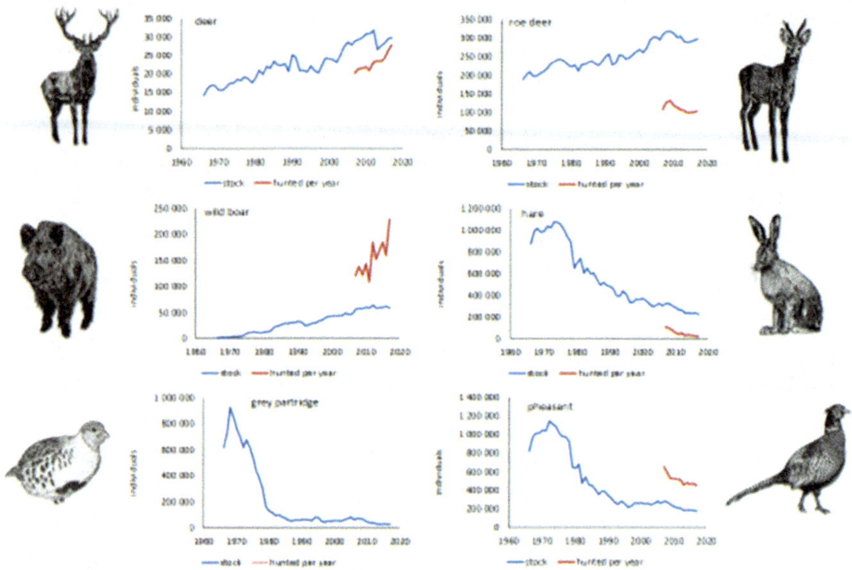

Fig. 3.33 Development of stock of selected game species in the Czech Republic. Data on hunting are only for the period up to 2000, Czech Statistical Office

Other important species of game are various duck and goose species. In Europe, important species are the mallard (*Anas platyrhynchos*), the tufted duck (*Aythya fuligula*), and the common pochard (*Aythya ferina*). Other waterfowl include the black coot (*Fulica atra*) and several species of geese, the most important of which is the great goose (*Anser anser*). The dynamics of stocks and hunting of some game species in Central Europe is summarized in Fig. 3.33. One may see a steady increase in some species, namely deer and wild boar, and a decrease in smaller game species such as hare, pheasant, and quail. Another interesting thing to note is that among European game species there is larger number of species that are either exotic or naturalized, introduced to Europe in pre-modern times.

3.5.2 Game Production, Hunting Strategies, Planning, and Management of Hunting

Recently besides hunting, ranching and other game kept in captivity are important as a meat source. Substantial part of game meat which is sold in markets is produced in more less controlled condition. Some animals are raised in controlled conditions (e.g. pheasants) and can then be released and hunted in the same way as wild animals. Some game species are kept in captivity in various forms of ranching. It has been shown that wild game ranching represents an economically viable option for land use in marginal land of Africa (Taylor et al., 2020). Wild animals kept in ranchland reach marketable or economically harvestable size earlier than cattle, and in about the same time as sheep. Weight gains vary from 0.06 kg per day for Thompson's gazelle (*Eudorcas thomsonii*) to 0.33 kg per day for eland (*Taurotragus oryx*), as compared with 0.14 kg per day for poorly managed cattle or 0.05 kg per day for sheep kept in similar conditions on the East African rangelands. Game has many advantages compared to cattle in marginal areas, importantly their efficient use of water, their resistance to disease, and also resistance to environmental stress due to their ability to adapt feeding strategies to the season and use lower quality forage (Ntiamoa-Baidu, 1997). In wildlife-only facilities, the average game meat production is about 4 kg ha^{-1} $year^{-1}$ but this can be intensified to level over 20 kg ha^{-1} $year^{-1}$, which is comparable to extensive cattle ranching (Taylor et al., 2020). Game ranching is also common in many European countries, North America, and New Zealand, as various cervids are widely kept in captivity, among them fallow deer being the most popular. In the USA bison (*Bison bison*) are raised on many farms.

The production of free living game is low in comparison with agriculture, and also in comparison with game species kept in semicaptive conditions because of the lower density of game and also because game has to spend more energy for maintenance in wild. For example, in Central Europe the game production per hectare of hunting ranges is, on average, between 0.7 and 0.8 kg ha^{-1} $year^{-1}$ in field dominated hunting grounds and up to 1.8 kg ha^{-1} $year^{-1}$ in forest hunting grounds (Sýkora, 2012). Planning of hunting is therefore focused on maintaining a stable population, rather than meat production.

In contrast to domesticated animals (and often even game kept in captivity), which are slaughtered to obtain the meat, wild game is obtained by hunting. Hunting, similar to other human activities presented so far, is based on many technologies. In terms of hunting, they are based on development of hunting weaponry and other hunting tools, the domestication of other animals that can assist in hunting, and cooperation between people not only in hunting but also in gathering of the knowledge that allows them to adjust hunting strategy. In terms of hunting weaponry, invention of long-distance projectile weaponry was particularly important. A study of hunting strategies of recent hunters suggests that before the invention of long-distance hunting weaponry, the hunting was limited to a smaller set of hunting strategies which also limited prey choice. The use of hand-delivered spears allows the hunting of large-bodied prey, but hunting techniques are dependent on specific features or terrain types which limit its uses. The invention of the spear-throwing lever or atlatl made hunting less dependent on specific terrain features. Finally, the use of bow and arrow allowed hunting of prey regardless their size and terrain features (Churchill, 1993). Although bow-propelled arrows are popular hunting tools even today, the appearance of firearms substantially increased hunting efficiency. Hames (1979) noted that one could kill monkeys and large birds in a neotropical forest from 43 m by rifle, from 25 m by arrow, and only from 17 m using a blowgun. The same author compared hunting efficiency of two tribes hunting either by arrow or by shotgun and concluded that shotgun hunters can hunt 38.6 kg of prey per 100 hour of hunting, while arrow hunters obtain only 6.8 kg of prey over the same time. One can only speculate that the difference could be even larger in terrain that is more open. In general, there are two most common firearms: rifles and shotguns. Rifles shoot one projectile from rifled barrels and allow great accuracy over large distances (commonly 200–300 m, although military sniper rifles are accurate far over 1 km) and are used to kill larger game. Shotguns usually have a smooth barrel and can be used either to shoot several small bullets at one time, ideal for hitting small, moving game, or it can be also used to shoot one projectile. Hunting communities and professional hunters also used various self-operating traps including cages, glue, and steel jaws traps to catch animals, but many of these are now legally limited in many countries. Also various tools and instruments were developed for observing and tracking game such as numerous kinds of optical devices, binoculars, night vision, and GPS technology that not only helps with hunter orientation, but may also inform hunters about the position of fellow hunters, or even hunting dogs. Besides weaponry and tools, the domestication of certain animals was important for hunting. It is likely that the domestication of dogs that could help with hunting and also transportation of prey may have helped with human success, resulting in megafauna extinction (Fiedel, 2005). Domestication of dogs and horses also allowed for the application of a wide range of hunting strategies. Falconry was another way of using animals for hunting. Finally, human used sophisticated tactics to hunt game; he not only used natural terrain advantages and a knowledge of game movement, but human also actively modified the terrain (e.g. by burning) to create suitable habitats where game aggregated and could be eventually hunted. Very likely, similar to modern hunters, human probably also constructed various shelters for observation and the safe hunting of prey, for instance, leaving a few solitary trees as hunting and observation posts in otherwise clear-cuts or burned areas. Development of human hunting strategy has made human hunting ability superior, and increased the risk of overhunting certain populations, which lead to regulation planning and the management of hunting.

Regulation of hunting has several forms, the main ones include the protection of certain game that is banned to hunt, the regulation of time when the hunting can be performed, which divides the year into times of game protection and hunting season, and finally, in the management of hunting in term of regulating the number of individuals that can be hunted in a certain area. The latter approach depends on knowledge of animal density in that area. The right to hunt is exercised within a certain area – the hunt, or hunting ground. The hunt is therefore land on which the right to hunt is allowed. The hunt can be either owned, that is, belonging to the person who exercises the right to hunt here, or the so-called social hunt that includes the land of several owners who have agreed to their hunting use. The minimum area of a hunt varies among European countries; for example, in Austria the minimum size of a hunt is 115 ha, in the Czech Republic, 500 ha, and in Slovakia, the minimum area of hunting grounds in areas for deer hunting is 3000 ha, in areas with roe deer it is 2000 ha, and in areas with small game it is 1000 ha. The principle of hunting regulation is based on estimates of the population of prey and its expected development; based on that we determine how the hunt will take place, and then after the hunt we estimate the state of prey again and the whole process is repeated.

As shown in Box 3.11, the size of the wildlife population can be estimated in various ways. In some countries, such as the Czech Republic and some other European countries, the line transect is used, and it is assumed that the transects are laid so that they cover the entire hunting ground, which is a somewhat optimistic assumption. This census of game is carried out on all hunting grounds in the Czech Republic on the same day. Of course, such an estimate is burdened with a number of errors (especially if we consider that these numbers are base relatively accurate the calculations). Despite this inaccuracy, the strength of this estimate is that it is carried out using a similar methodology in the long run (Fig. 3.33). However, it is recommended to correct it on the basis of other indicators, such as damage caused by game. Based on these values and on the basis of the quality class of the hunt (this will be determined by the state administration when recognizing the hunt), an estimate of the increase in game and its summer condition will be made. Based on this, the hunting manager decides how many animals should remain in the hunting grounds after the hunt (expected stock), and sets the hunting plan. In many countries, such as the USA, there is, in principle, no minimum land area on which hunting takes place, provided that it is safe to do so. In such a case, especially for small plots of land, of course, the owner does not manage the game population, but simply hunts what comes to his plot, in compliance with other regulations such as the conservation period, number or size of individuals allowed, or the laws for the protection of certain species. However, large owners, and in particular managers of public land on which hunting is allowed, plan and regulate hunting. The procedure is similar, but somehow more appreciate the fact that operate with approximate numbers. An estimate of the total number of animals is usually not used as a basis, but some indicator can be assumed to be related to the total number; for example, when planning deer hunt the sex ratio is often used as an indicator, when planning goose hunting it is the number of nests on selected indicator transects, etc. Based on these indicators, the degree of regulation of hunting is determined, for example, the number of stamps issued or sold to catch deer. The stamp is then used by hunters to identify the caught animal and is proof that the animal has been caught in accordance with the regulations in force. An understanding of how a certain hunting intensity affects the monitored indicator is based on the experience of previous years.

Box 3.11 Estimation of Wild Population Density

In Box 1.3, we dealt with the estimation of the biomass of organisms, namely plants. Biomass is important for the study of a number of environmental impacts, the role of organisms in biogeochemical cycles, and the provision of ecosystem services. On the other hand, to study population dynamics we often need to know the number of individuals, this is particularly true for animals. There are several basic methods for estimating the number of individuals (Fig. 3.34). The simplest method is a total inventory. Let us just count all the individuals. Such a procedure is possible if there are relatively few individuals, they are easy to count, and we are interested to explore a reasonably large area. For example, the number of trees in the park, the number of

(continued)

Box 3.11 (continued)

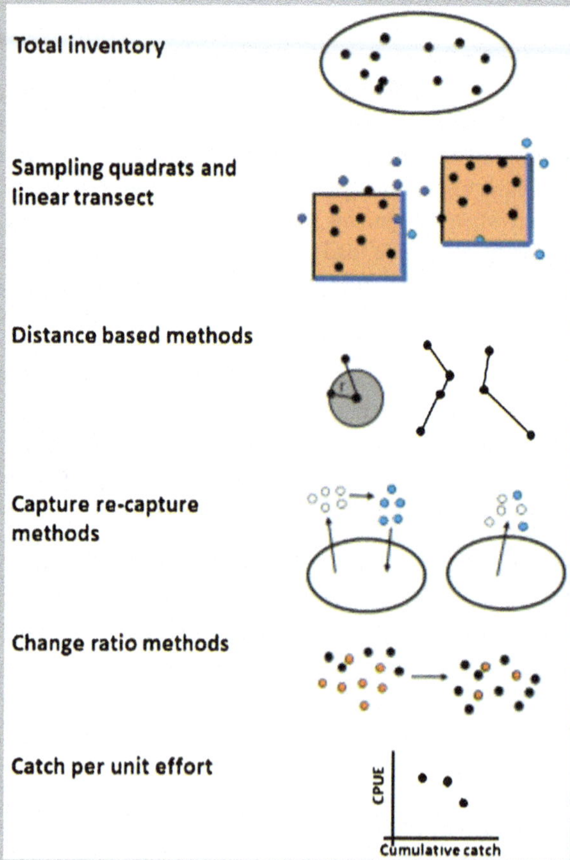

Fig. 3.34 Schematic representation of various methods for estimating the number of organisms

trees in a defined area of the forest, and so on. The counting of all individuals, even if they do not move, is not simple; we usually proceed by dividing the area into smaller sections and counting the individuals in them, while in some way marking the already counted individuals. This procedure is usually used in a situation where we want to obtain additional information about individuals, such as a map of trees, distribution of individual species, their dimensions, etc. We can also count individuals who move, but it is often a problem to mark them, so sometimes we can use characteristic features such as spots on the body, to distinguish them. With a sufficiently dense coverage of the area, for example, by photo traps and a longer time available, we could count all animals of certain species in certain area. As we have already said, the

(continued)

Box 3.11 (continued)
implementation of the total inventory is possible only where the area is reasonably small; we can certainly not count, for example, all the trees in the entire taiga. Here we would change the question a bit and ask ourselves how many trees there are, on average, per unit area, for example, per hectare. If we then determined the area of the taiga, for example, using aerial or satellite images, we could estimate the total number of trees. To determine the number of trees per unit area, we use the method of sampling squares. We proceed by defining sampling plots of known size and counting all units in them. Of course, we face the same problems we discussed in relation to the overall inventory, and some additional ones. This method is suitable for the study of immobile or slow moving individuals. The number of samples affects the accuracy of our estimate, which depends on the variability of the samples. For more details, see Krebs (1999). For organisms that move, we use a modification called a line transect. In a line transect, we go through the landscape and count the organisms on one or both sides of your path. We get a strip of the observed landscape, the area of which can be calculated as the length of our route multiplied by the width of the observed strip. The advantage is that organisms that run away from us usually run away from the surveyed area, so there is little chance that we will count them more than once. One concern is the determination of the width of the surveyed area, and a number of methods are applied here, the overview of which is given, for example, by Krebs (1999). For completeness, we will also briefly review methods based on distance. These methods measure the distance of organisms (e.g. trees) from a random point, or from each other, and assume that the area occupied by the individual corresponds to a circle of radius given by this measure. The number of individuals per area is then the inverse of this area.

The classic research method for monitoring the numbers or density of mobile individuals is the method of capture-recapture. We will explain the principle of the simplest of these techniques, which has a number of designations (e.g. Lincoln index or Petersen method). With this method, we capture a certain amount of undamaged individuals, mark them and subsequently release them back into nature. After some time we catch another group of animals in the same area and record total catch and the number of individuals that are marked among them. The number of individuals in the population is then calculated based on the assumption that the share of all individuals marked (in the first capture) in the whole population is the same as the share of marked individuals in the second capture. Therefore, if we mark 100 individuals in the first catch and in the subsequent catch we also catch 100 individuals, of which 10 are marked, we will assume that the total number of individuals marked in the first catch is one-tenth of the population, and the whole population would be 1000 individuals. This method is based on a number of assumptions. We assume that the released individuals will randomly disperse in the population,

(continued)

Box 3.11 (continued)

that the marked individuals will survive until the next capture, that marking does not change the probability of recatching them or cause mortality of the identified individuals, and finally, assumes that the population is demographically and geographically closed. In order for this to work, it is also necessary to label a sufficiently large part of the population, say at least 10%, and preferably 20%. So we should have some idea at the beginning of the estimate of how many individuals there may be. The disadvantage of Petersen's method is that the fulfillment of the above assumptions is difficult to verify. This shortcoming is eliminated by the Schnábel method, which has the same limitations and assumptions as Petersen's method, but some assumptions can be verified. The procedure is initially the same as for Petersen's method except that all individuals caught in the second catch are also tagged and released again, and then we continue with another third catch and possibly other catches. At each catch we record the number of captured individuals and the number marked, then the others are marked and released. The logic of the calculation is the same as for Petersen's method, but we can check that the share of tagged individuals in individual catches increases as the number of tagged individuals increases. Finally, we can proceed by marking the individual in each catch in unique way, for example, with a different color. This is the basis of Joli Seber's method. Unlike other methods, it does not require the population to be demographically and geographically closed and allows the turnover of the population to be monitored. When applying this method, in contrast to the methods described above, where the interval between catches is short, here we catch animals repeatedly, usually for longer often for several years.

Another group of methods are those designed to estimate the population of a harvested population. These include change ratio methods. Imagine that we have populations with significant dimorphism (e.g. deer and doe) and that we hunt exclusively or mostly only one sex (only deer). Then we can determine the size of the population from the knowledge of the sex ratio before and after hunting, and from the number of individuals we hunt. For instance, suppose we have a sex ratio of 1:1, then we catch 200 males and the new sex ratio is 1:2. In the beginning males represented 50% of the population, but now they are only 25% meaning that 25% of the population was caught. Since this 25% was 200 individuals, the entire original population had 800 individuals. The calculation is more complicated if we hunt both sexes, but one significantly more (Krebs, 1999). Finally, there are methods based on the conversion of catch to unit effort (catch per unit effort). The principle of the method is that as we hunt individuals the total number of individuals that can be caught, and thus the chances of catching them, is reduced. This is reflected in the fact that, with the total number of individuals caught, the number of individuals caught per unit of effort (one hunting expedition, number of hunters, time spent hunting, number of fishing boats, etc.) decreases. We proceed by drawing points describing the relationship of the catch per unit of effort, based on the total

(continued)

Box 3.11 (continued)
number of individuals caught (Fig. 3.34). Where this dependence intersects the x-axis (its value on the y-axis will be 0), we can subtract the number of catchable individuals in the population at the beginning of the hunt on the x-axis. Again, we assume that the population is demographically and geographically isolated. We can also use the catch per unit effort to compare the size of different stocks. This is the basis of a number of relative methods of monitoring abundance. While the goal of the other methods mentioned so far was to find out how many individuals we have, the goal of the relative methods is to compare the number of individuals between two or more habitats. We will not find out how many individuals are in the meadow and in the field, but we will find that, for example, there are three times more individuals in the meadow than in the field (better to say three time higher chance we catch them on meadow than on the filed), or that the number of individuals decreased by 50% after pesticide application, etc. In many practical applications we are more interested in such comparisons than in the total number, and in addition, relative methods are usually easier to implement than methods for determining the total number. These methods include comparing the number of individuals caught, or observed, in some standardized way (the number of individuals caught once a photo trap per day, caught per trap per unit of time, etc.).

3.5.3 Impact of Hunting on Game Populations and Ecosystems

A human hunter acts like a predator. Predators are a natural part of ecosystems, which raises the question of whether human influence on prey populations is similar to that of natural predators. If we compare the behavior of a predator and a human hunter, there is not only a fundamental similarity (both hunt prey), but also a number of differences in the way how it happens. The predator's hunting success is regulated by the catchability of prey. This is determined by the conditions, the type of prey, and its status. The predator therefore chooses prey which is in some way substandard; sick, old, young, wounded, etc. This preference for substandard prey is greater the more difficult it is for a predator to catch prey (Temple, 1987). It is assumed that the predator's catch are individuals who would have less ability to resist competition. Thus, predation pressure can alleviate competition and thus the loss of animals to predation will not be reflected in population decline. On the other hand, the human hunter often (not always) chooses trophy, above-standard, large, and strong individuals (Zeckhauser, 2017), probably because it will bring the hunter more admiration and respect in the community, or greater economic benefit. While predator-prey selection is determined by catchability, for humans it is determined by the social norm. Similar differences appear in the timing of the hunt. The predator depends on its prey as a food, so it must hunt constantly. But, for humans, the time of hunting, and the time when prey is not hunted are given by social norms

(Zeckhauser, 2017). At the same time, the time when social norms forbid, or at least condemn, prey hunting is often the time when prey is most easily caught by a predator. It could even be said that social norms preclude hunting at a time when the pressure of natural predators is greatest.

In addition to the effect on prey mortality, the predator has an effect on its behavior. Under the presence of a predator, prey chooses places where it cannot be caught so easily, or where the predator occurs less frequently, and at the same time, devotes time to patrolling to be informed about the predator's activity. All this is at the expense of foraging. The presence of a predator therefore has a lethal effect on prey due to the capture and killing of part of the prey population. In addition, there is a nonlethal effect on prey due to the fact that, in the presence of a predator, it devotes considerable effort to observing the environment in an attempt to identify the predator in time. This leads to a loss of time and opportunities to obtain food, and thus less available energy that can result in a slower growth, reduced offspring production, etc. These nonlethal effects are considered to be at least as significant, if not more significant, than the lethal effect due to predation-related mortality (Cresswell, 2008). Predators may, in a similar way, try to avoid areas visited by humans. The fact that humans do not hunt all the time, and not every human is a hunter, may thus contribute to creating areas where prey (or at least some prey species) can avoid the constant pressure of the predator. Human thus helps prey to win the game of space (Muhly et al., 2011) and thus indirectly supports the prey population.

The predator population depends on the prey population, and although it may switch to replacement prey the options are limited, so as the prey population decreases the predator population also decreases. In addition, with a declining prey population it is more difficult for a predator to hunt for that prey because it has to spend more time searching for it, and it is easier to switch to another prey. There is also no known case that suggests that a predator would exterminate the prey population without human input (Zeckhauser, 2017). On the contrary, human is so versatile in his resources that the extermination of one prey species will not endanger his population in any way. In addition, if there is a strong social or economic incentive, hunters can specialize in rare prey because its catch will bring more profit or prestige, and this supports a human ability to hunt down the prey population.

About 90–10,000 years ago, the expansion of modern people (*Homo sapiens*) took place outside Africa, and in this period modern human gradually settled all the continents. There has been a large loss of megafauna (animals over 44 kg) on all continents except Africa, where this effect was smallest. It is reported that 24 out of 46 (52%) species became extinct in Asia, as did 23 out of 39 (59%) in Europe, with more pronounced extinction observed on continents further from Africa, especially in Australia (19 out of 27, or 71% of species) and in North America (45 out of 61, or 74%), and South America (58 out of 71, or 82%). The conspicuous coincidence of human arrival with the extinction of the megafauna has led many researchers to believe that the megafauna became extinct due to human activity (Koch and Barnoski 2006). It is also often emphasized that the impact on the megafauna was smaller in places less distant from Africa, where the megafauna lived and evolved for a long time with humans and had more opportunities to adapt to this predator. On the contrary, in locations where human was a new player, his influence was more devastating. Recent research shows that there were more causes; and in addition to human activity, climate change at the end of the Ice Age, for example, has played a significant role, but the causes of extinction have probably differed from one animal to another. However, in many cases, human is very likely to have made a significant contribution to the extinction of these species, both directly by hunting them, and indirectly e.g. by changing their habitats.

However, even in historical times, there are known examples of complete or local extinction of some species by human activity in which hunting played an important place, although usually in combination with other factors. An example is the aurochs (*Bos taurus primigenius*) that was originally widespread in most of Europe, but gradually declined and then became extinct due to hunting and declining habitat. At the end of the sixteenth century, the aurochs survived only in the Białowieża Forest, where the last female died in 1627. Similarly, the American bison (*Bison bison*) was brought to the brink of extinction, primarily by human hunting activities. It is estimated that, in the sixteenth century, 20–30 million bison lived in North America of which only about a hundred remained in the late 1980s, and it has taken a concerted effort to restore the current population. Another example is the Steller's sea cow (*Hydrodamalis gigas*). This huge (up to 10 t) marine mammal from the group of ochechuls was discovered in 1741 by the expedition of Vitus Bering and was already exterminated in 1768, thus representing a record in the rate of extinction of a species by human since its discovery. A common feature of the above species is the fact that they were large, slowly reproducing animals. These strategists are among the animals most affected by hunting, especially if there is a targeted reason to systematically hunt these animals, such as trying to suppress indigenous populations in the case of bison hunting, or the belief in the healing ability of rhino horns, which is behind the current hunting and illegal poaching of these animals.

In addition to large animals with slow reproduction, hunting is a significant problem for migrating animals and for island populations. Probably the most textbook example of the extinction of island populations is the Mauritian Dronte (*Raphus cucullatus*) or dodo, a flightless bird, endemic to Mauritius, extinct as early as the end of the seventeenth century. Migrating animals are another group that is sensitive to hunting; they produce large flocks that occur in predictable places at predictable times, and therefore, hunting these can be very effective. As far as migratory birds are concerned, we have already mentioned the hunting of quail as they pass through Southern Europe. Many other birds were hunted in the south of Europe. Glutz von Blotzheim (2015) estimates that 11–35 million migratory birds are hunted in the Mediterranean each year, despite their declared protection. After all, the migratory pigeon (*Ectopistes migratorius*) was probably the most abundantly hunted bird. At the same time, it is estimated that this species once represented 40% of all birds in North America, and in 1873 an observer noted that a flock of this species passed over in a flight that lasted over 8 hours. Nevertheless, humans have managed to completely eradicate the species in nature through intensive hunting by the beginning of the 20th century, and during the first decade of the 20th century, individuals bred in captivity also became extinct.

With regard to the mechanism of action of hunting on wild populations, hunting reduces the number of individuals caught, both through the take and as a result of possible fatalities of injured individuals. In addition, there are a number of other indirect consequences for the hunted populations. For example, a Swedish study comparing two bear populations under weak and strong hunting pressure showed a 13.6% slower growth rate in the hunted population, mainly due to higher sexually specific male mortality and higher population turnover (Gosselin et al., 2015). However, paradoxically, hunting can also lead to an increase in population growth rate (Massei et al., 2014). An example is the effect of hunting on the social structure of wild boar (Bieber & Ruf, 2005). In feral pigs, the generation of this year's females is already fertile, which determines their very high reproduction rate. In the natural situation, males live solitarily and females form groups led by an old female leader. In these groups, the whole group has synchronized estrus. Subsequently, the males engage in combat and the winning male then mates preferentially with the dominant female, and further with other socially higher (older) females. Younger, socially inferior females are usually neglected, and at the same time the winner prevents other males from mating with the group. If, as a result of hunting, the dominant individuals of both sexes are eliminated, the group synchronization of estrus disappears and entire population which can reproduce does so, leading to a significant increase in population growth rate.

Other effects are those on the genetic structure of the population, as hunting can affect genetic drift and cause selection of certain genotypes (Harris et al., 2002). Besides hunting itself, the gene pollution by game introduction is also important. As mentioned earlier, importing exotic game in many areas of Europe, and also Asia, was popular among nobility throughout history (Lindemann, 1956). Some of the historical introductions have recently become naturalized, but in numerous cases these introductions bring problems including undesired hybridization with related local species. An example is the hybridization of the more recently introduced sika deer (*Cervus nippon*) with the European deer (Bartoš, 1993).

In addition to targeted introduction itself, there is a significant impact of artificial breeding and the release of wild animals for hunting. For many popular game species, such as pheasants or ducks, the natural reproduction of populations is not enough to cover the demand of hunters. Therefore, the breeding of individuals of the species in captivity is used to produce additional individuals, which are then released into the wild where they are the subject of hunting. These massive introductions can affect the genetic structure of wild populations. For example, it has been observed that due to the massive introduction of red-legged partridge (*Alectoris rufa*) in Spain in order to strengthen the hunted populations, individuals carrying genes of the related species *Alectoris chukar* have been introduced. Up to 28% of the wild population is now reported to carry *A. chukar* genes, with hybrids having lower survival.

Breeding wild animals for subsequent hunting also has negative consequences for animal welfare. Probably the most extreme example is the so-called canned hunting. Its goal is to grow attractive prey, such as lions, in captivity. Very often these animals are kept under very unsatisfactory conditions – small cages, unsuitable food, etc. Animals bred in this way are then released into the wild only for the purpose of being caught by a hunter who pays well for their hunting. The possibility of reintroduction of animals bred in this way into nature is difficult, almost impossible, for a number of reasons: they do not have hunting habits and experience, they cannot communicate with other individuals of their species, and they are not afraid of humans. The crowded conditions common in captive breeding also promotes the development and spread of diseases, which may then be transmitted to wild populations.

In addition to indirect effects at the level of the hunted population, the restriction of a certain species due to hunting has consequences for the functioning of the whole ecosystem. A classic textbook example is the suppression of sea otter (*Enhydra lutris*) populations due to their intensive hunting as prized fur animals (Mills et al., 1993). This led to a reduction in the pressure of the otters on their main prey, sea urchins. The subsequent increase in the sea urchin population, in turn, led to a reduction in the growth of kelp algae on which sea urchins feed, and thus to a decrease in the area covered by the underwater forest of kelp algae. This forest is home to a number of other organisms that have been affected, and it also deteriorate other kelp functions. For example kelp slows down the speed of water flow and thus reduces abrasion of the coast by the waves (Estes & Duggins, 1995). These effects differ depending on the position in the food cascade of the animal that was suppressed or eliminated as a result of the hunting. The removal of the top predator leads, on the one hand, to an increase in the herbivorous pressure by the herbivores hunted by it, and on the other hand, to an increase in the population density and predation pressure of the mesopredators. The increase in herbivorous pressure is related to an increase in their population density, but especially to a change in the behavior of herbivores that, in the presence of a predator, either completely avoid some risky localities, or act very cautiously and spend less time grazing, thereby having less impact on vegetation. The increase in the populations of mesopredators, such as foxes, can then lead to the suppression of small vertebrates, which are their prey. Suppression of dominant herbivores can lead to an increase in other, competing, herbivores, and this has implications for vegetation composition, as has already been shown in the sections on the effect of grazing. Elimination of predators and management of hunted population may also lead to an increase in populations of some game species (Fig. 3.33). Together with the absence of nonlethal effects of predators, as mentioned above, this increases grazing pressure on vegetation. This may be a particular problem for tree regrowth, as trees are more sensitive to grazing than are grasses and herbaceous vegetation. In some European countries, game grazing has become a significant problem that affects forest regeneration (Forest Europe, 2015) in a way that many newly planted patches require fencing as a protection against grazing.

Consumption of game by growing human populations is a motivation for intensive illegal hunting of animals, so-called bushmeat. Similarly, illegal or semi-legal hunting of animals and trade in their meat is a problem in many Asian countries. At the same time, bushmeat is not only sold in local markets, but it has been shown that due to the economic power of immigrants in many developed countries, and probably other "gourmets," there are large-scale illegal exports of meat caught in this way to many developed countries. It is estimated that more than 200 t of this illegal bushmeat can pass through one large European airport each year (Chaber et al.,

2010). It can only be speculated that seaports are an even larger center of such trade. Illegal wildlife trafficking, of which bushmeat is only a portion, has been estimated to be worth US $21 billion annually (Nellemann & INTERPOL, 2012). Such illegal imports not only contribute to the threat to hunted populations but also pose significant health risks. Consumption of the meat of wild animals, in particular, illegally hunted and imported meat that is not subject to any veterinary inspection, and certain groups of animals that host large quantities of zoonoses, can be a source of zoonotic transmission to humans. Viral diseases are particularly important in this regard; only recently, the emergence of a number of viral diseases (HIV, Ebola, SARS, HPAI) has been observed, which are thought to be related to the transmission of zoonoses to humans (Murray et al., 2016). After all, world pandemics such as COVID-19 can be a good example of these risks (Neupane, 2020).

However, bushmeat and/or obtaining luxury food is not the only driver, and probably not even the most significant driver, of poaching. One important driver is the medical use of many wild species, such as tigers, elephants, and rhinoceroses, in traditional medicine. Illicit trade in these commodities drives prices high and, thus, greatly increases the economic incentive for poaching which may be even more enhanced by poverty of local populations, despite increasing efforts to suppress poaching. Decreasing the demand for these commodities and reducing poverty may be as important for reducing poaching as poaching repression itself (Challender and MacMillan, 2014).

The above-described effects of hunting on ecosystems are generally negative. On the other hand, the interest in hunting leads people to not only to care for game but also to take an interest in what is happening in nature, which in turn has played an important role in conservation. For example, the interest in game hunting led to the creation of the first "protected areas." One of the largest complexes of the original forest in Europe, the Białowieża Forest, was initially protected as a hunting ground for Polish kings. In the colonial era, it was the aristocratic supporters of hunting who contributed to the creation of the first reserves in Africa. The decline in game and their habitats, which the US President Roosevelt observed as a passionate hunter, eventually prompted his significant efforts to protect nature. Hunting brings large financial resources that can be used for protection of wildlife, including species that are not hunted (Heffelfinger et al., 2013), and we could find many other similar examples. Our interest in hunting is also reflected in the great popularity of various safaris and other activities associated with game watching. Historically, game watching has been an essential part of hunting, and the popularity of this activity is still part of our culture, even if we do not hunt. The passion for hunting and, in particular, game watching, is one of the great drivers of conservation. This was clearly shown, among other things, during the COVID-19 pandemic. As a result of the absence of humans there was a reduction in wildlife disturbances in many areas, but the loss of hunters and tourists meant a major loss of income on which wildlife conservation in Africa and other countries is essentially dependent. This economic hardship has increased the risk of poaching. The risk of poaching has also increased due to the absence of people in the countryside, as in normal situations, a large part of the landscape and transport infrastructure is permanently controlled by hunters, tourists, and their guides (Neupane, 2020).

Chapter 4
Fisheries and Aquaculture

Fishery is human activity focused on harvesting or raising fish and other aquatic organisms. It can focus on harvesting a wild population and/or raising various organisms in captivity often referred to as aquaculture. These activities can happen in marine as well as in freshwater environments.

Fishing is one of the oldest human activities. Fish but also mollusks and crustaceans were an important source of food not only for modern people but also for early hominids such as *Homo habilis* and *Homo erectus* (Stewart, 1994). Many prehistoric artifacts of modern humans indicate the important role of fishing as a food source (Sahrhage & Lundbeck, 1992). Early fishing techniques include harpoons, fishing hooks, and various kinds of nests and traps. Gradually, fishing developed from selfsubsidence to specialized professions (Sahrhage & Lundbeck, 1992).

First trawlers appeared in the seventeenth century, but started to expand when sails became replaced by steam engines in the nineteenth century. Besides, fishery techniques itself also have developed modern preservation methods such as canning, refrigeration, or freezing that allow us to trade fish and other seafood over large distances, thereby substantially expanding the market and bringing important stimulus for the expansion of industrial fisheries. New technological innovation after World War II (such as sonar and others) increased the extraction capacity of modern fisheries tremendously, which brings a risk of overexploitation of the resources. This was even enhanced by the Cold War, leading to nations' territorial claims on ocean. The risk of resource damage by overfishing leads to new international law and the development of modern fishery regulations and management practices (Finley, 2016).

Aquaculture is also a very old practice; it dates back to Neolithic times where people stored the caught fish alive in fences or dammed in the shallow parts of lagoons, lakes, or rivers. Using fishponds for aquaculture was reported in ancient Egypt; it was widely distributed in the medieval period in Europe, China, or Hawaii.

While various organisms besides fish such as mollusks or crustaceans can be the subject of fisheries, fish are the most important target group.

© The Author(s), under exclusive license to Springer Nature Switzerland AG 2022

J. Frouz, J. Frouzová, *Applied Ecology*, https://doi.org/10.1007/978-3-030-83225-4_4

4.1 Biological Bases of Fisheries

4.1.1 Growth of Fish Populations and Methods of Optimizing Their Fishing

Most true fish spawn, females secrete eggs, and males containing sperm fertilize the eggs. Pelagic fish, such as tuna and sardine, spawn at the surface and their eggs float. Bottom fish, but also others, lay eggs on the bottom and other substrates. Many fish look for specific spawning sites or substrates, often different from the environment in which they normally live, such as some coral fish in mangroves and salmon in freshwater. Some fish (smelt, Pacific salmon) die after wiping. Small fish hatch from the eggs, first provided with a sac (the rest of the egg-containing nutrients); this stage is called the sac fry or alevin. The sac is gradually absorbed and the fry begins to feed itself. The fry often feeds on a different type of food than adult fish. Switching to adult fish is often associated with a change in growth dynamics (Fig. 4.1). For example, predatory fish that feed on plankton begin to grow faster once they switch to a piscivorous lifestyle. Fry mortality is huge. As the fish grows, the growth of the fish can be described by the von Bertalanffy growth curve (Box 4.1), which shows that the growth rate of the fish decreases with age and its size approaches some maximum value (more in Box 4.1). Growth rate and size vary with the amount of food, competition, and other environmental conditions that also affect fish mortality.

We can catch only a part of the fish population and fish that we can catch in the usual way (e.g., larger than those that pass through the mesh of the net) we call stock. Young fish that have just grown enough to join the population of adult fish are

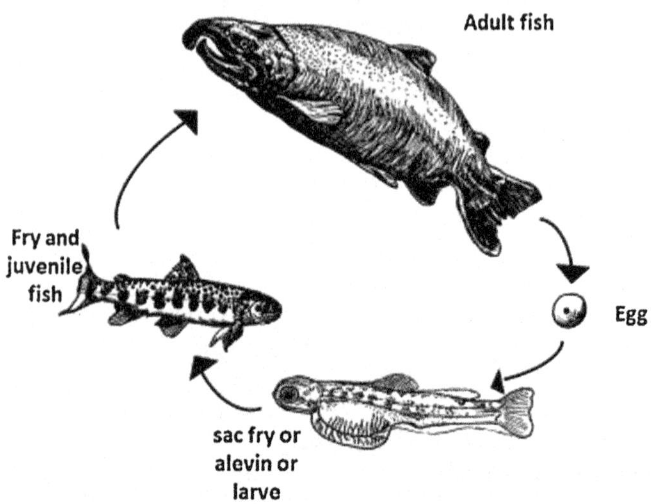

Fig. 4.1 Fish life cycle

called recruitment (or recruits, R). Conversely, the size of the fish we catch can be determined by regulation, such as the size of the net. There is an effort to regulate the size of nets so that the fish caught are adults, which means that it is very likely that we will not catch fish that have not yet reproduced massively. For most organisms, their population is established on the carrying capacity of the environment K. The value of K may not be constant and may vary depending on environmental conditions. Many species also oscillate very widely around the carrying capacity of the environment (Fig. 1.25). Such oscillations are very common in fish stocks, where the differences between long-term maxima and minima usually reach several times the minimum value. These fluctuations are due to the response of fish stocks to the variability of environmental conditions and can be significantly exacerbated by fishing (Shelton & Mangel, 2011).

In the introductory chapters, we said that if we manage a population of wild animals, it is optimal to take individuals so that the biomass of the population is kept at half the carrying capacity of the environment (K/2). Then, we can take a supplement in the long run, which will be equal to a quarter of K. We also call this point the maximum sustainable yield (MSY). This principle also applies to fishing. However, it is impractical to relate our considerations to the total biomass of fish stocks, for example, in the sea, because it is very difficult to estimate the total biomass.

We will now try to relate MSY, not to population size, but to fishing efforts (E). The dependence of the yield on the fishing effort gives a spherical curve, which shows that initially, as we increase our fishing effort, our catch will increase to a certain point, and then begins to decline (Fig. 4.2). This is due to the fact that initially, our fishing does not limit the size of the population because the removal of individuals by fishing reduces the competition of other individuals and, thus, increases their chances of survival and growth. If the number of individuals is not limiting, then form two units of effort we simply catch more fish than for one unit. However, as we increase the number of individuals caught, the population is insufficient to replenish and its size will decrease. The peak of the given curve is, therefore, the MSY, which corresponds to the EMSY fishing intensity. If the fishing intensity is greater than this value, the stock begins to be overfished and its abundance decreases over time. With a fishing effort equivalent to twice the EMSY, fishing leads to the extinction of the population as a result of overfishing. Mathematically, the relationship between fishing effort and yield is parabola.

$$Y = aE + bE^2$$

At the same time, as the fishing effort increases, the catch per unit of effort (CPUE) decreases. If we take the parabola equation described above and divide both sides by the effort (E) we get.

$$Y / E = a + Eb$$

At the same time, Y/E is the catch per unit of effort (CPUE). Therefore, if we have CPUE values for different levels of effort E, as in the example in Fig. 4.2, we

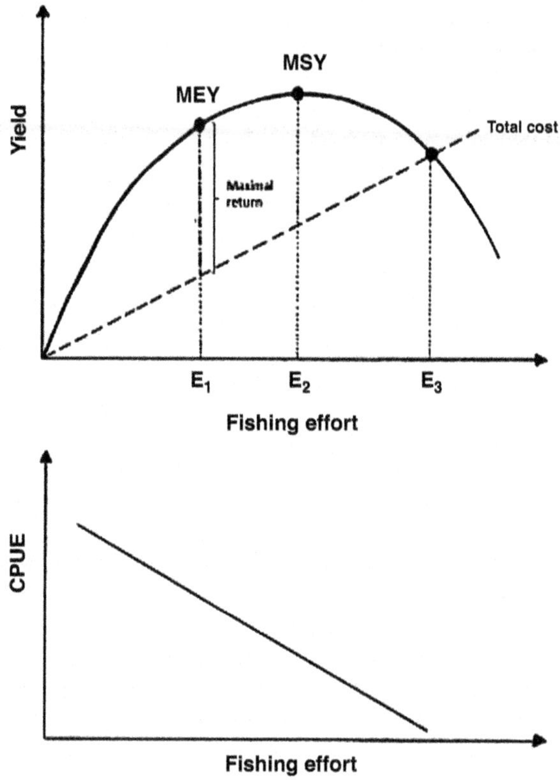

Fig. 4.2 Estimating MSY and maximum economic yield (MEY) and setting parameters for MSY estimations based on the relationship between fishing effort and CPUE

can estimate a, b, as coefficients of a line and, then, substitute the same coefficients into the equation of the parabola describing the dependence of Y on E and thus, mathematically describe the whole curve and, then, estimate the MSY from it. At the same time, we can express fishing efforts in various ways, such as the number of boats, the number of fishing expeditions, the length of fishing expeditions, the amount of fuel consumed during fishing, etc. The CPUE will then be the amount of fish, expressed as weight or number of fish, per unit effort. However, the estimation of individual parameters is burdened with a number of errors, and thus, there is a risk of overfishing even when using EMSY, so it is usually recommended to use a value lower than the estimated EMSY for regulatory purposes.

From the fisherman's point of view, another aspect is interesting, namely the profitability of fishing. Each unit of fishing effort incurs certain costs (fuel, maintenance, and depreciation of the ship, wear and tear of hunting equipment, crew salary, etc.). If we express the catch and these costs in the same units, then the costs at a certain point cross the yield curve (it can be before or after reaching the MSY). It is not economically viable to hunt beyond this point. On the contrary, the amount of yield above this line indicates the rate of profit; it allows us to determine the place where the profit is the largest (maximal economic yield MEY). MEY always comes at lower values of fishing effort than MSY, so it is also recommended as a target reference point (TRP) for fishery regulations. In addition, it has a clear interpretation from the point of view of fishermen. There are a large number of other types of TRP. As with MSY, their goal is to set up hunting efforts so that we can catch as many fish as possible while keeping our catch sustainable (does not reduce the size of the fish population in the long run).

Although the above-described method of estimating MSY is appealing in its simplicity, it has a number of drawbacks (Larkin, 1977). Recall that we assumed that all individuals multiply and, at the same time, population density affects their growth and mortality. However, in the real world, the density dependence is not the same throughout the life cycle, at the same time, the fish reproduce only as adults and we only catch adult fish. In addition, the K value fluctuates; populations often do not stick to the K value, but oscillate differently, and so on. We will, therefore, briefly mention some other possible TRPs, a more detailed overview is provided, for example, by Caddy and Mahon (1995) and Cadima (2003).

The age structure of the population must, therefore, be taken into account when optimizing fishing. The easiest way to take this age structure into account is to count on the so-called supplements (i.e., recruits – new individuals who are just entering the fished population). There are a number of practical reasons why we may not be able to catch younger juvenile stages in the normal way of fishing. They may live elsewhere, behave differently, being too small for fishing gear to trap them, etc. The dependence of the Y/R ratio on the total mortality of fish (sum of natural mortality and mortality given by fishing efforts) is used to estimate the TRP series (Fig. 4.3). At a low fishing effort, the Y/R ratio first increased with increasing fishing effort. This is due to the fact that as we remove part of the adult fish from the population, we reduce the competition for the next generation; it has lower mortality and individual fish can grow larger biomass at the same time. As a result, the yield achieved

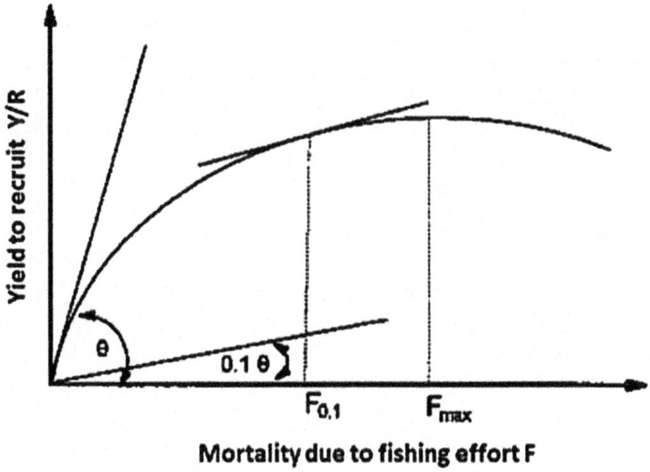

Fig. 4.3 Method of calculation of Fmax and F 0.1

per unit of the supplement will increase. However, as fishing efforts increase, we start to catch more and more recruits, and Y/R decreases.

On the curve describing the dependence between the mortality given by the fishing effort F and Y/R, we can observe a peak (maximum of the whole curve). Thus, Fmax corresponds to the fishing effort for this maximal Y/R. For Fmax, the first derivative of the function at this point is 0 while for lower F values, the derivative is positive (curve increases) and for higher F values, this derivative is negative (curve decreases). Fmax, like MSY, determines the maximum value, which due to various errors may already be beyond the edge leading to overfishing. This is one of the reasons why it is recommended to use a TRP at some more conservative points that can be found on the same curve, before Fmax. One of them is F0.1. It is the value when the slope of the first derivative of the Y/R curve reaches 10% of the maximum value. We said that the value of Y/R increases first, this increase is the largest at the beginning (for F close to 0). This steepest slope (the first derivative of the function at point 0) is used as 100%, of which the 10% is calculated. As F increases, the function becomes flatter, the slope of its increase (first derivative of the function) decreases. F0.1 is reached when this slope reaches 10% of the maximal slope of the curve (for F close to 0) that is, in simple terms, we could say that F0.1 is a conventionally defined point at lower values of F than Fmax. It, therefore, provides similar properties as Fmax but is more conservative and less likely to be overfished due to random estimation errors. Instead of F0.1, another point can be used (another percentage of the original directive 20% F0.2, etc.) based on a higher percentage of the original slope to provide a more conservative (more cautious) TRP. See Cadima (2003) for more detail and calculations.

MSY and other concepts discussed so far and the corresponding TRP assumed that we have a population where the main factor driving the development of the population is competition, which will be very stable without the absence of fishing. This is certainly a reasonable assumption for a number of K strategists. In contrast, external influences may be as significant or more significant than the competition for a number of R strategists located at the bottom of the food web, and these influences need to be taken into account when managing their hunt. There are, therefore, a number of methods for predicting the catchable part of a stock based on the evolution of environmental conditions that can be used to regulate the fishing of such stocks. It follows that for different species of fish with different ecology, different approaches are suitable for how to regulate their fishing (Table 4.1).

So far, we have dealt with the regulation of fishing of individual species, but species are not alone in their environment, fishing of one species can significantly change the living conditions of other species due to changes in competition or various bottom-up and top-down effects in the food web (Box 4.2). In this context, it is interesting to observe that humans prefer to catch fish that are at the top of the food web, large, predatory, long-lived fish (tuna, sailfish). When the fishing significantly reduces their stocks, fishing effort progress to lower trophic levels (cod) and from there gradually to small, short-lived fish at the bottom of the food pyramid (sprats, sardines, etc.). As a result, the average position of our catches in the food web is

Table 4.1 Distribution of ecophysiological properties of fish and possibilities of regulation of their catching according to Kawasaki (1980)

Parameter	R strategist		K strategist
	Subtype A	Subtype B	
Response to environmental conditions	Variable and difficult to predict		Stable predictable
	Rather short oscillations	Long-term oscillations	
Major investment of resources	Reproduction	Reproduction and maintenance	Growth and maintenance
Life expectancy	Short	Long	Long
Size	Small	Medium	Large
Age of reproduction	Very low	Low	High
Population growth rate	Very high	High	High
Representative	Anchovy	Herring	Cod
Recommended method of management	Catch forecast and its monitoring	MSY, MEY	Taking into account the age structure of the population, Fmax, F0.1

gradually decreasing (Pauly et al., 1998) (Fig. 4.4). At first glance, this does not look like alarming information. Fish at the top of the food chain make up only a small part of the total fish biomass. We might even expect that if we eliminate predators and shift our fishing efforts toward lower trophic levels, our catches will increase as we become more important predators of lower trophic levels and do not compete with natural predators. However, this is not the case. If we plot the total catch and the position of the catch in the food web, we see that the effect of catching top predators can be just the opposite and leads to a decrease in catches (Fig. 4.5). This is due to the fact that with the removal of the top predators, the energy flows in the food web will change and subsequently the whole community will change (ecosystem shift), which may result in reduced abundance and catch of commercially interesting fish species (more in Box 4.2 and Sect. 4.2.5).

A textbook example of ecosystem change due to fishing is the replacement of small pelagic fish by jellyfish. In Namibia and South Africa, sardines and anchovies (*Sardinops sagax* and *Engraulis encrasicolus*) are the subjects of intensive fishing, dependent on upwelling on the southwest coast of Africa. The dramatic overfishing of these small pelagic fish in Namibia has led to a shift in the composition of the ecosystem, which has begun to be dominated by jellyfish. This has not happened in South Africa, where fishing is more strictly regulated.

What is the cause of this shift? Simply put, small pelagic fish are important predators of the young ontogenetic stages of jellyfish, and at the same time, these small fish are prey for other organisms, which are also important predators of jellyfish. The suppression of small pelagic fish below a certain limit (tipping point) thus reduces the predation of jellyfish and this leads to their expansion. Jellyfish, once large, can reduce the abundance of small pelagic fish and prevent the ecosystem from returning to its original state (Roux et al., 2013).

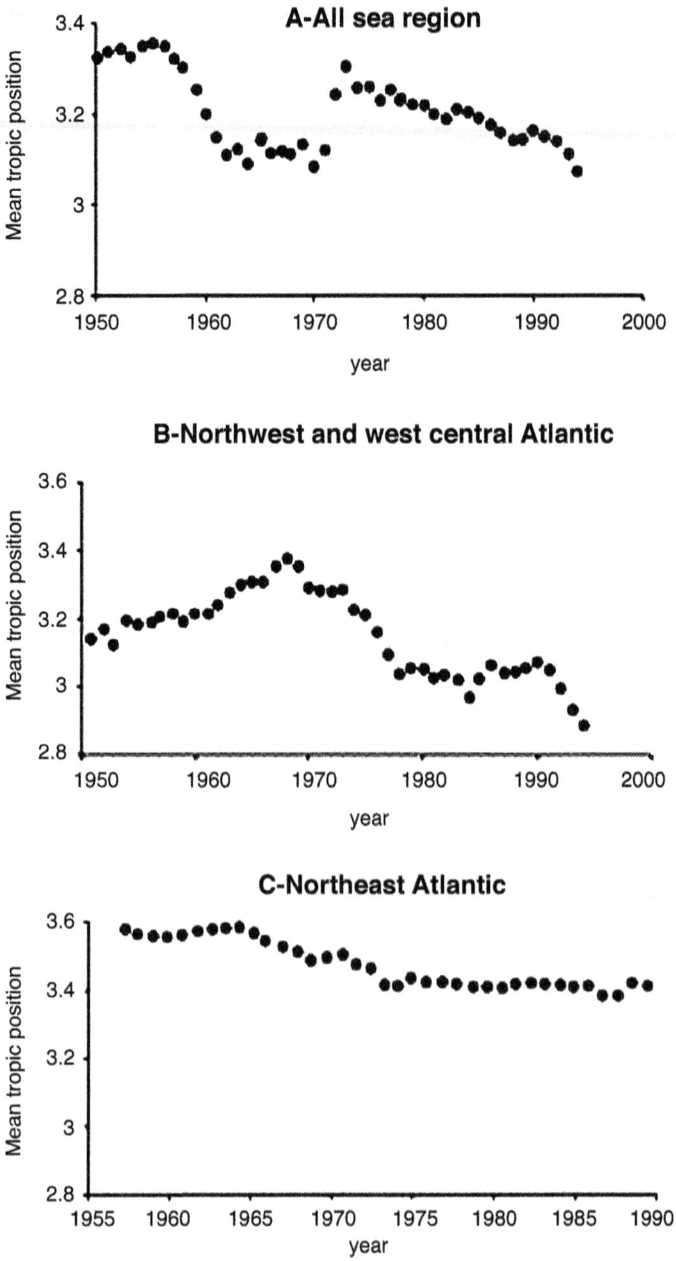

Fig. 4.4 Decrease in the average trophic level of catches in different fisheries according to Pauly et al. (1998)

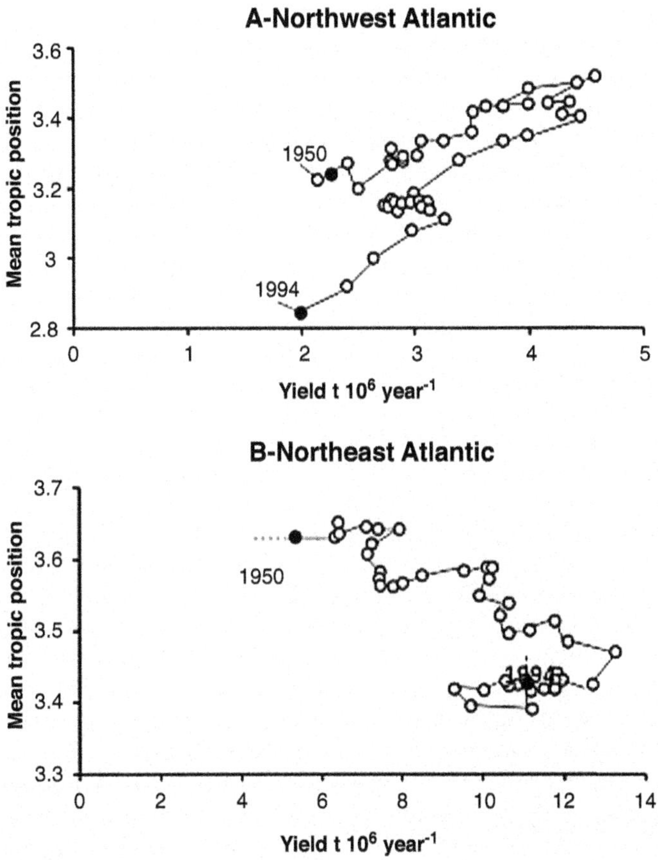

Fig. 4.5 Relationship between the average trophic level of catches and yield in different fisheries according to Pauly et al. (1998)

A comparison of Namibia and South Africa shows that regulating fisheries to maintain stocks that can continue to play a role in food web (predominantly suppressing jellyfish and supporting sufficient populations of other jellyfish predators) can prevent this undesirable shift in the community. This is the goal of the ecosystem approach to fisheries, its goal is no longer to regulate individual isolated stocks, but to keep stocks at a level that allows them to perform their functions in the ecosystem (Boxes 4.2 and 4.3). Revealing these ecologically relevant boundaries requires a far more complete study of marine ecosystems than the approaches described above based on the regulation of individual populations (Cury et al., 2016).

Box 4.1 Study of Fish Populations

In studying the key characteristics of fish stocks, important for fishing and its regulation, we are interested in how fish grow (i.e., how big they are and how old they are, where they are, and how many there are). We can study the growth of fish relatively comfortably in aquaculture, where we have a cohort of fish from which we can catch a sample and measure everything we need and we know when we put the fish in the tank (i.e., how old it is). If we catch fish from nature, we need to find out information about age and growth at once. A number of structures on the fish body can be used for this, which shows seasonal or even daily increments. Scales are one such structure. For many species, we can observe on the scales concentric circles (annulus), similar to annual rings, which are given by alternating slow growth in winter and faster growth in summer. In addition, the distance of individual annual circles is approximately proportional to fish increments. From the annual increments, we can find out not only how old the fish is, but if we know how big the fish is, we can also reconstruct how it grew (i.e., how big it was in individual previous years) (Fig. 4.6) and after approximating the parameters of von Bertalanffy function (Fig. 4.6), we can estimate the maximum size that could grow.

In some fish, the scales are unreadable, have inappropriate morphology, are not transparent enough to see the rings, etc. Other fish structures, most commonly otoliths, can be used here for a similar purpose. Otoliths are auditory "stones" of fish, on which annual increments are also visible, often after cutting and grinding. In small fish, even daily increments are observable, so we can determine the age of the fry within days, for example. The disadvantage is that although scales can be removed from live fish, we have to kill the fish to collect otoliths. Another important parameter we are interested in is the length to weight ratio, or in other words, an indication of how much a fish of a certain length weighs.

To find out where the fish are and how many there are, we can use a variety of observation and trapping methods. The most common methods of monitoring fish in the ocean, as well as in freshwater lakes and dams, are sonar and echo sounder. Active echo sounders, which use a transducer to convert electrical signals into the sound of a defined wavelength, are often used in fish surveys and fisheries. These sound waves then propagate through the water and strike the interface of two environments where the speed of sound varies (bones, scales, air bubbles, fish gas swim bladders, etc.), and part of the sound is reflected back. These reflected sound waves are then picked up by a transducer and converted back into electrical signals. From where and when the transducer was hit, the angle and time of return of the sound wave can be calculated and the distance and direction in which the object is located can be determined, and its size can be inferred from the strength of the acoustic reflection. In the use of echo sounders, waves of different frequencies are used, and the lower the frequency (longer wavelength), the better the waves penetrate the environment (water) but have a lower resolution.

Electrodes measuring the electrical resistance of the water are used to count fish in freshwater, especially draft fish in rivers, fish have less resistance

Box 4.1 (continued)

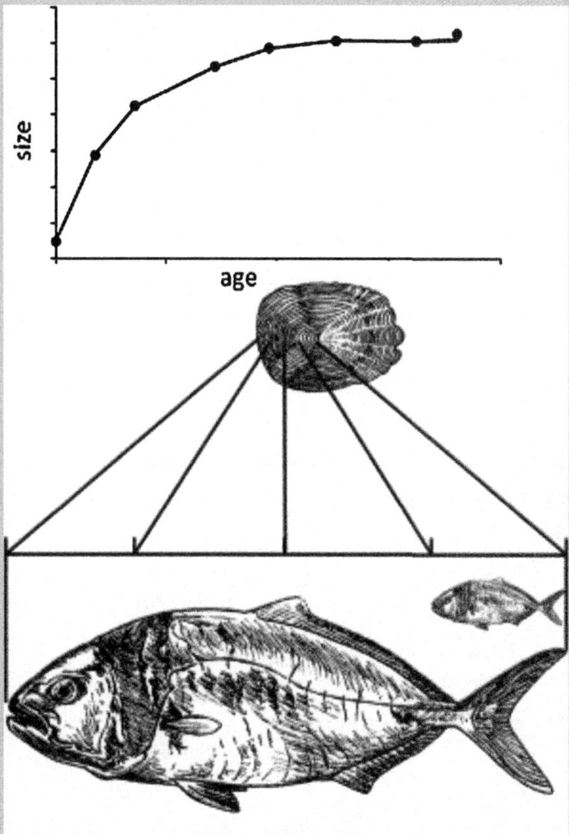

Fig. 4.6 Von Bertalanffy function and procedure of fish growth reconstruction from scale growth

than water, and the decrease in resistance is counted as the passage of the fish. A number of fish tagging methods and telemetry devices that send a signal about the position of fish are used for a detailed study of fish ecology. Specific to research purposes, electrofishing is worth mentioning.

4.1.2 Production of World Waters and Their Use

4.1.2.1 Sea and Oceans

Looking at the main habitats of the seas and oceans (Fig. 4.7), the area between low tide and high tide is called the tidal zone (intertidal) and is an important habitat for hunting a variety of commercially interesting animals (bivalves, cephalopods, and crustaceans). The coastal area where fresh land and salty seawater (brackish water)

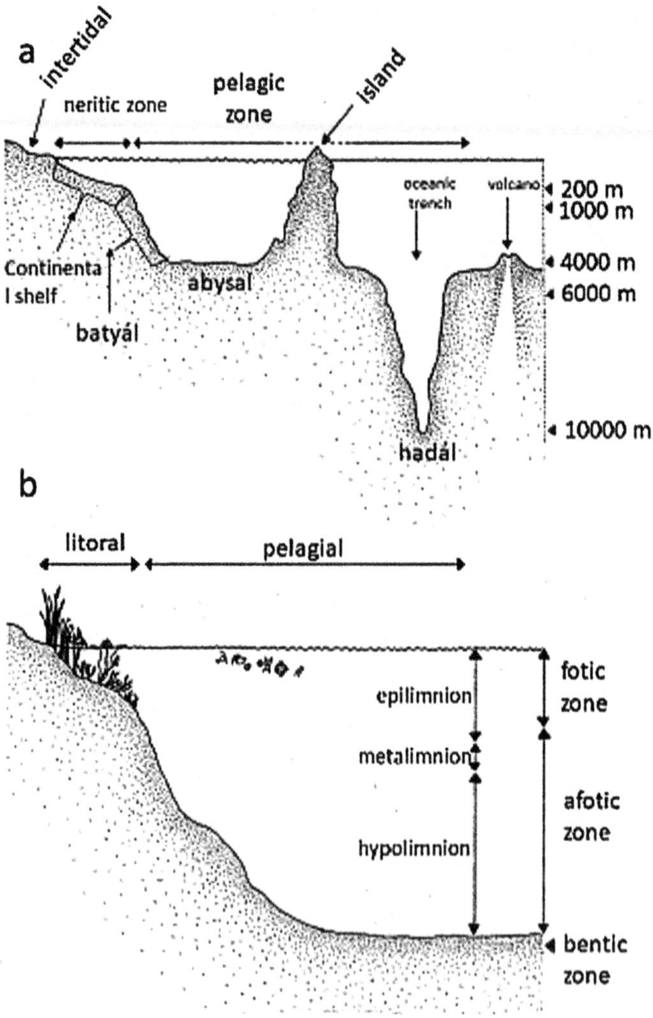

Fig. 4.7 Main habitats of seas (**a**) and lakes (**b**)

mix is called estuaries; they are often mouths of rivers, associated coastal wetlands, etc. For example, mangrove forests exist in the intertidal zone and estuary. The open water of the ocean forms pelagic habitats; they have several layers depending on the depth, the layer above the bottom where the bottom is affected is called demersal. The bottom is then formed by bottom habitats, we distinguish the continental shelves (there may be coral reefs or kelp forests on it) and several other benthic habitats (littoral, bathyal, abyssal, and hadal). The pelagial above the continental shelf (the neritic zone) is a very productive zone of the seas, important for fishing.

Horizontally, the world's oceans are divided into marine areas for fishery statistics. Marine fisheries (Fig. 4.8) can be divided into several zones (FAO, 2016): temperate regions, upwelling regions, tropical, and the area left around Antarctica, where fishery statistics are incomplete and, therefore, not mentioned in detail.

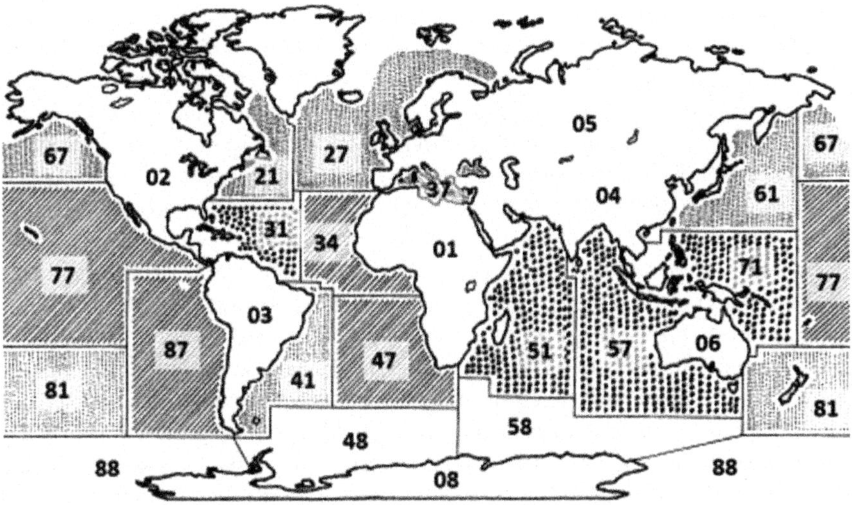

Fig. 4.8 Individual fishing areas and their aggregation into FAO zones, temperate areas – fine dots, upwelling areas, tropical thick dots, white-inland waters (1–8), and Arctic and Antarctic waters (two-digit numbers)

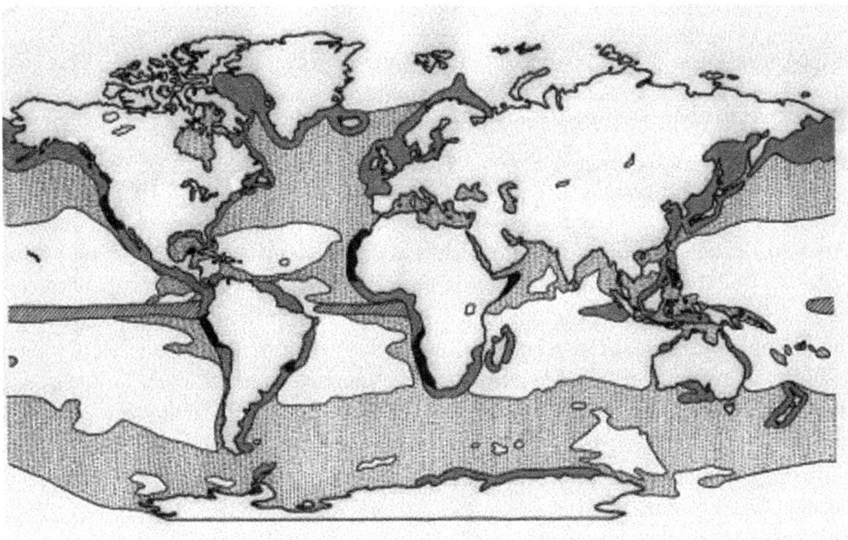

Fig. 4.9 Areas of sea with an upwelling, black colored is main offshore upwelling, grey is seasonal offshore upwelling, dots is oceanic upwelling, hatched area is equatorial upwelling

As we have said, production in the seas is severely limited by the supply of nutrients, both macroelements, and some microelements especially iron. The nutrients come to the surface layer seas from both land masses and from the deep layers of the ocean by upwelling (Fig. 4.9). The largest production is in coastal ecosystems, which includes tidal zones, kelp forests, mangroves, and coral reefs. Production is also high in the shelf seas, especially where there is an upwelling that brings nutrients. On the contrary, the smallest production is on the high seas in tropical waters, especially in the so-called ocean deserts, where there is a very small supply of nutrients, which is due to the great distance from land and very stable temperature stratification, which limits the supply of nutrients from greater depths.

In addition to the total primary production, the production of fish is determined by the further fate of this production at higher trophic levels. With sufficient nutrients, even in tropical seas, total phytoplankton production can be significant due to the high power of solar radiation. High and stable temperatures increase respiration, so less energy shifts to a higher trophic level (Fig. 4.10), and thus, the same primary production corresponds to a smaller catch in the tropics than in temperate or arctic waters. As the temperature rises, the share of benthos-feeding fish in the total catch also decreases. In temperate and subpolar waters, there is a strong seasonality due to fluctuations in temperature and light. As a result, zooplankton takes some time to increase its biomass with an increase in phytoplankton, followed by an increase in the biomass of pelagic fish. Then, with deteriorating conditions, some of the production dies and sinks to the bottom, where it becomes the food of benthic invertebrates and benthic fish. In contrast, in the tropics, production is stable at all trophic levels, most primary production is immediately consumed by zooplankton and subsequently, by higher trophic levels. As a result, only a small proportion of production falls to the bottom, which leads to a lower incidence of benthos, and thus, a small proportion of benthos-feeding fish in the total catch (Ursin, 1984). This may not be the case in ecosystems where nonphytoplankton organisms, such as coral reefs and mangroves, are the basis of primary production.

Not surprisingly, the species diversity of fish communities increases toward the equator and decreases toward the poles (Fig. 4.11) (Rohde, 1978, 1993); however, species-poor subpolar communities often reach high biomass. This is also reflected in the different reactions of temperate and tropical fish communities to fishing efforts. Unexploited temperate communities often achieve higher biomass and higher yields than tropical communities. As fishing effort increases, yield per recruit increases rapidly, peaks at relatively low levels of fishing effort and then declines rapidly. In contrast, tropical communities are characterized by lower biomass, and the increase in yield per recruit (Fig. 4.12) with increasing fishing effort is very gradual, peaking at significantly higher values of fishing effort than in temperate waters. Tropical waters are, therefore, less productive than temperate waters in terms of fish production, but they are less prone to overfishing. Subtropical waters with upwelling show great variability of production (and catch).

Looking at fish productivity in temperate seas, it is high and stable, although the trend is slightly stagnant (FAO, 2016). This is partly due to the high productivity of temperate seas, which have a high proportion of coastal areas (estuaries) and upwelling, and partly due to the technological superiority of the fishing industry (Fig. 4.13).

Tropical areas with coastal upwelling are very productive compared to other tropical waters, but their production is limited by significantly greater variability in tropical upwelling, disrupted by phenomena such as El Niño. Even for these

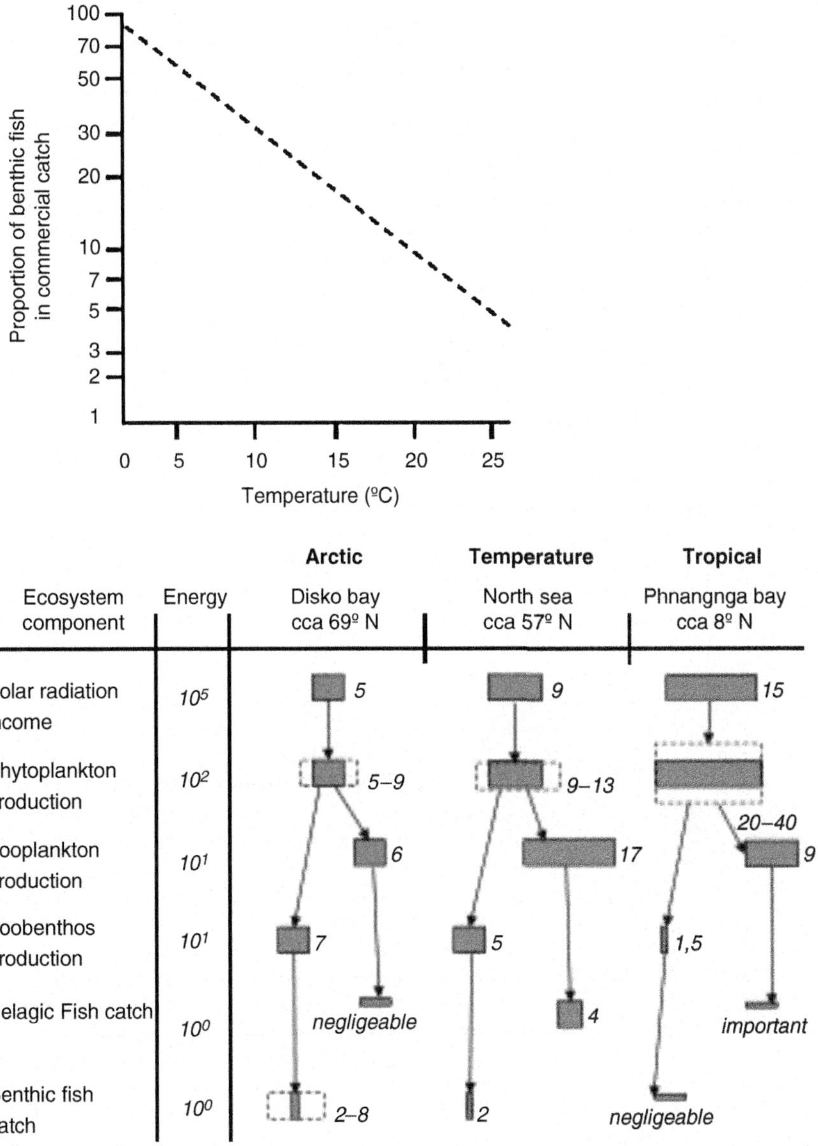

Fig. 4.10 Changes in the total catch and the proportion of different fish ecological groups in the total catch, depending on the latitude. It can be seen that in the direction from the polar regions to the tropics, the catch decreases and the catch of benthic fish decreases. Based on Ursin (1984)

reasons, production in tropical waters with upwelling stagnates. Remarkable is the increase in catches in other tropical areas (without upwelling) given mainly by the development of technology; the catch here consists mainly of not only small but also large (tuna) pelagic fish.

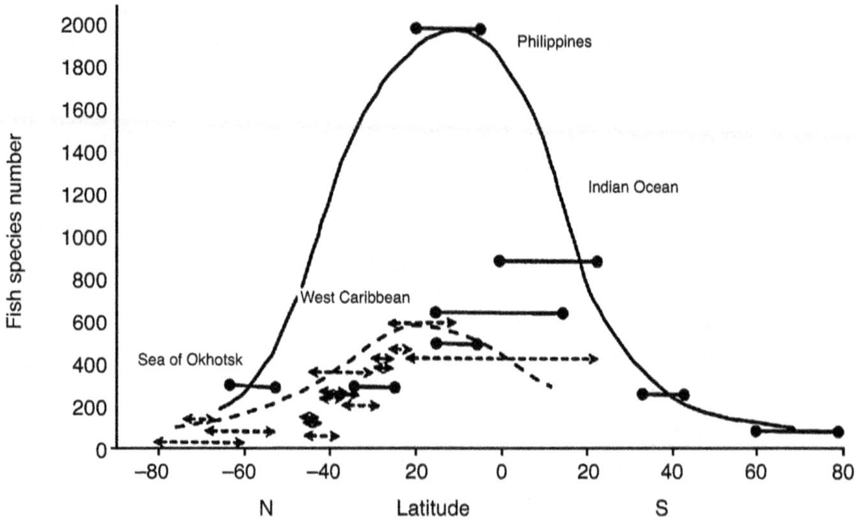

Fig. 4.11 Dependence of fish community diversity on latitude

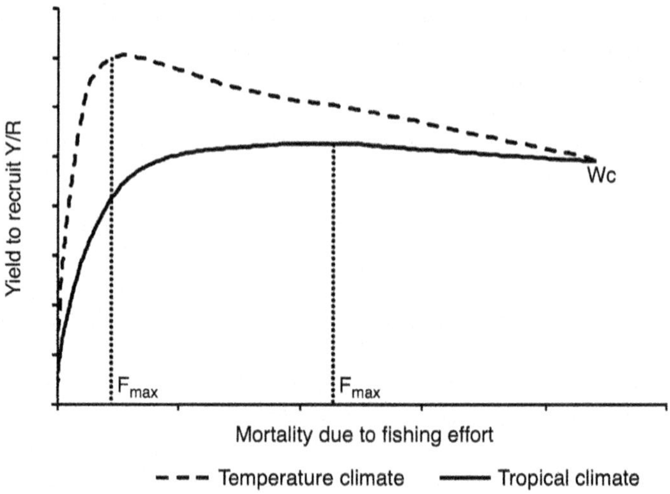

Fig. 4.12 Comparison of relationships between fishing effort and yield to recruit ratio in temperate and tropical climate, based on Ursin (1984)

4.1.2.2 Inland Waters

Worldwide, large amounts of surface freshwater are found in lakes, which are also important sites for commercial fisheries. The near-shore zone of lakes is called littoral, the open water zone is called a pelagial zone and the bottom zone is called a benthic zone. Depending on the water mixing during stratification and light penetration, the pelagial is divided into several further layers (Fig. 4.7).

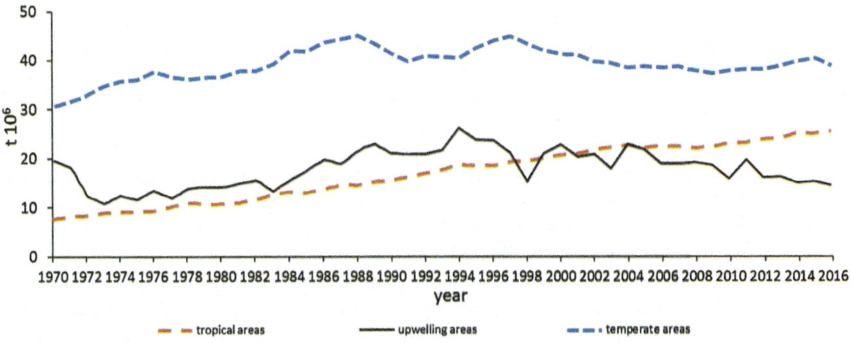

Fig. 4.13 Trend of fish catch in FAO's main fishing areas (FAO, 2016)

Besides natural lakes, man-made ponds and reservoirs are important bodies of standing water in many countries. Ponds are often used for aquaculture and will be mentioned in Sect. 4.3. Reservoirs usually serve more as a source of energy (hydropower), and water from reservoir is used as drinking water or for irrigation and other uses than for fishery and aquaculture. On the contrary, attempts are often made to manipulate the fish population in a reservoir in order to improve water quality (see Box 4.2).

We then divide running waters in Europe into several fish zones (Aarts & Nienhuis, 2003): trout, grayling, parma, and bream. We also sometimes refer to the first two zones as trout waters. The trout zone is located in the upper reaches of rivers; these are of high altitudes in mountainous fast-flowing streams with good oxygen saturation, but not very nutritious, supplied mainly with material from the vicinity of the river. The dominant fish found are trout and common bullhead. The grayling zone is represented by slower flowing sections of the upper reaches of rivers. Compared to the trout zone, it is more nutritious, still well saturated with oxygen; it is home to many species of invertebrates, and the dominant fish species is the grayling.

The barbel zone represents fast-flowing sections of the middle reaches of rivers in the foothills and their transition to the lowlands; the bottom of the flow is gravelly to sandy, the water already contains more organic matter and sludge and the oxygen content fluctuates. The dominant species here is the common barbel.

The bream zone occurs in slow sections of the middle and lower reaches of rivers, in the lowlands of natural rivers with numerous meanders, creating a gradient from stagnant to running water, the bottom is muddy, often freezing in winter, and nutrient and sludge content is highest here. This is sometimes mentioned also as a flounder zone, which includes estuaries into the sea, characterized by brackish water.

If we look at where commercial freshwater fishing is located in the world, it is mainly the lower reaches, deltas of rivers, and also the larger lakes. This is also due to the fact that the upper reaches of rivers are not very productive and are of fishing importance rather exceptionally, mainly when harbor few particular species of fish, especially salmon.

The fishing techniques for commercial fishing in lakes are similar to those described at seas. In the rivers, lift nets, gillnets, and various sets of traps are widely used. The fishing techniques in rivers often focus on migrating fish, which can be caught in various traps.

Box 4.2 Bottom-Up and Top-Down Effects in the Aquatic Food Web

The food web in aquatic ecosystems can be affected both by the availability of nutrients and changes in primary production, so-called bottom-up, and by top down by the effect of predators. One example of bottom-up control is the variability of nutrient supplies in areas with upwelling. A change in the amount of water coming from the deep layers of the ocean, due to changes in atmospheric circulation (e.g., El Niño) or changes in ocean temperature (higher ocean surface temperature strengthens stratification and reduces upwelling), leads to changes in the amount of nutrients coming to the surface. This affects the biomass of phytoplankton and subsequently the biomass of zooplankton, which feeds on phytoplankton and, thus, the amount of plankton-eating fish and higher levels of the trophic chain.

Similarly, the development of algae and cyanobacteria in connection with excessive nutrient intake (eutrophication) can affect the food web from below. Eutrophication can significantly affect the community and other influences; some species of cyanobacteria that multiply in this way can be toxic and can cause the death of other organisms higher up in the food web, not always being the primary consumers of cyanobacteria, but often consumers of organisms, which feed on algae. In addition, the death of multiplied algae and cyanobacteria and their subsequent decomposition can cause anaerobic conditions, which can then cause the death of other aquatic organisms. Conversely, some organisms may find refuge in the anaerobic zone from predators. These effects are, therefore, very complex and include both changes due to the amount of food in the food web and indirectly induced changes in living conditions.

In contrast, increasing the predation pressure of predatory fish on zooplankton eating fish, for example, in blind river branches, can lead to a decrease in the number of zooplankton eating fish and an increase in the number of large zooplankton species. These are effective filters and filter out phytoplankton from the water, which leads to a reduction in the amount of algae in the water and improves water quality. The described mechanism is a typical example of top-down control, which is used in many water reservoirs (recreational or water reservoirs) to improve water quality.

These top-down effects in marine ecosystems can be very complex, for example, the removal of peak predators can lead to an increase in mesopredators and this leads to, for example, the suppression of algae-eating fish from coral reefs and the subsequent suppression of coral, which in turn changes habitat quality. Here, these effects can be associated not only with direct effects on the amount of food but also with indirectly induced changes in the environment.

(continued)

Box 4.2 (continued)

Of course, in real situations, these interactions can be even more complex, for example, fishing pressure on one of the competing species can strengthen other competitors. Similarly, the elimination of a predator of one of the competing species may enhance that species. In addition, a number of other abiotic conditions are currently changing in connection with global climate change, such as water temperature and its salinity in the sea, which may again favor or disadvantage individual species.

4.1.3 Most Important Species of Caught Marine Animals

The FAO registers a total of almost 1600 taxa of marine animals that are the subject of commercial fishing; however, it is estimated that over 2500 species of fish and marine animals are caught, where the 20 most abundant species represent 39% of the total catch. These belong to several dominant groups (Table 4.2) (Fig. 4.14). From the order Clupeiformes, we mention two families of pelagic fish living near the seacoast. First is the family Herring (Clupeidae). Herring (*Clupea harengus*) lives in the Baltic and North Seas and off the North Atlantic coast, grows up to 45 cm, lives up to 20 years, feeds on zooplankton and ichthyoplankton, the fattest and, therefore, the tastiest in March and April, and has been an important fish since the Middle Ages. Sprat (*Sprattus sprattus*) occurs in flocks off the coast, often in brackish waters. It grows up to 16 cm, usually lives 4 years, and feeds on zooplankton. Sardine (*Sardina pilchardus*) lives in the subtropics of the Atlantic, Mediterranean and Black Sea, upper layers of coastal water, lives up to 14 years, grows up to 30 cm, feeds on zoo and ichthyoplankton, and is fished all year round. The genus *Sardina* is similar in genus and ecology to the genus *Sardinella*, which lives in tropical and subtropical seas, as is the Sardine California (*Sardinops melanostictus*). Family Anchovies (Engraulidae) such as the Peruvian anchovy *(Engraulis ringens)* is one of the most fished, which lives in schools in the southeastern Pacific at depths of 3–80 m feeding on phytoplankton. Japanese anchovies (*Engraulis japonicus*), which are used to produce fish oil, is also important. In Europe, the anchovy (*Engraulis encrasicolus)* is important, grows up to 20 cm, lives up to 3 years, feeds on zooplankton, lives in the Mediterranean, Black Sea, and Atlantic, in coastal waters up to 400-m deep.

The Gadiformes are a group of benthic fish, of which we shall mention two families. First is the cod family (Gadidae). One of the most commonly caught fish is the Alaska Pollock (*Theragra chalcogramma*)" measures 75 cm, lives in S. Pacific at a depth up to 500 m, and feeds on crustaceans. Another important species of cod is the *Micromesistius poutassou* – blue cod, which lives in the North Atlantic, but extends into the Mediterranean Sea. Cod (*Gadus macrocephalus*) grows to a maximum of 119 cm and weighs 22 kg, lives up to 18 years, at depths of 0–1250 m, at temperatures of 6–9 °C on the Pacific shelf, and feeds on fish, octopus, crustaceans, etc. Atlantic Cod (*Gadus morhua*) grows to a maximum size of 200 cm, weighs 96 kg,

Table 4.2 Overview of the most frequently caught taxa of marine animals; according to FAOstat data for 2016, the catch is in thousands of tons, representing total catch of all marine animals, not a sum of most frequently caught

Scientific name	Common name	Catch
Theragra chalcogramma	Alaska pollock	3476.15
Engraulis ringens	Peruvian anchovies	3192.48
Katsuwonus pelamis	Striped tuna	2829.93
Sardinella spp.	Sardine	2289.83
Trachurus spp.	Mackerel	1743.92
Clupea harengus	Herring	1639.76
Scomber japonicus	Japanese mackerel	1598.95
Thunnus albacares	Yellowfin tuna	1462.54
Gadus morhua	Cod	1329.45
Engraulis japonicus	Japanese anchovies	1304.48
Decapterus spp.	Crane	1298.91
Sardina pilchardus	Sardine	1281.39
Trichiurus lepturus	Atlantic lace	1280.21
Micromesistius poutassou	Blue whiting	1190.28
Scomber scombrus	Mackerel	1138.05
Scomberomorus spp.	Mackerel	918.97
Dosidicus gigas	Peruvian squid	747.01
Nemipterus spp.		683.21
Brevoortia patronus	Mexican menhaden	618.72
Sprattus sprattus	Sprat	584.58
Portunus trituberculatus	Crab (Japanese blue crab)	557.73
Total		79,276.85

lives for 25 years, is benthopelagic at depths of 0–600 m, lives in coastal waters to the Atlantic shelf at temperatures up to 15 °C, and feeds on crustaceans and other benthos. Second is the family of hake (Merluciidae), of which, for example, the European hake (*Merluccius merluccius*). It lives on the bottom at depths up to 600 m, with a preference for muddy bottoms.

The largest fish order is the ray-finned fish (Perciformes), which includes a number of representatives from the top 20 animals such as the Family scombrids (Scombridae). Mackerel (*Scomber scombrus*) grow up to 50 cm, live up to 20 years in the S. Atlantic, are fast-floating fish in the open water, form shoals at the surface, drive deeper in winter, and eat crustaceans, mollusks, and small fish. Other significantly hunted species are the Japanese mackerel (*Scomber japonicus*) and are representatives of the genera *Scomberomorus* spp. and *Trachurus* spp. Several genera of Tuna also belong to the same family. Tuna are large pelagic fast-floating fish which, thanks to a special system for heating blood entering the muscles, leaving the muscles with blood, are partially warm-blooded. This gives them the ability to swim quickly and at the same time gives their muscles a dark color. The most commercially important are yellowfin, skipjack, and bluefin tuna. *Thunnus albacares*

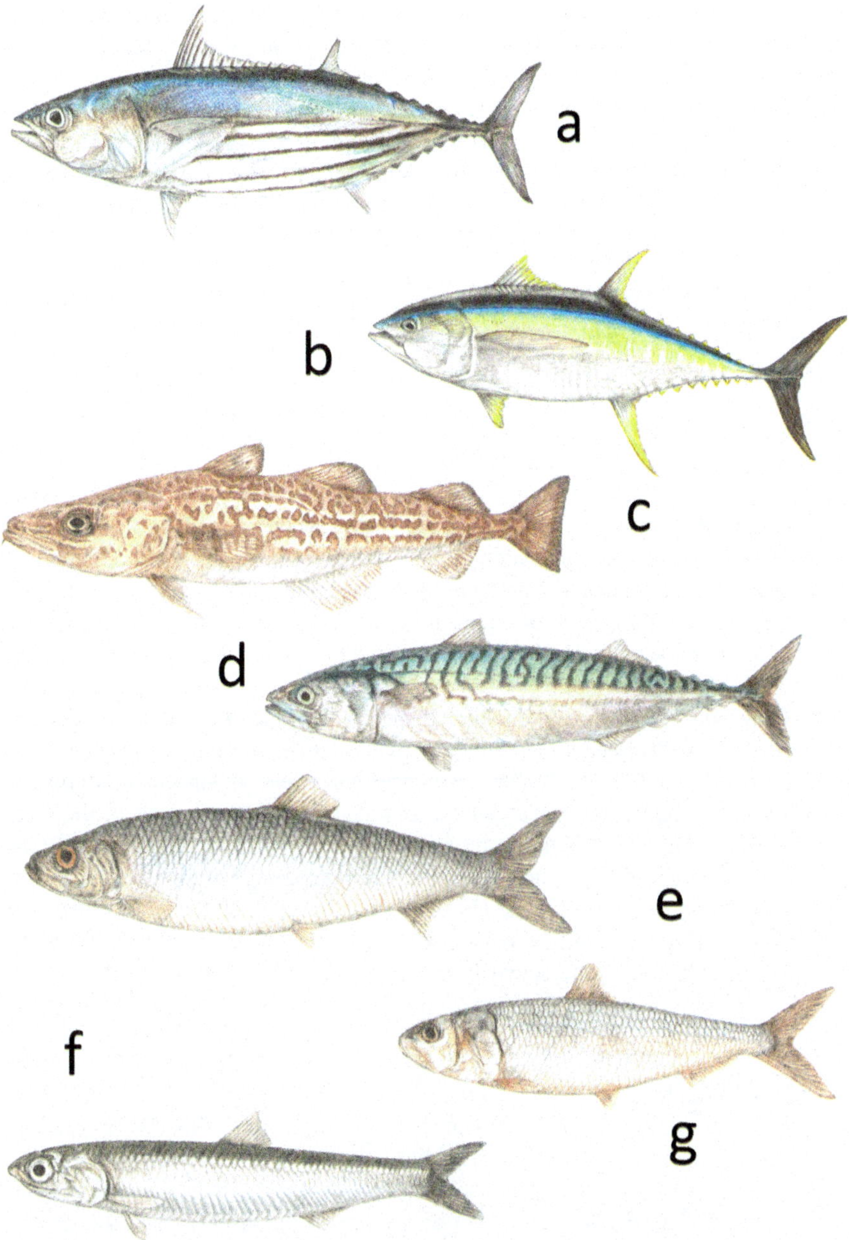

Fig. 4.14 Main species of fish that are subject of commercial fisheries, (**a**) Striped tuna (*Katsuwonus pelamis*), (**b**) yellowfin tuna (Thunnus albacares), (**c**) cod (*Theragra chalcogramma*), (**d**) mackerel (*Scomber scombrus*), (**e**) herring (*Clupea harengus*), (**f**) sardine (*Sardinella* sp.), (**g**) Peruvian anchovy (*Engraulis ringens*)

yellowfin tuna is almost endangered, occurs in most subtropical and tropical seas, and normally weighs about 60 kg. Skipjack tuna (*Katsuwonus pelamis)* is most often caught in the tropics and subtropics, grows to 115 cm, and weighs 35 kg. Northern bluefin tuna (*Thunnus thynnus*) reaches a size of up to 4.6 m, up to 640 kg, and lives up to 32 years. It makes great migrations for food, lives in the North Atlantic from Norway to North Africa and the Mediterranean. Their food consists of crustaceans, cuttlefish, and fish (mackerel, herring). Some species of tuna have been greatly decimated by fishing; these include bluefin tuna (*Thunnus thynnus*), Southern bluefin tuna (*Thunnus maccoyii),* and bigeye tuna (*Thunus obesus*). Tuna are similar to bonito (*Sarda* spp.). Significantly fished is the Largehead hairtail (*Trichiurus lepturus*), which lives in warm waters around the world, has a serpentine body, dorsal fin from head to tail, which is long and thin, is a predator, lives mainly in at depths and rises to the surface at night and includes sailfish (Istiophoridae). Although shrubs are not among the world's largest fish, they are a popular fish for both commercial and sport fishing.

Among the top 20 most hunted marine animals, there are only one mollusk and one crustacean (Fig. 4.15). The mollusk is a Peruvian cephalopod (*Dosidicus gigas*), up to 2-m-long predatory cephalopod; it is native to the South Pacific, but due to global warming, it is invading northern areas, especially California waters. In addition, another 20–30 species of squid are commercially important. Many bivalve mollusks are also caught, but they are often bred in aquaculture, so we will mention them there. Among the arthropods, the Japanese blue crab (*Portunus trituberculatus)* is one of the most abundantly hunted species, living at the bottom to a depth of 50 m in the coastal waters of Southeast Asia, feeding on other crustaceans, mollusks, and fish. Among other crustaceans, various species of shrimp are important, which will be mentioned in the chapter on aquaculture. Large crabs of the North Pacific, such as the Kamchatka crab (*Paralithodes camtschaticus*) or the snow crab (*Chionoecetes opilio*), are iconic. Other iconic crustaceans are lobsters and crawfish, such as the European lobster (*Homarus gammarus*) and the American lobster (*Homarus americanus*), but the most abundant is the Norwegian lobster (*Nephrops norvegicus*) is known as the crawfish or scampi.

Box 4.3 Sharks

Sharks are exploited for their meat, but mostly for their fins (Worm et al., 2013) and also for their cartilage, leather, oil, teeth, gill rakers, and jaws. The price of shark fins, especially dorsal fins, is about US $1100 per kg of fins. As a result, after the catch fishermen often just cut off the fins and leave the shark to its fate. It is estimated that about 100 million sharks are caught annually, which is about 1.4 million tons; however, official statistics are quite incomplete in this regard. Considering that the natural birth rate of sharks is low, fishing leads to a decrease in the population (Worm et al., 2013). In particular, some species of hammerhead sharks are seriously endangered. The reason for the high demand for shark fin is the popularity of shark fin soup. This is

(continued)

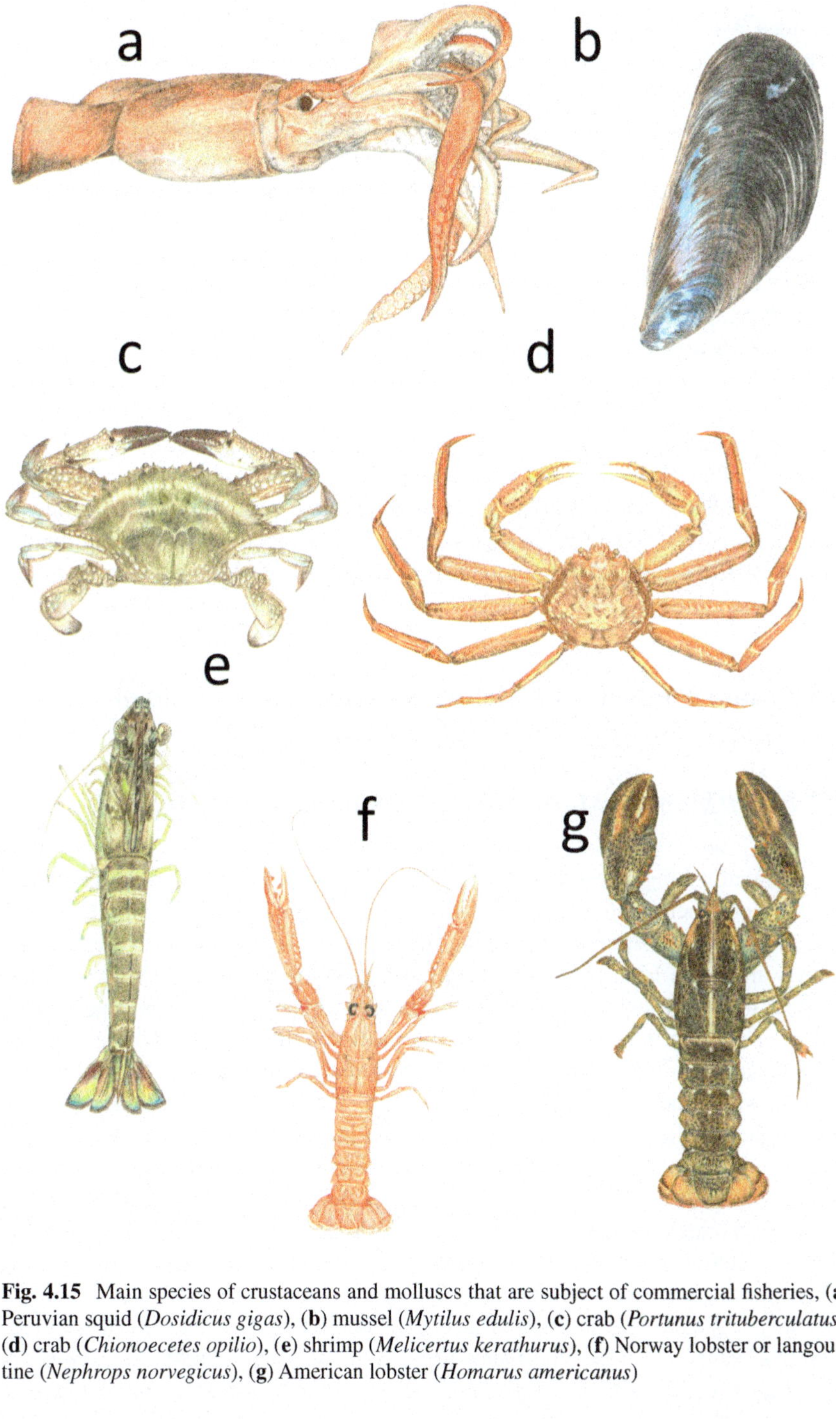

Fig. 4.15 Main species of crustaceans and molluscs that are subject of commercial fisheries, (**a**) Peruvian squid (*Dosidicus gigas*), (**b**) mussel (*Mytilus edulis*), (**c**) crab (*Portunus trituberculatus*), (**d**) crab (*Chionoecetes opilio*), (**e**) shrimp (*Melicertus kerathurus*), (**f**) Norway lobster or langoustine (*Nephrops norvegicus*), (**g**) American lobster (*Homarus americanus*)

Box 4.3 (continued)

basically a chicken broth in which shark fins are cooked. The popularity of the soup in China dates back to the time of the Empire. There are beliefs about the health benefits of shark soup consumption, and it contains symbolism of victory over the shark, making it one of the symbols of power and success. Its consumption increased with the economic growth of China and some other Asian countries. Shark fin soup became one of the ways to demonstrate power and wealth; thus, it was often offered at various banquets and receptions for both the government and large corporations. This has significantly increased the demand for shark fins. Media coverage of the problem is leading to gradual efforts to remedy the situation. In 2012, for example, the Chinese government banned the serving of shark fin soup at official banquets. The removal of sharks from the ecosystem is leading to an increase in mesopredators such as groupers and moray eels. The growth of mesopredators suppresses herbivores and fish that care for corals. The decline of herbivores leads to an increase in algae, with the corals not being able to compete, and this leads to coral decline. Finally, corals or underwater grasslands degrade, fish abandon them, and great harm is done to commercial fisheries (Ruppert et al., 2013).

4.2 Main Methods of Catching Fish and Other Animals and Their Environmental Impacts

4.2.1 Fishing for Marine Fish, Crustaceans, and Mollusks

Commercial fishing takes many forms (Box 4.4). There is a plethora of techniques used in commercial fishing; we will mention only the most important ones (Fig. 4.16). The most common fishing method is the use of various types of nets. Trawls are in the form of a large open bag that is towed forward by one or more vessels. In large trawls, the diameter of the open mouth of the trawl may exceed 100 m. The net can be towed either in the water column or on the bottom. Pelagic fish such as anchovies, tuna, mackerel, herring, and sardines are caught in the water column while cod are caught near the bottom. Trawls are very common in areas with high schooling densities of target fish species, especially in areas with coastal runoff (Fig. 4.17). Pelagic trawls are generally larger than benthic trawls. The trawl is nearly nonselective; anything in front of it is caught in the net and does not pass through the mesh, so the trawl has a significant amount of bycatch (Box 4.5). In addition, benthic trawls play an important role in bottom degradation. Especially where there are important structures for organisms, such as coral reefs, bottom-dwelling organisms are trapped or at least disturbed, and in the case of soft mud bottoms, sediment is scraped off and stirred up in the water column. An example of a species severely damaged by benthic trawls is the cold-water coral *Lophelia pertusa*.

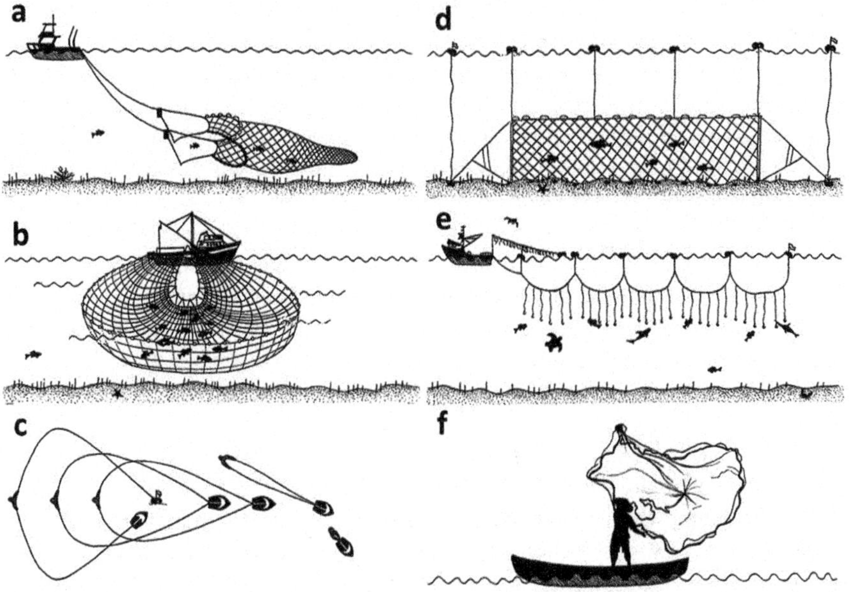

Fig. 4.16 Main types of fishing gear: (**a**) trawl, (**b**) purse seine, (**c**) Scottish net, (**d**) gillnet, (**e**) longline, (**f**) casting net

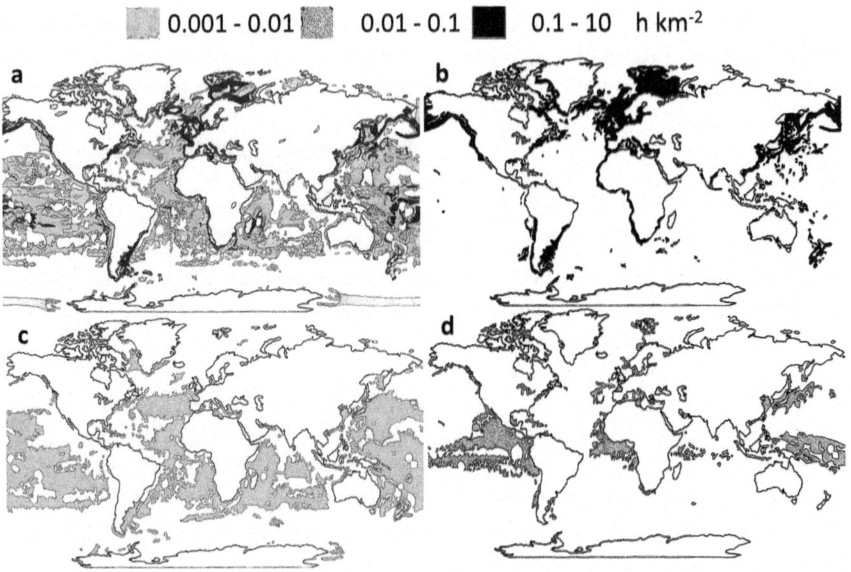

Fig. 4.17 A map of the total fishing effort in the world's seas as a whole (**a**) effort using trawls (**b**) and longlines (**c**) and purse seines (**d**), according to Kroodsma et al. (2018)

Dredging is similar to benthic trawling. The nets are smaller but more robust and they are used for shellfishing and have a similar effect on the bottom as benthic trawls, perhaps even more intense but more spatially limited.

Another type of net is the purse seine (lampara), which is used to catch fish that form shoals at the surface, such as tuna and marlin, sardines, and herring. The principle is to cast a long net around the shoal or fishing area, which the boat circles as it sets the net. This creates a wall of net from the surface (floats are attached to the upper line) down (the bottom line is loaded by weights). When the circle is complete, the bottom edge of the net drops and it is closed. The circle completed by large commercial purse seines can be up to 250 m in diameter and the net can be up to 200 m deep. Purse seines are somewhat more selective than trawls, especially when used on selected schools of fish, but can catch the species that accompany these schools. For example, dolphins are a common bycatch when fishing for tuna in this way.

Beach seines, used at the shore, are similar to purse seines. The net defines an area near the shore, which then gradually closes toward the shore.

At first glance, the hybrid between a purse seine and a towed net are nets of the Scottish seine, Danish seine, and anchor seine types. At first sight, they look like a trawl with long side wings or a trawl with a bag in the middle. They are used similarly to a trawl in coastal waters where the movement of tidal fish is used.

Another type of net we will mention are gillnets (tangle nets, trammel nets), they can be stationary or floating. Gillnets form a vertically oriented net wall in the water, the upper part of which has floats, and the lower part has weights. They can be located near the surface, in the water column, or on the bottom. They are designed so that when the fish tries to overcome this obstacle, it usually gets caught behind the gills or other protrusions on the body. Gill nets have a high degree of size selectivity. In terms of frequent bycatch, they pose a threat to sea turtles and mammals.

We will mention two other types of nets used for fishing in coastal waters. This includes various types of lift nets. These are flat nets attached at the corners of a structure by which they are lifted. The nets are submerged in the water or placed on the bottom. A bait can be placed in the middle of the net. When the fish swim over the net, it is lifted. Another type of net we will mention is the cast net. This is a circular net whose open end can be scrubbed into a large bag. The circle is equipped with weights around the circumference. The fisherman throws the net in front of him so that the circle opens like a large disc. When its edges sink deep enough, the fisherman closes the net and pulls it out.

In addition to nets, the most commonly used fishing technique is the so-called longlines. These are long lines on which shorter rigs with hooks are tied, which stretch in lengths of several kilometers behind the ship. They can be either epipelagic (e.g., for tuna fishing) or demersal for bottom fishing (e.g., for cod fishing). The hooks contain bait. They are relatively selective, but waterfowl, especially albatrosses or turtles, are a common bycatch of pelagic longlines. Fishing lines use a similar principle as angling, which is used only rarely in industrial fishing, for example, when fishing for some pelamid. Fishing lines are a very common fishing technique when fishing in international waters outside the economic zones of individual states (high seas) (Fig. 4.17). Economic zones are described in Sect. 4.2.6.

Fig. 4.18 Fishing boats (**a**) small fishing boat in Thailand, (**b**) small boat to tow Thailand, (**c**) whaling boat Iceland, (**d**, **e**) trawler Iceland, (**f**) long lines

When fishing for fish and other aquatic animals, various types of traps are used. These are designed to allow easy entry into the trap, where bait is usually placed, and, conversely, difficult escape. They are used to catch fish, but often also to catch crustaceans. The combination of bait, entry mechanisms, and selective escape holes makes them a relatively selective fishing gear. Sometimes these traps are designed to take advantage of fish migration.

A boat is needed to operate many of the fishing gear listed above. There are about 4.6 million fishing boats in the world (Fig. 4.18), most of them are small boats, 75% are smaller than 12 m, only 2% of them are larger than 24 m and only about 60% of them are equipped with an engine. Of course, different types of fishing vessels are needed for different types of fishing. Laying and retrieving smaller gillnets or traps is certainly possible even from ships without an engine. On the contrary, considerable power is required to tow the trawl. Small ships usually operate in coastal waters. In contrast, large ships operate on the high seas and are usually used for industrial fishing. Of course, the line between industrial and subsistence fishing is not entirely clear; we can understand them as two opposite positions of economic and technological approaches to fishing. They are very often combined during the development of fishing activities in a given locality (see Box 4.4).

Traditional artisanal fishing is based on the work of individuals or small family businesses. Production is usually used for own consumption or for the local market. Operators are often members of coastal or island ethnic groups. Fishing techniques include angling, the use of spears, harpoons, bows and arrows, traps, gillnets, throwing nets, or lift nets. This type of fishing often uses no boats or small boats only, often without an engine. It is, therefore, a technically and economically less demanding way of fishing, based on local knowledge of the environment. It is the dominant form of fishing in many developing countries.

Table 4.3 Comparison of selected parameters of subsistence and industrial fisheries in West Africa according to Belhabib et al. (2018)

Parameter	Self-sufficient fishing	Industrial fishing
Number of ships	252,000	3300
Length of ships (m)	2–24	10–100
Annual average power of the ship kW day^{-1}	14,748th most common	54,676th most common
Average number of crew members per ship	7.4	15
CPUE kg kW^{-1} day^{-1}	213	1.5

In contrast, industrial fishing uses larger vessels with high power, using fishing methods that allow this high power for trawling, the use of large purse seines, etc. State-of-the-art, highly sophisticated technical aids (sonar, satellite and aerial images, etc.) are used to search for fish. This is an economically very demanding activity, so the operators are large companies. The fish caught are usually processed on board or in downstream plants and are destined for the global market. Table 4.3 compares the parameters of industrial and subsistence fishing in West Africa (Belhabib et al., 2018), and it can be seen from Table 4.3 that the catch per kilowatt of energy put into propulsion is lower in industrial fishing than in subsistence fishing. It is not surprising that inshore (mostly subsistence) fisheries have a higher energy return (EROI = 1) than, usually industrial, offshore fisheries (EROI = 0.06) (Lough, 1999).

Box 4.4 Economic Development of the Fishing Locality

Fishing is economically and technically demanding. Industrial fishing, in particular, requires considerable investment and working capital. It turns out that the development of the fishing industry has certain common features in individual localities (FAO, 1984). *Preindustrial phase*: In this phase, recreational or subsistence fishing takes place, indicating the potential of a particular place to catch a particular species of fish. To ensure that such an opportunity is exploited economically, people should be able and willing to bring risk capital to start such activities. If they are commercially successful, their success will attract additional funding, whether from government, private, or international sources, which will put such an enterprise into a phase of expansion. For a number of reasons, first because decisions to invest in fishery development are made independently but in the same short time by a number of individual investors, and second because there is a time lag (sometimes significant) between the investment decision and the physical acquisition of new investments in the form of new vessels or processing facilities, fishing, and processing capacity is often oversized. Moreover, determining the correct fishing effort is not straightforward, as fish numbers are subject to a number of random fluctuations and more. In the subsequent stabilization phase (maturity

(continued)

Box 4.4 (continued)
phase), the number of fish in traditional localities begins to decline. This decline may be partially offset by improved knowledge of other fisheries, seasonal occurrence of fish, etc. The long-term relative sufficiency of fish or economic competition from financially strong entities may translate into a decline in prices at the end of this phase. This may have negative consequences for subsistence fisheries. At the end of this phase, catches by industrial fisheries decline significantly due to overfishing. The cycle continues into a phase of decline (senescent phase); this stage is the result of increasing the cost per unit of catch and low profit, thus lack of funds for the renewal of the fleet of old vessels are gradually worn out, and this leads to the decline of industrial fishing. A declining catch at this stage can lead to higher prices. This can lead to an increase in profits and, thus, in the capital of subsistence fishermen and can also lead to greater investment in subsistence fishing. This improved economic performance of small-scale fishers, the partial recovery of fish stocks due to declining fishing effort, may then lead to another investment cycle as the more successful small-scale fishers reinvest in larger industrial vessels in response to higher prices and more fish.

4.2.2 Whaling

Unlike fishing, where we try to catch several fish at once and often catch whole schools, with whales we catch single individuals. Whales are large, relatively long-lived, and reproduce slowly. This makes their populations vulnerable to hunting, and also makes them slow to recover after being overfished. In addition, whales make regular migrations, breeding in warmer waters and feeding in cold subarctic or subantarctic waters. Therefore, they occur at predicted times in predicted locations, which makes them easier to hunt.

The main catches are the minke whale (*Balaenoptera acutorostrata)*, giant sperm whale (*Physeter macrocephalus*), and gray whale (*Eschrichtius robustus*). The products and raw materials obtained by whaling are meat, fat, whalebones or spermaceti, and ambergris (sperm whales). The meat of whales is similar to the meat of other mammals; the fat of whales being used for lighting. The skeletons were used to make umbrellas, corsets, and the like. Spermaceti and ambergris were used in the cosmetics industry.

As mentioned earlier, whales are hunted one at a time, and the technique of hunting is based on finding the prey, chasing it, harpooning it, and then pulling it onto a boat or dragging it ashore and processing it. Originally, harpoons were hand-thrown from smaller boats, which involved considerable risk for hunters, and many whale species were very difficult to hunt this way. The breakthrough was the invention of the harpoon cannon and explosive harpoons, which made hunting very efficient. In the 1930s, 50,000 whales were caught per year. In 1946, the International

Commission for the Regulation of Whaling was established, in 1949, the International Whaling Commission, and in 1986, a moratorium was placed on commercial whaling, which has been extended to the present day. Current whale hunting only takes place in a few countries, and industrial whaling is most notable in Norway and Japan. Japan has long formally adhered to the moratorium and hunts about 600 whales annually for so-called research purposes. However, the meat from this hunt was commercially processed and sold; it was rather a cover for commercial hunting. Now, Japan has given its consent to the moratorium and it is questionable how this will be reflected in the further development of hunting. Norway caught 1286 whales in 2011 and Greenland 175; other countries involved in whaling are Canada, the Faroe Islands, Iceland, Russia, and the USA. In a number of countries such as the USA, Russia, and Canada, hunting is allowed mainly or only to Inuit indigenous communities for whom whaling is part of traditional culture and who fish in a way close to traditional methods. A moratorium on whaling has led to a gradual increase in populations, although not in all species and in all places of their original occurrence (Lotze et al., 2011).

4.2.3 Freshwater Fishing

In 2016, inland commercial fisheries amounted to 11.6 million tons. It was dominated by China, followed by India and Bangladesh. In addition to the countries of Southeast Asia, the countries of Africa are also represented in the top ten; of the countries that extend to Europe is only Russia. From the above, it follows that commercial fishing of wild fish stocks in freshwater (unlike fish farming) is not of great economic importance throughout Europe. Freshwater fishing is more widespread in some large lakes and in the lower reaches and deltas of larger rivers.

4.2.4 Environmental Impacts of Fishing

The impact of fishing on fish stocks can be divided into three categories: (1) the impact on harvested stocks, (2) the impact on stocks that were not fished but become incidental catch, and (3) the impact on ecosystems, which may feedback to fish stock.

Man is a very successful predator, as mentioned in the chapter on wildlife hunting. A significant part of the catch of most predators consists of the young. On the other hand, fisheries specialize in catching adults. Using commercially important fish species as examples, Darimont et al. (2015) showed that humans catch far more adult fish than their main natural enemies (Fig. 4.19). In addition, humans have been able to use some fish antipredation strategies to enhance hunting success. A typical example is the formation of schools or shoals. Swimming synchronously in a school protects against many predators because it is very difficult to focus on one target and catch it, which significantly increases the chances of survival for fish swimming in schools. However, there are specialized predators that can outsmart this behavior.

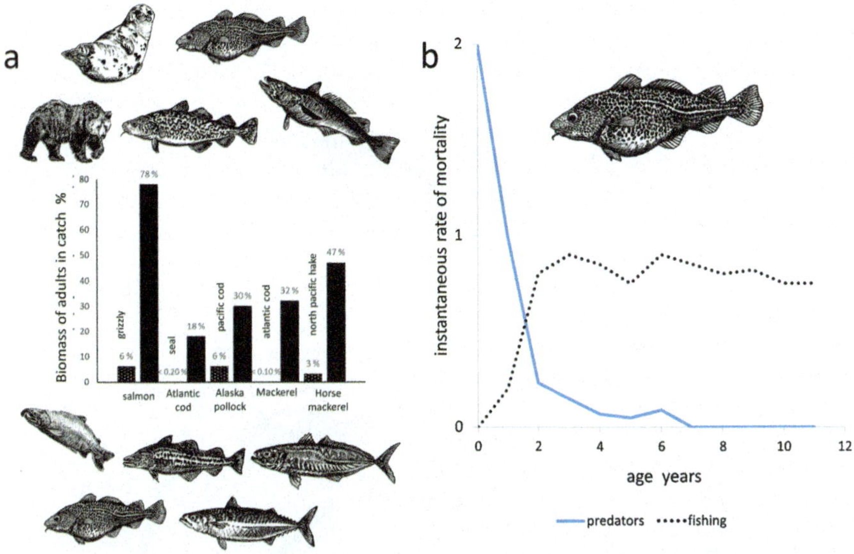

Fig. 4.19 Effect of predators and fishing on fish mortality. (**a**) The share of adult fish in the catch by humans (black) and by the most important natural predators (pictured above graph) of selected species of fish prey (hatched) (prey species depicted below graph), according to Darimont et al. (2015). (**b**) Instantaneous rate of mortality of Atlantic cod, given by natural predators, or fishing according to Kolding and van Zwieten (2011)

These include people who fish with nets. Nets can catch an entire school at once, so this antipredation strategy is rather disadvantageous for fish. Thus, if we remove individuals from the population faster than their replenishment (recruitment), it can reduce the size of the population and overfishing occurs. We know that there is overfishing from the fact that the catch per unit effort has decreased. Fish respond to predation pressure not only by reducing stock size but also individuals reach sexual maturity at a younger age and smaller size, growth rate slows and leads to a loss of genetic diversity (Jørgensen et al., 2007) (Fig. 4.20).

Fishing faster than the population is replenishing will result in a reduction in population size. If fishing pressure continues, the limit of the smallest viable population may be reached. At this limit, any disturbance can lead to the collapse of the entire fish population. The population is, as we said above, exposed to a number of natural influences that affect its survival. If the stock is small, then any significant loss threatens its existence. An illustrative example is the impact of El Niño on the populations of Peruvian anchovies, one of the most abundant species caught (Box 4.6). The anchovy population depends on rich sources of plankton, which depends on the abundance of nutrients from the offshore upwelling in the Peru area. The occurrence of the El Niño climatic phenomenon disrupts this upwelling, which brings nutrient-rich water and, thus, leads to a reduction in fish food.

Even if we stop fishing for the overfished population, pushed close to the minimum sustainable size of the population, its recovery to its original size may take a long time or may not be possible at all. Recovery is particularly slow in large

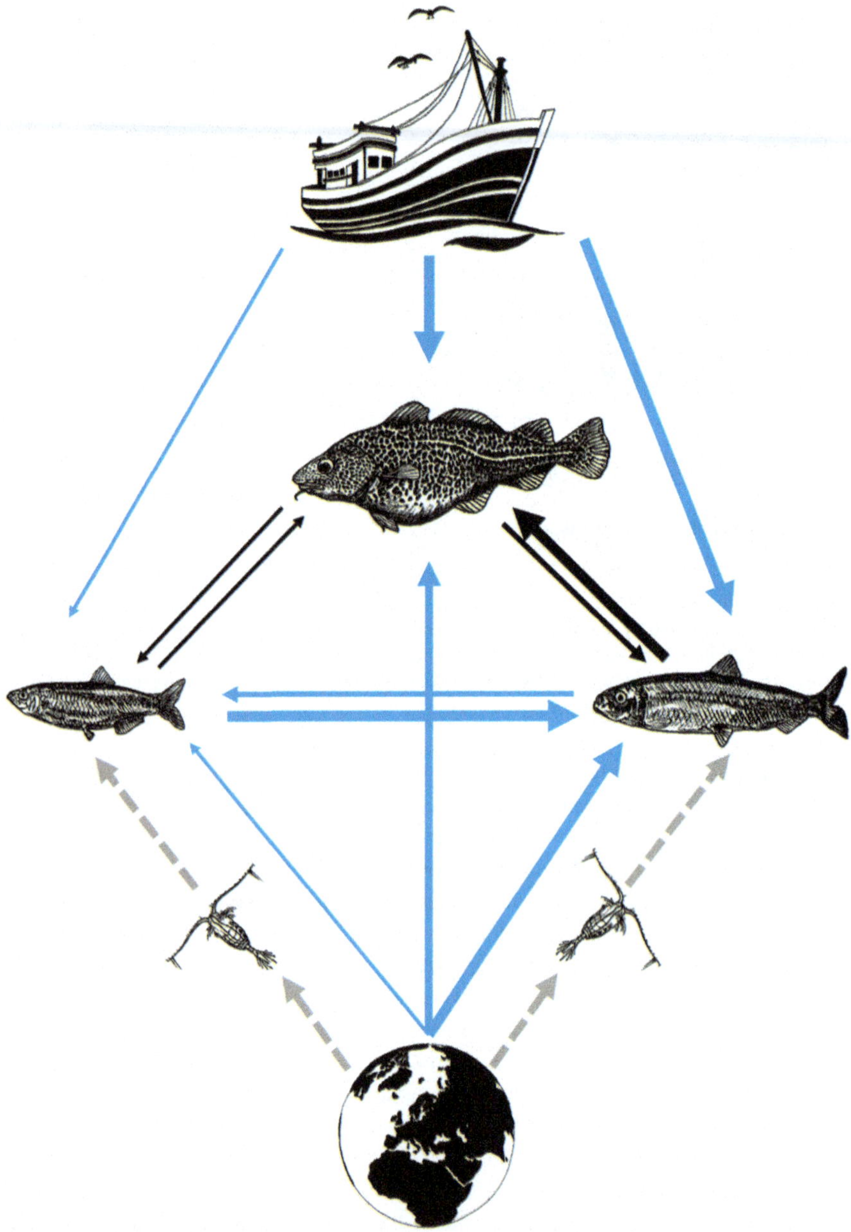

Fig. 4.20 Relationships between fishing, cod (top), sprats (left), herring (right), and global climate change (globe). The numbers indicate correlations between phenomena, negative correlations are blue, positive correlations are black, adjusted according to Jørgensen et al. (2007)

predatory fish (tuna), which take a long time to reach adulthood and, therefore, have little recruitment. In addition, it may happen that after the removal of a certain species from the community, the whole structure of the community will shift so that the restoration of the original populations is no longer possible (ecosystem shift).

An example of an ecosystem shift in commercially important fish is the suppression of cod in favor of sprat in the Baltic Sea. Massive fishing pressure has suppressed not only cod stocks but also herring stocks, the main food of cod. In addition, the increased nutrient supply in the Baltic Sea accelerated the growth of phytoplankton while at the same time promoting the development of anoxic zones. On the one hand, anoxia allowed parts of the phytoplankton to hide from predation pressure from zooplankton, and on the other hand, it had an indirect negative effect on cod. Under normal conditions, most cod eggs remain floating in the water above the anoxic zone after spawning. The decrease in salinity caused them to sink deeper, and the increase in anoxia landed most cod eggs in oxygen-deficient water where they could not develop. In contrast, the sprat population benefited from the increase in temperature and changes in zooplankton. This, in turn, further increased competitive pressure on herring, the food of cod, and worsened the living conditions of the cod fry. These complex interactions have led to the collapse of some cod populations in the Baltic Sea (Lindegren et al., 2009). Moreover, this situation may further deteriorate if temperatures continue to rise and salinity decreases. In order to restore cod stocks, it is important to reduce not only the cod fishery but also the herring fishery and to limit the input of nutrients to the Baltic Sea.

The FAO estimates that 33% of the world's fish stocks are overfished (caught more intensely than their stocks can replenish). The worst situation in this respect is in the Baltic and Mediterranean Seas, which are closely followed by the South-East Pacific (more than 60% of the population is overfished in these areas).

In addition to the target prey, unwanted prey or so-called bycatch can be affected in a similar way, which in many cases can represent a significant percentage of caught organisms. This unwanted catch can be not only other species of fish or fish of inappropriate sex or size and age but also invertebrates and other vertebrates such as turtles, birds, and mammals. For example, Read et al. (2006) estimated that fishing in the United States inadvertently catches over 6000 marine mammals each year. For more, see Box 4.5. To unintentional catch is related so-called ghost fishing. It is the catch of fish or other organisms in forgotten or lost fishing tackle. Many traps, nets, or fishing lines can hunt or damage marine organisms long after they have been lost or discarded.

Other fish species for which target fish species are prey are also threatened by overfishing. Cury et al. (2011) showed that the nesting success of bird populations is dependent on fisheries and on declines of their prey species. This response is very asymmetrical. With decreasing prey abundance (see Fig. 4.21) breeding success decreases first only slowly but then curve breaks sharply, for all monitored species, at approximately one-third of the maximum population size of the given species of fish prey. In other words, if we want to maintain nesting success and thus, bird populations dependent on fish, the populations of these fish must not fall below one-third of their maximum abundances (or otherwise below one-third of the population density achieved when the carrying capacity of the environment is filled).

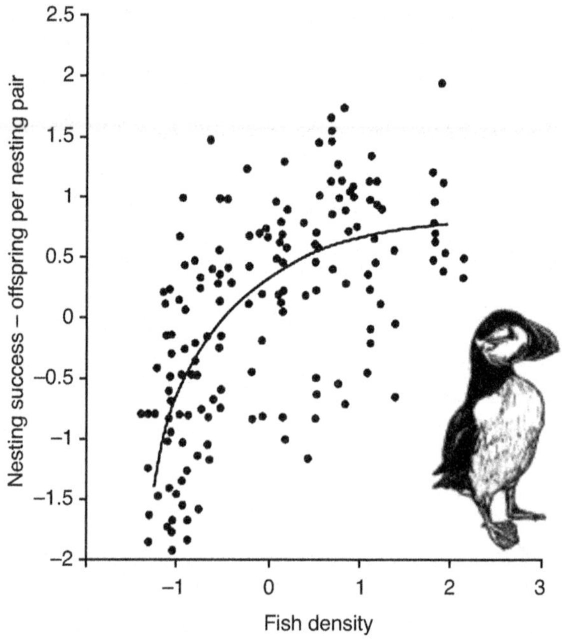

Fig. 4.21 Nesting success of birds in the North Sea as a function of fish abundance (both normalized as the number of standard deviations from the mean), according to Cury et al. (2011), modified

Finally, there is the impact on ecosystems. This stems from the effects of the removal of a given species from the ecosystem on the entire food web, see the ecosystem shift described above. Another example, described in the Box 4.3 sharks, shows how the removal of sharks as top predators, supports mesopredators and leads to the collapse of fish stocks that help keep coral reefs alive.

In addition to this effect via the trophic cascade, a number of fishing activities directly damage ecosystems. In particular, a number of bottom fishing and trapping activities play a role here and can cause significant mechanical damage both to the population of benthic organisms and to the various bottom structures on which the life of benthic organisms depends.

Another, indirect aspect of fisheries may be the contribution to sound pollution. Although also in this case, sound pollution in the ocean had many others, more serious sources such as container ship traffic, mining, or military activities. Nevertheless, sound propagates in seas on large distances and sound pollution can affect many marine mammals and fish, which, in turn, can cause a reduction of their population (Aguilar Soto et al., 2006; Warner, 2008).

Box 4.5 Bycatch and Its Reduction

It is estimated that bycatches worldwide account for about 35%, catches taken from fishing vessels in ports. This bycatch is usually dead or badly damaged and thrown back into the sea. Table 4.4 shows the bycatch ratio for selected groups of caught marine animals and Table 4.5 shows this ratio for the main fishing methods. We can also get a visual idea of the bycatch rate in Fig. 4.22.

(continued)

Box 4.5 (continued)

Table 4.4 Ratio (multiple) of bycatch to main catch in selected FAO catch group

Type of main catch	Bycatch ratio to main catch
Prawns	5.20
Crabs	2.49
Eels	0.84
Halibut	0.75
Perch	0.63
Lobsters	0.55
Horse mackerel, mullet	0.28
Cod	0.20
Tuna	0.18
Cephalopods	0.09
Salmon	0.05
Mackerel	0.03

Table 4.5 Ratio (multiple) of bycatch to main catch for selected FAO fishing methods

Method of fishing	Bycatch ratio to main catch
Shrimp trawls without bycatch reducing devices	14.7–10.3
Trawls for benthic fish	5.3–2.1
Pelagic trawls	0.1–1
Purse seines	0.8–0
Danish–Scottish networks	0.5–0.3
Fishing lines	1.1–0.2
Various traps	1.8–0.3

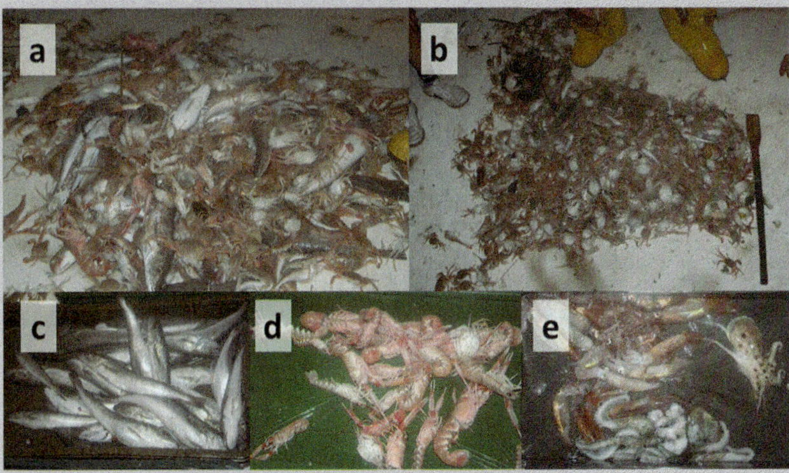

Fig. 4.22 Example of bycatch from benthic trawl in Croatia: (**a**) original total catch, (**b**) bycatch after sorting of commercially interesting catch, and (**c**–**e**) sorted, commercially interesting catch

(continued)

Box 4.5 (continued)

From Tables 4.4 and 4.5, it is seen that even if the variability is high, bycatch is on average 35% of the main catch. In other words, the total number of organisms caught (and killed) is more than a third higher than reported as the catch. If a species is caught intensively as a bycatch, it may be overfished, similar to fish that are the subject of commercial fishing. An example of such a situation is the decline of stingrays (*Raja batis*). This can lead to a change in the composition of the community. In addition, fish that are damaged as bycatches can be an important food for commercially important fish. Furthermore, fishermen often discard uninteresting sizes of market fish, which contributes even more to the overfishing of these fish stocks.

The undesirable consequences of fishing unintended, secondary prey have led to efforts to reduce it (Fig. 4.23). These efforts are basically in three directions: reducing bycatch, increasing the survival of bycatch, and, where appropriate, if bycatch are economically viable fish, utilizing those catches. Bycatch reduction techniques depend on the type of bycatch and the fishing method used. One of the largest bycatches occurred in shrimp trawling. In the 1990s, catches of marketable shrimp accounted for only 16% of the total catch, with the remainder being bycatches, especially fish (67%). This has led to the development of a number of mechanisms to reduce bycatches. In the case of shrimp trawls, these are the various openings in the net to allow fish or turtles to escape, to which the fish or turtles can be guided by means of coarse transverse ribs through which the shrimp pass easily.

The bycatch of dolphins and other marine mammals when fishing for tuna and other fish with purse seines also attracted a lot of attention. A number of techniques have been used to reduce their unwanted catches. These include acoustic jammers located on nets to discourage cetaceans from entering nets or the use of Medina panels (parts of the net where the cetaceans may occur after the net is closed are thicker and cause a drop of the upper line and allow the escape of animals), and back-down procedures for casting nets or using special boats to take dolphins out of nets before closing. Ultrasonic scarers are also placed on fishing lines. Holes in the upper part of the net with a guide grid are also used for trawl nets. The possibility of producing nets from materials detectable by echolocation of cetaceans has also been tested (Hamilton & Baker, 2019).

Another significant kind of unwanted bycatch is a catch of albatrosses and other aquatic birds, which are cast on bait placed on fishing cords (longlines). There another cord with tassels is used to discourage the birds from approaching the bait before it sinks out of their reach. Other strategies that can help are increasing the weight on the tow rope so that it sinks faster and fishing at night when no seabirds are active. Also, the casting of the bait from the side of the boat is effective, again equipped with another line with fringes, which will allow the bait to sink before the boat passes it. Gilman et al. (2008) showed that a combination of these approaches can significantly reduce bycatches of albatrosses and other seabirds.

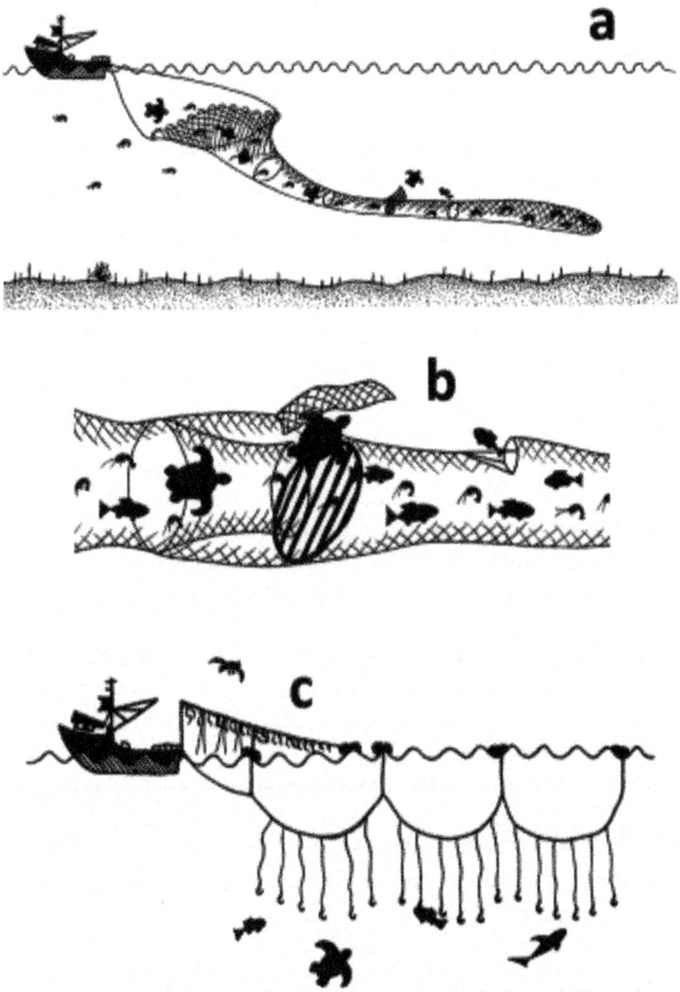

Fig. 4.23 Different methods of reducing bycatch, (**a**, **b**) using different escape holes in trawls, (**c**) using fringes around fishing lines to prevent birds from collecting bait before immersing it in depth

4.2.5 Sport Fishing and Its Environmental Impacts

Sport fishing may seem marginal, but in many inlands, water is not used for aquaculture; it is an important component of human fishing effort. Unlike industrial fishing, which focuses on the economic value of the catch, in sport fishing, in addition to the catch itself, is a significant fishing experience that many people do not hesitate to spend a lot of money on. Like hunting, angling is, therefore, a recreational activity with considerable economic potential. Many water bodies are, therefore, operated exclusively for sport fishing. In 2017, sport fishing in the United

States was valued at $4.9 billion in various permits and fees, $23.9 billion was directly related to fishing and fishing expeditions (bait, travel, subsistence, and other expenses), and $57.5 billion represented various purchases of equipment and other fishing-related supplies.

Cooke and Cowx (2004) estimated that recreational fishing may account for about 12% of global fish catches worldwide. Although the share of recreational fishing in global catches does not appear to be large, it can be significant. Recreational fishermen fish in freshwater and coastal sea areas. These areas often represent important habitats for the development of small fish or are important for fish migration. Post et al. (2002) point to a range of freshwater fish stocks declining due to overfishing by recreational fisheries. This has led to a number of regulations limiting the impact of fishermen on fish stocks. In addition to the closed season, when fishing for certain species is prohibited, it is the maximum daily catch or the number of fish a fisherman can keep (bag limit) and the setting of a minimum catch rate or minimum fish size that a fisherman can keep. Van Poorten et al. (2013) have shown that these two techniques can reduce fishing pressure on fish stocks. However, at the same time, comparisons between different US states have shown that these techniques have a more beneficial effect in some states than in others. This is due to the fact that in both cases the regulation can be filled in a way that will increase the mortality of the fish population (Woodward & Griffin, 2003). For example, if the legal quota is increased, fishermen may increase their fishing effort to catch fish of the required size. The difference between states may be due to the fact that people in different states or social groups have different tendencies to comply not only with the letter but also with the spirit of the law.

For many fish stocks, which are a popular target for sport fishermen, the natural replenishment of the stock is not enough to cover the loss associated with fishing pressure and their condition is maintained by stocking artificially farmed fish. The effect of supporting natural populations by stocking them with artificially bred fish can be illustrated by the example of the Czech Republic (Table 4.6).

Fish stocking has a number of pitfalls (Laikre et al., 2010). Nonnative species are often introduced, which can significantly suppress native species. For example, in Table 4.6, we can observe the introduction of two exotic species of American grayling

Table 4.6 Numbers of fish stocked and caught in thousands in the Czech Republic in 2018 according to statistics from the Czech Fishermen's Association

Species	Caught	Stocked
Common carp	945.8	2521.8
Eel	11.3	120.3
Catfish	8.6	37.5
Pike	51.0	693.7
Brown trout	28.7	820.1
Grayling	0.4	309.3
Siven	34.9	51.6
Rainbow trout	119.2	345.7

and rainbow trout in Czechia. Even the introduction of native species is not biological, without a problem. For example, the introduction of hatchery-reared trout has been shown to reduce genetic diversity in wild populations. There are studies showing differences in susceptibility to diseases and parasites, different dietary requirements, different choices of microhabitats, and other behavioral differences between wild and captive individuals. Therefore, the addition of hatchery fish does not necessarily lead to an increase in species abundance in the wild. On the contrary, there are studies showing that the introduction of farmed salmon leads to genetic changes that reduce the viability of such supported populations (Reisenbichler & Rubin, 1999). On the other hand, in the case of fish stocks in the lower parts of rivers, where fish stocks appear to have been enriched with the genetic material of pond stocks for a long time, restocking may have favorable consequences for the maintenance of the stock.

In addition to the targeted introduction of nonnative fish species or genotypes, sport fishing is involved in the adventitious spread of nonnative species of plants and animals that are transmitted by fishing gear, clothing, boats, and other equipment to sport fishermen. For example, in the Czech Republic, fishermen are thought to have contributed to the spread of American waterweed (*Elodea canadensis).* Sport fishing in the United States and Canada has also contributed, among other things, to the mass invasion of European earthworm species into North American forests. European species of earthworms are a popular bait for fishing, and fishermen usually buy them in shops where they are supplied by specialized companies. The part of the bait that fishermen do not use is then released into the wild, and at the end of the last century, there was a mass spread of European earthworm species in North American forests, which completely changed the soil conditions and functioning of local ecosystems.

As in the case of hunting, by describing the possible negative impacts of sport fishing on ecosystems, we certainly do not want to give the impression that the impact of sport fishermen on nature is entirely negative. This is certainly not true, fishing as a hobby attracts a large number of people to the water who are interested in sustaining aquatic ecosystems and that is certainly good for nature.

Box 4.6 Fisheries and Climate Change

Increasing the temperature of seawater has a lot of impacts, from the level of individual fish to complex effects on other environmental factors that affect the entire population and communities. At the level of individual fish, this concerns an acceleration of metabolism or movement, which can affect the success of fish in the community. However, it is generally assumed that with increasing ocean temperature, the biomass of marine organisms will decrease, and this effect will increase at higher trophic levels (Lotze et al., 2019).

The rising temperature of the sea surface is also related to the change in the global atmospheric circulation and the related changes in sea currents. This can have a number of effects, especially in areas with upwelling (Fig. 4.24).

The influence of El Niño, which is increasingly occurring in the context of the global warming process, is examined in detail on the example of the

(continued)

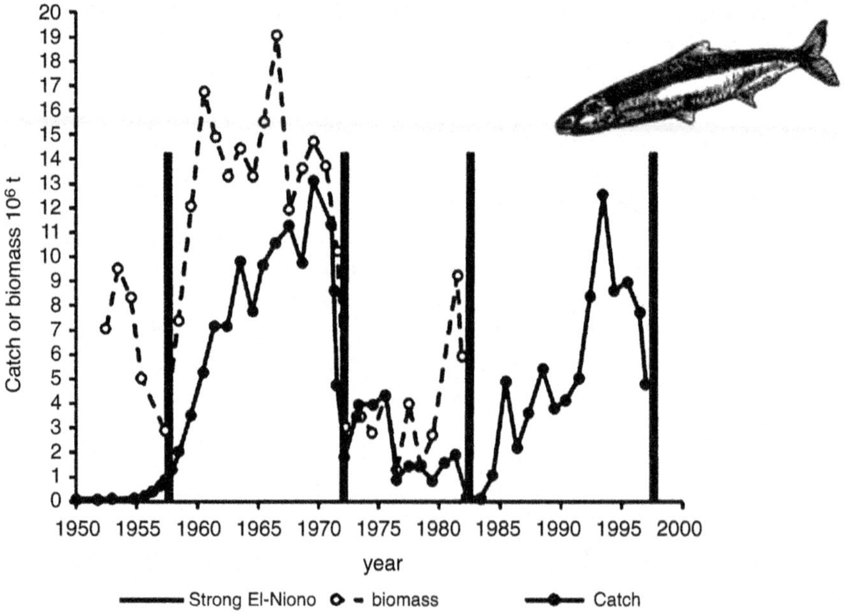

Fig. 4.24 Effect of El Niño on Peruvian anchovy populations according to Klyashtorin (2001)

Peruvian anchovy fishery. Peruvian fisheries have a dominant position in anchovy production. The productivity of this area is conditioned by two sea currents, the Humboldt current and the Peruvian coastal current, which create a coastal upwelling that brings a large amount of nutrients to the surface and, thus, supports the anchovy populations. Anchovy populations are negatively correlated with sea surface temperature. As the sea surface temperature increases, which happens with El Niño, the upwelling decreases with a decreased supply of nutrients. This affects the production of phytoplankton and consequently the abundance of the anchovy population. It must be said that the dependence of Peruvian anchovy and some other fish stocks on seawater temperature is nothing new (Klyashtorin, 2001). As the long-term reconstruction of the anchovy population from marine sediment wells shows, the anchovy population has responded to long-term changes in atmospheric circulation also in the past, but El Niño is now coming more often.

Rising sea temperatures also cause more frequent bleaching of corals and subsequent coral death, which is an important habitat for many fish species. The point is that at high temperatures, the polyps that corals form are unable to retain and expel green photosynthetic symbiotic algae in their bodies. This results in optical whitening of the corals. Polyps provide these algae with most of the energy needed for life; therefore, the coral will soon starve and die. The effect on corals will be further supported by acidification of water and a decrease in pH.

(continued)

Box 4.6 (continued)

Rising temperatures will also allow tropical and subtropical fish to shift to higher latitudes, which can have a variety of effects on local communities. From the point of view of fishing and fish consumption, it may be interesting that other organisms, such as *Dinoflagellate* protozoa of the genus *Gambierdiscus*, which potentially produce ciguatoxins and can cause shellfish toxicity to mollusks and fish, can potentially spread (Booth et al., 2018).

In addition to raising temperatures, climate change is accompanied by a number of climatic changes, such as torrential rains, floods, and conversely, drought. These lead to greater fluctuations in water inflow and increased supply of sediments to coastal waters, which can affect a number of habitats in coastal zones that are important for the development of small fish. Rising sea levels may further accelerate habitat degradation (Booth et al., 2018).

4.2.6 Possibilities of Fishing Regulation

In this chapter, we will focus mainly on the regulation of sea fishing we talked about the regulation of sportfishing in the relevant chapter. Here, it should be mentioned that the question of how much we can catch and how to catch was important for fishermen for centuries. As an example, Kolding and van Zwieten (2011) mentioned the petition forwarded in 1376 to the King of England, Edward III complaining about a new trawl, which signatories assume unsustainable.

The sea is large, fish stocks are moving, and fishing is done by many subjects who do not own the sea (unlike agricultural or forest land, which mostly has an owner), creating a problem known as the tragedy of commons. In other words, a situation where many individual users of a common resource follow their own interests, and since the resource is common, their intention to conserve the resource for the future may be constrained by the fact that someone else may use that conserved resource. This implies a need for regulation. First of all, it must be said that the possibilities of regulations differ within national waters (territorial waters) and within international waters. The right to fish in international waters is given by the 1982 United Nations Convention on the Law of the Sea (UNCLOS), which replaced the seventeenth century "freedom of the seas." This convention defines the rights and responsibilities of states, their use of the world's oceans and lays down rules for trade and resource management.

From the point of view of UNCLOS, we can divide the world's seas into three zones, the first of which are territorial waters within 12 nautical miles (22.2 km) from the coast (the coast is usually considered to be the average low-tide line). These are considered the territory of the state and it has complete sovereignty here. This is followed by the Exclusive Economic Zone (EEZ), which reaches 200 nautical miles from the coast. The EEZ is not a sovereign territory of states, they must allow ships of other countries to navigate freely, but they have control over the use of all resources both at the bottom and in the water column, including fishing. High seas (international waters) start further away than the EEZ. The third zone is the continental shelf boundary which, however, does not affect fisheries.

The use of fish stocks in the high seas, as well as stocks that migrate between national borders, is addressed in the 1995 UN Fish Stocks Agreement. In addition, there are a number of specific local associations for the protection of certain fish stocks [e.g., the International Convention on the Conservation of Atlantic Tuna (ICCAT) and the Convention on Indian Ocean Tuna (IOTC), the Convention on the Conservation of Antarctic Living Marine Resources (CCAMLR), various FAO regional institutions Fishery Committee for the Eastern Central Atlantic (CECAF) and Asia-Pacific Fishery Commission (APFIC)] in international waters. However, international waters contribute little to the total catch, with most commercially important fish being caught in areas encroaching on national EEZ, so most regulations take place at the national level or groupings of states (EU).

Cochrane (2002) discussed the general principles of regulation. They are to some extent, similar to the regulation of game hunting, however, due to the greater importance of fish in human nutrition, and thus, the greater importance of commercial fishing, there is a more complicated compromise between maximizing catch and sustainability of fish stocks. The main principles on which regulation should be based are summarized in Table 4.7.

Table 4.7 Guiding principles for Cochrane fisheries regulation (2002)

C	Principle	Implications for regulation
1	Fish stocks are limited and depletable and organic production limits the potential income from fishing	It is necessary to realistically estimate the biological possibilities of production
2	Organic fish production depends on the size of the population and the quality of the environment.	It is necessary to determine the TRP and monitor the quality of the environment
3	Human consumption and existing technologies make it possible to catch fish faster than stocks are recovering	Fisheries must be regulated
4	In multi-species fisheries, it is difficult to establish uniform regulation that suits all species	Realistic targets must be set for the whole ecosystem, taking into account interactions between species
5	The high dependence of society on fishing shifts priorities for greater exploitation of fish stocks at the expense of sustainability	Fishing cannot be regulated in isolation but must be considered in the context of the conditions of the whole society (diversification of economic opportunities, etc.)
6	The information is always limited	The greater the uncertainty, the more conservative the approach to fishing should be
7	A sense of ownership and a long-term share of resources are best suited to maintaining responsible fishing.	A system of effective and adequate access to resources must be established.
8	Real participation in the management process by fully informed users is in line with the democratic principle, facilitates the identification of acceptable management systems, and promotes compliance with laws and regulations.	Communication between the regulatory authority and the stackers at all stages of the regulatory process is crucial

In Sect. 4.1.1, we dealt with the theory related to regulatory points, in other words, determining how much to fish to make fishing sustainable. However, setting these points is only one part of the problem, the second part, no less important, is their implementation into laws and regulations, and ensuring that these regulations are accepted by society to fulfill their purposes. What can be regulated for individual species? The time when fishing is allowed can be regulated, the places where fishing is allowed, the way in which fishing is carried out, the number of fishing entities and techniques used (number of boats, type of net, etc.) can be regulated, and finally, the size of catches can be regulated (e.g., quotas per quantity caught fish). The first measures lay in reducing certain ways of fishing, for example, regulation of net mesh sizes, which reduces the chance of catching young fish. This type of regulation has, historically, been taken by fishermen since medieval times in order to suppress the catching of juveniles. Another way to support reproduction and prevent juveniles is the regulation of the fishing season. Another approach to regulate fishing is to regulate the number of fish that will be caught. For that, various TRP points were developed. To enforce the TRPs described above, various forms of quotas have been applied.

Fishing season and fishing gear regulations can also reduce bycatch. Bycatch can also be reduced by introducing bans on discarding less valuable parts of the catch back into the sea as is becoming part of EU policy.

Quotas can, however, only partly affect the fact that fishing is unbalanced, biased toward catching disproportionally more fish from a higher trophic level and adult fish. Most of the regulations are focused on protecting young fish. This may bring a problem as in fact, mature fish are crucial for fish stock reproduction, and also, they play a major role in food web interactions (Kolding & van Zwieten, 2011). One approach to solve this is linked to the effort to create marine protected areas (MPA) in places important for reproduction. An advantage of MPA is that all age classes of fish in whole food webs, including all ecological interactions, are protected. Most of MPA is in territorial waters. U.N.s World Database on Protected Areas, which summarizes records of individual countries, indicates that 27 million km^2 of the ocean (almost 7.5%) is somehow protected. Cabral et al. (2020) even claim that expanding the existing global MPA network by just 5%, when done in proper, strategic, places, can improve future catch by at least 20%. Many political declarations are aiming for an even larger increase in are of MPA, however, as pointed by Gill et al. (2017) success of MPA is closely related to investment into staff and protection measures. Hence, the global expansion of MPAs must be accompanied by adequate investment in human and financial capacity to achieve optimal results.

Also, there is a difference in the regulation of industrial fisheries and small-scale fisheries. Here, it should be mentioned that small-scale fisheries are often largely regulated by natural production. Because of technical limitations, fishermen often target different species with different gears, which generates a more balanced pressure on the fish community. Kolding and van Zwieten (2011) even believe that "small-scale unregulated, nonselective, adaptive fishing patterns could be healthier, and far more ecosystem conserving than the current imposed single species management strategies." Moreover, in small-scale fisheries, the fisherman also forms a more closed community using the same resort, which can little bit reduce the tragedy of commons problem.

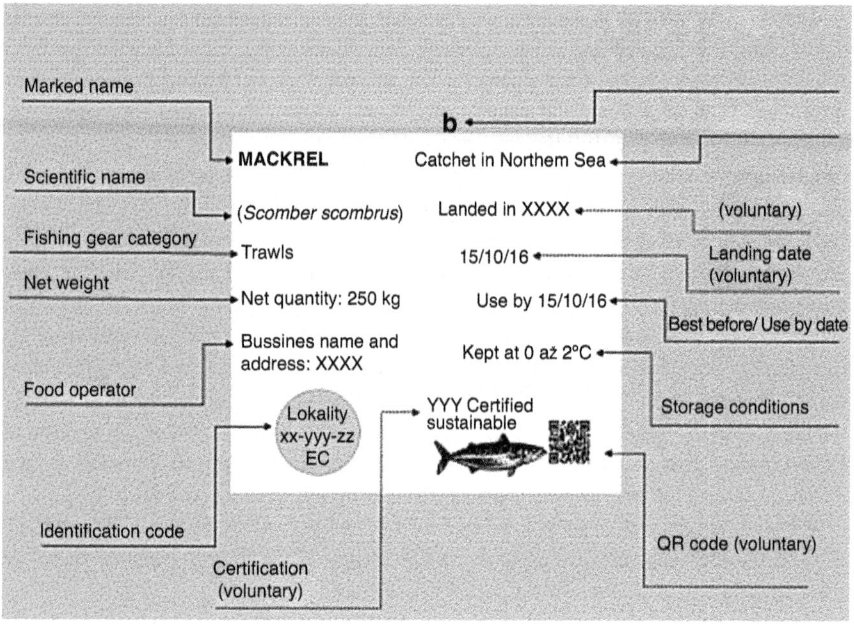

Fig. 4.25 Example of retail fish labeling

Last but not least, in addition to all sophisticated regulation, social perception is absolutely essential because it is basically impossible to have absolute control, which would enforce regulations, and there is always space for behavior that goes against the regulations. This may not be directly cheating, but the stakeholders can come up with many activities, which go against the spirit of the law without directly violating its letter. One from many examples of such behavior can be preemptive resource extraction in the areas intended for MPA before the conservation intervention is enforced. McDermott et al. (2019) showed that the fishing effort doubled once the area was designated for eventual protected status. Besides education and social perception of the regulations by fisherman, the same should apply also to the whole processing and supply chain as well as to customers. For example, part of the EU policy is an effort to increase the awareness of both stakeholders, experts and end customers (see Fig. 4.25), who can then reflect their opinion in their decision to purchase goods.

4.3 Aquaculture

4.3.1 Types of Aquaculture and Their Scope

By aquaculture, we generally mean the breeding of fish or other aquatic organisms in captivity or in some seminatural, but usually regulated conditions. The production of fish and aquatic animals from aquaculture is growing steadily and amounted to 80 million tons in 2016 (i.e., only slightly less than fisheries, which on the contrary have stagnated over the past 20 years). Unlike in fisheries, where we work with wild populations from which, we take caught individuals, in aquaculture, we usually work with cohorts of individuals of the same age. This gives us the opportunity to better regulate competition between individuals and the living conditions of the organisms in order to achieve maximum production. As individuals grow into competitors, we can either satisfy their needs with more food or give them more space and relocate them to a larger volume of water, constantly maximizing space use and yields. There are many types of aquaculture, the most widespread are inland aquaculture (i.e., systems more or less similar to pond farming) and in 2016, they produced 51.4 million tons (mostly in Asia). Marine aquaculture, which many people synonymies with the term aquaculture, is less widespread and produced 28.7 million tons in the same period.

4.3.2 Main Species of Fish and Other Organisms in Aquaculture

Among the fish reared in aquaculture (Figs. 4.26 and 4.27), freshwater fish from the carp family dominate (Table 4.8). Here, we start with the two species of herbivorous fish important in the aquaculture worldwide. The first feeds on aquatic vegetation Grass carp (*Ctenopharyngodon idella*), the second feeds on algae – Silver carp (*Hypophthalmichthys molitrix*), but they do not despise other food sources. Both come from the basin of the river Amur but they are bred in many countries. Silver carp is similar to reared Bighead carp *(Hypophthalmichthys nobilis)*, originating from China, it feeds on phytoplankton and zooplankton, and as with other species of herbivorous fish, transfers to plant food at higher water temperatures. Another species of fish that is widely grown is the carp (*Cyprinus carpio*). It is believed to have originated from the domestication of the wild form of the same species from the Danube basin (including the Black, Caspian, and Aral Sea basins). After domestication, it has spread to all continents. The carp is an omnivore, besides the larger zooplankton, it gathers food (insect larvae, etc.) on the bottom, where it digs. It can feed on plant food, grains, seeds, etc. Two other omnivorous cyprinids commonly found in aquaculture in Asia are the *Labeo rohita* and the *Catla catla*. Among the top ten aquaculture fish species were three representatives of other families, the first of which is the Nile tilapia (*Oreochromis niloticus*, Cichlidae), an omnivorous fish native to Africa, but bred in many parts of the world. The second is striped catfish (*Pangasianodon hypophthalmus)*, these catfish are omnivorous, come from Southeast Asia, among the catfish also include the North African catfish (*Clarias gariepinus*), originating from Africa, but reared in aquaculture in many places in the world, including in Europe.

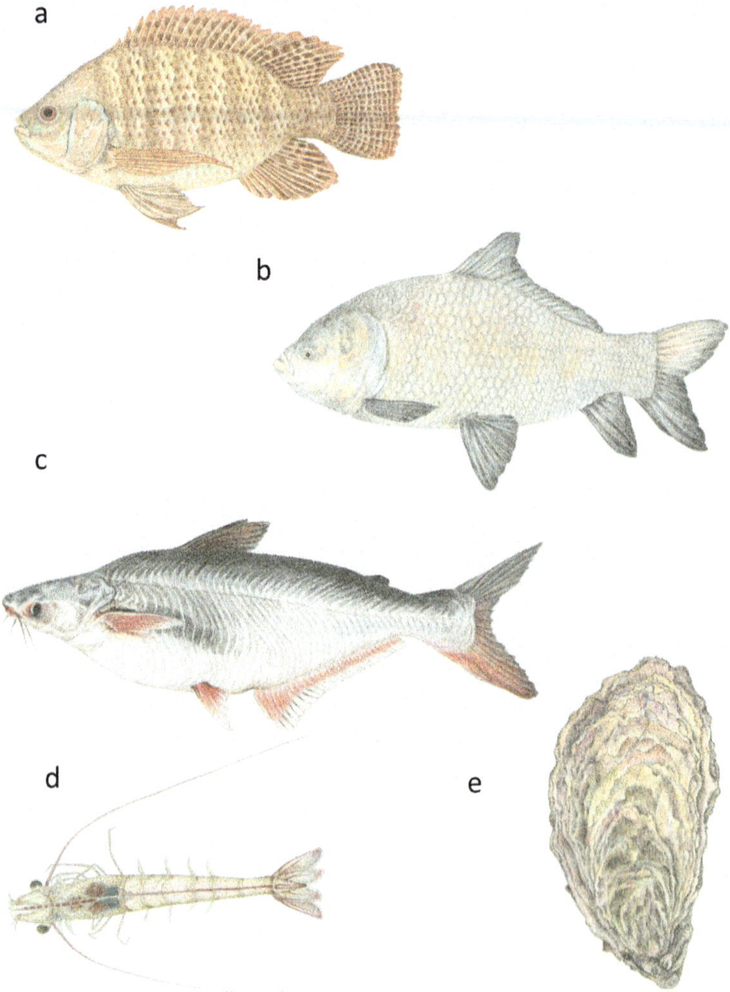

Fig. 4.26 Main species of fish, crustaceans, and mollusks grown in aquaculture, (**a**) Nile tilapia (*Oreochromis niloticus*), (**b**) catla (Indian carp) (*Catla catla*), (**c**) pangasius (*Pangasius* sp.), (**d**) King prawn (*Penaeus vannamei*), and (**e**) oyster (*Crassostrea* sp.)

The last important species farmed in aquaculture is the Atlantic salmon (*Salmo salar*), farmed in coastal cage aquaculture in Norway and elsewhere. It represents a large group of salmonid fish, important both in aquaculture and as well as for the fishing of wild populations. For others, we can mention Pink salmon (*Oncorhychus gorbuscha*) and Chum salmon (*O. keta*). Salmon meat has a pink to orange color that results from the accumulation of carotenoids derived from the consumption of marine crustaceans (krill), which, in turn, obtain pigments from marine algae. In

Fig. 4.27 Main species of fish grown in aquaculture II, (**a**) Silver carp (*Hypophthalmichthys molitrix*), (**b**) Grass carp (*Ctenopharyngodon idella*), (**c**) Common carp (*Cyprinus carpio*), (**d**) tench (*Tinca tinca*) salmon, (**e**) (*Salmo salar*), (**f**) trout (*Salmo trutta*), (**g**) pike (*Esox lucius*)

Table 4.8 Overview of organisms most frequently reared in aquaculture according to FAOstat data for 2016 in thousands tons

Grass carp	*Ctenopharyngodon idella*	6068
Silver carp	*Hypophthalmichthys molitrix*	5301
Common carp	*Cyprinus carpio*	4557
Nile tilapia	*Oreochromis niloticus*	4200
Bighead carp	*Hypophthalmichthys nobilis*	3527
Goldfish	*Carassius* spp.	3006
Catla (Indian carp)	*Catla catla*	2961
Atlantic salmon	*Salmo salar*	2248
Roho labeo	*Labeo rohita*	1843
Pangasius	*Pangasius* spp.	1741
Total fish		54,091
"King prawn"	*Penaeus vannamei*	4156
Red crayfish	*Procambarus clarkii*	920
River crab	*Eriocheir sinensis*	812
Tiger prawns	*Penaeus monodon*	701
Freshwater Shrimp	*Macrobrachium nipponense* and *M. rosenbergi*	507
Crustaceans, total		7862
Oyster	*Crassostrea spp.*	4864
Hearts	*Ruditapes philippinarum*	4229
Scallops	Pectinidae spp.	1861
Mussels	Mytilidae gen sp.	1100
Molluscs, total		17,139

farms, the fish are supplied with the synthetic carotenoid astaxanthin, which ensures the color of the meat. Among other salmonid fish in aquaculture, rainbow trout (*Oncorhynchus mykiss*) is interesting, unlike salmon, its meat is white, but by supplying astaxanthin, we can achieve the production of salmon-colored fish, such trout are then sold with the attribute salmon.

Of the crustaceans, the important shrimps *Penaeus vannamei* and *P. monodon* are widely reared in marine aquaculture in South-East Asia, as well as several species of freshwater prawns (*Macrobrachium nipponense* and *M. rosenbergi*) also reared in South-East Asia and the crab *(Eriocheir sinensis)*, which is mass-produced in China but occurs in Europe as an invasive species. Also, invasive in Europe is the Red crayfish (*Procambarus clarkii*) bred in the southeastern United States.

Of the mollusks, various species of *Crassostrea* spp. oysters are most often bred, as well as *Ruditapes philippinarum* on the petals and the so-called bongols or vongols, scallops (*Pectinidae* spp.), and mussels (*Mytilidae* spp.).

Among the more exotic animals, it is worth mentioning, the Softshell turtle (*Trionyx sinensis*), the Grass frog (*Rana* spp.), and the Chinese giant salamander (*Andrias davidianus*), which are mass-bred in China.

4.3.3 *Pond Farming and Other Inland Aquaculture and Their Environmental Impacts*

Inland aquaculture systems are very variable, from pond systems, through various cage systems similar to coastal marine aquaculture, to artificial tanks in the open air, or even covered recirculation systems with control of water quality and other environmental conditions. Here, we will mention in more detail pond farming, as the majority of the world's aquaculture production comes from fishponds.

Pond farming is a very old branch of human activity; it is described from China around 2300 BC, 700 BC from Egypt, and it was operated by the ancient Greeks and Romans. In Central Europe, already in the twelfth and thirteenth centuries, ponds are mentioned for fish farming purposes.

Using fishponds to rear fish is the dominant way of inland aquaculture in China. The most common are various species of the carp family. Pond fertilization and feeding fish namely with plant-derived food belongs to common intensification practices. Besides ponds, various other water bodies are used including dam reservoirs or drainage ditches. Specific aquaculture systems represent paddy fields, which are used for producing many species from the carp family such as black carp, grass carp, silver carp, bighead carp, common carp, crucian carp, Chinese bream, blunt-snout bream, and silver chub, however, also for other organisms such as freshwater shrimp, fresh-water mussel, river-snail, soft-shelled turtle, and river crab production (FAO, 2017).

Ponds are artificial reservoirs used for fish farming (Fig. 4.28). Ponds are also important elements of the landscape and they often perform other functions, such as water retention, affect the humidity in the region and compensate temperature differences. They are also involved in wastewater treatment, are part of an erosion

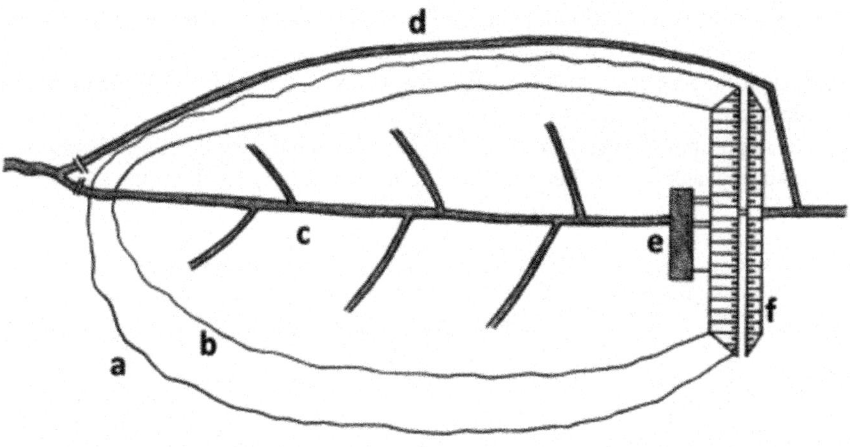

Fig. 4.28 Pond and its main components, and the perimeter of the pond (**a**) its cadastral territory, including coastal water-affected vegetation, (**b**) bottom or flooded area, (**c**) pond sewer, (**d**) perimeter sewers, (**e**) fishing grounds, and (**f**) dam

control system, and can be used for recreation, extraction of pond vegetation (reed, cattail), or for energy (hydropower) production. The basic element of a pond is a water-retaining dam, with a suitable drain enabling the handling of the water levels including complete drainage of the pond, which allows easy handling of the fish. In addition to the area flooded by water, the pond also includes the surrounding off-shore vegetation affected by water.

In general, fishponds need a certain suitable land configuration where fishponds can be built and it may be also limited by the availability of water. Fish density and consequently fish production can also be limited by the availability of oxygen; this can be improved by fishpond aeration. But, in nonaerated ponds production is limited by about 3500 kg ha^{-1} year^{-1} (Bosma & Verdegem, 2011). Fish production may also be limited by food availability in the pond. There are basically two mechanisms on how we can increase food availability: fertilize ponds to increase production of phytoplankton on which zooplankton and benthos grow, which can then serve as food for fish or feed fish directly. Often both methods are combined. Many of the most important fish species reared in aquaculture are completely or partly herbivorous or at least utilize plant feed well. Various feeds ranging from homemade mixtures to industrially prepared complex mixtures or granules are used (Ahmed, 2007) in various levels depending on the intensity of fish production (see Table 4.9). Both pond fertilization and excess of feeding, which turn in fish feces cause excess of nutrients, which can cause eutrophication not only in fishponds but also downstream (Bosma & Verdegem, 2011).

Fishpond aquaculture as already mentioned is very common in South-East Asia but also has a long tradition in central Europe. As we focused in this book on temperate ecosystems, we will describe European aquaculture in more detail below. Here, we describe fishpond technology and its effects on the environment using the example of Central Europe, which is an important hotspot of freshwater aquaculture in Europe and where these processes are well documented. The first significant appearance of pond aquaculture in Central Europe occurs in the twelfth and thirteenth centuries. In the second half of the nineteenth century, J. Šusta laid the foundations of modern pond farming with techniques such as manure fertilization and fish feeding.

An important parameter for fish farming is water temperature. Temperature is affected by water depth (the shallower the faster it heats up), flow (the smaller the

Table 4.9 Comparison of selected parameters of extensive, semi-intensive, and intensive pangas (*Pangasianodon hypophthalmus*) in Bangladesh based on the study of Ahmed (2007)

Parameter	Extensive	Semi-intensive	Intensive
Pond size, ha	0.12	0.21	0.36
Stocking density, fish ha^{-1}	12,065	23,575	35,900
Feed used, kg ha^{-1} year^{-1}	5790	13,010	22,370
Pangas production, kg ha^{-1} year^{-1}	3380	7705	13,945
Cost–benefit ratio	1.64	1.76	2.12
Annual profit, USD ha^{-1}	1188	2235	3649

flow, the more the water heats up), climate, exposure, and shading. Water temperature affects the speed of biological processes, including fish metabolism. At the same time, however, the solubility of gases (including oxygen) in water decreases with temperature. This may cause a lack of oxygen at a higher temperature, which can limit fish metabolism or even cause mortality. In addition to temperature, the pH of the water and the nutrient content are important. In addition to the source of water, the pond bottom is important; it retains and releases nutrients, retains water, and allows the development of benthos as food for fish. Key is its upper layer or the "active mud," which represents a 5–12 cm fine organic mud where decomposition processes and the formation of organic compounds take place with the participation of bacteria and benthos. It is easier to form on more fertile soils; on poorer soils, its formation is supported by fertilization with manure. The lower layer of mud represents a supply of nutrients, which is activated by wintering or summering of the pond and subsequent processing of the bottom. Under the mud, there are layers of permeable bedrock and soil, which is a supply of mineral nutrients. Below the bedrock, there is an impermeable layer important for retaining water in the pond.

The fish stock can be either single-species (monocultural) (e.g., just carp) or multispecies (polycultural). Polyculture is more common in extensive fishponds – especially with carp, which is supplemented with herbivorous fish Grass carp and Silver carp, then tench (*Tinca tinca*), which makes good use of space inshore vegetation, which is less usable for carp, and then predatory fish such as pike, zander, and catfish. Furthermore, the roach is stocked as food for predatory fish. The advantage of single-species stocks is simple fishing; the catch does not have to be sorted. The disadvantage is the incomplete use of the production potential of the pond, which is an advantage in polycultures. The more the requirements of individual species for the nature of food and the environment differ, the less competition there is between them, such as carp and tench. However, with increased farming intensity, monocultures have become more common.

However, most (up to 90%) of the harvest in Central Europe and particularly in the Czech Republic (which is the most important produces of fishpond-reared fish in the EU) comes from carp. Therefore, we will talk about carp breeding in more detail. Carp fry can be obtained either by natural spawning of generational fish in small ponds (Dubravius's ponds) or by artificial spawning and hatching of fry in special facilities (hatcheries). The bag fry is stocked in smaller, up to 5–ha, fertile nursery ponds, mainly with natural food. Up to 600 thousands individuals per ha are stocked. Mortality in this phase of life is high, up to 80%. Carp fry up to 5–10 g in size, 4–6 weeks old, it then moves to another, fingerling, pond where it grows to the size of 30–50 g and then overwinters. For overwintering, either water depth should be increased in a pond or specialized, deeper ponds can be used. One-year-old carp (30–50 g) are stocked in ongrowing ponds (up to 20 ha, up to 15 thousands individuals per ha). Two-year-olds (300–500 g) are placed in the main ponds (50 ha, 1500 individuals per ha), 3 to 4-year-old carp already represent the market size.

As can be seen, as the size of the fish increases, the number of fish in the hectarage of a pond decreases so that optimal conditions for fish growth can be achieved at all times. This can be achieved by either catching or sorting the fish every year and stocking it in other ponds. However, this requires frequent handling and higher costs. An alternative is to leave the fish in the same pond for 2 years. Here, however, either the first-year fish should be understocked or second-year fish should be supplied by additional food sources.

Fertilizing ponds and feeding fish are important intensification mechanisms that make it possible to increase the density of fish stocks and, thus, fish production. The purpose of fertilization is to support both the production of phytoplankton, on which zooplankton feed, and the production of macrozoenosis, which is also fish food. It is stated that for an increase of 0.5 t of carp, 3 t of zooplankton must be produced, which consumes about 21 t of phytoplankton.

A number of measures are used to supply nutrients to the pond. This includes wintering of ponds, during which the drained pond, or rather its bottom, freezes, and nutrients are mobilized in the deeper layers of the mud, a similar effect brings summering of ponds. In addition, during summering plants can grow at the bottom, including vetch-like N_2 fixing plants, and fertilization takes place in the form of flooded plant biomass. A significant amount of nutrients can be supplied by co-growing fish and waterfowl, especially ducks (the so-called duck-carp system), where nutrients are supplied in the form of duck excrements. Furthermore, we can supply nutrients by fertilization, most often in the form of organic fertilization with manure or slurry, or mineral fertilizers. Manure supports zooplankton from bacteria growing on the manure, which can be consumed by protozoans and they become prey for rotifera, which in turn can be consumed by zooplankton. Alternatively, microbial decomposition of manure releases nutrients, which support phytoplankton, which in turn can be consumed by zooplankton.

As we have seen in other cases of adding nutrients to the ecosystem, the increase in fish biomass is not linear with the addition of nutrients (Hartvich, 2012). With a smaller total dose of added nutrients, the relative increments per unit of nutrients supplied are greater and decrease with the amount of nutrients added. From a certain dose of nutrients, the total biomass no longer increases and may even decrease. This is due to several reasons. As we mention with increasing nutrient availability the amount of zooplankton increases. However, as nutrient load and fish stock increase, more zooplankton with low filtering rates and unpalatable algae prevail which reduces plankton palatably for fish (Potužák et al., 2007). Moreover, an overdose of fertilizer can also cause O_2 deficiency (Baxa et al., 2020), increased occurrence of free ammonia, and consequently overall weakening of fish, which can lead to deterioration of fish health, the development of diseases, or even the death of fish.

The supply of excess nutrients has a number of negative environmental impacts in addition to the potential negative impact on fish stocks. A significant amount of nutrients is discharged into the basin, which affects the quality of the water further downstream. For example, fishponds are responsible for the input of 49–53 t P per year into the Orlík reservoir in the Czech Republic, which is one of the most important nutrient inputs to the reservoir. These nutrients subsequently support the eutrophication of the reservoir and, thus, endanger the use of water in the reservoir (Hejzlar et al., 2010). Similarly, feeding fish with cereals has a negative effect on the P balance, because the use of P from cereals by fish is, in contrast to natural food, very small. In addition, the high fish population contributes to the remineralization of nutrients bound in the biomass, which is released back into the water column by the production of fish excrements.

Increasing nutrient supply and the associated increase in fish stock biomass have negative consequences for the biodiversity (birds, plants, and other groups of organisms) of pond ecosystems (Broyer & Curtet, 2012). Already at fish densities above 350–400 kg ha^{-1}, negative effects on bird populations were observed. These are related to the loss of food for their young, due to competition from fish for zooplankton and benthos, and to the decline of nesting habitats (macrophytic vegetation). Macrophyte vegetation is negatively correlated with the amount of nutrients supplied and the intensity of farming. The path to remediation is not easy and will be different for different ponds. The retention volume of ponds (clogged with sediments) should be increased, their technical condition should be improved so that the water surface can be better manipulated and the pond can be better prepared and, thus, reducing the loss of production, and at the same time optimizing fish stocks and nutrient supply.

In addition to fish farming in ponds, cage fish farming, which is similar to marine aquaculture, is used in some freshwaters. In addition, some more intensive methods of fish farming in tanks or even covered halls are used. These systems can be either flow-through or (semi-) closed recycling water. These are also very intensive systems with artificial feed and very well-regulated conditions. Flow systems have similar impacts on the ecosystem as ponds, especially in the runoff of nutrients from fish metabolism products and feed residues. In contrast, flow-through systems where water is recycled do not burden the environment with nutrients but have higher energy consumption. For example, in trout farming in a flow system, an energy consumption of 12–15 kWh kg^{-1} fish is reported, whereas in closed systems, it is 15–20 kWh kg^{-1} (Martins et al., 2010).

4.3.4 Marine Aquaculture and its Environmental Impacts

Coastal marine aquaculture takes many forms. Often, it is the cages for fish or crustaceans such as known, for example, from salmon farming in northern fjords (Fig. 4.29). Their principle is to place farmed animals in large cages made of nets, where they are fed industrially produced feed. Due to the high concentration of fish, the proportion of local natural food is negligible. In order to achieve sufficient growth, the diet must be rich in protein. For example, tilapia (*Oreochromis* spp.) needs up to 35% protein in the feed ration, whereas salmon (*Salmon* spp.) needs up to 2/3 of the diet consisting of fish meat and fats. Most of this food, or most of the raw material for its preparation, is obtained by fishing. Fish excrement falls out of the mesh bottom of the cage and the contact with the surroundings allows a constant exchange of water. Even so, due to the high concentration of organisms in one place, medication against some diseases may be necessary. As a rule, monocultures of individual species are grown in aquacultures; however, there are also aquacultures

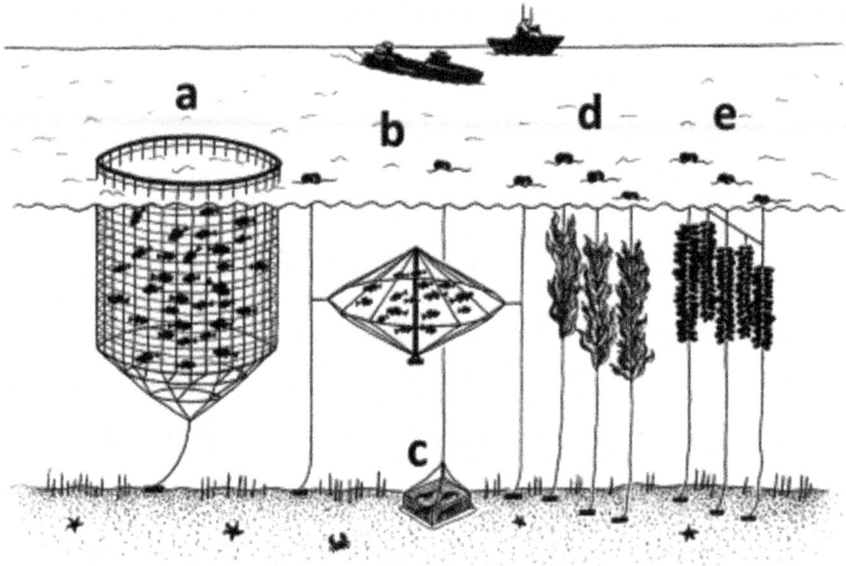

Fig. 4.29 Main types of marine aquaculture (**a**, **b**) cage fish farming, (**c**) cage farming of crustaceans, (**d**) growing algae, (**e**) shellfish

that use polycultures of several species. Integrated aquaculture, based on intensive synergistic cultivation of organisms at different trophic levels, seems promising. This allows the use of waste products of one organism by others as an example of such combinations is the joint cultivation of seaweeds and carnivores or salmon and mussels. Efficiency depends on management, technology, ratio of trophic levels, natural availability of food, temperature, and season. In addition to cage farming, a significant number of bivalve mollusks (e.g., oysters and cockles) are grown either in cages or as growths on artificial substrates.

The main environmental advantage of aquaculture is that the impacts of fishing described above, such as bycatches, impacts on the structure of populations, changes in habitats by bottom trawling, etc., are eliminated or minimized. However, aquaculture food is mostly obtained by fishing, so fishing problems can be transferred to fishing for aquaculture feed. In addition, in some aquacultures, we also obtain farmed stocks by fishing, and the problems of fishing are, therefore, passed on to its acquisition. This problem may be mitigated to some extent, by the use of bycatches for these purposes. Another disadvantage of aquaculture is the intensive use of various drugs and other food supplements that leak into the environment and can affect the surrounding ecosystem. The surrounding ecosystem can also be eutrophicated by the accumulation of food residues and the excrement of farmed fish. In addition, the large accumulation of fish leads to higher concentrations of predators, pathogens, and parasites, which can subsequently attack wild populations of fish and other animals. Farmed organisms can escape and, in the case of nonnative species, promote the spread of nonnative species. Even the escape of relayed individuals of domestic species can cause genetic erosion of wild populations. Similarly, as described in detail in the chapter on sportfishing, for the risks associated with the introduction of artificially bred fish. In some cases, for example, in the rearing of some shrimps in coastal aquaculture, the original habitats, especially mangroves, are eliminated at the expense of rearing cages.

There are, of course, efforts to minimize these negative impacts of aquaculture. One of the ways is hi-tech fish farms, based on completely closed cages located further from the coast in places with sea current, where local fish species are kept. The flowing stream dilutes the excrement and escaping food residues and, thus, mitigates their impact on the ecosystem, similarly mitigating a number of other impacts. In order to place these fish stations in an intense current, the stations are completely submerged and a number of improvements have had to be developed to allow them to withstand the current (see Fig. 4.29). Another option for mitigation already mentioned is polytrophic aquaculture based on a common culture of organisms at different trophic levels, so the waste products produced by one trophic level organisms are used by other trophic levels.

4.4 Impacts of Agriculture and Forestry on Fisheries and Aquatic Ecosystems

In previous chapters, we have addressed the environmental impacts of agriculture and forestry. In this chapter, we will mention in more detail the impacts of these two activities on fisheries. As we have already said, agricultural and forest land make up the majority of the land area worldwide. In addition, both of these activities have a major impact on the water regime of the landscape and biogeochemical cycles, so it is no wonder that they also have significant impacts on aquatic ecosystems and fisheries.

We will first focus on the impacts of agriculture and forestry on inland waters. Agriculture can adversely affect aquatic ecosystems and fish production in many ways. Pesticide and herbicide use can harm aquatic organisms. Excessive or improper use of chemical fertilizers can cause eutrophication. Soil erosion increases turbidity and sediment deposition in watercourses. Changes in sedimentation, river regulation, and wetland drainage alter the hydrologic regime of rivers, lakes, and other natural water bodies and damage fish habitats and migration routes. On the other hand, there are many opportunities for mutual integration of agriculture and aquaculture. Fish farming is part of the traditional way of growing rice in many East Asian countries. Similarly, alternating rice cultivation and freshwater shrimp farming are widespread in many East Asian countries (Nair et al., 2014).

Changes in forest management have a major impact on habitats, especially in stream species. For example, clear-cut logging leads to increased erosion, which is often accelerated by the movement of logging mechanization, and associated clogging of riverbeds, thanks to the supply of more deadwood (branches and parts of trunks). By removing the vegetation around, the temperature regime and flow change. The impacts of forest management on salmonid fish production have been studied in detail in British Columbia. It turned out to be a very complex process in which different factors can be applied on different spatiotemporal scales. Immediately after wood mining, the water temperature changes, and subsequently, the littoral vegetation is affected and the supply of litter to streams is reduced. The litter is the main basis of food webs in these streams, so its reduction will reduce the secondary production of benthos and, thus, the food supply of salmonids. In the medium term, there is an increased risk of erosion and sudden floods due to the lower water retention capacity of the basin, which can lead to a loss of fish eggs by leaching or clogging with fine sediment. Altered erosion conditions will also affect the composition of the riverbed, in particular, the increase in the proportion of fine particles in the surface layer of sediment, and this will affect the quality of the bottom as a habitat for laying eggs. Finally, in the even longer time (decades), the accumulation of trunk residues in the riverbed can cause changes in the flow of the stream and can lead to the formation of backwaters and pools, which are a suitable habitat for fish. However, their occurrence depends on the nature of the stream and the type of trees that the trunk remains there (Scrivener et al., 1998).

Management in the landscape with a significant contribution from agriculture also affects the ecosystems of the seas, especially the coastal parts. Agricultural activities that take place close to the coast are certainly more likely to affect the function of marine ecosystems in a number of ways (see Table 4.10) compared to upstream ecosystems. However, even agroecosystem upstream has a significant contribution, especially in the distribution of nutrients and possibly other agrochemicals. The supply of nutrients to coastal areas causes eutrophication of coastal areas with a number of negative impacts, the most well known of which are cases of mass cyanobacterial overgrowth. The most famous and, perhaps, the most striking is called Red tide due to mass overcrowding dinoflagellates (e.g., *R. Lingulodinium* or *Karenia*). Organisms producing red tide produce saxiotoxins, which can accumulate in the bodies of mollusks, which normally feed by filtering algae from seawater. Saxotoxins block sodium channels important for the transmission of nerve impulses and can cause poisoning of mollusks or other organisms that feed on mollusks. Although the causes of the red tide certainly occurred in the world's seas long before the industrial revolution, the massive occurrence of this phenomenon increased significantly with the advent of industrial fertilizers and the associated eutrophication of coastal seas and oceans.

Cyanobacterial toxins can be toxic to bathers, as in freshwater. Increased primary phytoplankton production due to eutrophication may have other consequences. The penetration of light through the water column is reduced, which reduces the depth at which submerged macrophytes can grow, and the proportion of relatively short-lived fibrous algae increases. These are important habitats for fish life, reproduction, and consequently, and their loss can also affect fish stocks. Increasing the concentration of nutrients often leads to deviations from the optimal so-called Redfield nutrient ratio (N:P 16:1), which can lead to a reduction in diversity and the support of species less sensitive to this ratio. Overall, the share of plankton in primary production is increasing compared to the benthic part of the ecosystem.

Table 4.10 Impacts of agriculture on aquatic ecosystems in coastal areas, fisheries, and ecosystem services provided, adjusted according to FAO (1998b)

Activity	Environmental impact	Socioeconomic impact
Flood control, regulation, and diversion of rivers	Increase in salinity, decrease in sediment contribution to coastal erosion	Reducing crop yields, reducing fish yields
Use of pesticides in coastal areas but also in higher parts of the river basin	Toxic substances in the food chain	Toxic substances in fish reduce fish yield
Use of fertilizers in coastal areas but also in higher parts of the basin	Eutrophication	Fish mortality, coral mortality, yield reduction, and other ecosystem services
Excessive plowing and grazing in coastal areas but also in higher parts of the basin	Increased erosion, increased sedimentation in canals and floodplains, increased turbidity, negative effect on benthos and organisms sedentary on rocks	Fish mortality, coral mortality, yield reduction, and other ecosystem services
Cattle breeding	Eutrophication	Fish mortality, coral mortality, yield reduction, and other ecosystem services
Irrigation from a seaside aquifer	Decline of a seaside aquifer snake	Penetration of saltwater into the aquifer, degradation of water in it
Drainage of wetlands and wet soils and regulation of rivers	Destruction of wetlands, loss of coastal protection against sea erosion, loss of sea protection against toxic substances, and eutrophication	Fish mortality, coral mortality, yield reduction, and other ecosystem services
Manufacturing industry	Eutrophication, toxic substances in the food chain	Fish mortality, coral mortality, yield reduction, and other ecosystem services

The composition of zooplankton is also changing, for example, in the Baltic Sea, due to eutrophication, the dominance of jellyfish has been observed at the expense of crustaceans. The shift to short-lived forms of primary producers and their mass increase then leads to more frequent and significant periods of their death, which, due to decomposition processes, leads to the consumption of oxygen and anoxia. This can lead to the extinction of many species of benthic organisms, as well as to the production of toxic hydrogen sulfide, which can poison other organisms. Furthermore, under anaerobic conditions, P is released into accessible forms, which may again stimulate further cyanobacterial growth and further occurrence of the aquatic flower.

A textbook example of eutrophication is the eutrophication of the North Sea, caused by an increased supply of nutrients from rivers, especially from the Elbe, reinforced in the 1970s and 1980s by local sources of pollution in coastal areas. In addition, the North Sea (North Sea and Waden Sea) belong to the parts of the seas where, due to natural conditions, the nature of the bottom and sea currents, nutrients naturally accumulated, which exacerbated the problem (Van Bennekom and Wetsteijn, 1990). This has led to increased phytoplankton production, an increase in cyanobacteria and the promotion of aquatic blooms, foam formation on beaches and the appearance of black spots, and the formation of large anoxic areas in tidal areas. These changes subsequently led to a decline in seabird populations. This decline sparked a series of political and technical measures. Finally, in 1990, at the International Conference on the Protection of the North Sea in London, a commitment was made to reduce the load of nutrients by 50% to the level of 1985, which was subsequently achieved. River pollution and other sources of pollution leading to eutrophication of coastal waters have also been the subject of several EU directives, in particular, the EC Urban Waste Water Treatment Directive, the EC Nitrates Directive, the EU Water Framework Directive, and the Marine Strategy Directive.

Chapter 5
Conclusion

In the conclusion section, we will try to summarize how humans have influenced and still influence nature by obtaining food and other raw materials. Table 5.1 explains how humans, as hunters and gatherers, have significantly influenced their surroundings. As a hunters and of fisherman humans affect age-dependent mortality patterns diferently then most of natural predators. As shown in examples in hunting and fisheries sections hunters or fisherman specialize in adult prey or adult fish (of reproductive age) and avoid exploiting younger individuals (e.g. Fig. 4.19). This may certainly affect population survival as exploitation is focused on reproducing individuals. While natural predator which preferentially exploited young individuals do not cause so much effect, as mortality is often naturally high in this stage and moreover removing individuals reduce competition and increase chance of others to survive. The expansion of modern human settlements from the African continent is closely correlated with the extinction of many species of megafauna in places where people have inhabited. However, it is not entirely clear to what extent humans have directly contributed to these extinctions. As hunters, people also contributed to the expansion of grassland ecosystems by burning forests in order to support the populations of certain groups of fauna, which they subsequently hunted. With climate change, people accumulated in the floodplains of large rivers, which supported the transition to agriculture and the domestication of plants and animals. Agriculture very likely originated independently in different parts of the Earth, so it is possible that the specific management mechanisms were different in different places. Interestingly, original plants, which are the ancestors of domestic plants, often came from multiple areas, genetic centers; today's crops were created by crossing these plants outside their area of origin. This shows that trade and thus, the exchange of goods between different agricultural centers was of great importance for the development of crops. In the case of domesticated animals, on the other hand, it seems that in many cases their domestication took place in one place, from where the already domesticated forms spread with humans all over the world; again the influence of trade is evident. Domestication, among other things, led to the loss of some

© The Author(s), under exclusive license to Springer Nature Switzerland AG 2022

J. Frouz, J. Frouzová, *Applied Ecology*, https://doi.org/10.1007/978-3-030-83225-4_5

Table 5.1 Summary of the main effects of hunting, fishing, agriculture and forestry on ecosystems in selected stages of human society development

Parameter	Hunters and gardeners	Preindustrial society	Industrial society
Temporal and geographical delimitation	Human society 5–10 thousand years BC. Today, some tribes in the rainforests of Papua New Guinea, Khoisan, etc.	European society of the early nineteenth century	European society of the twentieth to twenty-first centuries
Population density in populated areas	Tithes exceptionally up to tens of people per km^2	Tens of people per km^2	In densely populated areas, hundreds to thousands of people per km^2. Aggregation of people in cities
World population	5–10,000 years BC about 2.5–18 million	In 1850, 1.3 billion	2019, 7.7 billion
EROI	40 and more, man uses only the opportunities that nature offers and uses them with a minimum of supplied energy, but there are a limited number of such opportunities	Around 1, man makes intensive use of the landscape through his own work and the work of animals	Significantly less than 1, most energy is supplied from fossil sources
Work efficiency	The whole society participates in the production of food, in principle, the family unit feeds only itself	One family will feed several other people; this allows division of labor and the existence of cities	One producer will feed dozens of people, thanks to a large input of energy from fossil sources
Hunting efforts on land and in fresh waters	Impact similar to other large predators, however, can hunt some species to extinction especially K strategists. On the contrary, changes in the disturbance regime by burning were supported by large herbivores	In populated areas, suppression or extinction of top predators and large herbivores. Change in the structure of communities supported by changes in land use	Efforts to "breed" hauntingly attractive species lead to further changes in the composition of the community. This also has implications for the use of the landscape, e.g., overpopulation of game can negatively affect forest regeneration, etc. Fisheries and aquaculture are often important determinants of freshwater fish communities

(continued)

Table 5.1 (continued)

Parameter	Hunters and gardeners	Preindustrial society	Industrial society
Sea fisheries and whaling, etc.	Small effect	Suppression or extinction of individual K strategists, (especially marine mammals)	Population overfishing in 30% of species. Suppression of higher trophic levels (top predators). The growth of mesopredators and changes in communities, supported in addition by changing nutrient cycles
Landscape use biodiversity and Nutrient cycling	Changes in the landscape by burning in swidden gardening comparable in extent to natural disturbances. For hunting communities, burning the landscape can lead to a change in the area of the main types of ecosystems. Grassland expands at the expense of the forest. The gardens of swidden farmers have little effect on the nutrient cycle. Larger for large-scale indigenous burning. Humans likely contributed to extinction of important parts of megafauna	The intensity of land use limited by sufficient nutrients, especially N and P, led to the suppression of the forest and to the effort to obtain feed and litter of biomass by fresh grazing and as a rake or turf. The nutrients contained in this biomass were then supplied to arable land, which led to local mobilization of nutrients at the cost of oligotroph cation of the entire landscape. There was also support for nonforest habitats, the species composition of forests significantly influenced by humans. Expansion of arable land increased erosion and affected the erosion-sedimentation regime of the landscape, drainage of wet areas, tillage, and erosion also led to the loss of organic matter Changes in habitat extensively affect biodiversity	The availability of fossil fuels has made limiting nutrients available, both through Haber–Bosch synthesis and from the extraction of apatite and other nutrient-producing minerals The availability of fossil fuels and mechanization have reduced need for human labor. The intensive use of mechanization has led to an increase in field size, with negative consequences for biodiversity and erosion. It accelerated the previously initiated changes associated with the loss of organic matter and increased the water runoff from the landscape The availability of nutrients leads to their greater loss and runoff into rivers. This supports the eutrophication of the landscape, the spread of fast-growing species, including invasive ones, and the eutrophication of freshwater in the coastal seas Habitat loss, global change and other factors described above caused substantial loss of biodiversity

properties important for survival in nature in domesticated forms, and these were replaced by properties advantageous for coexistence with humans (synchronized germination and maturation in plants, taming animals, etc.). From an evolutionary point of view, human orientation was advantageous for them, domesticated forms have become the dominant species on the planet, for example, the biomass of domesticated animals is several times higher than the biomass of wild animals.

The development of agriculture has enabled the emergence of cities as centers of trade and crafts, thereby increasing labor productivity. Urban development has placed great demands on the development of agriculture, fisheries, and aquaculture. The rural population had to produce food not only for themselves, but also for the people in the cities. In addition, the urban population had greater demands on the consumption of more luxurious foods, more meat, eggs, oil, fats, milk, etc. Increasing the consumption of animal products makes it necessary to grow more biomass, as the conversion of plant biomass to animal biomass results in significant energy losses. Increasing the consumption of animal products therefore places greater demands on ensuring even greater crop production. This could be further increased by the trade in luxury items, high demands on food for war or religious purposes, etc., which further increased the supply intensity. It is believed that the decline of many ancient empires was due to the gradual increase in demands on the countryside, which had to supply to urban centers, and then climate change or war was enough to collapse the whole system. However, despite these declines, cities reappeared. Urban development has enabled the development of crafts, the division of labor, and specialization in certain activities and thus greater work efficiency. With the beginning of the industrial revolution, there was a further outflow of people from the countryside to the cities, as well as the discovery of a number of machines that facilitated tillage, plant cultivation (improvement of the plow and seeder), and animal husbandry. All this increased the efficiency of human work in agriculture. This effort continued with the development of the industry and the use of fossil fuels as a source of energy, which not only allows substantially reduced manpower needed in agriculture but also increased the amount of additional energy supplied to grow crops. This additional energy was realized not only in work operations done in fields (fuel machinery used for tillage and other operation), but also in products manufactured elsewhere and applied in the field (machines, fertilizers, pesticides, etc.).

An important step was the discovery of the Haber–Bosh synthesis, which enabled the industrial fixation of nitrogen from the atmosphere. This significantly increased agricultural production and consequently led to an explosion of the human population. At the same time, the development of this "industrial agriculture" required a greater division of labor, specialization in individual subprofessions not only in agriculture as such, but also in the production of machinery, fertilizers, and other inputs to agricultural production. This further strengthened the shift of the population to cities and placed even greater demands on the supply of food and industry to raw materials. This has increased the demands on food storage and processing. On the one hand, for logistical reasons, because food was transported over longer distances and stored for a longer period of time, and on the other hand, because of the convenience of the urban customer, food was increasingly sold in the form of a product. Bread or croissants instead of grains, processed meat instead of fresh, canned, or frozen instead of live animals, etc. This has gradually led to a greater and greater separation of consumers from food production, not only technically and logistically, but also economically, socially, and emotionally. The fact that the urban population has been buying food in the processed form for generations also contributed to a feeling that food is often not associate with living organisms.

In the hunter gardeners' stage, forests were a source of many goods, game, fruits, etc. With the development of agriculture, it progressively became more of a source of wood, pasture, and litter. At the beginning of the industrial revolution, forests were mainly a source of fuel, especially charcoal, which could be used in smelters, glassworks, etc. With the development of fossil-fuel combustion, interest in charcoal decreased, but more wood was needed for logging and other construction work. This clearly shows how market demands modify the behavior of primary producers in agriculture and forestry. In forestry, this shift is particularly notable, as forests will often be harvested in the next generation, so market adjustment decisions are made for a very long time (for which market developments are very difficult to predict). At the same time, however, a by-product of these decisions is the level of social ecosystem services provided, which are also affected by this decision for a long time.

It should be emphasized that this development of agricultural and forestry production, as well as the intensification of fisheries, has been accompanied by great intellectual efforts and a number of discoveries in applied research (development of mechanization, fertilizers, pesticides, breeding of varieties and breeds, development of economic thinking about production, theory of net soil yield, prediction of fish, sonar, etc.). After all, this can also be seen in a number of Nobel Prize winning scientists mentioned in the text of this book.

If we look at the way in which this intensification has been achieved, then instead of populations from which we would take suitable individuals, intensive agriculture, forestry, and fishing have focused on growing or breeding cohorts of organisms with a limited number of species of excellent production characteristics. In agriculture this was even increased by extensive use of annual plants. Using cohorts instead of populations makes it possible to increase production by optimizing competition between individuals and supporting their growth by optimizing the number of individuals, supplying nutrients or food, especially if we work within a certain area with a monoculture or species-poor mixture. The biggest advantage of even-aged cohorts is the ability to synchronize certain types of the above interventions and this gives the ability to use machines to perform these tasks. In next step, method of production was modified even further to maximize potential for use of machinery an efficiency of machinery use.

The development of mechanization, especially in agriculture, but also in forestry, was made possible by the development of industrial production. This massive use of mechanization is connected with the loss of people in rural areas and thus the loss of labor and sufficient fossil fuels. Mechanized production is more energy intensive per unit of production than without it and this energy demand is usually covered by the need for fossil fuels (Woods et al., 2010). The energy return on production (EROI) is therefore decreasing with its intensification. At the same time, the importance of recycling natural biomass in energy flows and biogeochemical cycles is declining, and it is being replaced by additional energy and nutrient addition in form of fertilizers. This extra energy brings two environmental problems: one is the dependence of agriculture on fossil fuels. This can possibly be improved by moving to a more sustainable energy mix. The second problem is that the way the additional energy is used ends up reducing the role of natural processes in biogeochemical cycles, carbon, and energy turnover through a series of complex interactions (including homogenization and simplification of the system). This can threaten system stability and the provision of ecosystem services. To reduce these negative impacts, we would need profound changes in technologies, not just a change in energy sources.

The intensification of the production and use of cohorts has had a major impact on the flow of energy and nutrients to ecosystems due to adjusting the conditions for plant growth (drainage, ploughing, or fertilization, etc.). In addition, the use of cohorts combined with extensive using of annuals in agriculture promotes a higher incidence of initial succession stages in the landscape. This causes a situation where large areas of soil are without vegetation or only with sparse vegetation. It promotes the occurrence of erosion, affects the water cycle, and energy flows in the landscape. This is further reinforced by the fact that most of the species we grow in our landscape come from significantly different climatic conditions. The original species from which the dominant cultivated plants (for which we adjust the conditions in the agricultural landscape) originated from the Mediterranean or from the mountainous areas of the subtropics. However, the most important production areas occur in temperate climates. On the contrary, Norway spruce, the most abundant tree in many European forests (forming the conditions of forest soils), comes from the mountainous regions or northern Europe. Many approaches such as ploughing and summering ponds aim to accelerate the turnover of nutrients in ecosystems and provide cultivated organisms with more nutrients or food. This effort has become significantly more effective with the discovery of industrial fertilizers, which has led to a large-scale change in biogeochemical cycles at both the global and local levels.

In addition, the use of cohorts, specialization in certain species and the effort to make effective use of mechanization often leads to specialization in a certain type of vegetation and homogenization of the landscape, which is covered by large homogeneous units instead of small, diverse areas. These influences such as the homogenization of the landscape, the cultivation of cohorts, extensive use of annuals in agriculture, the modification of the dominant vegetation, and our efforts to adjust the conditions for this vegetation have affected the size of erosion, the radiation balance, and the flow of nutrients in the landscape. This has affected the landscape's ability to regulate the flow of water, nutrients, and other processes on which many ecosystem services depend.

Thanks to the specialization at the landscape level, many of these effects have been enhanced by trade and a number of environmental impacts have begun to be exported to other countries (e.g., loss of cerado and rainforests in connection with soybean cultivation in Brazil, which is then exported to many countries, etc.).

It would be very unfair to identify food and raw material producers as the main culprits in this situation. Rather, its roots lie in the detachment of consumers from the production of individual commodities. This detachment was triggered by urban development, specialization in certain professions, and division of labor, which at the same time, enabled greater efficiency of work and technological development. The consumer usually tries to buy products at the lowest possible price and is often not even able to assess the other consequences of their decision. This motivates processors and sellers of products to put pressure on producers to reduce primary production costs. This is usually achieved by intensifying production and this often reduces the ability of ecosystems to provide ecosystem services, especially regulatory services such as climate regulation, regulation of water quantity, and quality in the landscape, etc. All this is complicated by the fact that although reducing the ability of ecosystems to provide services has great economic value, no one specifically pays for these services. In addition, many of these damages are exported elsewhere; for example, tiger prawns that we buy in a supermarket do not endanger ecosystem services provisioning in counties where thez are marketed. However, in the countries of their production their breeding cause loss of mangroves, which cauese loss of many ecosystem services mangrove provide, incluing higher tsunami susceptibility, due to the loss of mangroves. In addition, impacts on ecosystem services are often not immediste and proportional to man induced increase in production. In many cases ecosystems damage appear with some delay when some tipping points are reached and thus, not manifest itself immediately, and so many of these damages will not be paid until the next generation (climate change can be an example).

Massive habitat changes, which to a large extent, are caused by agriculture and forestry interact with other anthropogenic stressors, such as contamination, urbanization, and development of infrastructure and global climatic changes. Global climatic changes are particularly important in this context. They not only strongly affect agriculture, forestry, and fisheries, but at the same time, these human activities strongly affect the impact global changes have on biodiversity and ecosystem functioning. For example, the ability of biota to respond to global change by migration to climatically more suitable areas is strongly affected by habitat changes. Changes in rainfall distribution, longer periods of drought followed by more intensive rain, may enhance the negative effect of cultivation of soil on erosion. Last but not the least, land-use changes driven by agriculture and forestry may also strengthen climate changes itself; for example, by greenhouse gas production or effects in the radiation budget.

In several places in this book, we saw that "business as usual" exacerbates ecosystem change driven by global change and also makes ecosystems more vulnerable to some indirect effects of global change. In agriculture, we focused on growing cohorts of monocultures of annual crops. To use our machinery efficiently, we do this in large fields. This leaves large areas of soil without vegetation cover for large parts of the year. This increases the sensitivity to erosion. Growing annual crops and associated cultivation also reduces soil organic matter content. This reduces the ability of soil to be a sink of atmospheric CO_2 (or even made it a source), decreases the ability to hold water, increases sensitivity to erosion, etc. All of this increases the sensitivity of ecosystems to irregular rainfall, prolonged drought, and heavy rain. In forestry, we often shift forest composition toward higher proportion of conifers, which often store more carbon in the forest floor compared to many broadleaf trees, which tend to store carbon more in mineral soil. This can potentially make forests more sensitive to fire in drought periods. Also, efforts to increase wood production by achieving higher tree density may increase sensitivity to drought and consequent, bark beetle attack. All of these activities reduce the ability of the ecosystem to be carbon sink and increases the risk that it will become a carbon source as a result of drought, fire, or insect outbreaks. Interestingly, some of our efforts to mitigate global change of climate such as attempts to replace savannas with exotic trees may have similar negative effects. In the ocean, the global change impact is less understood. However, one of the most often mentioned impacts of global change is the effect of increasing CO_2 concentration on ocean acidification combined with increasing temperatures, which threaten coral reefs (Richardson et al., 2012). Our focus on fishing at the top of the food web, which supports mesopredators and hence suppresses fish helping coral reefs to live, is another example where our common practice magnifies global change impact.

The question is to what extent natural ecosystems are needed and necessary to maintain the functioning of our planetary ecosystem. The approach to this question is a complex ethical, social, and scientific problem that transcends a number of disciplines. There are opinions that we should best tame nature and then steer this domesticated nature consistently and judiciously (stewardship) (Kareiva et al., 2007). In contrast, many researchers point out that we are able to influence, but not control, a number of natural processes. The failure of the Biosphere 2 project is, after all, a demonstration of how difficult it is even on a much smaller scale than a planetary scale. These researchers point to a large information deficit or to the fact that natural energy flows are many orders of magnitude larger than those governed by our society (Table 5.1). Wilson (2016) even suggests preserving half the globe for natural processes. This resulted in an effort not to reduce the size of natural ecosystems, but to support their restoration where possible. The Global Deal for Nature (Dinerstein et al., 2019) also believes that in order to protect diversity and achieve climate neutrality, at least 30% of the planet's surface needs to be protected and another 20% should be subject to climate mitigation. This was reflected by many political initiatives. However, besides the debate about how much natural ecosystem we need, we should also discuss how much natural or close to natural processes we should use in a productive landscape to allow reasonable provisioning or ecosystem services. Another approach that does not address the issue of natural or anthropogenic ecosystems but rather the capacity of individual key processes of the terrestrial planetary ecosystems are the so-called planetary boundaries. A group of experts led by Rockström et al. (2009) tried to define critical values of selected indicators of human activity, including CO_2 concentrations, nitrogen and phosphorus fluxes, agricultural areas of the planet, etc.

Despite the conceptual contradictions of the approaches described above, their recommendations on what we should do in the cultural landscape and where we use natural resources will be similar and will include approaches such as increasing diversity, landscape mosaics, optimizing the distribution of cultures and technologies used in ecosystem services, to support research in this direction, and to use products with less impact on the functions of ecosystems. Many of these things are, in principle, already happening to some extent. But often the intensity of more environmentally friendly production is lower than conventional one. This makes the products more expensive, while the benefit for consumers to improve ecosystem services is less visible.

There is a big role for research and education of producers, traders, and especially customers. Research is needed both to develop new environmentally friendly production methods and, above all, to understand how our behavior and market decisions affect ecosystems. Many companies have sensed a shift in customers' environmental behavior. They understand that selling products at the same price and quality as the competition, but also environmentally friendly, increases their competitiveness. For many companies, this behavior is slowly but surely becoming part of the corporate culture. At the same time, the good news is that many companies became interested about the environmental impacts of their supply chain. However, companies and their social responsibility cannot be expected to completely solve the problem (Kotecký, 2019).

Another option is to pay producers to improve the state of ecosystem services from some central source (e.g., taxes). An example of this approach is part of EU Common agricultural policy payments and other subsidies. However, these approaches still face problems in terms of the effectiveness of their real impact on ecosystem services. This also applies to the various certifications of environmentally friendly production within individual industries.

The impacts of our activities on the functioning of ecosystems are not only large but also complex and take place on various spatiotemporal scales. Most of our knowledge is gained by experiments with simple systems on small spatial scales over a short time. Their extrapolation to large and more complex spatiotemporal scales is not trivial at all. The question of how best to ensure the key functions of ecosystems, while ensuring adequate production and how to reach a social consensus on how to pay the potential costs of such a change, is certainly a key question we need to answer to ensure sustainable well-being of the human population.

References

Aarts, B. G. W., & Nienhuis, P. H. (2003). Fish zonnations and guilds as the basis for assessment of ecological integrity of large rivers. *Hydrobiologia, 500*, 157–178. https://doi.org/10.1023/A:1024638726162

Adams, R., Rosenzweig, C., Peart, R., et al. (1990). Global climate change and US agriculture. *Nature, 345*, 219–224.

Aguilar Soto, N. A., Johnson, M., Madsen, P. T., Tyack, P. T., Bocconcelli, A., & Borsani, J. F. (2006). Does intense ship noise disrupt foraging in deep-diving Cuvier's beaked whales (*Ziphius cavirostris*)? *Marine Mammal Science, 22*(3), 690–699.

Ahmed, N. (2007). Economics of aquaculture feeding practices: Bangladesh. In M. R. Hasan (Ed.), *Economics of aquaculture feeding practices in selected Asian countries,* FAO fisheries technical paper 505 (pp. 33–64). Food and Agriculture Organization (FAO) of the United.

Aizen, M. A., & Harder, L. D. (2009). The global stock of domesticated honey bees is growing slower than agricultural demand for pollination. *Current Biology, 19*(11), 915–918.

Alaphilippe, A., Boissy, J., Simon, S., & Godard, C. (2016). Environmental impact of intensive versus semi-extensive apple orchards: Use of a specific methodological framework for Life Cycle Assessments (LCA) in perennial crops. *Journal of Cleaner Production, 127*, 555–561.

Albert, M., Nagel, R. V., Nuske, R. S., Sutmöller, J., & Spellmann, H. (2017). Tree species selection in the face of drought risk – Uncertainty in forest planning. *Forest, 8*, 363.

Allen, P. A., & White, E. P. (2003). Effects of range size on species-area relationships. *Evolutionary Ecology Research, 5*(4), 493–499.

Allen, M., Brunner, A., Antón-Fernández, C., & Astrup, R. (2020). The relationship between volume increment and stand density in Norway spruce plantations. *Forestry*. https://doi.org/10.1093/forestry/cpaa020

Andermann, T., Faurby, S., Turvey, S. T., Antonelli, A., & Silvestro, D. (2020). The past and future human impact on mammalian diversity. *Science Advances, 6*, eabb2313.

Angilletta, M. J., Steury, T. D., & Sears, M. W. (2004). Temperature, growth rate, and body size in ectotherms: Fitting pieces of a life-history puzzle. *Integrative and Comparative Biology, 44*(6), 498–509.

Aponte, A., Samarappuli, D., & Berti, M. T. (2019). Alfalfa–grass mixtures in comparison to grass and alfalfa monocultures. *Agronomy Journal, 111*, 628–638.

Ardyna, M., Lacour, L., Sergi, S., d'Ovidio, F., Sallée, J. B., Rembauville, M., Blain, S., Tagliabue, A., Schlitzer, R., Jeandel, C., Arrigo, K. R., & Claustre, H. (2019). Hydrothermal vents trigger massive phytoplankton blooms in the Southern Ocean. *Nature Communications, 10*, 1.

© The Author(s), under exclusive license to Springer Nature Switzerland AG 2022

J. Frouz, J. Frouzová, *Applied Ecology*, https://doi.org/10.1007/978-3-030-83225-4

Armbruster, M., Kohler, H., & Feger, K. H. (2000). Chemical composition of first-order forest streams in the upper Black Forest – Variability related to runoff dynamics and soil liming. *Forstwissenschaftliches Centralblatt, 119*(5), 249–262.

Arvidsson, J., & Hakansson, L. (1991). A model for estimating crop yield losses caused by soil compaction. *Soil and Tillage Research, 20*, 319–332.

Asner, G. P., Knapp, D. E., Broadbent, E. N., Oliveira, P. J. C., Keller, M., & Silva, J. N. (2005). Selective logging in the Brazilian Amazon. *Science, 310*(5747), 480–482.

Assmann, E. (1961). *Waldertragskunde. Organische Produktion, Struktur, Zuwachs und Ertrag von Waldbeständen*. BLV Verlagsgesellschaft.

Augustine, D. J. (2003). Long-term, livestock-mediated redistribution of nitrogen and phosphorus in an East African savanna. *Journal of Applied Ecology, 40*, 137–149.

Babin, S. M., Carton, J. A., Dickey, T. D., & Wiggert, J. D. (2004). Satellite evidence of hurricane-induced phytoplankton blooms in an oceanic desert. *Journal of Geophysical Research, 109*, C03043. https://doi.org/10.1029/2003JC001938

Baker, B. P., Benbrook, C. M., Groth, E., III, & Lutz Benbrook, K. (2002). Pesticide residues in conventional, IPM-grown and organic foods: Insights from three U.S. data sets. *Food Additives and Contaminants, 19*(5), 427–446.

Báldi, A. (1996). Edge effects in tropical versus temperate forest bird communities: Three alternative hypotheses for the explanation of differences. *Acta Zoologica Academiae Scientiarum Hungaricae, 42*(3), 163–172.

Baling, R. C. (2005). Observational surface temperature records versus model prediction. In P. J. Michaels (Ed.), *Shattered consensus: The true state of global warming* (pp. 50–71). Rowman & Littlefield.

Barrow, C. (1987). *Water resources and agricultural development in the tropics* (356 pp). Longman.

Bartoš, L. (1993). Sika red deer hybridization – Recent status and main problems. *Development in Animal and Veterinary Sciences, 26*, 415–416.

Battisti, A., Benvegnu, I., Colombari, F., & Haack, R. A. (2013). Invasion by the chestnut gall wasp in Italy causes significant yield loss in *Castanea sativa* nut production. *Agricultural and Forest Entomology, 16*(1), 75–79.

Baxa, M., Musil, M., Kummel, M., Hanzlík, P., Tesařová, B., & Pechar, L. (2020). Dissolved oxygen deficits in a shallow eutrophic aquatic ecosystem (fishpond) – Sediment oxygen demand and water column respiration alternately drive the oxygen regime. *Science of the Total Environment, 142647*. https://doi.org/10.1016/j.scitotenv.2020.142647

Bayala, J., & Prieto, I. (2020). Water acquisition, sharing and redistribution by roots: Applications to agroforestry systems. *Plant and Soil, 453*, 17–28.

Bayle, G. K. (2019). Ecological and social impacts of eucalyptus tree plantation on the environment. *Journal of Biodiversity Conservation and Bioresource Management, 5*, 93–104. https://doi.org/10.3329/jbcbm.v5i1.42189

Begon, M., Harper, J. L., & Townsend, C. R. (1986). *Ecology. Individuals, populations and communities*. Blackwell Scientific Publications.

Belhabib, D., Greer, K., & Pauly, D. (2018). Trends in industrial and artisanal catch per effort in West African fisheries. *Conservation Letters, 11*, e12360.

Belnap, J., & Gillette, D. A. (1998). Vulnerability of desert biological soil crusts to wind erosion: The influences of crust development, soil texture, and disturbance. *Journal of Arid Environments, 39*(2), 915–918.

Betts, R. A. (2000). Offset of the potential carbon sink from boreal forestation by decreases in surface albedo. *Nature, 408*, 187–190.

Beylich, A., Oberholzer, H. R., Schrader, S., Höper, H., & Wilke, B. M. (2010). Evaluation of soil compaction effects on soil biota and soil biological processes in soils. *Soil and Tillage Research, 109*, 133–143.

Bharati, L., Lee, K. H., Isenhart, T. M., & Schultz, R. C. (2002). Soil-water infiltration under crops, pasture, and established riparian buffer in Midwestern USA. *Agroforestry Systems, 56*, 249–257.

Bieber, C., & Ruf, T. (2005). Population dynamics in wild boar Sus scrofa: Ecology, elasticity of growth rate and implications for the managementof pulsed resource consumers. *Journal of Applied Ecology, 42*, 1203–1213.

BIO Intelligence Service. (2011). *Disturbances of EU forests caused by biotic agents*. Final report prepared for European Commission, BIO Intelligence Service.

Birkby, J. (2016). *Vertical farming*. ATTRA sustainable agriculture – NCAT, Butte, IP516.

Björnerfeldt, S., Webster, M. T., & Vilà, C. (2006). Relaxation of selective constraint on dog mitochondrial DNA following domestication. *Genome Research, 16*(8), 990–994. https://doi.org/10.1101/gr.5117706

Blankenship, R. E., Tiede, D. M., Barber, J., et al. (2011). Comparing photosynthetic and photovoltaic efficiencies and recognizing the potential for improvement. *Science, 332*(6031), 805–809. https://doi.org/10.1126/science.1200165

Blum, W. H. E. (2013). Soil and land resources for agricultural production: General trends and future scenarios A worldwide perspective. *International Soil and Water Conservation Research, 1*, 1–14.

Bohn, U., Gollub, G., Hettwer, C., Neuhäuslová, Z., Raus, T., Schlüter, H., & Weber, H. (2003). *Map of the natural vegetation of Europe*. Federal Agency for Nature Conservation (BfN).

Booth, D. J., Feary, D., Kobayashi, D., Luiz, O., & Nakamura, Y. (2018). Tropical marine fishes and fisheries and climate change. In B. F. Phillips & M. Pérez-Ramírez (Eds.), *Climate change impacts on fisheries and aquaculture: A global analysis II* (1st ed.). Wiley.

Bosma, R. H., & Verdegem, M. C. J. (2011). Sustainable aquaculture in ponds: Principles, practices and limits. *Livestock Science, 139*, 58–68.

Bourn, D., & Prescott, J. (2002). A comparison of the nutritional value, sensory qualities, and food safety of organically and conventionally produced foods. *Critical Reviews in Food Science and Nutrition, 42*(1), 1–34.

Boyd, W. (2001). Making meat science, technology, and American poultry production. *Technology and Culture, 42*, 631–664.

Brockman, C. F. (1968). *Trees of North America: A guide to field identification*. St Martins Press.

Brooker, R. W., Bennett, A. E., Cong, W. F., Daniell, T. J., George, T. S., Hallett, P. D., Hawes, C., Iannetta, P. P. M., Jones, H. G., Karley, A. J., Li, L., McKenzie, B. M., Pakeman, R. J., Paterson, E., Schöb, C., Shen, J., Squire, G., Watson, C. A., Zhang, C., … White, P. J. (2015). Improving intercropping: A synthesis of research in agronomy, plant physiology and ecology. *The New Phytologist, 206*, 107–117.

Brouwer, C., & Prins, K. (1989). *Irrigation water management: Irrigation scheduling*. FAO.

Broyer, J., & Curtet, L. (2012). Biodiversity and fish farming intensification in French fishpond systems. *Hydrobiologia, 694*(1), 205–218.

Brunner, P. H., & Rechberger, H. (2002). Anthropogenic metabolism and environmental legacies. In *Encyclopedia of global environmental change*. Wiley.

Buchwald, E. (2005, January 11–19). A hierarchical terminology for more or less natural forests in relation to sustainable management and biodiversity conservation. In *Proceedings. Third expert meeting on harmonizing forest-related definitions*. Edition: ftp://ftp.fao.org/docrep/fao/008/j4959e/j4959e.pdf. Chapter: Annex VIPublisher: FAO, Editors: FAO, Anonymous.

Burgett, M. (2010). Pacific northwestpolination surwey. *USDA Honey Market News*.

Buri, P., Humbert, J.-Y., & Arlettaz, R. (2014). Promoting pollinating insects in intensive agricultural matrices: Field-scale experimental manipulation of hay-meadow mowing regimes and its effects on bees. *PLoS One, 9*(1), e85635.

Butler, R. W. (1980). The concept of a tourism area cycle of evolution: Implication for management of resources. *Canadian Geographer, 24*(1), 5–12.

Cabral, R. B., Bradley, D., Mayorga, J., Goodell, W., Friedlander, A. M., Sala, E., Costello, C., & Gaines, S. D. (2020). A global network of marine protected areas for food. *Proceedings of the National Academy of Sciences, 117*(45), 28134–28139.

Caddy, J. F., & Mahon, R. (1995). *Reference points for fisheries management*. FAO fisheries technical paper 347 (p. 83). FAO.

Cadima, E. L. (2003). *Fish stock assessment manual*. FAO fisheries technical paper 393 (p. 161). FAO.

Calder, I., Hofer, T., Vermont, S., & Warren, P. (2007). Towards a new understanding of forests and water. *Unasylva, 58*, 3–10.

Callaway, E. (2018). CRISPR plants now subject to tough GM laws in European Union. Top court's ruling threatens research on gene-edited crops in the bloc. *Nature, 560*, 16. https://doi.org/10.1038/d41586-018-05814-6

Cambi, M., Certini, G., Neri, F., & Marchi, E. (2015). The impact of heavy traffic on forest soils: A review. *Forest Ecology and Management, 338*, 124–138.

Carlsson, G., & Huss-Danell, K. (2003). Nitrogen fixation in perennial forage legumes in the field. *Plant and Soil, 253*, 353–372.

Carson, R. (1962). *Silent spring*. Houghton Mifflin.

Cejpek, J., Kuráž, V., Vindušková, O., & Frouz, J. (2017). Water regime of reclaimed and unreclaimed post-mining sites. *Ecohydrology, 11*, e1911.

Chaber, A. L., Allebone-Webb, S., Lignereux, Y., Cunningham, A. A., & Rowcliffe, J. M. (2010). The scale of illegal meat importation from Africato Europe via Paris. *Conservation Letters, 3*(5), 317–321.

Challender, D. W. S., & MacMillan, D. C. (2014). Poaching is more than an enforcement problem. *Conservation Letters, 7*, 484–494.

Chapin, F. S., III, Matson, P. A., & Vitousek, P. (2011). *Principles of terrestrial ecosystem ecology*. Springer.

Chassé, P., Pelosi, C., Lata, J. C., & Barot, S. (2019). Impact of crop genetic diversity on a litter consumer. *Basic and Applied Ecology, 36*, 1–11.

Chaudhary, A., Burivalova, Z., Koh, L. P., & Hellweg, S. (2016). Impact of forest management on species richness: Global meta-analysis and economic trade-offs. *Scientific Reports, 6*, 23954.

Chave, J., Coomes, D., Jansen, S., Lewis, S. L., Swenson, N. G., & Zanne, A. E. (2009). Towards a worldwide wood economics spectrum. *Ecology Letters, 12*, 351–366. https://doi.org/10.1111/j.1461-0248.2009.01285.x

Chen, X., Cui, Z., Fan, M., et al. (2014). Producing more grain with lower environmental costs. *Nature, 514*(7523), 486–489. https://doi.org/10.1038/nature13609

Chimeli, A. B., & Soares, R. R. (2017). The use of violence in illegal markets: Evidence from mahogany trade in the Brazilian Amazon. *American Economic Journal: Applied Economics, 9*(4), 30–57.

Churchill, S. E. (1993). Weapon technology, prey size selection, and hunting methods in modern hunter-gatherers: Implications for hunting in the palaeolithic and mesolithic. *Archeological Papers of the American Anthropological Association, 4*, 11–24.

Clark, D. A., & Clark, D. B. (1984). Spacing dynamics of a tropical rainforest tree: Evaluation of the Janzen-Connel model. *The American Naturalist, 124*, 769–788.

Clayton, S. (2007). Domesticated nature: Motivations for gardening and perceptions of environmental impact. *Journal of Environmental Psychology, 27*(3), 215–224.

Cleveland, C. C., Townsend, A. R., Schimel, D. S., Fisher, H., Howarth, R. W., Hedin, L. O., Perakis, S. S., Latty, E. F., Von Fischer, J. C., Elseroad, A., & Wasson, M. F. (1999). Global patterns of terrestrial biological nitrogen (N_2) fixation in natural ecosystems. *Global Biogeochamical Cycles, 13*, 623–645.

Cochrane, K. L. (2002). *A fishery manager's guidebook – Management measures and their application*. Fisheries technical paper 424. FAO.

Connell, J. H. (1970). On the role of natural enemies in preventing competitive exclusion in some marine animals and in rain forest trees. In P. J. Den Boer & G. R. Gradwell (Eds.), *Dynamics of population*. Pudoc.

Cooke, S. J., & Cowx, I. G. (2004). The role of recreational fishing in global fish crises. *Bioscience, 54*(9), 857–859.

Corre, M. D., Beese, F. O., & Brumme, R. (2003). Soil nitrogen cycle in high nitrogen deposition forest: Changes under nitrogen saturation and liming. *Ecological Applications, 13*(2), 287–298.

Costanza, R., d'Arge, R., de Groot, R., Farber, S., Grasso, M., Hannon, B., Limburg, K., Naeem, S., O'Neill, R. V., Paruelo, J., Raskin, R. G., Sutton, P., & van den Belt, M. (1997). The value of the world's ecosystem services and natural capital. *Nature, 387*(6630), 253–260.

Costanza, R., Sutton, P., van der Ploeg, S., Anderson, A. J., Kubiszewski, I., Farber, S., & Turner, R. K. (2014). Changes in the global value of ecosystem services. *Global Environmental Change, 26*, 152–158.

Cresswell, W. (2008). Non-lethal effects of predation in birds. *Ibis, 150*, 3–17.

Crews, T. E., & Peoples, M. B. (2004). Legume versus fertilizer sources of nitrogen: Ecological tradeoffs and human needs agriculture. *Ecosystems and Environment, 102*, 279–297.

Crutzen, P. J., & Stoermer, E. F. (2000). The antropocene. *Global Change Newsletters, 41*, 17–18.

Culman, S. W., Snapp, S. S., Ollenburger, M., Basso, B., & DeHaan, L. R. (2013). Soil and water quality rapidly responds to the perennial grain an Kernza wheatgrass. *Agronomy Journal, 105*(3), 735–744.

Cunningham, S. A., Floyd, R. B., & Weir, T. A. (2005). Do *Eucalyptus* plantations host an insect community similar to remnant *Eucalyptus* forest? *Austral Ecology, 30*, 103–117.

Cury, P. M., Boyd, I. L., Bonhommeau, S., Anker-Nilssen, T., Crawford, R. J. M., Furness, R. W., Mills, J. A., Murphy, E. J., Österblom, H., Paleczny, M., Piatt, J. F., Roux, J. P., Shannon, L., & Sydeman, W. J. (2011). Global seabird response to forage fish depletion – One-third for the birds. *Science, 334*, 1703–1706.

Cury, P., Bertrand, A., Bertrand, S., Coll, M., Gros, P., Kifani, S., Le Loch, F., Maury, O., Menard, F., Renaud, F., Shannon, L., & Shin, Y. J. (2016). The ecosystem approach to fisheries: Reconciling conservation and exploitation. In A. Monaco & P. Prouzet (Eds.), *Tools for oceanography and ecosystemic modeling*. ISTE Ltd/Wiley.

Da Ronch, F., Caudullo, G., & de Rigo, D. (2016). *Pseudotsuga menziesii* in Europe: Distribution, habitat, usage and threats. In J. San-Miguel-Ayanz, D. de Rigo, G. Caudullo, T. Houston Durrant, & A. Mauri (Eds.), *European atlas of forest tree species*. European Comission.

Dahlgren, R. A., & Singer, M. J. (1994). *Nutrient cycling in managed and non-managed oak woodland-grass ecosystems*. Final report: Integrated hardwood range management program. Land, air and water resources paper 100028. University of California.

Daniels, R. F., & Burkhart, H. E. (1988). An integrated system of forest stand models. *Forest Ecology and Management, 23*, 159–177.

Darimont, C. T., Fox, C. H., Bryan, H. M., & Reimchen, T. E. (2015). The unique ecology of human predators. *Science, 349*(6250), 858–860.

de Bertoldi, M., Vallini, G., & Pera, A. (1983). The biology of composting: A review. *Waste Management & Research, 1*(2), 157–176. https://doi.org/10.1016/0734-242X(83)90055-1

de Souza, J. G., Schaan, D. P., Robinson, M., et al. (2018). Pre-Columbian earth-builders settled along the entire southern rim of the Amazon. *Nature Communications, 9*, 1125.

de Vries, F. T., Thébault, E., Liiri, M., Birkhofer, K., Tsiafouli, M. A., Bjornlund, L., Jorgensen, H., Bracht, B., Mark, V., Christensen, S., de Ruiter, P. C., d'Hertefeldt, T., Frouz, J., Hedlund, K., Hemerik, L., Hol, W. H. G., Hotes, S., Mortimer, S. R., Setala, H., … Bardgett, R. D. (2013). Soil food web properties explain ecosystem services across European land use systems. *Proceedings of the National Academy of Sciences of the United States of America, 110*, 14296–14301.

DeFoliart, G. (1994). Edible insects as minilivestock. *Biodiversity and Conservation, 4*(3), 306–321.

Dias, L. C. P., Macedo, M. N., Costa, M. H., Coe, M. T., & Christopher, N. C. (2015). Effects of land cover change on evapotranspiration and streamflow of small catchments in the Upper Xingu River Basin, Central Brazil. *Journal of Hydrology: Regional Studies, 4*, 108–122.

Dinerstein, E., Vynne, C., Sala, E., Joshi, A. R., Fernando, S., Lovejoy, T. E., Mayorga, J., Olson, D., Asner, G. P., Baillie, J. E. M., Burgess, N. D., Burkart, K., Noss, R. F., Zhang, Y. P., Baccini, A., Birch, T., Hahn, N., Joppa, L. N., & Wikramanayake, E. (2019). A global deal for nature: Guiding principles, milestones, and targets. *Science Advances, 5*, eaaw2869.

Dixon, R. K., Solomon, A. M., Brown, S., Houghton, R. A., Trexier, M. C., & Wisniewski, J. (1994). Carbon pools and flux of global forest ecosystems. *Science, 263*(5144), 185–190.

Dolan, C., & Humphrey, J. (2000). Governance and trade in fresh vegetables: The impact of UK supermarkets on the African horticulture industry. *The Journal of Development Studies, 37*, 147–176.

Donald, P. F., Pisano, G., Rayment, M. D., & Pain, D. J. (2002). The common agricultural policy, EU enlargement and the conservation of Europe's farmland birds. *Agriculture, Ecosystems and Environment, 89*, 167–182.

Doughty, C. E., Roman, J., Faurby, S., Wolf, A., Haque, A., Bakker, E. S., Malhi, Y., Dunning, J. B., & Svennin, J. C. (2016). Global nutrient transport in a world of giants. *PNAS, 113*(4), 868–873.

Edixhoven, J. D., Gupta, J., & Savenije, H. H. G. (2013). Recent revisions of phosphate rock reserves and resources: Reassuring or misleading? An in-depth literature review of global estimates of phosphate rock reserves and resources. *Earth System Dynamics, 5*(2), 491–507.

Edwards, C. A., & Subler, S. (2011). Human pathogen reduction during vermicomposting. In C. A. Edwards, N. Q. Arancon, & R. Sherman (Eds.), *Vermiculture technology* (pp. 249–261). CRC Press/Taylor & Francis Group.

Ellis, E. C., & Ramankutty, N. (2008). Putting people in the map: Anthropogenic biomes of the world. *Frontiers in Ecology and the Environment, 6*, 439–447.

Ellis, E. C., Kaplan, J. O., Fuller, D. Q., Vavrus, S., Klein Goldewijk, K., & Verburg, P. H. (2013). Used planet: A global history. *Proceedings of the National Academy of Sciences, 110*(20), 7978–7985.

Erisman, J. W., Sutton, M. A., Galloway, J., Klimont, Z., & Winiwarter, W. (2008). How a century of ammonia synthesis changed the world. *Nature Geoscience, 1*(10), 636–639.

Esper, J., Büntgen, U., Frank, D. C., Nievergelt, D., & Liebhold, A. (2007). 1200 years of regular outbreaks in alpine insects. *Proceedings of the Royal Society B-Biological Sciences, 274*(1610), 671–679.

Estes, J. A., & Duggins, D. O. (1995). Sea otters and kelp forest in Alaska: Generality and variation in a community ecological paradigm. *Ecological Monographs, 65*, 75–100.

EU Nitrogen Expert Panel. (2015). *Nitrogen use efficiency (NUE) – An indicator for the utilization of nitrogen in agriculture and food systems*. Wageningen University.

Evans, J., McNeill, A. M., Unkovich, M. J., Fettell, N. A., & Heenan, D. P. (2001). Net nitrogen balances for cool-season grain legume crops and contributions to wheat nitrogen uptake: A review. *Australian Journal of Experimental Agriculture, 41*(3), 347–359.

Fahrig, L., Girard, J., Duro, D., Pasher, J., Smith, A., Javorek, S., King, D., Lindsay, K. F., Mitchell, S., & Tischendorf, L. (2015). Farmlands with smaller crop fields have higher within-field biodiversity. *Agriculture, Ecosystems and Environment, 200*, 219–234.

Falge, E., Baldocchi, D., Tenhunen, J., Aubinet, M., Bakwin, P., Berbigier, P., Bernhofer, C., Burba, G., Clement, R., Davis, K., Elbers, J., Goldstein, A. H., Grelle, A., Granier, A., Gudmundsson, J., Hollinger, D., Kowalski, A., Katul, G., Law, B., & Suyker, A. (2002). Seasonality of ecosystem respiration and gross primary production as derived from FLUXNET measurements. *Agricultural Forest Meteorology, 113*(1–4), 53–74.

FAO. (1984). Papers presented at the expert consultation on the regulation of fishing effort (fishing mortality). In *A preparatory meeting for the FAO World Conference on fisheries management and development* (2nd ed., p. 214). FAO fishery report 289.

FAO. (1998a). *State of the world plant genetic resourses for food and agriculture*. FAO.

FAO. (1998b). *Integrated coastal area management and agriculture, forestry and fisheries*. FAO.

FAO. (2008). *The state of food insecurity in the world.* Annual report.

FAO. (2010). *Global Forest Resources Assessment 2010*. FAO.

FAO. (2011). *Global food losses and food waste: Extent, causes and prevention*. FAO.

FAO. (2016). *The state of world fisheries and aquaculture 2016. Contributing to food security and nutrition for all*. FAO.

FAO. (2017). *Fishery and aquaculture country profiles. China. Country profile fact sheets*. FAO Fisheries Division

Ferreira, V., Boyero, L., Calvo, C., et al. (2019). A global assessment of the effects of eucalyptus plantations on stream ecosystem functioning. *Ecosystems, 22*, 629–642.

Fiedel, S. J. (2005). Man's best friend – Mammoth's worst enemy? A speculative essay on the role of dogs in Paleoindian colonization and megafaunal extinction. *World Archaeology, 37*(1), 11–25.

Filková, V., Kolár, T., Rybníček, M., Gryc, V., Vavrčík, H., & Jurčík, J. (2015). Historical utilization of wood in southeastern Moravia (Czech Republic). *iForest-Biogeosciences and Forestry, 8*(1), 101–107.

Finley, C. (2016). The industrialization of commercial fishing, 1930–2016. In *Oxford research encyclopedia of environmental science*. https://doi.org/10.1093/acrefore/9780199389414.013.31

Fischer, J., Abson, J. D., Butsic, V., Chappell, M. J., Ekroos, J., Hanspach, J., Kuemmerle, T., Smith, H. G., & von Wehrden, H. (2014). Land sparing versus land sharing: Moving forward. *Conservation Letters, 7*, 149–157.

Fitzherbert, E. B., Struebig, M. J., Morel, A., Danielsen, F., Brül, C. A., Donald, P. F., & Phalan, B. (2008). How will oil palm expansion affect biodiversity*? Trends in Ecology & Evolution*, 993.

Floate, K. D., Düring, R. A., Hanafi, J., Jud, P., Lahr, J., Lumaret, J.-P., Scheffczyk, A., Tixier, T., Wohde, M., Römbke, J., Sautot, L., & Blanckenhorn, W. U. (2016). Validation of a standard field test method in four countries to assess the toxicity of residues in dung of cattle treated with veterinary medical products. *Environmental Toxicology and Chemistry, 35*, 1934–1946.

Foglar-Deinhardstein, A., Piribauer, V. C., & Prem, J. (2015). *Sustainable forest management in Austria – Austrian forest report 2015*. Republic of Austria, Federal Ministry of Agriculture, Forestry, Environment and Water Management.

Foley, J. A., Prentice, I. C., Ramankutty, N., Levis, S., Pollard, D., Sitch, S., & Haxeltine, A. (1996). An integrated biosphere model of land surface processes, terrestrial carbon balance and vegetation dynamics. *Global Biogeochemical Cycles, 10*(4), 603–628.

Foley, J. A., DeFries, R., Asner, G. P., Barford, C., Bonan, G., Carpenter, S. R., Chapin, F. S., Coe, M. T., Daily, G. C., Gibbs, H. K., Helkowski, J. H., Holloway, T., Howard, E. A., Kucharik, C. J., Monfreda, C., Patz, J. A., Prentice, I. C., Ramankutty, N., & Snyder, P. K. (2005). Global consequences of land use. *Science, 309*, 570–574.

Fonseca, T. F., Abreu, C. G., & Parresol, B. R. (2004). Soil compaction and chestnut ink disease. *Forest Pathology, 34*, 273–283.

FOREST EUROPE. (2015). *State of Europe's forests 2015, ministerial conference on the protection of forests in Europe*. FOREST EUROPE Liaison Unit.

Forestry Agency Japan. (2019). *State of Japan's forests and forest management*. Forestry Agency Japan.

Franklin, J. F., Berg, D. R., Thornburgh, D. A., & Tappeiner, J. C. (1997). Alternative silvicultural approaches to timber harvesting: Variable retention harvest systems. In K. A. Kohn & J. F. Franklin (Eds.), *Creating a forestry for the 21st century: The science of ecosystem management* (p. 111). Island Press.

Friel, S., Schram, A., & Townsend, B. (2020). The nexus between international trade, food systems, malnutrition and climate change. *Nature Food, 1*, 51–58.

Frost, R. A., & Launchbaugh, K. L. (2003). Prescription grazing for rangeland weed management. *Rangelands, 25*, 43–47.

Frouz, J. (2018a). Effects of soil macro- and mesofauna on litter decomposition and soil organic matter stabilization. *Geoderma, 332*, 161–172.

Frouz, J. (2018b). Changes of water budget during ecosystem development in post-mining sites at various spatiotemporal scales: The need for controlled systems. In J.-F. Liu & W.-Z. Gu (Eds.), *Hydrology of artificial and controlled experiments*. IntechOpen.

Frouz, J. (2021). Chapter 6: Soil recovery and reclamation of mined lands. In J.-A. Stanturf & M. A. Callaham (Eds.), *Soils and landscape restoration* (pp. 161–191). Academic Press.

Frouz, J., & Franklin, J. A. (2014). Vegetation and soil development in planted pine and naturally regenerated hardwood stands 48 years after mining. *Journal American Society of Mining and Reclamation, 3*(2). https://doi.org/10.21000/JASMR14020021

Frouz, J., & Vindušková, O. (2018). Soil organic matter accumulation in postmining sites: Potential drivers and mechanisms. In M. Á. Munoz & R. Zornoza (Eds.), *Soil management and climate change. Effects on organic carbon, nitrogen dynamics and greenhouse gas emissions*. Academic Press.

Frouz, J., Šantrůčková, H., & Kalčík, J. (1997). The effect of wood ants (*Formica polyctena* Foerst) on the transformation of phosphorus in a spruce plantation. *Pedobiologia, 41*, 437–447.

Frouz, J., Keplin, B., Pižl, V., Tajovský, K., Starý, J., Lukešová, A., Nováková, A., Balík, V., Háněl, L., Materna, J., Düker, C., Chalupský, J., Rusek, J., & Heinkele, T. (2001). Soil biota and upper soil layers development in two contrasting post-mining chronosequences. *Ecological Engineering, 17*, 275–284.

Frouz, J., Prach, K., Pižl, V., Háněl, L., Starý, J., Tajovský, K., Materna, J., Balík, V., Kalčík, J., & Řehounková, K. (2008). Interactions between soil development, vegetation and soil fauna during spontaneous succession in post mining sites. *European Journal of Soil Biology, 44*, 109–121.

Frouz, J., Kalčík, J., & Syrovátka, O. (2010a). The effect of pipe drainage on peat meadow soil: Physical and chemical soil properties. *Environment, 24*, 83–90.

Frouz, J., Pižl, V., Tajovský, K., & Syrovátka, O. (2010b). The effect of pipe drainage on peat meadow soil: Soil macrofauna. *Environment, 24*, 91–108.

Frouz, J., Jílková, V., Cajthaml, T., Pižl, V., Tajovský, K., Háněl, L., Burešová, A., Šimáčková, H., Kolaříková, K., Franklin, J., Nawrot, J., Groninger, J. W., & Stahl, P. D. (2013a). Soil biota in post-mining sites along a climatic gradient in the USA: Simple communities in shortgrass prairie recover faster than complex communities in tallgrass prairie and forest. *Soil Biology and Biochemistry, 67*, 212–225.

Frouz, J., Livečková, M., Albrechtová, J., Chroňáková, A., Cajthaml, T., Pižl, V., Háněl, L., Starý, J., Baldrian, P., Lhotáková, Z., Šimáčková, H., & Cepáková, Š. (2013b). Is the effect of trees on soil properties mediated by soil fauna? A case study from post-mining sites. *Forest Ecology and Management, 309*, 87–95.

Frouz, J., Špaldoňová, A., Fričová, K., & Bartuška, M. (2014). The effect of earthworms (*Lumbricus rubellus*) and simulated tillage on soil organic carbon in a long-term microcosm experiment. *Soil Biology and Biochemistry, 78*, 58–64.

Frouz, J., Vobořilová, V., Janoušová, I., Kadochová, Š., & Matějíček, L. (2015a). Spontaneous establishment of late successional tree species English oak (Quercus robur) and European beech (Fagus sylvatica) at reclaimed alder plantation and unreclaimed post mining sites. *Ecological Engineering, 77*, 1–8.

Frouz, J., Dvorščík, P., Vávrová, A., Doušová, O., Kadochová, Š., & Matějíček, L. (2015b). Development of canopy cover and woody vegetation biomass on reclaimed and unreclaimed post-mining sites. *Ecological Engineering, 84*, 233–239.

Frouz, J., Jílková, V., & Sorvari, J. (2016a). Contribution of wood ants to nutrient cycling and ecosystem functionings. In S. J. A. Robinson & E.J.H. (Eds.), *Wood ant ecology and conservation*. Cambridge University Press.

Frouz, J., Toyota, A., Mudrák, O., Jílková, V., Filipová, A., & Cajthaml, T. (2016b). Effects of soil substrate quality, microbial diversity and community composition on the plant community during primary succession. *Soil Biology and Biochemistry, 99*, 75–84.

Frouz, J., Mudrák, O., Reitschmiedová, E., Walmsley, A., Váchová, P., Šimáčková, H., Albrechtová, J., Moradi, J., & Kučera, J. (2018). Rough wave-like heaped overburden promotes establishment of woody vegetation while leveling promotes grasses during unassisted post mining site development. *Journal of Environmental Management, 205*, 50–58.

Gajic, B., Dugalic, G., Sredojevic, Z., & Tomic, Z. (2008). Effect of different vegetation types on infiltration and soil water retention. *Cereal Research Communications, 36*, 991–994.

Galloway, J. N., Townsend, A. R., Erisman, J. W., Bekunda, M., Cai, Z., Freney, J. R., Martinelli, L. A., Seitzinger, S. P., & Sutton, M. A. (2008). Transformation of the nitrogen cycle, recent trends, questions, and potential solutions. *Science, 20*(5878), 889–892.

Gaston, K. (2000). Global patterns in biodiversity. *Nature, 405*, 220–227.

Gereffi, G. (1994). The organisation of buyer-driven global commodity chains: How U.S. retailers shape overseas production networks. In G. Gereffi & M. Korzeniewicz (Eds.), *Commodity chains and global capitalism* (pp. 95–122). Praeger Publishers.

Gerten, D., Hoff, H., Bondeau, A., Lucht, W., Smith, P., & Zaehle, S. (2005). Contemporary "green" water flows: Simulations with a dynamic global vegetation and water balance model. *Physics and Chemistry of the Earth, Parts A/B/C, 30*(6–7), 334–338.

Gill, D., Mascia, M., Ahmadia, G., et al. (2017). Capacity shortfalls hinder the performance of marine protected areas globally. *Nature, 543*, 665–669.

Gilman, E., Kobayashi, D., & Chaloupka, M. (2008). Reducing seabird bycatch in the Hawaii longline tuna fishery. *Endangered Species Research, 5*, 309–323.

Gizicki, Z. S., Tamez, V., Galanopoulou, A. P., Avramidis, P., & Foufopoulos, J. (2017). Long-term effects of feral goats (*Capra hircus*) on Mediterranean island communities: Results from whole island manipulations. *Biological Invasions, 20*(1), 1537–1552.

Gleason, K. E., Bradford, J. B., Bottero, A., D'Amato, A. W., Fraver, S., Palik, B. J., Battaglia, M. A., Iverson, L., Kenefic, L., & Kern, C. C. (2017). Competition amplifies drought stress in forests across broad climatic and compositional gradients. *Ecosphere, 8*(7), e01849. https://doi.org/10.1002/ecs2.1849

Glutz von Blotzheim, U. N. (2015). Illegal bird hunting in the EU: Discrepancy between written words and effective action. *GNOR Info, 123*, 18–23.

Goded, S., Ekroos, J., Domínguez, J., Azcárate, J. G., Guitián, J. A., & Smith, H. G. (2019). Effects of eucalyptus plantations on avian and herb species richness and composition in North-West Spain. *Global Ecology and Conservation, 19*, e00690.

Goklany, I. M. (2002). The ins and outs of organic farming. *Science, 298*(5600), 1889–1890.

Golley, F. B., Petrides, G. A., Rauber, E. L., & Jenkins, J. H. (1965). Food intake and assimilation by bob-cats under laboratory conditions. *Journal of Wildlife Management, 29*, 442–447.

Gonzalez-Sanchez, E. J., Veroz-Gonzalez, O., Blanco-Roldan, G. L., Marquez-Garcia, F., & Carbonell-Bojollo, R. (2015). A renewed view of conservation agriculture and its evolution over the last decade in Spain. *Soil and Tillage Research, 146*, 204–212.

Goosse, H., Barriat, P. Y., Lefebvre, W., Loutre, M. F., & Zunz, V. (2010). *Introduction to climate dynamics and climate modeling*. Online textbook available at http://www.climate.be/textbook

Gosselin, J., Zedrosser, A., Swenson, J. E., & Pelletier, F. (2015). The relative importance of direct and indirect effects of hunting mortality on the population dynamics of brown bears. *Proceedings of the Royal Society B, 282*, 20141840.

Gray, D. M. (Ed.). (1973). *Handbook on the principles of hydrology*. Water Information Center.

Green, R. E., Cornell, S. J., Scharlemann, J. P. W., & Balmford, A. (2005). Farming and the fate of wild nature. *Science, 307*, 550–555.

Griscom, B. W., Adams, J., Ellis, P. W., Houghton, R. A., Lomax, G., Miteva, D. A., Schlesinger, W. H., Shoch, D., Siikamäki, J. V., Smith, P., Woodbury, P., Zganjar, C., Blackman, A., Campari, J., Conant, R. T., Delgado, C., Elias, P., Gopalakrishna, T., Hamsik, M. R., … Fargione, J. (2017). Natural climate solutions. *Proceedings of the National Academy of Sciences, 114*(44), 11645–11650.

Grove, A. T., & Rackham, O. (2003). *The nature of Mediterranean Europe: An ecological history*. Yale University Press.

Guerrero, I., Morales, M. B., Onate, J. J., Aavik, T., Bengtsson, J., Berendse, F., Clement, L. W., Dennis, C., Eggers, S., Emmerson, M., Fischer, C., Flohre, A., Geiger, G., Hawro, V., Inchausti, P., Kalamees, A., Kinks, R., Liira, J., Melendez, L., … Weisser, W. W. (2011). Taxonomic and functional diversity of farmland bird communities across Europe: Effects of biogeography and agricultural intensification. *Biodiversity and Conservation, 20*, 3663–3681.

Güsewell, S. (2004). N: P ratios in terrestrial plants: Variation and functional significance. *New Phytologist, 164*, 243–266.

Gustafsson, L., Kouki, J., & Sverdrup –Thygeson, A. (2010). Tree retention as a conservation measure in clear-cut forests of northern Europe: A review of ecological consequences. *Scandinavian Journal of Forest Research, 25*(4), 295–308.

Haberl, H., et al. (2009). Using embodied HANPP to analyze teleconnections in the global land system: Conceptual considerations. *Danish Journal of Geography, 109*(2), 119–130.

Hák, T., et al. (2015). *Metabolismus společnosti. Materiály, energie, území*. Karolinum.

Hamberger, J., & von Carlowitz, H. C. (2013). *Sylvicultura oeconomica oder Haußwirthliche Nachricht und Naturmäßige Anweisung zur Wilden Baum-Zucht Gebundene Ausgabe*. Oekom.

Hames, R. B. (1979). A comparison of the efficiencies of the shotgun and the bow in neotropical forest hunting. *Human Ecology, 7*, 219–252.

Hamilton, S., & Baker, G. B. (2019). Technical mitigation to reduce marine mammal bycatch and entanglement in commercial fishing gear: Lessons learnt and future directions. *Reviews in Fish Biology and Fisheries, 29*, 223–247. https://doi.org/10.1007/s11160-019-09550-6

Harrington, J. L., & Conover, M. R. (2006). Characteristics of ungulate behavior and mortality associated with fences. *Wildlife Society Bulletin, 34*, 1295–1305.

Harris, R., Wall, W., & Allendorf, F. (2002). Genetic consequences of hunting: What do we know and what should we do? *Wildlife Society Bulletin, 30*(2), 634–643.

Harris, N. L., Brown, S., Hagen, S. C., Saatchi, S. S., Petrova, S., Salas, W., Hansen, M. C., Potapov, P. V., & Lotsch, A. (2012). Baseline map of carbon emissions from deforestation in tropical regions. *Science, 336*, 1573–1576.

Hart, T. (2016). *Summary of pre-industrial forest condition of English River, Caribou, Black Spruce and Dog River-Matawin Forests, Ontario*. Resolution, Forest Product.

Hartvich, P. (2012). *Význam organických hnojiv pro výživu rybniční biocenózy. Intenzivní metody chovu ryb a ochrana kvality vody*. Dada.

Hati, K. M., Swarup, A., Mishra, B., Manna, M. C., Wanjari, R. H., Mandal, K. G., & Misra, A. K. (2008). Impact of long-term application of fertilizer, manure and lime under intensive cropping on physical properties and organic carbon content of an Alfisol. *Geoderma, 148*, 173–179.

Haws, I. D., Inoue, H., Tanaka, A., & Yoshida, S. (1983). Comparison of crop productivity in the tropics and temperate zone. In *Potential productivity of field corops under different environment* (pp. 403–414). International Rice Research Institute.

Heal, O. W., & MacLean, S. F. (1975). Comparative productivity in ecosystems – Secondary productivity. In W. H. van Dobben & R. H. Lowe-McConnell (Eds.), *Unifying concepts in ecology* (pp. 89–108). Springer.

Heffelfinger, J. R., Geist, V., & Wishart, W. (2013). The role of hunting in North American wildlife conservation. *International Journal of Environmental Studies, 70*(3), 399–413.

Heitefuss, R. (1989). *Crop and plant protection: The practical foundations*. Ellis Hotwood Ltd. ISBN 0 7458 0523 X.

Hejzlar, J., Borovec, J., Polívka, J., Turek, J., & Volková, A. (2010). Bilanční studie zdrojů živin v povodí nádrže Orlík 2. scénářová analýza pro návrh strategie snižování odnosu fosforu (A mass balance study of nutrient sources in the catchment of Orlík Reservoir 2. Scenario analysis for proposal of strategy to decrease ph). In J. Borovec & I. Očásková (Eds.), *Sborník příspěvků Revitalizace Orlické nádrže 2010* (pp. 67–80). Hydrobiologický ústav Biologické Centrum AV ČR.

Hettelingh, J. P., Posch, M., & Potting, J. (2005). Country-dependent characterisation factors for acidification in Europe – A critical evaluation. *International Journal of Life Cycle Assessment, 10*(3), 177–183.

Hindar, A., Wright, R. F., Nilsen, P., Larssen, T., & Hogberget, R. (2003). Effects on stream water chemistry and forest vitality after whole-catchment application of dolomite to a forest ecosystem in southern Norway. *Forest Ecology and Management, 180*, 509–525.

Hippert, C. (2018). Agriculture and Colonialism. In P. Thompson & D. Kaplan (Eds.), *Encyclopedia of food and agricultural ethics*. Springer.

Hoekstra, J. M., Boucher, T. M., Ricketts, T. H., & Roberts, C. (2005). Confronting a biome crisis: Global disparities of habitat loss and protection. *Ecology Letters, 8*, 23–29.

Hoersrna, G. J., Kannenberg, L. W., & Christine, B. R. (1985). Grain yield comparison of pure stands and equal proportion mixtures for seven hybrids of maize. *Canadian Journal of Plant Science, 65*, 471–479.

Hofmeister, J., Oulehle, F., Krám, P., & Hruška, J. (2008). Loss of nutrients due to litter raking compared to the effect of acidic deposition in two spruce stands, Czech Republic. *Biogeochemistry, 88*(2), 139–151.

Hole, D. G., Perkins, A. J., Wilson, J. D., Alexander, I. H., Grice, P. V., & Evans, A. D. (2005). Does organic farming benefit biodiversity? *Biological Conservation, 122*, 113–130.

Hommeltenberg, J., Schmid, H. P., Drösler, M., & Werle, P. (2014). Can a bog drained for forestry be a stronger carbon sink than a natural bog forest? *Biogeosciences, 11*, 3477–3493.

Hooper, D., Adair, E., Cardinale, B., et al. (2012). A global synthesis reveals biodiversity loss as a major driver of ecosystem change. *Nature, 486*, 105–108.

Horn, R., Fleige, H., Peth, S., & Peng, X. (Eds.). (2006). Soil management for sustainability. *Advances in Geoecology, 38*, 436.

Houghton, R. A. (2007). Balancing the global carbon budget. *Annual Review of Earth and Planetary Sciences, 35*, 313–347.

Howarth, R. W. (2008). Coastal nitrogen pollution: A review of sources and trends globally and regionally. *Harmful Algae, 8*, 14–20.

Hüblová, L., & Frouz J. (2021). Contrasting effect of coniferous and broadleaf trees on soil carbon storage during reforestation of forest soils and afforestation of agricultural and post-mining soils. *Journal of Environmental Management, 290*, 112567. https://doi.org/10.1016/j.jenvman.2021.112567

Hubbard, R. K., Newto, L., & Hill, G. M. (2004). Water quality and the grazing animal. *Journal of Animal Science, 82*(E-Suppl), E255–E263.

Hudson, R. J. M., Gherini, S. A., & Goldstein, R. A. (1994). Modeling the global carbon cycle: Nitrogen fertilization of the terrestrial biosphere and the "missing" CO_2 sink. *Global Biogeochemical Cycles, 8*, 307–333.

Huggins, D. R., Buyanovsky, G. A., Wagne, G. H., Brown, J. R., Darmody, R. G., Peckc, T. R., Lesoingd, G. W., Vanottie, M. B., & Bundyf, L. G. (1998). Soil organic C in the tallgrass prairie-derived region of the corn belt: Effects of long-term crop management. *Soil & Tillage Research, 47*, 219–234.

Huntley, B. (1993). Species-richness in north-temperate zone forests. *Journal of Biogeography, 20*, 163–180.

Huston, M. A., & Wolverton, S. (2009). The global distribution of net primary production: Resolving the paradox. *Ecological Monographs, 79*(3), 343–377.

IGN. (2019). *Le Mémemto inventaire forestier*. Institut national de l'information géographique et forestière.

INFC. (2009). I caratteri quantitativi 2005 – Parte 1, versione 2. In P. Gasparini, F. De Natale, L. Di Cosmo, C. Gagliano, G. Salvadori, G. Tabacchi, & V. Tosi (Eds.), *Inventario nazionale delle foreste e dei serbatoi forestali di carbonio. MiPAAF – Ispettorato Generale Corpo Forestale dello Stato*. CRA-MPF.

Ingerslev, M. (1997). Effects of liming and fertilization on growth, soil chemistry and soil water chemistry in a Norway spruce plantation on a nutrient-poor soil in Denmark. *Forest Ecology and Management, 92*, 55–66.

ISAAA. (2018). ISAAA Brief 54-2018.

IUSS Working Group WRB. (2015). *World reference base for soil resources 2014, update 2015 International soil classification system for naming soils and creating legends for soil maps*. World soil resources reports no. 106. FAO.

Janssens, I. A., Freibauer, A., Ciais, P., Smith, P., Nabuurs, G. J., Folberth, G., Schlamadinger, B., Hutjes, R. W. A., Ceulemans, R., Schulze, E. D., Valentini, R., & Dolman, A. J. (2003). Europes terrestrial biosphere absorbs 7 to 12% of European anthropogenic CO_2 emissions. *Science, 300*(5625), 1538–1542.

Janzen, D. H. (1970). Herbivores and the number of tree species in tropical forests. *The American Naturalist, 104*(940), 940. https://doi.org/10.1086/282687

Jedrzejewski, W., Schmidt, K., Theuerkauf, J., & Kowalczyk, R. (2007). Territory size of wolves Canis lupus: Linking local (Białowieża Primeval Forest, Poland) and Holarctic-scale patterns. *Ecography, 30*(1), 66–76.

Jing, J., Søegaard, K., Cong, W. F., & Eriksen, J. (2017). Species diversity effects on productivity, persistence and quality of multispecies swards in a four-year experiment. *PLoS One, 12*(1), e0169208.

Johnson, P. T. J., et al. (2010). Linking environmental nutrient enrichment and disease emergence in humans and wildlife. *Ecological Applications, 20*, 16–29.

Jolliffe, P. (1997). Are mixed populations of plant species more productive than pure stands? *Oikos, 80*(3), 595–602. https://doi.org/10.2307/3546635

Jørgensen, C., Enberg, K., Dunlop, E. S., Arlinghaus, R., Boukal, D. S., Brander, K., Ernande, B., et al. (2007). Managing the world's evolving fish stocks. *Science, 318*, 1247–1248.

Jucker, T., Avăcăriţei, D., Bărnoaiea, I., Duduman, G., Bouriaud, O., & Coomes, D. A. (2016). Climate modulates the effects of tree diversity on forest productivity. *Journal of Ecology, 104*, 388–398.

Jucker, T., Bouriaud, O., & Coomes, D.A. (2015). Crown plasticity enables trees to optimize canopy packing in mixed-species forests. *Funct Ecology, 29*, 1078–1086. https://doi.org/10.1111/1365-2435.12428.

Just, T., Matoušek, V., Dušek, M., Fišer, D., & Karlík, P. (2005). *Vodohospodářské revitalizace a jejich uplatnění v ochraně před povodněmi*. ZO ČSOP Hořovicko se spolupráci se spol. Ekologické služby s.r.o., AOPK ČR a MŽP.

Kabrda, J., & Bičík, I. (2010). Dlouhodobé změny rozlohy lesa v Česku i ve světě. *Geografické rozhledy, 20*(1), 2–5.

Kapička, J., Lang, J., Petrus, D., Novotný, I., Brant, V., & Kroulík, M. (2017). *Metodický postup optimalizace velikosti zemědělských pozemků*. VUMOP, CZU.

Kaplan, J. O., Krumhardt, K. M., & Zimmermann, N. (2009). The prehistoric and preindustrial deforestation of Europe. *Quaternary Science Reviews, 28*, 3016–3034.

Kareiva, P., Watts, S., McDonald, R., & Boucher, T. M. (2007). Domesticated nature: Shaping landscapes and ecosystems for human welfare. *Science, 316*, 1866–1869.

Kawasaki, T. (1980). Fundamental relations among the selections of life history in the marine teleosts. *Bulletin. Japanese Society of Scientific Fisheries, 46*(3), 289–293.

Kemper, N. (2008). Veterinary antibiotics in the aquatic and terrestrial environment. *Ecological Indicators, 8*, 1–13.

Kiær, L. P., Skovgaard, I. M., & Østergård, H. (2009). Grain yield increase in cereal variety mixtures: A meta-analysis of field trials. *Field Crops Research, 114*(3), 361–373.

Kjellander, P., Hewison, A. J. M., Liberg, O., Angibault, J. M., Bideau, E., & Cargnelutti, B. (2004). Experimental evidence for density-dependence of home-range size in roe deer (*Capreolus capreolus* L.): A comparison of two long-term studies. *Oecologia, 139*, 478–485.

Kleiber, M. (1961). *The fire of life. An introduction to animal energetics*. Wiley.

Klimo, E., Midriak, R., Pelikán, J., Buček, A., & Materna, J. (1994). *Lesnická ekologie*. MZLU.

Klyashtorin, L. B. (2001). *Climate change and long-term fluctuations of commercial catches – The possibility of forecasting*. FAO fisheries technical paper 410. FAO.

Koch, P. L., & Barnosky, A. D. (2006). Late quaternary extinctions: State of the debate. *Annual Review of Ecology Evolution and Systematics, 37*, 215–250.

Kolding, J., & van Zwieten, P. A. M. (2011). The tragedy of our legacy: How do global management discourses affect small scale fisheries in the south? *Forum for Development Studies, 38*(3), 267–297.

Konvička, M., Beneš, J., & Poláková, S. (2016). Smaller fields support more butterfliees: Comparing two neighbouring European countries with different socioeconomic heritage. *Journal of Insect Conservation, 20*, 1113–1118.

Kotecký, V. (2019). Nenápadný půvab kapitalizmu. *Vesmír, 98*, 438.

Král, K., Daněk, P., Janík, D., Krůček, M., & Vrška, T. (2018). How cyclical and predictable are Central European temperate forest dynamics in terms of development phases? *Journal of Vegetation Science, 29*, 84–97.

Krebs, C. J. (1999). *Ecological methodology*. Benjamin/Cummings.

Kreutzweiser, D. P., Hazlett, P. W., & Gunn, J. M. (2008). Logging impacts on the biogeochemistry of boreal forest soils and nutrient export to aquatic systems: A review. *Environmental Reviews, 16*, 157–179.

Kroodsma, D. A., Mayorga, J., Hochberg, T., Miller, N. A., Boerder, K., Ferretti, F., Wilson, A., Bergman, B., White, T. D., Block, B. A., et al. (2018). Tracking the global footprint of fisheries. *Science, 359*, 904–908. https://doi.org/10.1126/science.aao5646

Kukla, J., Whitfeld, T., Cajthaml, T., Baldrian, P., Veselá-Šimáčková, H., Novotný, V., & Frouz, J. (2019). The effect of traditional slash-and-burn agriculture on soil organic matter, nutrient content and microbiota in tropical ecosystems of Papua New Guinea. *Land Degradation and Development*, 1–12.

Kuneš, P., Odgaard, B. V., & Gaillard, M.-J. (2011). Soil phosphorus as a control of productivity and openness in temperate interglacial forest ecosystems. *Journal of Biogeography, 38*, 2150–2164.

Kuuluvainen, T., Tahvonen, O., & Aakala, T. (2012). Even-aged and uneven-aged forest management in boreal Fennoscandia: A review. *Ambio, 41*(7), 720–737. https://doi.org/10.1007/s13280-012-0289-y

Kuzyakov, Y., Siniakina, S. V., Ruehlmann, J., Domanski, G., & Stahr, K. (2002). Effect of nitrogen fertilisation on below-ground carbon allocation in lettuce. *Journal of the Science of Food and Agriculture, 82*, 1432–1441.

Laikre, L., Schwartz, M. K., Waples, R. S., & Ryman, N. (2010). Compromising genetic diversity in the wild: Unmonitored large-scale release of plants and animals. *Trends in Ecology and Evolution, 25*, 520–529.

Lair, R. C. (1999). *Gone astray – The care and management of the Asian elephant in domesticity*. FAO Regional Office for Asia and the Pacific.

Lal, R. (2004). Soil carbon sequestration impacts on global climate change and food security. *Science, 304*(5677), 1623–1627.

Lal, R. (2017). *Enciklopedia of soil science*. CRC Press.

Lammerts van Bueren, E. T., Jones, S. S., Tamm, L., Murphy, K. M., Myers, J. R., Leifert, C., et al. (2011). The need to breed crop varieties suitable for organic farming, using wheat, tomato, and broccoli as examples: A review. NJAS Wagening. *NJAS – Wageningen Journal of Life Sciences, 58*, 193–205.

Landers, A. (2000). *Farm machinery: Selection, investment and management*. Farming Press.

Larkin, P. A. (1977). An epitaph for the concept of maximum sustained yield. *Transactions of the American Fisheries Society, 106*, 1–11.

Lassaletta, L., Billen, G., Grizzetti, B., Anglade, J., & Garnier, J. (2014). 50 year trends in nitrogen use efficiency of world cropping systems: The relationship between yield and nitrogen input to cropland. *Environmental Research Letters, 9*, 105011.

Laurance, W. F., Nascimento, H. E. M., Laurance, S. G., Andrade, A., Ewers, R. M., et al. (2007). Habitat fragmentation, variable edge effects, and the landscape-divergence hypothesis. *PLoS One, 2*(10), e1017. https://doi.org/10.1371/journal.pone.0001017

Laurance, W. F., Camargo, J. L., Fearnside, P. M., Lovejoy, T. E., Williamson, G. B., Mesquita, R. C., Meyer, C. F., Bobrowiec, P. E., & Laurance, S. G. (2018). An Amazonian rainforest and its fragments as a laboratory of global change. *Biological Reviews, 93*(1), 223–247.

Lavalle, P., et al. (2005). Chapter 12: Nutrient cycling. In R. Hassan, R. Scholes, & N. Ash (Eds.), *Ecosystems and human well-being: Current state and trends: Findings of the condition and trends working group* (331–353 pp). Island Press.

LeBauer, D. S., & Treseder, K. K. (2008). Nitrogen limitation of net primary productivity in terrestrial ecosystems is globally distributed. *Ecology, 89*, 371–379.

Ledgard, S. F., & Steele, K. W. (1992). Biological nitrogen fixation in mixed legume/grass pastures. *Plant and Soil, 141*, 137–153.

Lehrman, S. (1994). Organic farm face dilema over new gene technology. *Nature, 367*(6459), 106–106.

Lepš, J. (2005). Chapter 8: Diversity and ecosystem function. In E. van der Maarel (Ed.), *Vegetation ecology*. Blackwell Publishing.

Lindegren, M., Möllmann, C., Nielsen, A., & Stenseth, N. C. (2009). Preventing the collapse of the Baltic cod stock through an ecosystem-based management approach. *Proceedings of the National Academy of Sciences of the United States of America, 106*(34), 14722–14727.

Lindemann, W. (1956). Transplantation of game in Europe and Asia. *The Journal of Wildlife Management, 20*(1), 68–70.

Loss, S. R., Will, T., & Marra, P. P. (2013). The impact of free-ranging domestic cats on wildlife of the United States. *Nature Communications, 4*, 1396.

Lotze, H. K., Coll, M., Magera, A. M., Ward-Paige, C., & Airoldi, L. (2011). Recovery of marine animal populations and ecosystems. *Trends in Ecology & Evolution, 26*, 595–605.

Lotze, H. K., Tittensor, D. P., Bryndum-Buchholz, A., Eddy, T. D., Cheung, W. W. L., Galbraith, E. D., Barange, M., Barrier, N., Bianchi, D., Blanchard, J. L., Bopp, L., Büchner, M., Bulman, C. M., Carozza, D. A., Christensen, V., Coll, M., Dunne, J. P., Fulton, E. A., Jennings, S., … Worm, B. (2019). Global ensemble projections reveal trophic amplification of ocean biomass declines with climate change. *Proceedings of the National Academy of Sciences, 116*(26), 12907–12912.

Lough, T. S. (1999). Energy, agriculture, patriarchy and ecocide. *Human Ecology Review, 6*(2), 100–111.

Lowe, K. F., Hume, D. E., & Fulkerson, W. J. (2016). Forages and pastures: Perennial forage and pasture crops – Species varieties. In *Reference module in food science*. Elsevier.

Lumaret, J. P. (1986). Toxicite de certains helminthicides vis-a-vis des insectes coprophages et consequences sur la disparition des excrements de la surface du sol. *Acta Oecologica. Oecologia Applicata, 7*(4), 313–324.

Lumaret, J. P., & Kadiri, N. (1995). The influence of the first wave of colonizing insects on cattle dung dispersal. *Pedobiologia, 39*, 506–517.

Lundström, U. S., van Breemen, N., & Bain, D. C. (2000). The podzolization process. A review. *Geoderma, 94*(2–4), 91–107.

Lützow, M. V., Kögel-Knabner, I., Ekschmitt, K., Matzner, E., Guggenberger, G., Marschner, B., & Flessa, H. (2006). Stabilization of organic matter in temperate soils: Mechanisms and their relevance under different soil conditions – A review. *European Journal of Soil Science, 57*, 426–445.

Lyons, S. K., Amatangelo, K. L., Behrensmeyer, A. K., Bercovici, A., Blois, J. L., Davis, M., DiMichele, W. A., Du, A., Eronen, J. T., Faith, J. T., Graves, G. R., Jud, N., Labandeira, C., Looy, C. V., McGill, B., Miller, J. H., Patterson, D., Munoz, S. P., Potts, R., … Gotelli, N. J. (2016). Holocene shifts in the assembly of plant and animal communities implicate human impacts. *Nature, 529*, 80–83.

MacArthur, R. H., & Wilson, E. O. (1967). *The theory of island biogeography*. Princeton University Press.

MacDonald, S. E., Landhaeusser, S. M., Skousen, J., Franklin, J., Frouz, J., Hall, S., Jacobs, D. F., & Quideau, S. (2015). Forest restoration following surface mining disturbance: Challenges and solutions. *New Forests, 46*, 703–732.

Mader, P., Fliessbach, A., Dubois, D., Gunst, L., Fried, P., & Niggli, U. (2002). Soil fertility and biodiversity in organic farming. *Science, 296*(5573), 1694–1697.

Maestre, F. T., Callaway, R. M., Valladares, F., & Lortie, C. J. (2009). Refiningthe stress-gradient hypothesis for competition and facilitation in plant communities. *Journal of Ecology, 97*, 199–205.

Makarieva, M. A., Gorshkov, V. G., & Li, B. L. (2008). Energy budget of the biosphere and civilization: Rethinking environmental security of global renewable and non-renewable resources. *Ecological Complexity, 5*, 281–288.

Man, M. (1994). Phosporic acid by wet proces, heavy metal in phosphoric acid and their removal. In C. A. Hoge & N. N. Popovichi (Eds.), *Pollution control in fertilizer production*. Marcel Dekker.

Marco, I., Padró, R., Cattaneo, C., Caravaca, J., & Tello, E. (2018). From vineyards to feedlots: A fund-flow scanning of sociometabolic transition in the Vallès County (Catalonia) 1860–1956–1999. *Regional Environmental Change, 18*, 981–993.

Marshall, E. J., & Moonen, A. C. (2002). Field margins in northern Europe: Their functions and interactions with agriculture. *Agriculture Ecosystems & Environment, 89*(1), 5–21.

Marshall, C. H., Roger, A., Pielke, S. R., Steyaert, L. T., & Willard, D. A. (2004). The impact of anthropogenic land-cover change on the Florida Peninsula sea breezes and warm season sensible weather. *Monthly Weather Review, 138*, 28–52.

Martins, C. I. M., Eding, E. H., Verdegem, M. C. J., Heinsbroek, L. T. N., Schneider, O., Blancheton, J. P., Roque d'Orbcastel, E., & Verreth, J. A. J. (2010). New developments in recirculating aquaculture systems in Europe: A perspective on environmental sustainability. *Aquacultural Engineering, 43*, 83–93.

Masiero, M., Pettenella, D., Boscolo, M., Barua, S. K, Animon, I., & Matta, J. R. (2019). *Valuing forest ecosystem services: A training manual for planners and project developers*. Forestry Working paper no. 11 (216 pp). FAO.

Massei, G., Kindberg, J., Licoppe, A., Gačic, D., Sprem, N., Kamler, J., Baubet, E., Hohmann, U., Monaco, A., Ozolins, J., Cellina, S., Podgorski, T., Fonseca, C., Markov, N., Pokorny, B., Rosell, C., & Nahlik, A. (2014). Wild boar populations up, numbers of hunters down? A review of trends and implications for Europe. *Pest Managemen Science, 71*, 492–500.

Matějíček, J., & Zádrapa, R. (2014). *Oceňování lesa*. Mendelova univerzita v Brně.

Matson, P. A., Parton, W. J., Power, A. G., & Swift, M. J. (1997). Agricultural intensification and ecosystem properties. *Science, 277*(5325), 504–509.

McDermott, G. R., Meng, K. C., McDonald, C. G., & Costello, C. J. (2019). The blue paradox: Preemptive overfishing in marine reserves. *Proceedings of the National Academy of Sciences, 116*(12), 5319–5325.

McGrath, M. J., Luyssaert, S., Meyfroidt, P., Kaplan, J. O., Bürgi, M., Chen, Y., Erb, K., Gimmi, U., McInerney, D., Naudts, K., Otto, J., Pasztor, F., Ryder, J., Schelhaas, M. J., & Valade, A. (2015). Reconstructing European forest management from 1600 to 2010. *Biogeosciences, 12*, 4291–4316.

McTavish, E. J., Decker, J. E., Schnabel, R. D., Taylor, J. F., & Hillis, D. M. (2013). New World cattle show ancestry from multiple independent domestication events. *Proceedings of the National Academy of Sciences, 110*(15), E1398–E1406. https://doi.org/10.1073/pnas.1303367110

Micklin, P. (2014). Irrigation in the Aral Sea basin. In P. Micklin, N. Aladin, & I. Plotnikov (Eds.), *The Aral Sea* (Springer earth system sciences) (Vol. 10178). Springer.

Mie, A., Andersen, H. R., Gunnarsson, S., Kahl, J., Kesse-Guyot, E., Rembiałkowska, E., Quaglio, G., & Grandjean, P. (2017). Human health implications of organic food and organic agriculture: A comprehensive review. *Environmental Health, 16*, 111. https://doi.org/10.1186/s12940-017-0315-4

Miklosi, A., & Soproni, K. (2006). A comparative analysis of animals' understanding of the human pointing gesture. *Animal Cognition, 9*(2), 81–93.

Miko, L., & Storch, D. (2015). Biodiversity conservation under energy limitation: Possible consequences of human productivity appropriation for species richness, ecosystem functioning, and food production. *Ecosystem Services, 16*, 146–149.

Milesia, C., Elvidge, C. D., Dietz, J. B., Tuttle, B. T., Nemani, R. R., & Running, S. W. (2005). A strategy for mapping and modeling the ecological effects of US lawns. In M. Moeller & E. Wentz (Eds.), *Proceedings of joint symposia URBAN – URS 2005*, XXXVI-8/W27.

Milewski, W. (2017). *Forests in Poland*. The State Forests Information Centre.

Millennium Ecosystem Assessment. (2005). *Ecosystems and human well-being: Opportunities and challenges for business and industry*. World Resources Institute.

Mills, L. S., Soule, M. E., & Doak, D. F. (1993). The keystone-species concept in ecology and conservation. *Bioscience, 43*, 219–224.

Minson, D. J. (1980). Nutritional differences between tropical and temperate pastures. In F. H. W. Morey (Ed.), *Grazing animals* (pp. 167–182). C. A. B. Furnham Royal.

Moreno-Mateos, D., Barbier, E. B., Jones, P. C., Jones, H. P., Aronson, J., López-López, J. A., McCrackin, M. L., Meli, P., Montoya, D., & Benayas, J. M. R. (2017). Anthropogenic ecosystem disturbance and the recovery debt. *Nature Communications, 8*, 14163.

Mosquera-Losada, M. R., Moreno, G., Pardini, A., McAdam, J. H., Papanastasis, V., Burgess, P. J., Lamersdorf, N., Castro, M., Liagre, F., & Rigueiro-Rodríguez, A. (2012). Past, present and future of agroforestry systems in Europe. In P. K. R. Nair & D. Garrity (Eds.), *Agroforestry – The future of global land use* (pp. 285–312). Springer.

Moya, B., Alison Parker, A., & Sakrabani, R. (2019). Challenges to the use of fertilisers derived from human excreta: The case of vegetable exports from Kenya to Europe and influence of certification systems. *Food Policy, 85*, 72–78.

Muhly, T. B., Semeniuk, C., Massolo, A., Hickman, L., & Musiani, M. (2011). Human activity helps prey win the predator-prey space race. *PLoS One, 6*(3), e17050. https://doi.org/10.1371/journal.pone.0017050

Mukherjee, A., Speh, D., Dyck, E., & Diez-Gonzalez, F. (2004). Preharvest evaluation of coliforms, *Escherichia coli*, salmonella, and *Escherichia coli* O157:H7 in organic and conventional produce grown by Minnesota farmers. *Journal of Food Protection, 67*(5), 894–900.

Mulder, J., de Wit, H. A., Boonen, H. W. J., & Bakken, L. R. (2001). Increased levels of aluminium in forest soils: Effects on the stores of soil organic carbon. *Water, Air, and Soil Pollution, 130*, 989–994.

Murray, K. A., Allen, T., Loh, E., Machalaba, C., & Daszak, P. (2016). Emerging viral zoonoses from wildlife associated with animal-based food systems: Risks and opportunities. In M. Jay-Russell & M. Doyle (Eds.), *Food safety risks from wildlife* (Food microbiology and food safety). Springer.

Nabuurs, G. J., & Schelhaas, M. J. (2002). Carbon profiles of typical forest types across Europe assessed with CO_2FIX. *Ecological Indicators, 1*, 213–223.

Nair, C., Salin, K., Joseph, J., Aneesh, B., Geethalakshmi, V., et al. (2014). Organic rice–prawn farming yields 20% higher revenues. *Agronomy for Sustainable Development, 34*(3), 569–581.

National Research Council. (2005). *Radiative forcing of climate change: Expanding the concept and addressing uncertainties*. The National Academies Press.

Nellemann, C., & INTERPOL Environmental Crime Programme (Eds.). (2012). *Green carbon, Black trade: Illegal logging, tax fraud and laundering in the worlds tropical forests. A rapid response assessment*. United Nations Environment Programme, GRIDArendal.

Něměček, J., Smolíková, L., & Kutílek, M. (1990). *Pedologie a paleopedologie*. Academia.

Neupane, D. (2020). How conservation will be impacted in the COVID-19 pandemic. *Wildlife Biology, 2020*(2). https://doi.org/10.2981/wlb.00727

Ni, B. J., Pan, Y., Guo, J., Virdis, B., Hu, S., Chen, X., & Yuan, Z. (2017). Denitrification processes for wastewater treatment. In I. Moura, J. J. G. Moura, S. R. Pauleta, & L. B. Maia (Eds.), *Metalloenzymes in denitrification: Applications and environmental impacts* (RSC metallobiology series) (Vol. 9, pp. 368–418). Royal Society of Chemistry.

Niehl, A., Soininen, M., Poranen, M. M., & Heinlein, M. (2018). Synthetic biology approach for plant protection using dsRNA. *Plant Biotechnology Journal, 16*, 1679–1687.

Nohrstedt, H. O. (2001). Response of coniferous forest ecosystems on mineral soils to nutrient additions: A rewiew of Swedish experiences. *Scandinavian Journal of Forest Research, 16*(6), 555–573.

Nørgaard, D. S., Stærmose, M. D., & Jensen, P. R. (2018). Protein from green biomass as a food resource. In *Sustain conference 2018: Creating technology for a sustainable society*. Technical University of Denmark.

Nowacky, G. J., & Abrams, M. D. (2008). The demise of fire and "mesophication" of forests in the eastern United States. *BioScience, 58*, 123–138.

Ntiamoa-Baidu, Y. (1997). *Wildlife and food security in Africa*. FAO conservation guide 33, FAO.

Nussbaumer, T. (2003). Combustion and co-combustion of biomass: Fundamentals, technologies, and primary measures for emission reduction. *Energy & Fuels, 17*, 1510–1521.

Oberson, A., Bunemann, E. K., Friesen, D. K., Rao, I. M., Smithson, P. C., Turner, B. L., & Frossard, E. (2006). Improving phosphorus fertility in tropical soils through biological interventions. In N. Uphoff, A. S. Ball, E. Fernandes, H. Herren, O. Husson, M. Laing, C. Palm, J. Pretty, P. Sanchez, N. Sanginga, & J. Thies (Eds.), *Biological approaches to sustainable soil systems*. CRC Tailor Francis.

Oinam, S. S., Rawat, Y. S., Kuniyal, J. C., Vishvakarma, S. C., & Pandey, D. C. (2008). Thermal supplementing soil nutrients through biocomposting of night-soil in the northwestern Indian Himalaya. *Waste Management, 28*(6), 1008–1019.

Oldfield, E. E., Bradford, M. A., & Wood, S. A. (2019). Global meta-analysis of the relationship between soil organic matter and crop yields. *The Soil, 5*, 15–32.

Oliver, C. D., Osawa, A., & Camp, A. (1998). Forest dynamics and resulting animal and plant population changes at the stand and landscape levels. *Journal of Sustainable Forestry, 6*(3/4), 281–312.

Olsen, J., Griepentrog, H., Nielsen, J., & Weiner, J. (2012). How Important are Crop Spatial Pattern and Density for Weed Suppression by Spring Wheat? *Weed Science, 60*(3), 501–509. https://doi.org/10.1614/WS-D-11-00172.1

Oswald, S. N., & Smith, W. B. (2014). *US Forest resurce fact and historical trend*. USDA.

Page, S., Siegert, F., Rieley, J., et al. (2002). The amount of carbon released from peat and forest fires in Indonesia during 1997. *Nature, 420*, 61–65.

Palma, J. H. N., Graves, A. R., Bunce, R. G. H., Burgess, P. J., de Filippi, R., Keesman, K. J., van Keulen, H., Liagre, F., Mayus, M., Moreno, G., Reisner, A. Y., & Herzog, F. (2007). Modeling environmental benefits of silvoarable agroforestry in Europe. *Agriculture, Ecosystems and Environment, 119*, 320–334.

Pan, Y., Birdsey, R. A., Fang, J., Houghton, R., Kauppi, P. E., Kurz, W. A., Phillips, O. L., Shvidenko, A., Lewis, S. L., Canadell, J. G., Ciais, P., Jackson, R. B., Pacala, S. W., McGuire, A. D., Piao, S., Rautiainen, A., Sitch, S., & Hayes, D. (2011). A large and persistent carbon sink in the world's forests. *Science, 333*, 988–993.

Parcerisas, L.,& Dupras, J. (2018). From mixed farming to intensive agriculture: energy profiles of agriculture in Quebec, Canada, 1871–2011. *Reg Environ Change 18*, 1047–1057. https://doi.org/10.1007/s10113-018-1305-y

Pang, J. F., Kluetsch, C., Zou, X. J., Zhang, A. B., Luo, L. Y., Angleby, H., Ardalan, A., Ekström, C., Sköllermo, A., Lundeberg, J., Matsumura, S., Leitner, T., Zhang, Y. P., & Savolainen, P. (2009). mtDNA data indicate a single origin for dogs south of Yangtze River, less than 16,300 years ago, from numerous wolves. *Molecular Biology and Evolution, 26*(12), 2849–2864. https://doi.org/10.1093/molbev/msp195

Pangala, S. R., Enrich-Prast, A., Basso, L. S., Peixoto, R. B., Bastviken, D., Hornibrook, E. R. C., Gatti, L. V., Marotta, H., Calazans, L. S. B., Sakuragui, C. M., Bastos, W. R., Malm, O., Gloor, E., Miller, J. B., & Gauci, V. (2017). Large emissions from floodplain trees close the Amazon methane budget. *Nature, 552*, 230–234.

Patrício, M. S., Monteiro, M. L., & Tomé, M. (2005). Biomass equations for castanea sativa high forest in the northwest of Portugal. In C. G. Abreu, E. Rosa, & A. A. Monteiro (Eds.), *Proceedings of the III International Chestnut Congress*. ISHS. Acta Hort. 693.

Pauly, D., Christensen, V., Dalsgaard, J., Froese, R., & Torres, F., Jr. (1998). Fishing down marine food webs. *Science, 279*(5352), 860–863.

Peltoniemi, A., & Niemi, J. (2016). Price margins in the finnish food chain. In *Proceedings in system dynamics and innovation in food networks*, pfsd.2016.1615.

Peltzer, D. A., Wardle, D. A., Allison, V. J., Baisden, W. T., Bardgett, R. D., Chadwick, O. A., Condron, L. M., Parfitt, R. L., Porder, S., Richardson, S. J., Turner, B. L., Vitousek, P. M., Walker, J., & Walker, L. R. (2010). Understanding ecosystem retrogression. *Ecological Monographs, 80*(4), 509–529.

Peng, S. X., Hansen, M. C., Stehman, S. V., Potapov, P. V., Tyukavina, A., Vermote, E. F., & Townshend, J. R. (2018). Global land change from 1982 to 2016. *Nature, 560*(7720), 639–643.

Peterson, A. T., & Brisbin, I. L. (1998). Genetic endangerment of wild Red Junglefowl Gallus gallus. *Bird Conservation International, 8*, 387–394.

Phalan, B., Onial, M., Balmford, A., & Green, R. E. (2011). Reconciling food production and biodiversity conservation: Land sharing and land sparing compared. *Science, 333*, 1289–1291.

Pimentel, D., Oltenacu, P. A., Nesheim, M. C., Krummel, J., Allen, M. S., & Chick, S. (1980). The potential for grass-fed livestock: Resource constraints. *Science, 207*, 843–848.

Pimentel, D., Dazhong, W., & Giampietro, M. (1990). Technological changes in US agricultural energy use. In S. R. Gliessman (Ed.), *Agroecology: Researching the ecological basis for sustainable agriculture* (pp. 305–321). Springer.

Pimentel, D., Stachow, U., Takacs, A., Brubaker, W., Dumas, A. R., Meaney, J. S., et al. (1992). Conserving biological diversity in agricultural/forestry systems. *Bioscience, 42*, 354–362.

Polovina, J. J., Howell, E. A., & Abecassis, M. (2008). Ocean's least productive waters are expanding. *Geophysical Research Letters, 35*, L03618. https://doi.org/10.1029/2007GL031745

Pommerening, A., & Murphy, S. T. (2004). A review of the history, definitions and methods of continuous cover forestry with special attention to afforestation and restocking. *Forestry, 77*, 27–44.

Ponge, J. F. (2003). Humus forms in terrestrial ecosystems: A framework to biodiversity. *Soil Biology and Biochemistry, 35*(7), 935–945.

Porder, S., & Ramachandran, S. (2013). The phosphorus concentration of common rocks-a potential driver of ecosystem P status. *Plant and Soil, 367*, 41–55.

Post, J. R., Sullivan, M., Cox, S., Lester, N. P., Walters, C. J., Parkinson, E. A., Paul, A. J., Jackson, L., & Shuter, B. J. (2002). Canada's recreational fishery: The invisible collapse? *Fisheries, 27*, 6–17.

Potter, K. M., Escanferla, M. E., Jetton, R. M., & Man, G. (2019). Important insect and disease threats to United States – Tree species and geographic patterns of their potential impacts. *Forests, 10*(4), 304. https://doi.org/10.3390/f10040304

Potužák, J., Hůda, J., & Pechar, L. (2007). Changes in fish production effectivity in eutrophic fishponds – Impact of zooplankton structure. *Aquaculture International, 15*(3), 201–210. https://doi.org/10.1007/s10499-007-9085-2

Pretzsch, H. (2009). Forest dynamics, growth, and yield. In *Forest dynamics, growth and yield*. Springer.

Pretzsch, H., Schütze, G., & Uhl, E. (2013). Resistance of European tree species to drought stress in mixed versus pure forests: Evidence of stress release by inter-specific facilitation. *Plant Biology, 15*(3), 483–495.

Quiroga, A., Funaro, D., Noellemeyer, E., & Peinemann, N. (2006). Barley yield response to soil organic matter and texture in the Pampas of Argentina. *Soil & Tillage Research, 90*, 63–68.

Ramage, M. H., Burridge, H., Busse-Wicher, M., Fereday, G., Reynolds, T., Shah, D. U., Wu, G., Yu, L., Fleming, P., Densley-Tingley, D., Allwood, J., Dupree, P., Linden, P. F., & Scherman, O. (2017). Thewoodfromthetrees: Theuse of timber in construction. *Renewable and Sustainable Energy Reviews, 68*(1), 333–359.

Ramos-Elorduy, J. (2008). Energy supplied by edible insects from Mexico and their nutritional and ecological importance. *Ecology of Food and Nutrition, 47*, 280–297.

Rea, A. W., & Munns, W. R., Jr. (2007). The value of nature: Economic, intrinsic, or both? *Integrated Environmental Assessment and Management, 13*(5), 953–955. https://doi.org/10.1002/ieam.1924

Read, A. J., Drinker, P., & Northridge, S. (2006). Bycatch of marine mammals in U.S. and global fisheries. *Conservation Biology, 20*(1), 163–169.

Redfield, A. C., Ketchum, G. H., & Richards, F. A. (1963). The influence of organisms on the composition of sea water. In M. N. Hill (Ed.), *The sea* (pp. 26–77). Wiley-Interscience.

Reganold, J. P., Elliot, L. F., & Unger, Y. L. (1987). Long term effect of organic and conventional farming on soil erosion. *Nature, 330*(6146), 370–372.

Reid, E. M. (1935). British floras antecedent to the Great Ice Age. Discussions on the origin and relationship of the British flora. *Proceedings of the Royal Society of London. Series B, Biological Sciences, 118*, 197–241.
Reif, J., & Vermousek, Z. (2018). Collapse of farmland bird populations in an Eastern European country following its EU accession. *Conservation Letters*. https://doi.org/10.1111/conl.1258
Reineke, L. H. (1933). Perfecting a stand-density index for even-aged forests. *Journal of Agricultural Research, 46*, 627–638.
Reisenbichler, R. R., & Rubin, S. P. (1999). Genetic changes from artificial propagation of Pacific salmon affect the productivity and viability of supplemented populations. *ICES Journal of Marine Science, 56*(4), 459–466.
Rekha, Naik, S. N., & Prasad, R. (2006). Pesticide residue in organic and conventional food-risk analysis. *Journal of Chemical Health and Safety, 13*(6), 12–19.
Remeš, J., Pulkrab, K., Bilek, L., & Podrazsky, V. (2020). Economic and production effect of tree species change as a result of adaptation to climate change. *Forests, 11*(4), 431.
Rengasamy, P. A. C., & Marchuk, A. (2011). Cation ratio of soil structural stability (CROSS). *Soil Research, 49*, 280–285.
Richardson, D. M. (Ed.). (1998). *Ecology and biogeography of Pinus*. Cambridge University Press.
Richardson, A. J., Brown, C. J., Brander, K., Bruno, J. F., Buckley, L., Burrows, M. T., Duarte, C. M., Halpern, B. S., Hoegh-Guldberg, O., Holding, J., Kappel, C. V., Kiessling, W., Moore, P. J., O'Connor, M. I., Pandolfi, J. M., Parmesan, C., Schoeman, D. S., Schwing, F., Sydeman, W. J., & Poloczanska, E. S. (2012). Climate change and marine life. *Biology Letters, 8*, 907–909.
Richman, S. (1958). The transformation of energy by Daplinia pulex. *Ecological Monographs, 28*, 273–291.
Rifai, S. W., Girardin, C. A. J., Berenguer, E., et al. (2018). ENSO Drives interannual variation of forest woody growth across the tropics. *Philosophical Transactions of the Royal Society B: Biological Sciences, 373*(1760), rstb20170410.
Rockström, J., Steffen, W., Noone, K., Persson, Å., Chapin, F. S., Lambin, E. F., Lenton, T. M., Scheffer, M., et al. (2009). A safe operating space for humanity. *Nature, 461*(7263), 472–475.
Rodríguez, C., & Wiegand, K. (2009). Evaluating the trade-off between machinery efficiency and loss of biodiversity-friendly habitats in arable landscapes: The role of field size. *Agriculture Ecosystems & Environment, 129*(4), 361–366.
Rohde, K. (1978). Latitudinal gradients in species diversity and their causes. II. Marine parasitological evidence for a time hypothesis. *Biologisches Zentralblatt, 97*, 405–418.
Rohde, K. (1993). *Ecology of marine parasites* (2nd ed.). CABI.
Roman, J., & McCarthy, J. J. (2010). The whale pump: Marine mammals enhance primary productivity in a coastal basin. *PLoS One, 5*(10), e13255.
Rosenzweig, M. L. (1968). Net primary productivity of terrestrial communities: Prediction from climatological data. *American Naturalist, 102*, 67–74.
Rosenzweig, C., & Liverman, D. (1992). Predicted effects of climate change on agriculture: A comparison of temperate and tropical regions. In S. K. Majumdar (Ed.), *Global climate change: Implications, challenges, and mitigation measures* (pp. 342–361). The Pennsylvania Academy of Sciences.
Roser, M. (2013). *Employment in agriculture*. Published online at OurWorldInData.org. https://ourworldindata.org/employment-in-agriculture [online resource].
Roubalová, M. (2017). *Situlační a výhledová zpráva. Krmiva*. MZe.
Roux, J. P., van der Lingen, C. D., Gibbons, M. J., Moroff, N. E., Shannon, L. J., Smith, A. D. M., & Cury, P. M. (2013). Jellyfication of marine ecosystems as a likely consequence of overfishing small pelagis fishes: Lessons from the Benguela. *Bulletin of Marine Science, 89*(1), 249–284.
Roy, E., Richards, P., Martinelli, L., et al. (2016). The phosphorus cost of agricultural intensification in the tropics. *Nature Plants, 2*, 16043.
Ruppert, J. L. W., Travers, M. J., Smith, L. L., Fortin, M. J., & Meekan, M. G. (2013). Caught in the middle: Combined impacts of shark removal and coral loss on the fish communities of coral reefs. *PLoS One, 8*(9), e74648.

Russell, T. (2017). *The illustrated encyclopedia of trees of Britain and Europe*. Southwater.
Sahrhage, D., & Lundbeck, J. (1992). *A history of fishing*. Springer-Verlag.
Šálek, M., Hula, V., Kipson, M., Daňková, R., Niedobová, J., & Gamero, A. (2018). Bringing diversity back to agriculture: Smaller fields and non-crop elements enhance biodiversity in intensively managed arable farmlands. *Ecological Indicators, 90*, 65–73.
Salusová, D. (2018). *České zemědělství očima statistiky 1918–2017*. Český statistický úřad.
Samojlik, T., & Kuijper, D. P. J. (2013). Grazed wood pasture versus browsed high forests: Impact of ungulates on forest landscapes from the perspective of the Białowieża Primeval Forest. In I. D. Rotherham (Ed.), *Trees, forested landscapes and grazing animals – A European perspective on woodlands and grazed treescapes* (pp. 143–162). Routledge.
Sanchez, P. (2019). Organic carbon. In *Properties and management of soils in the tropics* (pp. 259–306). Cambridge University Press.
Sanchez, P. A., & Buol, S. W. (1975). Soils of the tropics and the world food crisis. *Science, 188*(4188), 598–603.
Sanderson, E. W., Jaiteh, M., Levy, M. A., Redford, K. H., Wannebo, A. V., & Woolmer, G. (2002). The human footprint and the last of the wild. *Bioscience, 52*(10), 891–904.
Saner, P., Loh, Y. Y., Ong, R. C., & Hector, A. (2012). Carbon stocks and fluxes in tropical lowland dipterocarp rainforests in Sabah, Malaysian Borneo. *PLoS One, 7*(1), e29642.
San-Miguel-Ayanz, J., de Rigo, D., Caudullo, G., Houston Durrant, T., & Mauri, A. (Eds.). (2016). *European atlas of forest tree species*. European Comission.
Santamaría-Fernández, M., Molinuevo-Salces, B., Kiel, P., Steenfeldt, S., Uellendahl, H., & Lübeck, M. (2017). Lactic acid fermentation for refining proteins from green crops and obtaining a high quality feed product for monogastric animals. *Journal of Cleaner Production, 162*, 875–881.
Satterthwaite, D., McGranahan, G., & Tacoli, C. (2010). Urbanization and its implications for food and farming. *Philosophical Transactions of the Royal Society B, 365*, 2809–2820. https://doi.org/10.1098/rstb.2010.0136
Schiermeier, Q. (2007). Ocean circulation noisy, not stalling. *Nature, 448*, 844–845.
Schmid, M., Pautasso, M., & Holdenrieder, O. (2014). Ecological consequences of Douglas fir (*Pseudotsuga menziesii*) cultivation in Europe. *European Journal of Forest Research, 133*, 13–29.
Schneider, K. D., Thiessen Martens, J. R., Zvomuya, F., Reid, D. K., Fraser, T. D., Lynch, D. H., O'Halloran, I. P., & Wilson, H. F. (2019). Options for improved phosphorus cycling and use in agriculture at the field and regional scales. *Journal of Environmental Quality, 48*, 1247–1264.
Schulp, C. J. E., Thuiller, W., & Verburg, P. H. (2014). Wild food in Europe: A synthesis of knowledge and data of terrestrial wild food as an ecosystem service. *Ecological Economics, 105*, 292–305.
Scott, A. C., & Glasspool, I. J. (2006). The diversification of Paleozoic fire systems and fluctuations in atmospheric oxygen concentration. *Proceedings of the National Academy of Sciences, 103*(29), 10861–10865.
Scrivener, J. C., Tschaplinski, P. J., & Macdonald, J. S. (1998). An introduction to the ecological complexity of salmonid life history strategies and of forest harvesting impacts in Coastal British Columbia. In D. L. Hogan, P. J. Tschaplinski, & S. Chatwin (Eds.), *Land management handbook* (Vol. 41, pp. 23–28). Brithish Columbia Ministry of Forests.
Sekercioglu, C. H., & Robinson, S. K. (2011). Effects of habitat fragmentation on tropical birds. In N. S. Sodhi, Ç. H. Sekercioglu, J. Barlow, & S. K. Robinson (Eds.), *Conservation of tropical birds*. Blackwell Publishing Ltd.
Senici, D., Chen, H. Y. H., Bergeron, Y., & Cyr, D. (2010). Spatiotemporal variations of fire frequency in central boreal forest. *Ecosystems, 13*, 1227–1238.
Senici, D., et al. (2013). Multi-millennial fire frequency and tree abundance differ between xeric and mesic boreal forests in Central Canada. *Journal of Ecology, 101*, 356–367.
Senici, D., Chen, H. Y. H., & Bergeron, Y. (2015). The effects of forest fuel connectivity on spatiotemporal dynamics of holocene fire regimes in the central boreal forest of North America. *Journal of Quarternary Science, 30*(4), 365–375.

Seobi, T., Anderson, S. H., Udawatta, R. P., & Gantzer, C. J. (2005). Influence of grass and agroforestry buffer strips on soil hydraulic properties for an Albaqualf. *Soil Science Society of America Journal, 69*, 893–901.

Seufert, V., Ramankutty, N., & Foley, J. A. (2012). Comparing the yields of organic and conventional agriculture. *Nature, 485*, 229–232.

Sheil, D. (2018). Forests, atmospheric water and an uncertain future: The new biology of the global water cycle. *Forest Ecosystems, 5*, 19. https://doi.org/10.1186/s40663-018-0138-y

Shelton, A. O., & Mangel, M. (2011). Fluctuations of fish populations and the magnifying effects of fishing. *Proceedings of the National Academy of Sciences, 108*(17), 7075–7080.

Shorter, R., & Frey, K. J. (1979). Relative yields of mixtures and monocultures of oat genotypes. *Crop Science, 19*, 548–553.

Siddique, K. H. M., Tennant, D., Perry, M. W., & Belford, R. K. (1990). Water-use and water use efficiency of old and modern wheat cultivars in a mediterranean environment. *Australian Journal of Agricultural Research, 41*, 431–447.

Sikström, U. (2001). Effects of pre-harvest soil acidification, liming and N-fertilization on the survival, growth and needle eement concentrations of Picea abies L. Karst. seedlings. *Plant and Soil, 231*, 255–266.

Simola, H., Pitkänen, A., & Turunen, J. (2012). Carbon loss in drained forestry peatlands in Finland, estimated by re-sampling peatlands surveyed in the 1980s. *European Journal of Soil Science, 63*, 798–807.

Sjostrom, E. (1993). *Wood chemistry*. Academic Press.

Skovsgaard, J. P., & Vancay, J. K. (2008). Forest site productivity: A review of the evolution of dendrometric concepts for even-aged stands. *Forestry, 81*, 7–31.

SLU. (2020). *Forest statistics 2020, official statistics of Sweden*. Swedish University of Agricultural Sciences.

Šmelko, S., Bošeľa, M., Merganič, J., & Jankovič, J. (2006). *National forest inventory and monitoring of the Slovak Republic 2005–2006. Basic concept and selected summary information*. National Forest Centre – Forest Research Institute.

Smil, V. (2004). *Enriching the earth: Fritz Haber, Carl Bosch, and the transformation of world food production*. MIT Press.

Smil, V. (2011). Harvesting the biosphere: The human impact. *Population and Development Review, 37*, 613–636.

Smith, G. C. (1976). Ecological energetics of three species of ectothermic vertebrates. *Ecology, 57*(2), 252–264.

Song, X. P., Hansen, M. C., Stehman, S. V., Potapov, P. V., Tyukavina, A., Vermote, E. F., & Townshend, J. R. (2018). Global land change from 1982 to 2016. *Nature, 560*(7720), 639–643. https://doi.org/10.1038/s41586-018-0411-9

Soons, M. B., Hefting, M. T., Dorland, E., Lamers, L. P. M., Versteeg, C., & Bobbink, R. (2017). Nitrogen effects on plant species richness in herbaceous communities are more widespread and stronger than those of phosphorus. *Biological Conservation, 212*, 390–397.

Sorokin, J. I., & Panov, D. A. (1966). The use of C1″ for the quantitative study of the nutrition of fish larvae. *Internationale Revue der Gesamten Hydrobiologie, 51*, 743–756.

Špaldoň, E. (1982). *Rostlinná výroba. [Plant production]*. SZN.

Spiecker, H., Lindner, M., & Schuler, J. (Eds.). (2019). Douglas-fir – An option for Europe. EFI what science can tell us 9. The European Forest Institute, Joensu.

Steffen, W., Sanderson, A., Tyson, P. D., Jäger, J., Matson, P. A., Moore, B., III, Oldfield, F., Richardson, K., Schellnhuber, H. J., Turner, B. L., II, & Wasson, R. J. (2004). *Global change and the earth system. A planet under pressure*. Springer.

Steinfeld, H., Gerber, P., Wassenaar, T., Castel, V., Rosales, M., & De Haan, C. (2006). *Livestock's long shadow – Environmental issues and options*. Food and Agriculture Organisation.

Stewart, K. M. (1994). Early hominid utilisation of fish resources and implications for seasonality and behaviour. *Journal of Human Evolution, 27*(1–3), 229–245.

Stockan, J. A., & Robinson, E. J. H. (2016). *Wood ant ecology and conservation*. Cambridge University Press.

Strakhov, N. M. (1967). *Principles of lithogenesis* (Vol. 1). Oliver and Boyd.

Stringham, S. A., Mulroy, E. E., Xing, J., Record, D., Guernsey, M. W., Aldenhoven, J. T., Osborne, E. J., & Shapiro, M. D. (2012). Divergence, convergence, and the ancestry of feral populations in the domestic rock pigeon. *Current Biology, 22*(4), 302–308.

Sýkora, I. (2012). Produkce zvěřiny v jednotlivých typech honiteb. *Myslivost, 3*, 26.

Tagliabue, A., Bowie, A., Boyd, P., et al. (2007). The integral role of iron in ocean biogeochemistry. *Nature, 543*, 51–59.

Tainter, J. A. (1988). *The collapse of complex societies*. Cambridge University Press.

Taylor, W. A., Lindsey, P. A., Nicholson, S. K., Relton, C., & Davies-Mostert, H. T. (2020). Jobs, game meat and profits: The benefits of wildlife ranching on marginal lands in South Africa. *Biological Conservation, 245*, 108561.

Temple, S. A. (1987). Do predators always capture substandard individuals disproportionately from prey populations? *Ecology, 68*, 669–674.

Thomas, R. Q., Canham, C. D., Weathers, K. C., & Goodale, C. L. (2010). Increased tree carbon storage in response to nitrogen deposition in the US. *Nature Geoscience, 3*, 13–17.

Tigchelaar, M., Battisti, D. S., Naylor, R. L., & Ray, D. K. (2018). Future warming increases probability of globally synchronized maize production shocks. *Proceedings of the National Academy of Sciences, 115*(26), 6644–6649.

Tilman, D., Reich, P. B., Knops, J., Wedin, D., Mielke, T., & Lehman. (2001). Diversity and productivity in a long-term grassland experiment. *Science, 29*(5543), 843–845.

Tilman, D., Balzer, C., Hill, J., & Befort, B. L. (2011). Global food demand and the sustainable intensification of agriculture. *PNAS, 108*(5), 20260–20264.

Tinner, W., Conedera, M., Ammann, B., & Lotter, A. F. (2005). Fire ecology north and south of the Alps since the last ice age. *The Holocene, 15*, 1214–1226.

Toyota, A., Hynšt, J., Cajthaml, T., & Frouz, J. (2013). Soil fauna increase nitrogen loss in tilled soil with legume but reduce nitrogen loss in non-tilled soil without legume. *Soil Biology and Biochemistry, 60*, 105–112.

Trenbath, B. R. (1974). Biomass productivity of mixtures. *Advances in Agronomy, 26*, 177–210.

Trenberth, K., Fasullo, J., & Kiehl, J. (2009). Earth's global energy budget. *Bulletin of the American Meteorological Society, 90*(3), 311–323.

Tscharntke, T., Klein, A. M., Kruess, A., Steffan-Dewenter, I., & Thies, C. (2005). Landscape perspectives on agricultural intensification and biodiversity – Ecosystem service management. *Ecology Letters, 8*, 857–874.

Tuomisto, H. L., Hodge, I. D., Riordan, P., & Macdonald, D. W. (2012). Does organic farming reduce environmental impacts? – A meta-analysis of European research. *Journal of Environmental Management, 112*, 309–320.

UKZUZ. (2020). *Metodický návod pro hnojení plodin*. UKZUZ.

Ulrich, R. S. (1986). Human responses to vegetation and landscapes. *Landscape and Urban Planning, 13*, 29–44.

Unger, N. (2014). Human land-use-driven reduction of forest volatiles cools global climate. *Nature Climate Change, 4*, 907–910.

Ursin, E. (1984). The tropical, the temperate and the arctic seas as media for fish production. *Dana, 3*, 43–60.

Vacek, S., & Podrázský, V. (2006). *Přírodě blízké lesní hospodářství v podmínkách střední Evropy: pěstování lesů*. ČZU v Praze, Fakulta lesnická a environmentální.

Vacek, S., Hejcman, M., Semelová, V., et al. (2009). Effect of soil chemical properties on growth, foliation and nutrition of Norway spruce stand affected by yellowing in the Bohemian Forest Mts., Czech Republic. *European Journal of Forest Research, 128*, 367–375.

Vacek, S., Vacek, Z., Ulbrichová, I., Remeš, J., Podrázský, V., Vach, M., Bulušek, D., Král, J., & Putalová, T. (2019). The effects of fertilization on the health status, nutrition and growth of

Norway spruce forests with yellowing symptoms. *Scandinavian Journal of Forest Research, 34*, 267–281.

van Bennekom, A. J., & Wetsteijn, F. J. (1990). The winter distribution of nutrients in the southern bight of the North Sea (1961–1978) and in the estuaries of the Scheldt and the Rhine/Meuse. *Netherlands Journal of Sea Research, 25*, 75–87.

van Capelle, C., Schrader, S., & Brunotte, J. (2012). Tillage-induced changes in the functional diversity of soil biota – A review with a focus on German data. *European Journal of Soil Biology, 50*, 165–181.

van der Waal, C., Kool, A., Meijer, S. S., Kohi, E., Heitkönig, I. M., de Boer, W. F., van Langevelde, F., Grant, R. C., Peel, M. J., Slotow, R., de Knegt, H. J., Prins, H. H., & de Kroon, H. (2003). Large herbivores may alter vegetation structure of semi-arid savannas through soil nutrient mediation. *Oecologia, 165*(4), 1095–1107. https://doi.org/10.1007/s00442-010-1899-3

van Mourik, J., & Ligtendag, W. (2015). Relicts of a peat cover in the Westerkoggepolder (West Friesland, North-Holland, The Netherlands): The genesis of an eluvial clay soil. *Catena, 135*, 105–113.

van Poorten, B. T., Cox, S. P., & Cooper, A. B. (2013). Efficacy of harvest and minimum size limit regulations for controlling short-term harvest in recreational fisheries. *Fisheries Management and Ecology, 20*, 258–267.

Vavilov, N. I. (1926). Studies on the origin of cultivated plants. *(Russian) Bulletin of Applied Botany and Plant Breeding, 14*, 1–245.

Vayssières, J., & Rufino, M. C. (2012). Managing nutrient cycles in crop and livestock systems with green technologies. In J. I. Boye & Y. Arcand (Eds.), *Green Technologies in Food Production and Processing* (pp. 151–182). Springer Science.

Veldman, J. W., Overbeck, G. E., Negreiros, D., Mahy, G., Le Stradic, S., Fernandes, G. W., et al. (2015). Where tree planting and forest expansion are bad for biodiversity and ecosystem services. *Bioscience, 65*, 1011–1018. https://doi.org/10.1093/biosci/biv118

Vera, F. W. M. (2000). *Grazing ecology and forest history*. CAB International.

Vesterdal, L., Schmidt, I. K., Callesen, I., Nilsson, L. O., & Gundersen, P. (2008). Carbon and nitrogen in forest floor and mineral soil under six common European tree species. *Forest Ecology and Management, 255*, 35–48.

Vigani, M., Parisi, C., Rodríguez-Cerezo, E., Barbosa, M. J., Sijtsma, L., Ploeg, M., & Enzing, C. (2015). Food and feed products from micro-algae: Market opportunities and challenges for the EU. *Trends in Food Science & Technology, 42*(1), 81–92.

Viitala, E. J. (2013). The discovery of the Faustmann formula in natural resource economics. *History of Political Economy, 45*, 521–548.

Viitala, E. J. (2016). Faustmann formula before Faustmann in German territorial states. *Forest Policy and Economics, 65*, 47–58.

Vilà, M., Carrillo-Gavilàn, A., Vayreda, J., Bugmann, H., Fridman, J., et al. (2013). Disentangling biodiversity and climatic determinants of wood production. *PLoS One, 8*(2), e53530.

Vindušková, O., & Frouz, J. (2013). Soil carbon accumulation after open-cast coal and oil shale mining in Northern Hemisphere: A quantitative review. *Environmental Earth Sciences, 69*(5), 1685–1698.

Vindušková, O., Pánek, T., & Frouz, J. (2019). Soil C, N and P dynamics along a 13 ka chronosequence of landslides under semi-natural temperate forest. *Quaternary Science Reviews, 213*, 18–29.

Vinton, M. A., Hartnett, D. C., Finck, E. J., & Briggs, J. M. (1993). The interactive effects of fire, bison (Bison bison) grazing and plant community composition in tallgrass prairie. *American Midland Naturalist, 129*(1), 10–18.

Vitkova, L., Bače, R., Svoboda, M., & Kjucukov, P. (2018). Deadwood management in Central European forests: Key considerations for practical implementation. *Forest Ecology and Management, 429*, 394–405.

Vitousek, P. M., & Howarth, R. W. (1991). Nitrogen limitation on land and in the sea: How can it occur? *Biogeochemistry, 13*, 87–115.

Vitousek, P. M., Ehrlich, P. R., Ehrlich, A. H., & Matson, P. A. (1986). Human appropriation of the products of photosynthesis. *Bioscience, 36*(6), 368–373.

Vitousek, P. M., Aber, J. D., Howarth, R. W., Likens, G. E., Matson, P. A., Schindler, D. W., Schlesinger, W. H., & Tilman, D. G. (1997). Human alteration of the global nitrogen cycle: Sources and consequences. *Ecological Applications, 7*, 737–750.

Vos, C. C., & Stumpel, A. H. E. (1995). Comparison of habitat-isolation parameters in relation to fragmented distribution patterns in the tree frog (*Hyla arborea*). *Landscape Ecology, 11*(4), 203–214.

Vrška, T., Adam, D., Hort, L., Kolar, T., & Janık, D. (2009). European beech (*Fagus sylvatica* L.) and silver fir (*Abies alba* Mill.) rotation in the Carpathians – A developmental cycle or a linear trend induced by man? *Forest Ecology and Management, 258*, 347–356.

Waldrop, M. P., Zak, D. R., Sinsabaugh, R. L., Gallo, M., & Lauber, C. (2004). Nitrogen deposition modifies soil carbon storage through changes in microbial enzymatic activity. *Ecological Applications, 14*, 1172–1177.

Walmsley, A., & Sklenicka, P. (2017). Various effects of land tenure on soil biochemical parameters under organic and conventional farming – Implications for soil quality restoration. *Ecological Engineering, 107*, 134–143.

Wang, X., Edwards, R., Auler, A., et al. (2017). Hydroclimate changes across the Amazon lowlands over the past 45,000 years. *Nature, 541*, 204–207.

Wardle, D. A., Hörnberg, G., Zackrisson, O., Kalela-Brundin, M., & Coomesc, D. A. (2003). Long-term effects of wildfire on ecosystem properties across an island area gradient. *Science, 300*(5621), 972–975.

Wardle, D. A., Bardgett, D. R., Walker, L., Peltzer, A. D., & Lagerström, A. (2008). The response of plant diversity to ecosystem retrogression: Evidence from contrasting long-term chronosequences. *Oikos, 117*, 93–103.

Warner, R. (2008). Protecting the diversity of the depths: Environmental regulation of bioprospecting and marine scientific research beyond national jurisdiction. *Ocean Yearbook, 22*, 411–443.

Warren, J. M., Jensen, A. M., Medlyn, B. E., Norby, R. J., & Tissue, D. T. (2015). Carbon dioxide stimulation of photosynthesis in *Liquidambar styraciflua* is not sustained during a 12-year field experiment. *AoB Plants, 7*, plu074. https://doi.org/10.1093/aobpla/plu074

Weinzettel, J., Hertwich, E. G., Peters, G. P., Steen-Olsen, K., & Gallic, A. (2013). Affluence drives the global displacement of land use. *Global Environmental Change-Human and Policy Dimensions, 23*(2), 433–438.

Welch, H. E. (1968). Relationships between assimilation efficiencies and growth efficiencies for aquatic consumers. *Ecology, 49*, 755–759.

West, P. C., Gibbs, H. K., Monfreda, C., Wagner, J., Barford, C. C., Carpenter, S. R., & Foley, J. A. (2010). Trading carbon for food: Global comparison of carbon stocks vs. crop yields on agricultural land. *PNAS, 107*(46), 19645–19648.

Williams, A. G., Audsley, E., & Sandars, D. L. (2006). Energy and environmental burdens of organic and non-organic agriculture and horticulture. *Aspects of Applied Biology, 79*(2006), 19–23.

Williamson, M., & Fitter, A. (1996). The varying success of invaders. *Ecology, 77*(6), 1661–1666.

Wilson, E. O. (2016). *Half-earth: 'Our planet's fight for life*. Norton & Company.

Wilson, J. D., Morris, A. J., Arroyo, B. E., Clark, S. C., & Bradbury, R. B. (1999). A review of the abundance and diversity of invertebrate and plant foods of granivorous birds in northern Europe in relation to agricultural change. *Agriculture, Ecosystems and Environment, 75*, 13–30.

Winsberg, M. D. (1980). Concentration and specialization in United States agriculture, 1939–1979. *Economic Geography, 56*(3), 183–189.

Wolynetz, M. S., & Sibbald, I. R. (1990). Estimates of body components in broiler chickens from body weight and dry matter. *Poultry Science, 69*, 1318–1324.

Wood, H. B. (1977). Hydrologic differences between selected forested and agricultural soils in Hawaii. *Soil Science Society of America Journal, 41*, 132–136.

Woodcock, B. A., Bullock, J. M., Shore, R. F., Heard, M. S., Pereira, M. G., Redhead, J., Ridding, L., Dean, H., Sleep, D., Henrys, P., Peyton, J., Hulmes, J. S., Hulmes, L., Sárospataki, M., Saure,

C., Edwards, M., Genersch, E., Knäbe, S., & Pywell, R. F. (2017). Country-specific effects of neonicotinoid pesticides on honey bees and wild bees. *Science, 356*(63450), 1393–1395.

Woods, J., Williams, A., Hughes, J. K., Black, M., & Murphy, R. (2010). Energy and the food system. *Philosophical Transactions of the Royal Society B-Biological Science, 365*(1554), 2991–3006.

Woodward, R. T., & Griffin, W. L. (2003). Size and bag limits in recreational fisheries: Theoretical and empirical analysis. *Marine Resource Economics, 18*, 239–262.

World Resources Institute. (2005). *Millenium ecosystem assessment, ecosystems and human well-being*. World Resources Institute.

Worm, B., Davis, B., Kettemer, L., Ward-Paige, C. A., Capman, D., Heithaus, M. R., Kessel, S. T., & Gruber, S. H. (2013). Global catches, exploitation rates, and rebuilding options for sharks. *Marine Policy, 40*, 194–204.

Wright, D. H. (1983). Species-energy theory: An extension of species-area theory. *Oikos, 41*, 496–506.

Wright, R. F., Alewell, C., Cullen, J. M., Evans, C. D., Marchetto, A., Moldan, F., Prechtel, A., & Rogora, M. (2001). Trends in nitrogen deposition and leaching in acid-sensitive streams in Europe. *Hydrology and Earth System Sciences, 5*, 299–310.

Wunderlin, P., Mohn, J., Joss, A., Emmenegger, L., & Siegrist, H. (2012). Mechanisms of N_2O production in biological wastewater treatment under nitrifying and denitrifying conditions. *Water Research, 46*(4), 1027–1037.

Yamamoto, S. I. (2000). Forest gap dynamics and tree regeneration. *Journal of Forest Research, 5*(4), 223–229.

Zaks, D. P. M., & Kucharik, C. J. (2011). Data and monitoring needs for a more ecological agriculture. *Environmental Research Letters, 6*, 014017, (10 pp). https://doi.org/10.1088/1748-9326/6/1/014017

Zeckhauser, R. (2017). Human hunters and nonhuman predators: Fundamental differences. *Proceedings of the National Academy of Sciences of the United States of America, 114*(1), 13–14.

Zeebe, R. E. (2012). History of seawater carbonate chemistry, atmospheric CO^2, and ocean acidification. *Annual Review of Earth and Planetary Sciences, 40*(1), 141–165.

Zhang, C., & Noguchi, N. (2017). Development of a multi-robot tractor system for agriculture field work. *Computers and Electronics in Agriculture, 142*, 79–90.

Zhang, L., Dawes, W. R., & Walker, G. R. (2001). Response of mean annual evapotranspiration to vegetation changes at catchment scale. *Water Resource Research, 37*, 701–708.

Zhao, M. S., Heinsch, F. A., Nemani, R. R., & Running, S. W. (2005). Improvements of the MODIS terrestrial gross and net primary production global data set. *Remote Sensing of Environment, 95*, 164–176.

Zuidhof, M. J., Schneider, B. L., Carney, V. L., Korver, D. R., & Robinson, F. E. (2014). Growth, efficiency, and yield of commercial broilers from 1957, 1978, and 2005. *Poultry Science, 93*, 2970–2982.

Index

© The Author(s), under exclusive license to Springer Nature Switzerland AG 2022

J. Frouz, J. Frouzová, *Applied Ecology*, https://doi.org/10.1007/978-3-030-83225-4

The manufacturer's authorised representative in the EU is Springer Nature Customer Service Centre GmbH, Europaplatz 3, 69115 Heidelberg, Germany. If you have any concerns regarding our products, please contact ProductSafety@springernature.com

Printed and bound by CPI Group (UK) Ltd, Croydon, CR0 4YY
22/09/2026
02220946-0016